Methods in Molecular Biology

For further volumes:
http://www.springer.com/series/7651

For over 35 years, biological scientists have come to rely on the research protocols and methodologies in the critically acclaimed Methods in Molecular Biology series. The series was the first to introduce the step-by-step protocols approach that has become the standard in all biomedical protocol publishing. Each protocol is provided in readily-reproducible step-by step fashion, opening with an introductory overview, a list of the materials and reagents needed to complete the experiment, and followed by a detailed procedure that is supported with a helpful notes section offering tips and tricks of the trade as well as troubleshooting advice. These hallmark features were introduced by series editor Dr. John Walker and constitute the key ingredient in each and every volume of the Methods in Molecular Biology series. Tested and trusted, comprehensive and reliable, all protocols from the series are indexed in PubMed.

Manuscript Instructions https://drive.google.com/file/d/1K4Az5vpAFVzYadJ3jKx2-QSo7Dqa_SNA/view?usp=sharing

Gangliosides

Methods and Protocols

Edited by

Koichi Furukawa

College of Life and Health Sciences, Chubu University, Kasugai, Aichi, Japan

Humana Press

Editor
Koichi Furukawa
College of Life and Health Sciences
Chubu University
Kasugai, Aichi, Japan

ISSN 1064-3745 ISSN 1940-6029 (electronic)
Methods in Molecular Biology
ISBN 978-1-0716-5287-9 ISBN 978-1-0716-5288-6 (eBook)
https://doi.org/10.1007/978-1-0716-5288-6

This Humana imprint is published by the registered company Springer Science+Business Media, LLC, part of Springer Nature.
The registered company address is: 1 New York Plaza, New York, NY 10004, U.S.A.

Preface

Since we started to make the concept and plan the specific lineup, almost two and a half years have passed, but we are now completing a wonderful book of *Methods in Molecular Biology: Gangliosides* with the help of many authors.

In this book, more than 30 groups have participated, ranging from medical scientists, basic structure analysts, imaging specialists, MS specialists, microbiologists, and immunologists. There are 89 authors in total.

The contents of this book consist of:

1. Synthesis and degradation of gangliosides, including chemical synthesis, substrate preparation, and synthetic and degradation enzymes.
2. Structure analysis of gangliosides, including LC-MS, MS-imaging, NMR, ceramide analysis, and immunochemical analysis.
3. Interacting molecules and functions of gangliosides, including proximity labeling, toxin trafficking, interaction with hormone receptor, ganglioside-interacting lectins, ganglioside–bacteria interactions, and molecular complex formation.
4. Ganglioside behaviors and biology, including neutral glycolipids, single particle tracking, lipid raft analysis, metabolic analysis in foods, and flow cytometry of extracellular vesicles.
5. Disease and tissue development, including bone metabolism, melanocyte gangliosides, autoimmune neuropathies, and amyloid fibril formation.
6. Application of gangliosides, including ganglioside-reconstituted liposomes, and intracerebral administration of gangliosides for in vivo function and neural stem cell analysis.

Medical biology has been advancing rapidly in recent years. Glycoscience has been also developing, and dramatic progress based on new discovery and novel technology is being made.

At this point, we need both newest technology and approaches, and also classical methodology in order to correctly lead genuine glycoscience and ganglioside science.

Among all glycoscience fields, gangliosides may be recognized as a model and pioneer science, considering its historical accumulation and detailed development of researches.

I believe that this issue can speak to all young and veteran glycoscience researchers regardless of their specialty and experiences in this field.

Kasugai, Aichi, Japan *Koichi Furukawa*

Acknowledgment

I would like to thank all the authors and the co-authors for their kind cooperation and also Prof. John M. Walker, Series Editor/Professor, University of Hertfordshire, Mr. Patrick Marton, Executive Editor, Springer, Ms. Anna Rakovsky, Associate Editor, Springer and Ms. Catherine A., Production Supervisor - Books Production, MPS Limited, for their patient help. I would also like to thank Ms. M. Kojima for her excellent secretarial assistance.

Contents

Part V Diseases and Tissue Development

Part VI Application of Gangliosides

Contributors

S. M. ABE KAWSAR • *Department of Chemistry, Faculty of Science, University of Chittagong, Chittagong, Bangladesh*

KAZUNARI AKIYOSHI • *Department of Immunology, Graduate School of Medicine, Kyoto University, Kyoto, Japan*

HIROMUNE ANDO • *Institute for Glyco-core Research (iGCORE), Gifu University, Gifu, Japan*

HIDEMI ARAKAWA • *Department of Biosciences, Graduate School of Science and Engineering, Soka University, Hachioji, Tokyo, Japan*

TOSHIO ARIGA • *Department of Neuroscience and Regenerative Medicine, Medical College of Georgia at Augusta University, Augusta, GA, USA*

ROBIUL HASAN BHUIYAN • *Department of Biomedical Sciences, Chubu University College of Life and Health Sciences, Kasugai, Aichi, Japan; Department of Biochemistry and Molecular Biology, University of Chittagong, Chattogram, Bangladesh; Department of Otolaryngology-Head and Neck Surgery, New York University, New York, USA*

SHINICHI CHISADA • *Kyorin University School of Medicine, Mitaka-shi, Tokyo, Japan; Department of Bioscience and Biotechnology, Faculty of Agriculture, Kyushu University, Nishiku, Fukuoka, Japan; Department of Bioscience and Biotechnology, Faculty of Agriculture, Kyushu University, Fukuoka, Japan*

TAKAHIRO FUCHIGAMI • *Department of Neuroscience and Regenerative Medicine, Medical College of Georgia at Augusta University, Augusta, GA, USA; Departmet of Molecular Diagnosis, Graduate School of Medicine Chiba University, Chiba, Japan*

YUKI FUJII • *Graduate School of Pharmaceutical Sciences, Nagasaki International University, Sasebo, Nagasaki, Japan*

HIDEAKI FUJITA • *Graduate School of Pharmaceutical Sciences, Nagasaki International University, Sasebo, Nagasaki, Japan*

KEIKO FURUKAWA • *Department of Biomedical Sciences, Chubu University Graduate School of Life and Health Sciences, Kasugai, Aichi, Japan; Department of Biomedical Sciences, Chubu University College of Life and Health Sciences, Kasugai, Aichi, Japan*

KOICHI FURUKAWA • *Department of Biomedical Sciences, Chubu University College of Life and Health Sciences, Kasugai, Aichi, Japan*

MARCO GERDOL • *Department of Life Sciences, University of Trieste, Trieste, Italy*

KAZUNORI HAMAMURA • *Department of Pharmacology, School of Dentistry, Aichi Gakuin University, Nagoya, Japan*

IMTIAJ HASAN • *Department of Microbiology, Faculty of Biological Science, University of Rajshahi, Rajshahi, Bangladesh; Department of Biochemistry and Molecular Biology, Faculty of Science, University of Rajshahi, Rajshahi, Bangladesh*

NOBORU HASHIMOTO • *Biochemistry II, Nagoya University Graduate School of Medicine, Nagoya, Japan; Tissue Regeneration, Tokushima University Graduate School of Biomedical Sciences, Tokushima, Japan*

MOHAMMAD ABUL HASNAT • *Department of Biomedical Sciences, Chubu University Graduate School of Life and Health Sciences, Kasugai, Aichi, Japan; Department of Biochemistry and Molecular Biology, Shahjalal University of Science and Technology, Sylhet, Bangladesh*

TETSUYA HIRABAYASHI • *Biomembrane Group, Tokyo Metropolitan Institute of Medical Science, Tokyo, Japan*

KOICHI HONKE • *Department of Biochemistry, Kochi University Medical School, Nankoku, Kochi, Japan*

MELANIE HORKEY • *Department of Pharmacology, University of Alberta, Edmonton, AB, Canada*

NICOLAS HUMBERT • *Laboratoire de Bioimagerie et Pathologies, UMR 7021 CNRS, Faculté de Pharmacie, Université de Strasbourg, Illkirch, Strasbourg, France*

KAZUTAKA IKEDA • *Laboratory of Biomolecule Analysis, Department of Applied Genomics, Kazusa DNA Research Institute, Kisarazu, Chiba, Japan; Laboratory of Omics and Informatics, Department of Molecular and Chemical Life Sciences, Graduate School of Life Sciences, Tohoku University, Sendai, Japan*

AKIHIRO IMAMURA • *Institute for Glyco-core Research (iGCORE), Gifu University, Gifu, Japan; Department of Applied Bioorganic Chemistry, Gifu University, Gifu, Japan*

ARIFUL ISLAM • *Department of Cellular and Molecular Anatomy, Hamamatsu University School of Medicine, Hamamatsu, Shizuoka, Japan; Department of Biochemistry and Microbiology, School of Health and Life Sciences, North South University, Dhaka, Bangladesh*

TATSUKI ISOGAI • *The United Graduate School of Agricultural Science, Gifu University, Gifu, Japan*

MAKOTO ITO • *Department of Bioscience and Biotechnology, Faculty of Agriculture, Kyushu University, Nishiku, Fukuoka, Japan; Department of Bioscience and Biotechnology, Faculty of Agriculture, Kyushu University, Fukuoka, Japan*

SHIZUKA ITO • *Biochemistry II, Nagoya University Graduate School of Medicine, Nagoya, Japan*

YUTAKA ITOKAZU • *Department of Pharmacology and Toxicology, Medical College of Georgia at Augusta University, Augusta, GA, USA; Department of Neuroscience and Regenerative Medicine, Medical College of Georgia at Augusta University, Augusta, GA, USA*

KAZUHISA IWABUCHI • *Institute for Environmental and Gender-Specific Medicine, Graduate School of Medicine, Juntendo University, Urayasu, Chiba, Japan*

KAZUYA KABAYAMA • *Interdisciplinary Research Center for Radiation Sciences, Institute for Radiation Sciences, The University of Osaka, Suita, Osaka, Japan; Department of Chemistry, Graduate School of Science, The University of Osaka, Toyonaka, Osaka, Japan; Forefront Research Center, Graduate School of Science, The University of Osaka, Toyonaka, Osaka, Japan*

TSUNEYASU KAISHO • *Department of Immunology, Institute of Advanced Medicine, Wakayama Medical University, Wakayama, Wakayama, Japan*

SHIORI KAJI • *Second Department of Internal Medicine, Wakayama Medical University, Wakayama, Wakayama, Japan*

KENICHI KAMATA • *Department of Materials and Applied Chemistry, College of Science and Technology, Nihon University, Tokyo, Japan; Graduate School of Biomedical Sciences, Yokohama City University, Yokohama, Japan*

MIKI KANNO • *Graduate School of Natural Science and Technology, Gifu University, Gifu, Japan*
KOHJI KASAHARA • *Biomembrane Group, Tokyo Metropolitan Institute of Medical Science, Tokyo, Japan*
TATSUYA KAWASAKI • *Graduate School of Pharmaceutical Sciences, Nagasaki International University, Sasebo, Nagasaki, Japan*
NORIHITO KIKUCHI • *Biomembrane Group, Tokyo Metropolitan Institute of Medical Science, Tokyo, Japan*
KEN KITAJIMA • *Integrated Glyco-Biomedical Research Center (iGMED), Institute for Glyco-core Research (iGCORE), Nagoya University, Nagoya, Aichi, Japan*
TOSHIHIDE KOBAYASHI • *Laboratoire de Bioimagerie et Pathologies, UMR 7021 CNRS, Faculté de Pharmacie, Université de Strasbourg, Illkirch, Strasbourg, France*
KEISUKE KOMATSUYA • *Biomembrane Group, Tokyo Metropolitan Institute of Medical Science, Tokyo, Japan; Clinical Proteomics and Molecular Medicine, St Marianna University Gaduate School of Medicine, Kawasaki, Kanagawa, Japan*
NAOKO KOMURA • *Institute for Glyco-core Research (iGCORE), Gifu University, Gifu, Japan*
HISATAKA KONDO • *Department of Pharmacology, School of Dentistry, Aichi Gakuin University, Nagoya, Japan; Department of Dental Hygiene, Aichi Gakuin University Junior College, Nagoya, Japan*
NORIHIRO KOTANI • *Department of Pharmacology, Saitama Medical University, Iruma-gun, Saitama, Japan; Integrated Research Administration Center (IRAC), Saitama Medical University, Iruma-gun, Saitama, Japan; Department of Biochemistry, Saitama Medical University, Iruma-gun, Saitama, Japan*
HIROKA KOYANAGI • *Department of Biomedical Sciences, Chubu University Graduate School of Life and Health Sciences, Kasugai, Aichi, Japan*
SUSUMU KUSUNOKI • *Department of Neurology, Kindai University Faculty of Medicine, Minami-ku, Sakai, Osaka, Japan*
MOTOI KUWAHARA • *Department of Neurology, Kindai University Faculty of Medicine, Minami-ku, Sakai, Osaka, Japan*
DONGPEI LI • *Department of Pharmacology and Toxicology, Medical College of Georgia at Augusta University, Augusta, GA, USA*
MATTHEW S. MACAULEY • *Department of Chemistry, University of Alberta, Edmonton, Alberta, Canada; Department of Medical Microbiology and Immunology, University of Alberta, Edmonton, Alberta, Canada; Neuroscience and Mental Health Institute, University of Alberta, Edmonton, Alberta, Canada*
TERUHIKO MATSUBARA • *Department of Biosciences and Informatics, Keio University, Yokohama, Kanagawa, Japan*
NAOYUKI MATSUNAGA • *Bizen Chemical Co. Ltd., Akaiwa-shi, Okayama, Japan; Department of Bioscience and Biotechnology, Faculty of Agriculture, Kyushu University, Nishiku, Fukuoka, Japan; Department of Bioscience and Biotechnology, Faculty of Agriculture, Kyushu University, Fukuoka, Japan*
LAURA MAURI • *Department of Medical Biotechnology and Translational Medicine, University of Milan, Milan, Italy*
YVES MÉLY • *Laboratoire de Bioimagerie et Pathologies, UMR 7021 CNRS, Faculté de Pharmacie, Université de Strasbourg, Illkirch, Strasbourg, France*
ARISA MIYAGAWA-YAMAGUCHI • *Department of Biochemistry, Kochi University Medical School, Nankoku, Kochi, Japan*

TAEKO MIYAGI • *Cancer Chemotherapy, Miyagi Cancer Center Research Institute, Miyagi, Japan*
JOHN MONYROR • *Department of Pharmacology and Glycomics Institute of Alberta, University of Alberta, Edmonton, AB, Canada*
MOTOHIDE MURATE • *Laboratoire de Bioimagerie et Pathologies, UMR 7021 CNRS, Faculté de Pharmacie, Université de Strasbourg, Illkirch, Strasbourg, France; Support Unit for Electron Microscopy Techniques, Research Resource Division, Center for Brain Science, RIKEN, 2-1 Hirosawa, Wako, Saitama, Japan*
MAYU NAGAO • *Department of Pharmacology, School of Dentistry, Aichi Gakuin University, Nagoya, Japan*
YUKIKO OGAWA • *Graduate School of Pharmaceutical Sciences, Nagasaki International University, Sasebo, Nagasaki, Japan*
YUKI OHKAWA • *Department of Glyco-Oncology and Medical Biochemistry, Osaka International Cancer Institute, Chuo-ku, Osaka, Japan*
YUHSUKE OHMI • *Department of Clinical Technology, Chubu University College of Life and Health Sciences, Kasugai, Aichi, Japan; Department of Biomedical Sciences, Chubu University Graduate School of Life and Health Sciences, Kasugai, Aichi, Japan*
NOZOMU OKINO • *Department of Bioscience and Biotechnology, Faculty of Agriculture, Kyushu University, Nishiku, Fukuoka, Japan; Department of Bioscience and Biotechnology, Faculty of Agriculture, Kyushu University, Fukuoka, Japan*
TETSUYA OKUDA • *Molecular Biosystems Research Institute, National Institute of Advanced Industrial Science and Technology (AIST), Ikeda, Osaka, Japan*
TAKAYUKI OMOTO • *Integrated Glyco-Biomedical Research Center (iGMED), Institute for Glyco-core Research (iGCORE), Nagoya University, Nagoya, Aichi, Japan*
YASUHIRO OZEKI • *Graduate School of NanoBio Sciences, Yokohama City University, Yokohama, Japan*
SULTANA RAJIA • *Center for Interdisciplinary Research, Varendra University, Rajshahi, Bangladesh*
LUDOVIC RICHERT • *Laboratoire de Bioimagerie et Pathologies, UMR 7021 CNRS, Faculté de Pharmacie, Université de Strasbourg, Illkirch, Strasbourg, France*
GYOSUKE SAKASHITA • *Department of Pharmacology, School of Dentistry, Aichi Gakuin University, Nagoya, Japan*
IZUMI SASAKI • *Department of Immunology, Institute of Advanced Medicine, Wakayama Medical University, Wakayama, Wakayama, Japan*
CHIHIRO SATO • *Integrated Glyco-Biomedical Research Center (iGMED), Institute for Glyco-core Research (iGCORE), Nagoya University, Nagoya, Aichi, Japan; Laboratory of Glyco-Life Science, Graduate School of Agricultural Sciences, Nagoya University, Nagoya, Aichi, Japan*
EDWARD N. SCHMIDT • *Department of Chemistry, University of Alberta, Edmonton, Alberta, Canada*
MITSUTOSHI SETOU • *Department of Cellular and Molecular Anatomy, Hamamatsu University School of Medicine, Hamamatsu, Shizuoka, Japan; International Mass Imaging Center, Hamamatsu University School of Medicine, Hamamatsu, Shizuoka, Japan; Department of Systems Molecular Anatomy, Institute for Medical Photonics Research, Preeminent Medical Photonics Education and Research Center, Hamamatsu, Shizuoka, Japan*

KOHEI SHIMIZU • *Takenaka Co., Manufacturing & Logistics Engineering Division, Tokyo, Japan; Department of Bioscience and Biotechnology, Faculty of Agriculture, Kyushu University, Nishiku, Fukuoka, Japan*

ASAKO SHIMODA • *Department of Immunology, Graduate School of Medicine, Kyoto University, Kyoto, Japan; The Hakubi Center for Advanced Research, Kyoto University, Kyoto, Japan*

KAZUHIRO SHIOZAKI • *Faculty of Fisheries and The United Graduate School of Agricultural Science, Kagoshima University, Kagoshima, Japan*

SIMONETTA SIPIONE • *Department of Pharmacology, Neuroscience and Mental Health Institute and Glycomics Institute of Alberta, University of Alberta, Edmonton, AB, Canada*

SANDRO SONNINO • *Department of Medical Biotechnology and Translational Medicine, University of Milan, Milan, Italy*

KENICHI G. N. SUZUKI • *The United Graduate School of Agricultural Science, Gifu University, Gifu, Japan; Graduate School of Natural Science and Technology, Gifu University, Gifu, Japan; Institute for Glyco-core Research (iGCORE), Gifu University, Gifu, Japan; Division of Advanced Bioimaging, National Cancer Center Research Institute (NCCRI), Tokyo, Japan*

ORIE TAJIMA • *Department of Biomedical Sciences, Chubu University Graduate School of Life and Health Sciences, Kasugai, Kasugai Aichi, Japan*

AKIRA TOGAYACHI • *Department of Biosciences, Graduate School of Science and Engineering, Soka University, Hachioji, Tokyo, Japan; Glycan and Life Systems Integration Center (GaLSIC), Soka University, Hachioji, Tokyo, Japan*

NARIO TOMISHIGE • *Laboratoire de Bioimagerie et Pathologies, UMR 7021 CNRS, Faculté de Pharmacie, Université de Strasbourg, Illkirch, Strasbourg, France*

NAOKO WAKAKI-NISHIYAMA • *Department of Immunology, Institute of Advanced Medicine, Wakayama Medical University, Wakayama, Wakayama, Japan*

DI WU • *Integrated Glyco-Biomedical Research Center (iGMED), Institute for Glyco-core Research (iGCORE), Nagoya University, Nagoya, Aichi, Japan; Bioscience and Biotechnology Center, Nagoya University, Nagoya, Aichi, Japan*

MASAO YAMADA • *Graduate School of NanoBio Sciences, Yokohama City University, Yokohama, Japan*

YOSHIKI YAMAGUCHI • *Division of Structural Biology, Institute of Molecular Biomembrane and Glycobiology, Tohoku Medical and Pharmaceutical University, Sendai, Japan*

AKIHITO YAMAMOTO • *Tissue Regeneration, Tokushima University Graduate School of Biomedical Sciences, Tokushima, Japan*

KOJI YAMAMOTO • *Faculty of Health and Medical Care, Saitama Medical University, Saitama, Japan*

FARHANA YESMIN • *Department of Biomedical Sciences, Chubu University College of Life and Health Sciences, Kasugai, Aichi, Japan; Department of Otolaryngology-Head and Neck Surgery, New York University, New York, USA*

NORIKO YOKOYAMA • *Institute for Environmental and Gender-Specific Medicine, Graduate School of Medicine, Juntendo University, Urayasu, Chiba, Japan*

ASIFA K. ZAIDI • *Department of Pharmacology and Glycomics Institute of Alberta, University of Alberta, Edmonton, AB, Canada*

Part I

Synthesis and Degradation

Chapter 1

Structure and Structural Properties of the Gangliosides

Laura Mauri and Sandro Sonnino

Abstract

We describe here the primary and secondary structure of gangliosides, focusing on their structural variety that leads gangliosides to be an incredibly large family of amphiphilic compounds. Each ganglioside, due to the structural differences, displays a specific amphiphilicity and all this allows a different behavior of the ganglioside monomer inserted into the membrane. The structural differences allow gangliosides to exhibit different interaction and organizing properties at the cell surface. Some aspects of the ganglioside chemical, physicochemical, and biological properties are discussed in this chapter.

Keywords Gangliosides, Ceramide, Sphingosine, Lipid rafts

1 Introduction

Gangliosides are glycosphingolipids that contain in their structure one or more negatively charged sialic acid. The pKa of sialic acid is 10^{-2} M, and this confers a strong amphiphilic property to the molecule and gives gangliosides an optimal ability to be components of the membranes.

Gangliosides are inserted into plasma membranes with the lipid moiety, strongly interacting with the membrane lipid components, and with the hydrophilic moiety protruding into the extracellular medium. The hydrophilic carbohydrate chain of gangliosides is mobile and has several possible conformations. This allows a number of possible interactions with membrane neighboring components, with extracellular soluble proteins or ions, and with components of the membranes of approaching cells.

Gangliosides are defined as complex membrane lipids belonging to the larger family of sphingolipids and represent the most abundant class of glycolipids in eukaryotic membranes. The name

Koichi Furukawa (ed.), *Gangliosides*, Methods in Molecular Biology, vol. 3035,
https://doi.org/10.1007/978-1-0716-5288-6_1,

"gangliosides" was introduced by Ernst Klenk [1] for a mixture of compounds isolated from human brain. He later showed that the gangliosides contained an acidic carbohydrate, which was named "neuraminic acid" [2].

The field attracted an outstanding, but limited, group of scientists. This together with the structural complexity of gangliosides and the limitation in the chemistry analytical procedures, required to wait up to 1963 [3] to obtain the first total structure of a single purified ganglioside.

Also from this knowledge, gangliosides remained a field of biochemistry not capable to attract the critical mass of scientists necessary to rapidly understand their role in the cells. They remained a "little sister" to glycerophospholipids, proteins and nucleic acids for at least another three decades.

Then, the first information on the association of gangliosides to the transduction of the information throughout the membrane, and their participation to cell signaling processes fundamental for the correct biology of cells, progressively attracted more and more scientists working in several and different fields of science.

2 The Nomenclature of Gangliosides

The most used nomenclature of gangliosides is that established by Lars Svennerholm [4]. Even though it was considered trivial, this nomenclature was used worldwide so that it has been accepted as an alternative to that recommended by the IUPAC-IUB commission for the nomenclature of carbohydrate [5].

The Svennerholm's nomenclature identifies and names gangliosides with a combination of letters and numbers and is applied mainly to brain gangliosides (see Table 1). The ganglioside name starts with the capital letter G, for ganglioside, and is followed by the capital letter M, D, T, Q, or P indicating the number of sialic acid residues 1, 2, 3, 4, or 5, respectively. Then, there is a number, 1, 2, 3, or 4, that indicates the neutral series ganglio-4, ganglio-3, Lac or 4, respectively. At the end, for the gangliosides belonging to the ganglio-series and with more than one sialic acid residue, there is the lowercase letter a, b, or c. The letter a indicates that only one sialic acid is linked to the internal galactose, the letter b indicates that two sialic acids are linked to the internal galactose, and the letter c indicates that three sialic acids are linked to the internal galactose. Gangliosides GM1 with the sialic acid linked to the internal or external galactose are named GM1a and GM1b, respectively.

The IUPAC-IUB nomenclature is more explicative in writing but is not useful in the in speaking. All the ganglioside series are used according to the abbreviation reported in Table 1. Sugars are defined with Roman numbering from I to V starting from the

Table 1
Structures of the main neutral chains

Series	Abbreviation	Structure
4	Gal	ß-Gal-
	Glc	ß-Glc-
Lactose	Lac	ß-Gal-(1-4)-ß-Glc-
Ganglio-3	Gg_3	ß-GalNAc-(1-4)-ß-Gal-(1-4)-ß-Glc-
Ganglio-4	Gg_4	ß-Gal-(1-3)-ß-GalNAc-(1-4)-ß-Gal-(1-4)-ß-Glc-
Ganglio-5	Gg_5	ß-GalNAc-(1-4)-ß-Gal-(1-3)-ß-GalNAc-(1-4)-ß-Gal-(1-4)-ß-Glc-
Globo-3	Gb_3	α-Gal-(1-4)-ß-Gal-(1-4)-ß-Glc-
Globo-4	Gb_4	ß-GalNAc-(1-3)-α-Gal-(1-4)-ß-Gal-(1-4)-ß-Glc-
Globo-5	Gb_5	ß-Gal-(1-3)-ß-GalNAc-(1-3)-α-Gal-(1-4)-ß-Gal-(1-4)-ß-Glc-
Isoglobo-3	iGb_3	α-Gal-(1-3)-ß-Gal-(1-4)-ß-Glc-
Lacto	Lc_4	ß-Gal-(1-3)-ß-GlcNAc-(1-3)- ß-Gal-(1-4)-ß-Glc-
Neolacto-4	nLc_4	ß-Gal-(1-4)-ß-GlcNAc-(1-3)- ß-Gal-(1-4)-ß-Glc-
Neolacto-6	nLc_6	ß-Gal-(1-4)-ß-GlcNAc-(1-3)-ß-Gal-(1-4)-ß-GlcNAc-(1-3)-ß-Gal-(1-4)-ß-Glc-

saccharide linked to ceramide. This allows to identify to which saccharide is linked an additional residue; the linkage position of the additional glycose(s) is coded with an apical Arabic number near to the Roman number (see Table 2 for some examples using both the two nomenclature).

3 Structure of Gangliosides

Since these are amphiphilic compounds, two portions with opposite physical–chemical properties, one hydrophilic, the other hydrophobic, represent the ganglioside structure. In the hydrophilic portion, one or more pairs of torsional angles, thus allowing for different ganglioside conformers, may represent each glyosidic bond.

3.1 The Ganglioside Hydrophilic Portion

Sugar units in varying numbers compose the hydrophilic portion: both neutral sugar(s) and sialic acid residue(s) range from one to six. Neutral sugars are glucose, galactose, *N*-acetylated glycosamines and fucose. Both α and β configurations are available. Sialic acid is the trivial name of all the derivative of neuraminic acid, that according to the IUPAC-IUB nomenclature is coded as

Table 2
Structures of the main gangliosides of human brain nervous system

Abbreviations		Structure
According to Svennerholm	According to the IUPAC-IUB.	
GM4	Neu5AcGalCer	α-Neu5Ac-(2-3)-ß-Gal-(1-1)-Cer
GM3	II^{3}Neu5AcLacCer	α-Neu5Ac-(2-3)-ß-Gal-(1-4)-ß-Glc-(1-1)-Cer
GD3	II^3(Neu5Ac)$_2$LacCer	α-Neu5Ac-(2-8)-α-Neu5Ac-(2-3)-ß-Gal-(1-4)-ß-Glc-(1-1)-Cer
GM2	II^{3}Neu5AcGg$_3$Cer	ß-GalNAc-(1-4)-[α-Neu5Ac-(2-3)]-ß-Gal-(1-4)-ß-Glc-(1-1)-Cer
GM1	II^{3}Neu5AcGg$_4$Cer	ß-Gal-(1-3)-ß-GalNAc-(1-4)- [α-Neu5Ac-(2-3)]-ß-Gal-(1-4)-ß-Glc-(1-1)-Cer
GD1a	IV^{3}Neu5AcII^{3}Neu5AcGg$_4$Cer	α-Neu5Ac-(2-3)-ß-Gal-(1-3)-ß-GalNAc-(1-4)-[α -Neu5Ac-(2-3)]-ß-Gal-(1-4)-ß-Glc-(1-1)-Cer
GalNAc-GD1a	IV^{3}Neu5AcII^{3}Neu5AcGg$_5$Cer	ß-GalNAc-(1-4)-[α-Neu5Ac-(2-3)]-ß-Gal-(1-3)-ß-GalNAc-(1-4)-[α -Neu5Ac-(2-3)]-ß-Gal-(1-4)-ß-Glc-(1-1)-Cer
GD1b	II^3(Neu5Ac)$_2$Gg$_4$Cer	ß-Gal-(1-3)-β-GalNAc-(1-4)-[α-Neu5Ac-(2-8)-α -Neu5Ac-(2-3)]-ß-Gal-(1-4)-ß-Glc-(1-1)-Cer
GD1b-lactone	II^3[Neu5Ac-(2-8,1-9)-Neu5Ac]Gg$_4$Cer	ß-Gal-(1-3)-β-GalNAc-(1-4)-[α-Neu5Ac-(2-8,1-9)-α -Neu5Ac-(2-3)]-ß-Gal-(1-4)-ß-Glc-(1-1)-Cer
GT1b	IV^{3}Neu5AcII^3(Neu5Ac)$_2$Gg$_4$Cer	α-Neu5Ac-(2-3)-ß-Gal-(1-3)-β-GalNAc-(1-4)-[α-Neu5Ac-(2-8)-α -Neu5Ac-(2-3)]-ß-Gal-(1-4)-ß-Glc-(1-1)-Cer
GQ1b	IV^3(Neu5Ac)$_2$II^3(Neu5Ac)$_2$Gg$_4$Cer	α-Neu5Ac-(2-8)-α-Neu5Ac-(2-3)-ß-Gal-(1-3)-β-GalNAc-(1-4)-[α-Neu5Ac-(2-8)-α-Neu5Ac-(2-8)-α-Neu5Ac-(2-3)]-ß-Gal-(1-4)-ß-Glc-(1-1)-Cer

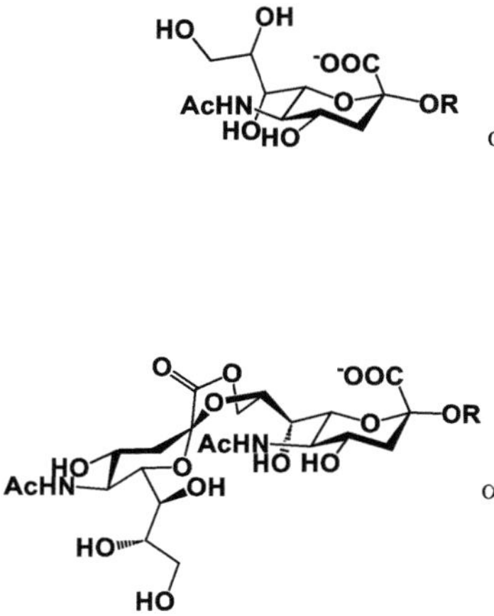

α-Neu5Ac-R

α-Neu5Ac-(2-8,1-9)-α-Neu5Ac-R

Fig. 1 Structure of the *N*-acetyl neuraminic acid, Neu5Ac, and of the lactone ring between two Neu5Ac

5-amino-3,5-dideoxy-D-*glycero*-D-*galacto*-non-2-ulopyranosonic acid. A few tens of sialic acid structure have been found in nature [6], but the Neu5Ac, the Neu5Gc, and the Neu5,9Ac_2 result to be the most abundant. Neu5Gc is not present in healthy humans but has been found in some human tumors [7]. In a few cases, lactone rings between two sialic acids have been described [8] (Fig. 1). The biosynthesis of the oligosaccharide chain occurs in the Golgi by sequential activity of glycosyltransferases that requires nuclear nucleotide activated sugars (see Fig. 2). The glycosyltransferases are specific for both the nucleotide activated sugar and the acceptor. Remodeling of the Golgi structures can occur at the surface of the cells by membrane associated glycosyltransferases and glycohydrolases.

Figure 3 shows the structure of five different human gangliosides, GM3, GM1, GalNAc-GD1a, GD1b-Lactone, and GQ1b. GM3 is the most abundant ganglioside in all mammals being component of all cells [9]. GM1 is one of the main brain gangliosides [10]. GalNAc-GD1a is a minor brain ganglioside [11]. GD1b-Lactone has been found in human brain in growing quantity with aging [12]. GQ1b has been described to be specifically associated to the Miller–Fisher syndrome [13].

Table 1 shows the structure of the main neutral chains present in gangliosides together with their abbreviations.

3.2 The Ganglioside Hydrophobic Portion

The hydrophilic head of gangliosides is linked to a double-tailed hydrophobic portion named ceramide (Fig. 4). Ceramide is constituted by a long chain amino alcohol (LCB) [14], linked *via* an amide bond with a fatty acyl chain. The most abundant long chain amino alcohol is the (2*S*,3*R*,4*E*)-2-amino-1,3-dihydroxy-octadec-4-ene, trivially known as sphingosine or C18-sphingosine. In the nervous system, there is also the (2*S*,3*R*,4*E*)-2-amino-1,3-dihydroxy-eicos-4-ene,

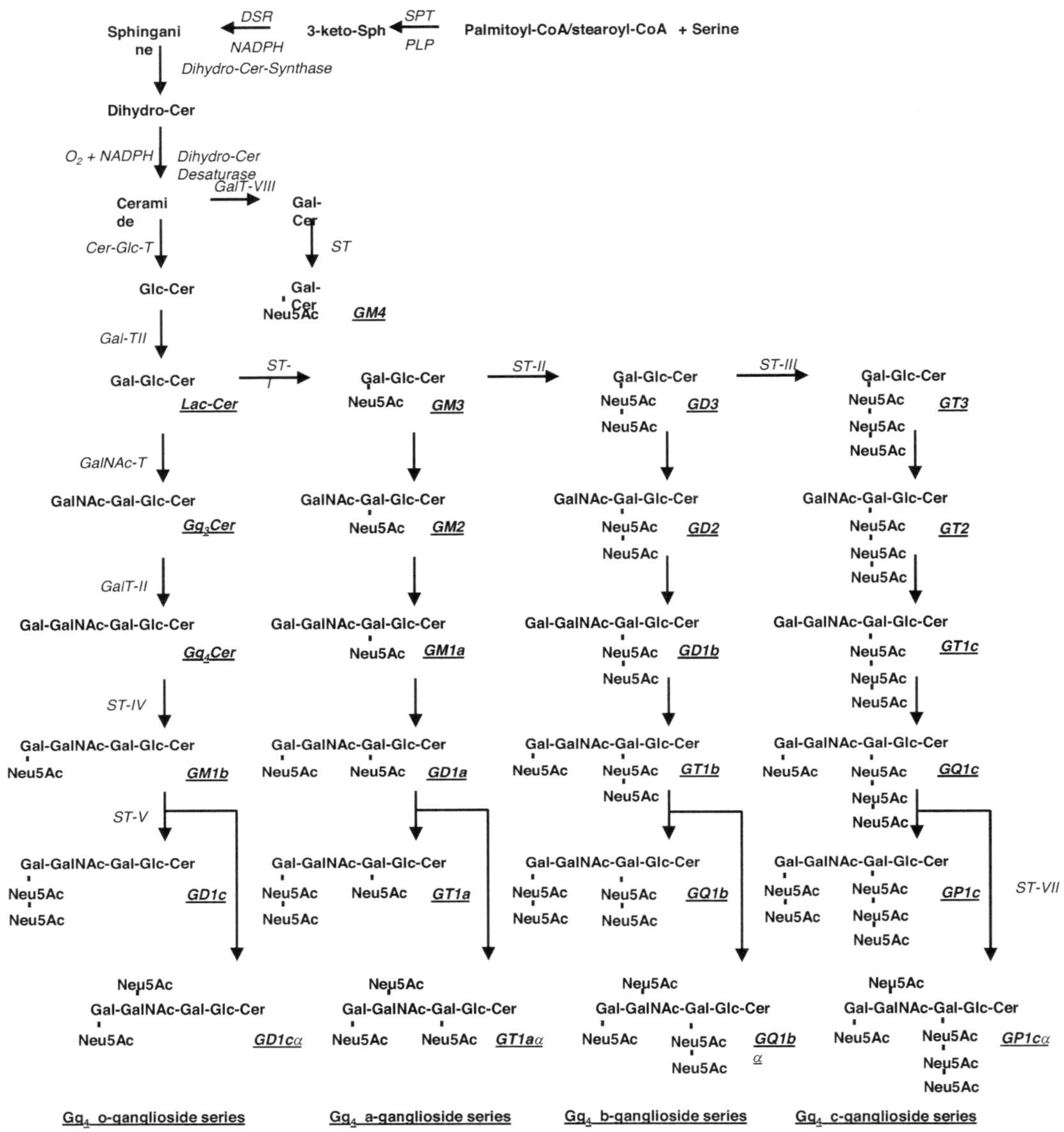

Fig. 2 Scheme of the sequential steps that lead to the synthesis of different series of gangliosides

GM3

GM1

GalNAc-GD1a

GD1b-Lactone

GQ1 b

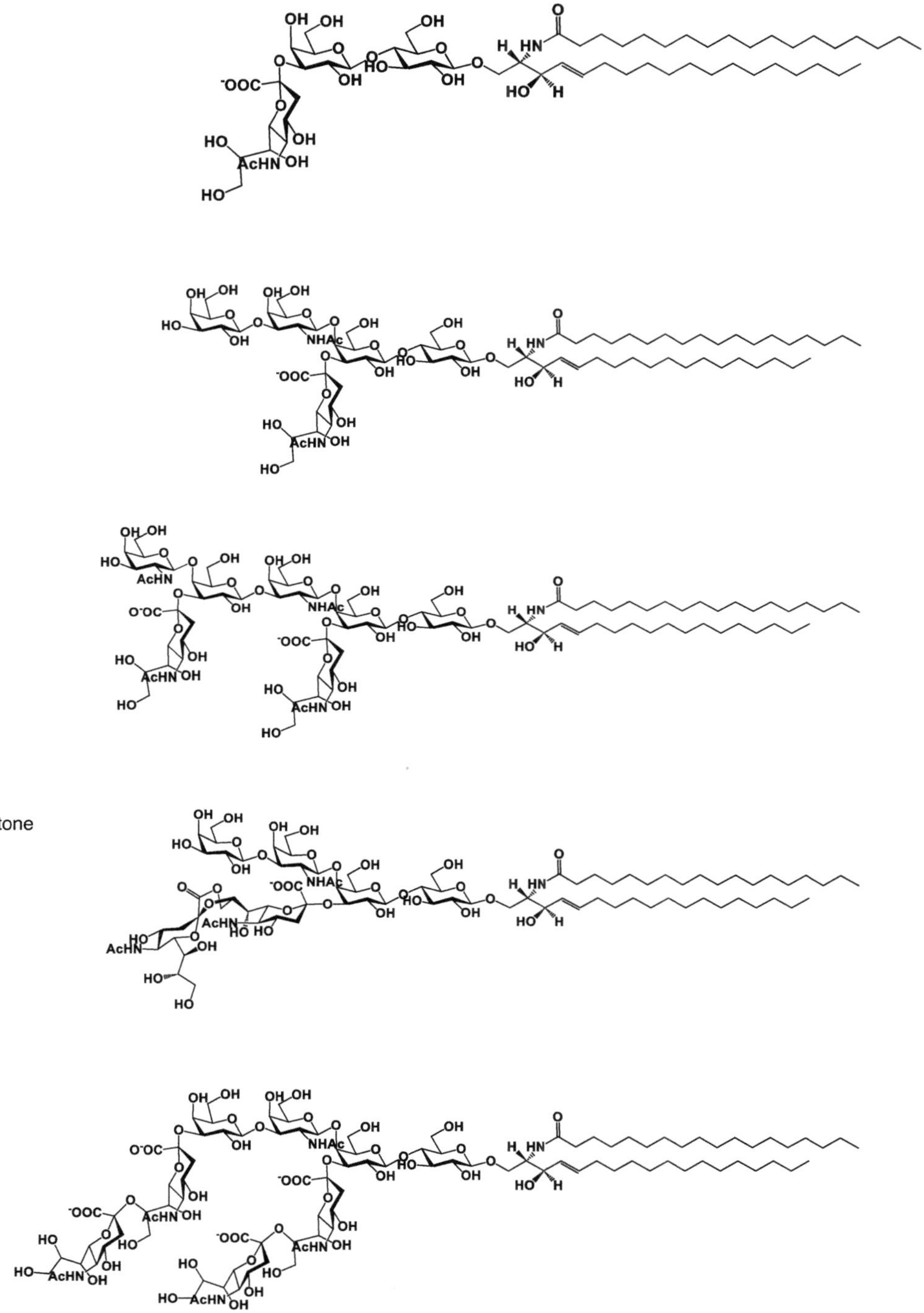

Fig. 3 Structure of gangliosides GM3, GM1, GalNAc-GD1a, GD1b-Lactone, and GQ1b

sphingosine

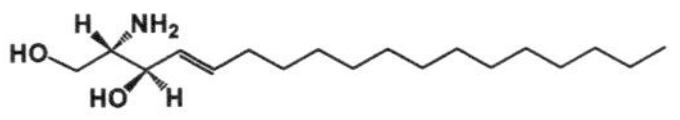

ceramide

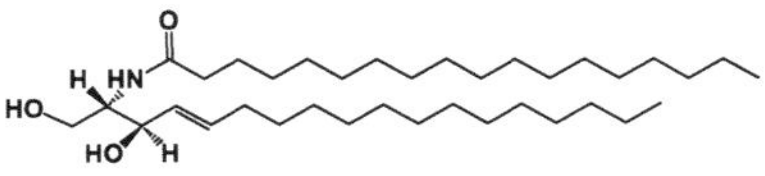

Fig. 4 Structure of C18-sphingosine and of Ceramide containing C18-sphingosine and stearic acid

or C20-sphingosine. The double bond is in trans configuration, which is quite unusual in mammalian long hydrophobic chains. The C20-sphingosine is absent or difficult to detect in human fetuses, but after birth it appears and then progressively increases, becoming abundant in aged people. A minor quantity of the LCB, 2%–4% of the total, is as unsaturated species. We recall that the word "sphingosine" is often used for any type of structure.

The fatty acid is variable in the structure, but the main species are from C14 to C24, with or without double bonds. In some cases, α-hydroxyl-acids have been described [15]. In the brain, stearic acid covers over the 90% of total fatty acids.

The ceramide has a perpendicular orientation with respect to the axes of the two hydrocarbon chains, a parallel orientation that is stabilize by the four to five unsaturation of sphingosine [16]. This makes ceramide a rigid structure. In addition, in the amide linkage the NH is a proton donor, and the carbonyl is a proton acceptor allowing the formation of cross hydrogen bonds. Both these features are in favor of ganglioside segregation and reduction of membrane fluidity [17].

The ceramide contributes to the ganglioside packing parameter [18] by mean of its volume and length of the alkyl chains. Thus, the length of LCB, 18 or 20 carbons, and the length of fatty acid contribute to the ganglioside properties and, indirectly, to membrane organization and properties. Increasing the LCB chain length by two carbons leads to an increased LCB hydrophobic volume (13%). The membrane must be flatter to accept the monomer, thus affecting the organization of the membrane environment in which the monomer is inserted [18].

3.3 The Three-Dimensional Structure of Brain Gangliosides

In gangliosides, the spatial arrangement of the neutral disaccharide -ß-Gal-(1-4)-ß-Glc-linked to ceramide fluctuates in a large energy minimum. A low motional freedom is predicted for the Glc residue, fully extended away from the bilayer normal due to the movement restriction imposed on the saccharide-lipid bond by the surrounding membrane surface [19]. Glucose, as well as the lipid portion inserted into the membrane, shows an averaging wobbling motion about the glucosidic bond, the director axis being the bilayer normal; however, the motional averaging effects seem not be large [20]. Instead, by adding sialic acid to the disaccharide, as in GM3, the α-Neu5Ac-(2-3)-β-Gal- linkage exists, at least, in two main conformations (Fig. 5).

By increasing the length of the chain, as in gangliosides GM2, GM1, GD1a, and GalNAc-GD1a, the trisaccharide sequence (-)β-GalNAc-(1-4)[α-Neu5Ac-(2-3)-]-β-Gal-, is considered a very rigid super unit due to several interresidual interactions [21-23] occurring between the Neu5Ac and GalNAc units. This trisaccharide super unit is stabilized by a hydrogen bond between the GalNAc amide proton and the Neu5Ac carboxyl group. Figure 6 shows the three-dimensional representation of the β-GalNAc-(1-4)[α-Neu5Ac-(2-3)-]-β-Gal trisaccharide. In contrast to this, the addition of galactose to hexosamine makes the disaccharide β-Gal-(1-3)-β-GalNAc- very mobile, allowing the existence of, at least, 2 main conformations of the glycosidic linkage (Fig. 7). Thus, excluding the two linkages nearest to ceramide, GM2 is present only in one preferred

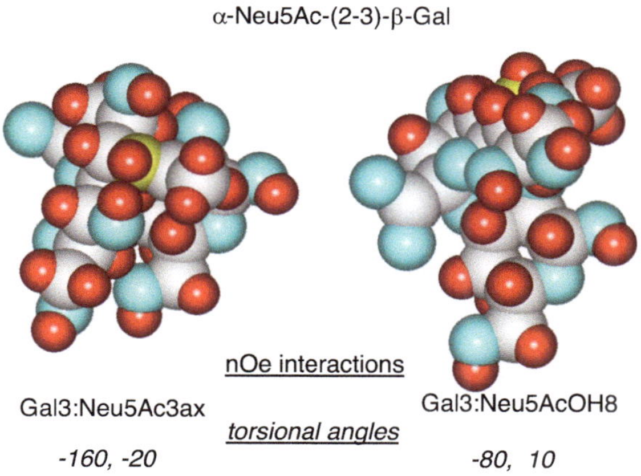

Fig. 5 The two conformations of the α-Neu5Ac-(2-3)-β-Gal linkage

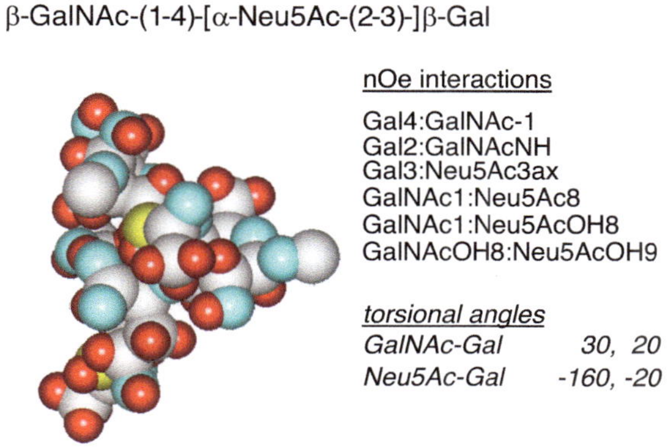

Fig. 6 The single rigid conformation of the trisaccharide β-GalNAc-(1-4)[α-Neu5Ac-(2-3)-]-β-Gal

conformation, while GM1 carrying the external mobile linkage β -Gal-(1-3)-β-GalNAc, is in two and GD1a carrying the external α -Neu5Ac-(2-3)-β-Gal-(1-3)-β-GalNAc in four. In GalNAc-GD1a, the addition of a GalNAc to GD1a gives a second β-GalNAc-(1-4) [α-Neu5Ac-(2-3)-]-β-Gal-fragment directly bound to the first one, this reducing the number of conformers from four to two. Figure 8 shows a scheme for the number of ganglioside conformers.

In GD1b and GT1b, the addition of a sialic acid residue to the sialic acid of the -β-GalNAc-(1-4)[α-Neu5Ac-(2-3)-]-β-Gal-trisaccharide, leads to the loss of the interresidual contacts between GalNAc and Neu5Ac linked to Gal. But interactions between GalNAc

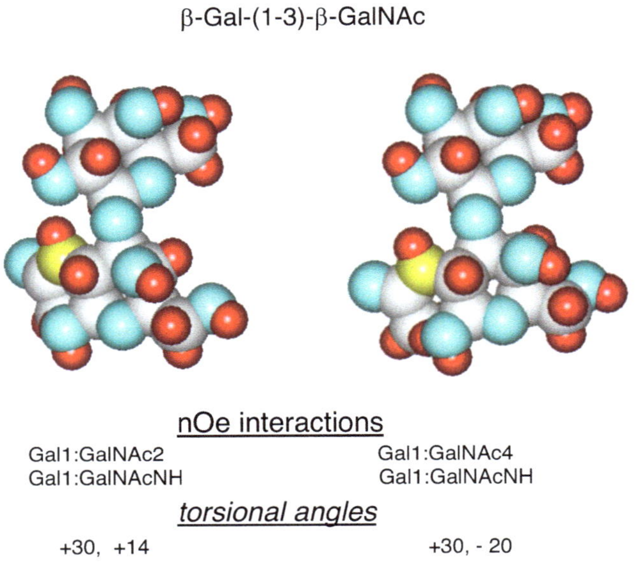

Fig. 7 The two conformations of the β-Gal-(1-3)-β-GalNAc linkage

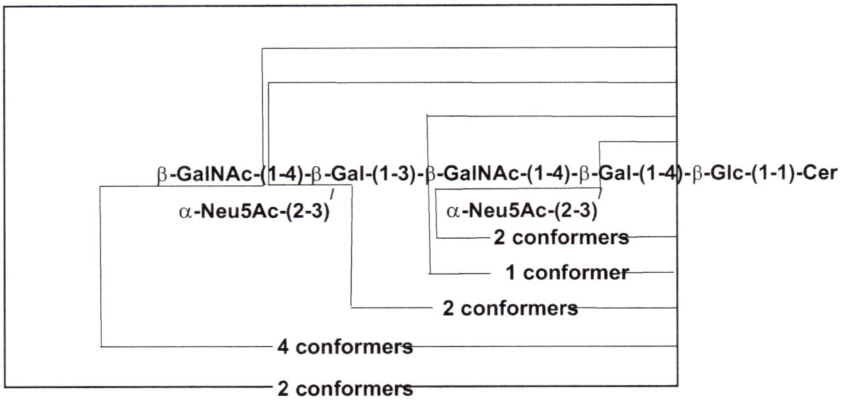

Fig. 8 Scheme showing the number of ganglioside conformers due to saccharide composition

and external Neu5Ac occur in the two gangliosides. According to these constraints, the tetrasaccharide chain -β-GalNAc-(1-4)[α-Neu-5Ac-(2-8)-α-Neu5Ac-(2-3)-]-β-Gal- is organized in a circle with a hole that has been calculated to be about 3 Å wide in the center [24]. Such a conformation confers a bulkier character to the ganglioside portion closer to the hydrophobic–hydrophilic interface. Interestingly, the solid angle required by the disialosyl chain is wide enough to accommodate an additional sialic acid unit in an external position. In fact, despite carrying an extra sugar linked to the external galactose unit, GT1b requires the same surface area as GD1b [18]. When the external sialic acid carboxyl group of the disialyl chain esterifies the inner sialic acid residue, as in GD1b-lactone, the compact conformation of the -β-GalNAc-(1-4)[α-Neu5Ac-(2-3)-]-β-Gal- observed in GM1 and GD1a is restored. This forces a better lining up of the disialosyl chain with the neutral oligosaccharide chain, reducing the angle between the neutral chain and the inner sialic acid axis [18].

4 Physical Properties of Gangliosides

4.1 Aggregation Properties of Gangliosides

Gangliosides, due to the extended double tail, are amphiphilic molecules, but large differences in the hydrophilic chain size compared to minor differences in maximum length and volume of ceramide result in large differences on their physicochemical properties. Thus, the hydrophilic–hydrophobic balance is the competition between two opposite strong requirements; any variation in such requirements affects their behavior when inserted in the cell plasma membranes. This has been determined by studying the aggregative behavior of gangliosides dispersed in water solutions. For this purpose, we may considered the ganglioside series, GM3, GM2, GM1, GD1a, GalNAc-GD1a, GD1b, GD1b-L, and GT1b, all prepared to have the same hydrophobic portion [18, 25]. Thus, inside an aggregate, the gangliosides require an interfacial area large enough to provide the in hydrophilic layer a place wide enough to host the oligosaccharide chain together with its hydration water. According, larger interfacial areas require smaller aggregates with lower aggregation numbers.

Table 3 shows the aggregate physical parameters of different gangliosides determined by static and dynamic laser light scattering measurements, confirmed for some gangliosides by neutron and X-ray scattering experiments [18, 25].

GM3, with the smallest headgroup in the studied series, forms vesicles. The ganglioside vesicles form spontaneously, with no need for external energy supply (sonication, pH jump, etc.). This suggests that the energy required for ganglioside vesicle formation is quite low, probably due to the head groups that require a surface area exceeding the lipid chains lateral dimension. This is supported by the fact that

Table 3
The molecular weight M, the aggregation number N, the hydrodynamic radius Rh, the axial ratio Ra/Rb, the charge Q of the ganglioside aggregate, the surface area a_0 calculated with a rigid lipid chain length l = 23.64 Å, and the packing parameter P of the monomer in the aggregate

	Aggregate	Rh Å	N	M KDa	Ra/ Rb	a_0 Å^2	P	Q e.u.
GM3	ves	~250	~14,000	~16,700		~80	>0.5	
GM2	mic	66.0	529	740	3.1	92.0	0.440	100
GM1	mic	58.7	301	470	2.3	95.4	0.428	48
GD1a	mic	58.0	226	418	2.0	98.1	0.416	60
GalNAc-GD1a	mic	60.0	246	509	2.1	97.0	0.421	
GD1b	mic	52.0	168	311	1.8	100.8	0.405	
GD1b-lactone	mic	57.0	229	424	2.1	97.6	0.418	
GT1b	mic	53.2	176	378	1.8	100.8	0.405	

Form of aggregation: *ves* vesicle, *mic* micelle. Gangliosides contained a ceramide moiety constituted by a mixture of C18 and C20 sphingosine (about 1:1) and stearic acid (>90% of the total fatty acid content)

ganglioside vesicles are interdigitated, so that a larger surface area per monomer is available than in noninterdigitated situation.

From GM2 to GT1b, the sugar head group starts to be much more extended so that micelles are formed, not vesicles. In micelles, even small changes in the hydrophilic portion change the headgroup repulsive contribution and the aggregation properties.

Considering the three gangliosides GM3, GM2, and GM1 (Table 3), that contain a sole sialic acid linked to position 3 of the same galactose II, the progressively increase of neutral sugar units enhances the hydrophilic character of the ganglioside together with a stronger hydrophilic repulsive contribution. Thus, the ganglioside requires a larger interfacial area and this leads to the decrease of the aggregation number and to a more curved and smaller micelle.

The gangliosides GM1, GD1a and GT1b carry the ganglio-4 chain but differ in the number and disposition of the sialic acid residues. The progressive increase of sugar number, from five to seven, leads to a progressive greater hydrophilic repulsive contribution, leading to progressively smaller and more spherical aggregates with lower aggregation numbers. Nevertheless, GD1b, which has the same saccharide composition as GD1a, behaves similarly to GT1b, a more hydrophilic ganglioside. GalNac-GD1a, that carries seven sugar units like GT1b packs tighter than GD1a, which carries only six.

Thus, we can say that in front to an extended and ramified structure of the ganglioside head group, together with a general hydrophilicity evaluation, it is necessary to consider some additional rules.

In GD1b, the second sialic acid residue, which in GD1a is linked to the external galactose unit, is linked to the first sialic acid, forming a disialyl chain close to the hydrophilic–hydrophobic interface. Thus, the monomer needs more room by at this level and a wider solid angle per GD1b molecule is required inside the micelle; this is reflected in a smaller aggregation number. Such a conformation confers a bulkier character to the ganglioside portion closer to the hydrophobic–hydrophilic interface. This is confirmed comparing the micellar properties of GD1b and GT1b. Although GT1b carries one more sugar, has a sialic acid linked to the external galactose unit and a higher hydrophilic character, it forms micelles with the same aggregation number as GD1b. This means that the solid angle required by the disialosyl chain is wide enough to host an additional sialic acid unit in an external position. When we compare GD1b-lactone with GD1b, we observe that the aggregation number of GD1b-L micelles is higher than GD1b and close to GD1a; this means that it is no longer necessary to increase the solid angle to locate the second sialic acid. In fact, the presence of the lactone ring in GD1b-L induces changes in the three-dimensional structure (see the above chapter), that is more similar to that of GD1a.

Far from following the number-of-sugars rules, the aggregation number of GalNAc-GD1a is slightly higher than GD1a. As stated earlier, the trisaccharide GalNAc-Gal-Neu5Ac linked to the glucose unit forms a strongly interacting bulky superunit able to influence the final ganglioside 3D-structure. GalNAc-GD1a differs from GD1a in that it has a second GalNAc unit at the end of the neutral oligosaccharide chain. Therefore, GalNAc-GD1a carries a second GalNAc-Gal-Neu5Ac trisaccharide rigid fragment directly bound to the first one (see the above chapter). This reduced mobility could explain why GalNAc-GD1a monomers occupy a reduced solid angle, in comparison with GD1a.

4.2 The Critical Aggregation Concentration of Gangliosides

It is important to spend some words on the critical aggregation concentration of gangliosides, critical liposome concentration (clc) for GM3 and GM4, and critical micellar concentration (cmc) for the other gangliosides. Only a very low value can explain how gangliosides remain stably inserted into the plasma membrane, presenting a very scant process of monomer shedding [26] and how few are the monomers that become components of plasma membranes when gangliosides are added to the cell culture medium [27].

The cmc of GM1, GD1a, GD1b, GT1b, and GQ1b has been calculated according to the suggestion of Charles Tanford [28] and knowing the surface area occupied by the gangliosides as determined by laser light scattering measurements.

The attraction contributions due to the hydrocarbon chains are almost the same for brain gangliosides [28], and depends on the contribution due to sphingosine, considering the double bond in position 4–5 of the chain and the contribution due to a C18 fatty acid (the main acyl chain). From these contributions, knowing the ganglioside surface areas, and from previous values determined using two distinct procedures, 2×10^{M8} M [18], and 3.6×10^{-10} M [29], it is in the range of 3.4×10^{-9} M - 3.9×10^{-8} M.

Applying the same procedure to the synthetic GM1-C2, where acetic acid replaces stearic acid [30], the calculated cmc resulted 3.8×10^{-5} M, which is very close to 2.3×10^{-5} M, determined by surface tension measurement. See [18] for details on the procedure for calculate the c.m.c. according to [28].

Results indicating that the ganglioside c.m.c.(s) are very low were yet obtained 50 years ago, when we studied the behavior of gangliosides on dialysis, as a function of ganglioside concentration and dialysis time [31]. However, at that time, the results have neither been fully understood nor discussed properly. Solutions of the ganglioside mixture containing the 5 main gangliosides GM1, GD1a, GD1b, GT1b, and GQ1b down to up 10^{-8} M was submitted to 4 days dialysis, the water being changed several times a day: at the end the mixture was analyzed to assess the ganglioside content. Both the ganglioside distribution and the total concentration were unchanged, an indication that no detectable amount of any species was lost during the dialysis. This is further indirect proof that the exchange rates of the five gangliosides are slow and quite similar and that the c.m.c.(s) cannot be too different.

4.3 Plasma Membrane Segregation of Gangliosides

Gangliosides, with their structure display characteristics considered as the most relevant driving forces for the formation of membrane lipid domains, now worldwide coded as lipid rafts.

Gangliosides are component of the external leaflet of plasma membranes, with the oligosaccharide chain protruding toward the extracellular space, and interacting with a variety of cell surface and extracellular molecules. Thus, their plasma membrane concentration is locally relatively high, and their influence on the membrane organization can represent an important driving force for the creation of the lateral order.

Gangliosides exhibit a hydrophilic bulky structure compared to that of phospholipids. The difference in volume of the hydrophilic groups among the membrane lipids is the main driving force for clustering and segregation of the different membrane components. This, in fact reduces the interfacial free energy. The oligosaccharide chains of gangliosides show great differences, due to the number of sugars units, the number of ramifications, the number of negative charges, and the number of isoforms. All this determines the surface area necessary to host the monomer at the water/lipid interface and this increases with the increase of the volume occupied by the head group. Larger is the interfacial area required by ganglioside oligosaccharide structure, and more pronounced is the segregation of the ganglioside within the phospholipids, and more positive is the curvature of the resulting ganglioside-rich membrane microenvironment [32].

Concerning the volume occupied by the oligosaccharide chain, it is necessary to recall the role of molecules of water associated to the sugar units [33].

The number of water molecules tightly associated with a given oligosaccharide chain ranges from 22 to 60, depending on the structure of the oligosaccharide. And some of these interactions are very strong [23]. Water mediates hydrogen bridges and avoid repulsive effects between the negatively charged oligosaccharide chains, thus significantly contributing to the stabilization of the membrane cluster.

In addition to the features of the hydrophilic portions, we must recall those of the ceramide. Gangliosides with a high content of saturated acyl chains, stearic and palmitic acid, display a high melting point and can be tightly packed with a high degree of order having a transition temperature higher than glycerophospholipids. In addition to this, the lateral phase separation of gangliosides within the phospholipid bilayer is directly correlated with the acyl chain length and inversely correlated with the degree of unsaturation [34]. In addition, the amide group between sphingosine and fatty acid, involving the nitrogen as donor and the oxygen as acceptor, allows cross linkages.

Conflict of Interest

The authors have no conflicts of interest to declare that are relevant to the content of this chapter.

References

1. Klenk E (1935) Über die natur der phophatide und anderer lipoide des gehirns und der leber bei der niemann-pickschen krankheit. Z Phys Chem 23:24–36
2. Klenk E (1939) Beitrdge zur Chemie der Lipiodosen. Z. Phys Chem 262:128–143
3. Kuhn R, Wiegandt H (1963) Die konstitution der Ganglio-N-Tetraose und des gangliosids GI. Chem Ber 96:866–880
4. Svennerholm L (1980) Ganglioside designation. Adv Exp Med Biol 125:11
5. IUPAC-IUB Joint Commission (1999) IUPAC-IUB Joint Commission on biochemical nomenclature (JCBN) nomenclature of glycolipids recommendations 1997. J Mol Biol 286(3):963–970
6. Schauer R (1982) Sialic acids. Chemistry, metabolism and function. Springer-Verlag, Wien, New York
7. Marquina G, Waki H, Fernandez LE, Kon K, Carr A, Valiente O, Perez R, Ando S (1996) Gangliosides expressed in human breast cancer. Cancer Res 56(22):5165–5171
8. Mauri L, Sonnino S (2023) Alkali-labile gangliosides. Glycoconj J 4:269–276
9. Mauri L, Casellato R, Kirschner G, Sonnino S (1999) A procedure for the preparation of GM3 ganglioside from GM1-lactone. Glycoconj J 16:197–203
10. Aureli M, Mauri L, Ciampa MG, Prinetti A, Toffano G, Secchieri C, Sonnino S (2016) GM1 ganglioside: past studies and future potential. Mol Neurobiol 53:1824–1842
11. Svennerholm L, Månsson JE, Li YT (1973) Isolation and structural determination of a novel ganglioside, a disialosylpentahexosylceramide from human brain. J Biol Chem 248:740–742
12. Riboni L, Sonnino S, Acquotti D, Malesci A, Ghidoni R, Egge H, Mingrino S, Tettamanti G (1986) Natural occurrence of ganglioside lactones. Isolation and characterization of GD1b

inner ester from adult human brain. J Biol Chem 261:8514–8519
13. Willison HJ, Veitch J, Paterson G, Kennedy PG (1993) Miller Fisher syndrome is associated with serum antibodies to GQ1b ganglioside. J Neurol Neurosurg Psychiatry 56(2):204–206
14. Karlsson KA (1970) On the chemistry and occurrence of sphingolipid long-chain bases. Chem Phys Lipids 5:6–43
15. Riboni L, Acquotti D, Casellato R, Ghidoni R, Montagnolo G, Benevento A, Zecca L, Rubino F, Sonnino S (1992) Changes of the human liver GM3 ganglioside molecular species during aging. Eur J Biochem 203:107–113
16. Pascher I (1976) Molecular arrangements in sphingolipids. Conformation and hydrogen bonding of ceramide and their implication on membrane stability and permeability. Biochim Biophys Acta 455:433–451
17. Bertoli E, Masserini M, Sonnino S, Ghidoni R, Cestaro B, Tettamanti G (1981) Electron paramagnetic resonance studies on the fluidity and surface dynamics of egg phosphatidylcholine vesicles containing gangliosides. BBA-Biomembranes 647:196–202
18. Sonnino S, Cantu' L, Corti M, Acquatti D, Venerando B (1994) Aggregative properties of gangliosides in solution. Chem Phys Lipids 71:21–45
19. Nyholm PG, Pascher I (1993) Orientation of the saccharide chains of glycolipids at the membrane surface: conformational analysis of the glucose-ceramide and the glucose-glyceride linkages using molecular mechanics (MM3). Biochemistry 32:1225–1234
20. Aubin Y, Prestegard JH (1993) Structure and dynamics of sialic acid at the surface of a magnetically oriented membrane system. Biochemistry 32:3422–3428
21. Acquotti D, Poppe L, Dabrowski J, von der Lieth GW, Sonnino S, Tettamanti G (1990) Three-dimensional structure of the oligosaccharide chain of GM1 ganglioside revealed by a distance-mapping procedure: a rotating and laboratory frame nuclear overhauser enhancement invastigation of native glycolipid in dimethyl sulfoxide and in water-dodecylphosphocholine solutions. J Am Chem Soc 112:7772–7778
22. Poppe L, van Halbeek H, Acquotti D, Sonnino S (1994) Carbohydrate dynamics at a micellar surface: GD1a headgroup transformations revealed by NMR spectroscopy. Biophys J 66:1642–1652
23. Brocca P, Berthault P, Sonnino S (1998) Conformation of the oligosaccharide chain of GM1 ganglioside in a carbohydrate enriched surface. Biophys J 74:309–318
24. Acquotti D, Fronza G, Ragg E, Sonnino S (1991) Three dimensional structure of GD1b and GD1b-monolactone gangliosides in dimethylsulphoxide: a nuclear overhauser effect investigation supported by molecular dynamics calculations. Chem Phys Lipids 59:107–125
25. Sonnino S, Prinetti A, Mauri L, Chigorno V, Tettamanti G (2006) Dynamic and structural properties of sphingolipids as driving force to the formation of membrane domains. Chem Rev 2006(106):2111–2125
26. Chigorno V, Giannotta C, Ottico E, Sciannamblo M, Mikulak J, Prinetti A, Sonnino S (2005) Sphingolipid uptake by cultured cells: complex aggregates of cell sphingolipids with serum proteins and lipoproteins are rapidly catabolized. J Biol Chem 280:2668–2675
27. Chigorno V, PittoM, Cardace G, Acquatti D, Kirschner G, Sonnino S, Guidoni R, Tettamanti G (1985) Association of gangliosides to fibroblasts in culture: a study performed with GM1(14C)-labelled at the sialic acid acetyl group. Glycoconjugate J 2:279–291
28. Tanford, C, 1980. The hydrophobic effect: formation of micelles and biological membranes. John Wiley and Sons Inc, New York
29. Hall DG, Pethica BA (1967) In: Shick MJ (ed) Nonionic surfactants. Marcel Dekker, New York
30. Sonnino S, Cantù L, Corti M, Acquotti D, Kirschner G, Tettamanti G (1990) Aggregation properties of semisynthetic GM1 ganglioside (II3Neu5AcGgOse4Cer) containing an acetyl group as acyl moiety. Chem Phys Lipids 56:49–57
31. Ghidoni R, Sonnino S, Tettamanti G (1978) Behaviour of gangliosides on dialysis. Lipids 13:820–822
32. Brocca P, Sonnino S (1997) Dynamic and spatial organization of surface gangliosides. Trends Glycosci Glyc 9:433–445
33. Bach D, Sela B, Miller IR (1982) Compositional aspects of lipid hydration. Chem Phys Lipids 31:381–394
34. Masserini M, Palestini P, Freire E (1989) Influence of glycolipid oligosaccharide and long-chain base composition on the thermotropic properties of dipalmitoylphosphatidylcholine large unilamellar vesicles containing gangliosides. Biochemistry 28:5029–5034

Chapter 2

Chemical Synthesis of Gangliosides: Synthesis of Uniform Ganglioside

Hiromune Ando and Akihiro Imamura

Abstract

Chemical synthesis of fine gangliosides will benefit biological and therapeutic studies because of their major roles in biological processes. However, the chemical synthesis of gangliosides needs considerable effort to construct α-glycoside of sialic acid and the glycosidic linkage between oligosaccharide moiety and ceramide. This chapter focuses on recent progress in the chemical synthesis of gangliosides, regarding the stereoselective glycosidation of sialic acid and the construction of glycolipid framework.

Keywords Chemical synthesis, Glycosidation, Ganglioside, Sialic acid, Ceramide

1 Introduction

Gangliosides are a large and diverse family of glycolipids composed of a branched lipid called ceramide (Cer) and glycans terminated with sialic acid. In the outer leaflet of the plasma membrane, gangliosides are anchored by their lipid moieties through hydrophobic interactions with other membrane molecules, thereby presenting their glycan moieties to the outer milieu. Sialic acid exposes an array of polar functional groups to the outer milieu because of its outermost positioning and the characteristic structure, which comprises a ω -amino acid-like moiety and glycerol branch. Therefore, gangliosides participate in unique interactions with sugar-binding proteins. Gangliosides in mammals are involved in biological processes, including cell-type-specific adhesion and the entry of toxins, viruses, and bacteria into cells [1–4]. Additionally, gangliosides modulate the function of protein receptors in the plasma membrane [5–7]. Gangliosides produced in echinoderms (e.g., starfish and sea cucumbers) exhibit neuritogenic activity toward model neuron cells (PC-12) in the presence of neural growth factors [8]. Some echinodermatous gangliosides (EGs) show greater activity than the representative neuritogenic mammalian ganglioside GM1, making EGs appealing

Koichi Furukawa (ed.), *Gangliosides*, Methods in Molecular Biology, vol. 3035,
https://doi.org/10.1007/978-1-0716-5288-6_2,

molecules as potential drug seeds for neural disorder therapy. EGs differ from mammalian gangliosides in their unique glycan and Cer moieties. The glycan moieties of EGs contain partially modified sialic acid congeners and their oligomerized congeners. In addition, sialic acids are often seen as internal residues in EGs, which is not observed in mammalian gangliosides. More than 15 types of EGs have been structurally identified, and their neuritogenic activity has been examined. However, the structure–activity relationships and molecular mechanisms underlying the potentiation of neurite generation by EGs are unknown.

1.1 Issues in the Chemical Synthesis of Gangliosides

The structural diversity of gangliosides has been a major problem in understanding and harnessing their biological functions and activities; thus, the chemical synthesis of gangliosides is important. The major difficulties in synthesizing diverse gangliosides are to incorporate sialic acids into oligosaccharide moieties in a stereoselective manner and to combine sialyl oligosaccharides with the Cer moiety (Fig. 1).

α-Glycosides of sialic acid (α-sialosides) can be synthesized through the reaction of an oxocarbenium ion generated from the sialic acid donor with the hydroxyl group of the glycosyl partner. However, the coupling yield and stereoselectivity are complicated by the structural features of sialic acid. First, the electron-withdrawing carboxyl group on the anomeric center makes the tertiary oxocarbenium ion intermediate unstable and susceptible to 2,3-elimination in combination with its 3-deoxy structure. Second, owing to the deoxy structure of the C3 position, the glycosidation reaction of sialic acid preferentially generates β-glycoside by the anomeric effect during glycosidation. In the past 50 years of research on sialic acid chemistry, extensive efforts have been devoted to overcoming the issue of stereocontrol of sialic acid glycosidation [9].

The issue of the conjugation of sialyloligosaccharide and Cer is mostly due to the poor reactivity of the 1-OH group of Cer. This poor reactivity can be attributed to the C3 amide moiety, which forms a hydrogen bond with 1-OH. In addition, Cer is likely to aggregate in organic solvents owing to hydrogen bonding between the head parts carrying an amide and a hydroxyl group and hydrophobic interaction

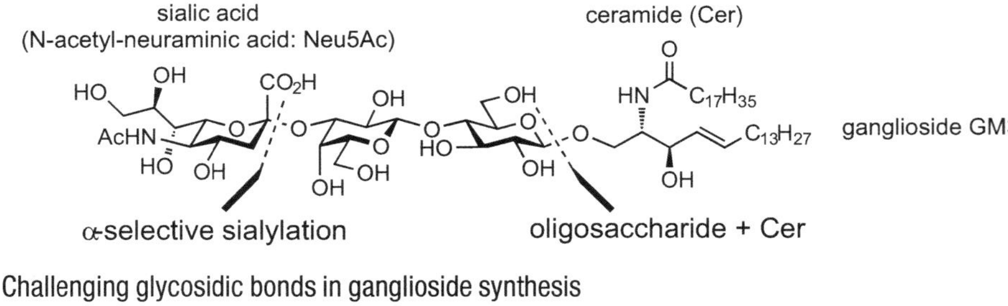

Fig. 1 Challenging glycosidic bonds in ganglioside synthesis

between long alkyl chain parts. This aggregation hampers the glycosylation reaction between the glycosyl donor and C1 hydroxyl group of Cer. Therefore, the glycosylation yield of Cer tends to decrease as the length of the glycosyl donor increases.

In this chapter, recently developed solutions to these issues in the chemical synthesis of gangliosides are described.

1.2 Sialylation

To predominantly produce α-glycoside during the glycosidation reaction of sialic acid, the β-face of oxocarbenium ion needs to be shielded from the nucleophilic attack by a hydroxyl group. To date, acetonitrile participation and participation of a nucleophilic substituent equipped at the C1 or C3 position have been employed for producing α-sialosides predominantly. Although the mechanism is not fully understood, the introduction of 4*O*,5*N*-oxazolizinone into a sialic acid donor significantly increases the α-selectivity. However, the profound effect of the substrate structure and reaction conditions on stereoselectivity is a common issue in the reported methods.

We examined the possibility of the substrate-based stereocontrol of sialidation via bridgehead oxocarbenium ions [10]. Inspired by the bridgehead carbocation (anti-Bredt carbocation) chemistry developed by Kraus et al. [11], we envisaged that an anti-Bredt oxocarbenium ion generated from a bicyclic α-sialic acid would restrict the attack of the nucleophile on the α-face via steric hindrance of the bicyclic system (Fig. 2). We also anticipated that a bridgehead oxocarbenium ion with a high ring strain would be more likely to form a tetrahedral sp_3 anomeric center with a glycoside bond.

The initial examination of the glycosidation reaction using macrocyclized sialyl donors with bicylco [8.2.2] to [13.2.2] systems revealed that bicyclic sialyl donors were completely α-selective, and sialyl donor **5** with a bicyclo [12.2.2] system provided the highest yield of α-glycoside (83%) among the bicyclic donors tested. Although donor **6**, with one carbon longer macroring than **5**, appeared comparable to **5** in terms of reactivity, **6** was more amenable to 1,2-elimination, yielding a dehydro derivative as the major byproduct, which may stem from the higher tolerance of the more flexible **6** to a planar structure around the anomeric center as an sp_2 carbon (Fig. 3).

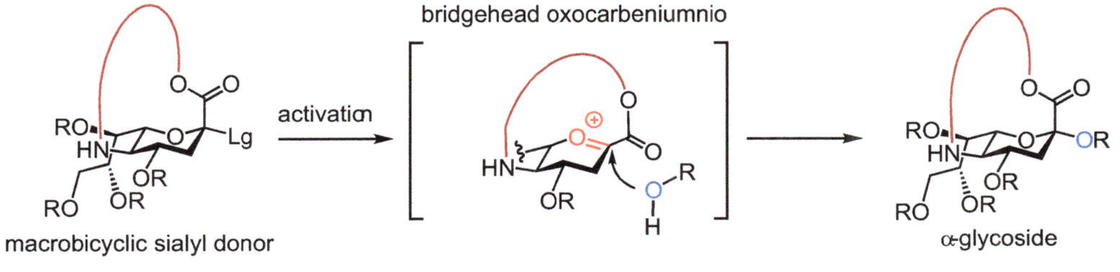

Fig. 2 Bridgehead oxocarbenium ion strategy for α-selective glycosidation of sialic acid

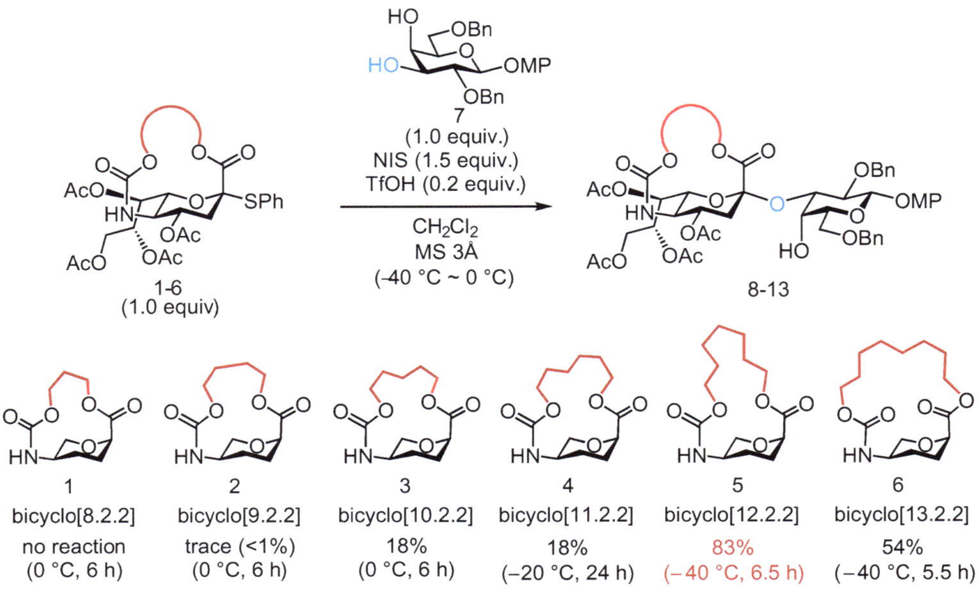

Fig. 3 Examination of macrobicyclic sialy donors

Next, we introduced a 2,2-dichloroethoxycarbonyl moiety into the tether to enable selective cleavage of the C5 amine group and subsequent modification into natural congeners, yielding dichlorosialyl donors **14** and **15**. These derivatives are glycosidated by a variety of hydroxylated molecules, including simple alcohols (**22**), monosaccharides (**23, 24**), polycyclic compounds (**25**), cholesterol (**26**), and oligosaccharides (**27**). In all cases, the sialoside was produced as a single stereoisomer with high yield, indicating the broad substrate scope of the bicyclic sialyl donors (Fig. 4). Phosphate **15** was also used to form *C*-glycoside bonds with carbon nucleophiles under acidic conditions.

1.3 Late-Stage Sialylation of Glycolipid to Produce Ganglioside

To overcome the issue of low efficiency in the conjugation of oligosaccharide and Cer, we devised building blocks known as "glucosyl ceramide (Glc-Cer) cassette" in which glucose was linked with Cer and protected at the C2, C3, and C6 hydroxyl groups. For example, Glc-Cers (**29** and **31**) were successfully coupled with complex sialyloligosaccharyl donors (**28** and **30**), providing high yields of the ganglioside frameworks, which were converted to gangliosides GQ1b and GP3 via two-step global deprotection, respectively [12, 13] (Fig. 5).

However, this approach using the Glc-Cer cassette requires early-stage sialylation for the preparation of sialyloligosaccharyl donor owing to the incomplete stereoselectivity of sialylation. Having obtained a fully stereoselective bicyclic sialyl donor, we explored late-stage sialylation in the ganglioside synthesis to increase the

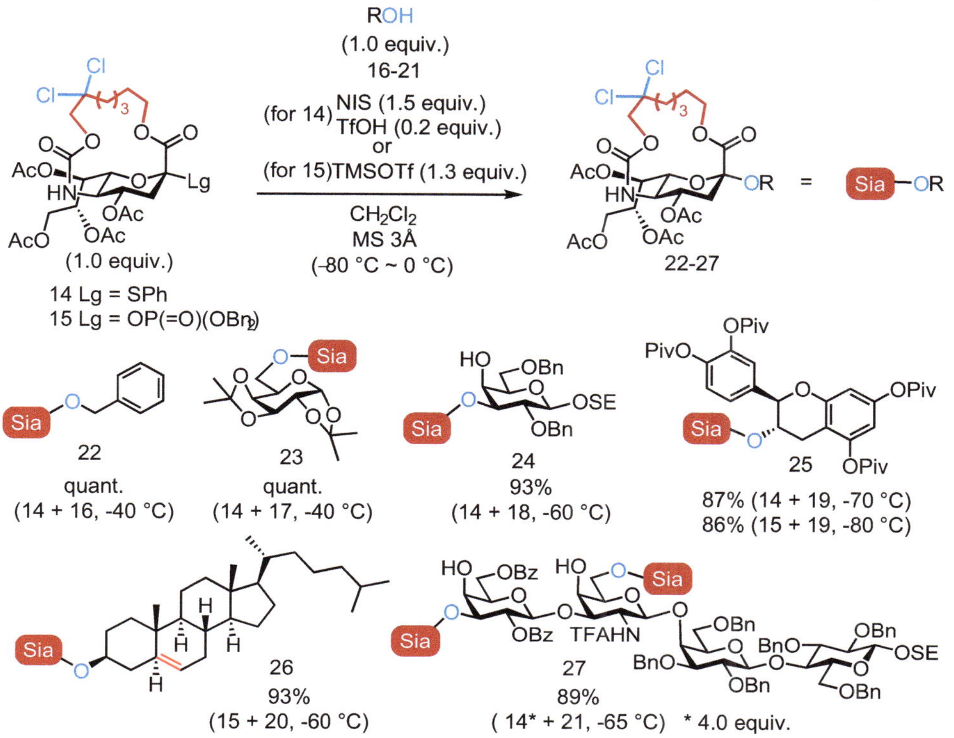

Fig. 4 Substrate scope study of bicyclic sialy donors

degree of accessibility toward ganglioside analogs bearing modified sialic acids. As a coupling partner of bicyclic sialyl donor, we have developed *tert*-butylbenzoylated (TBBz) oligosaccharyl Cer acceptors [14, 15]. The TBBz group increases the solubility in organic solvent by inhibiting π-π or CH-π interaction with the bulky *tert*-butyl group [16]. We confirmed that the oligosaccharyl Cer acceptors were dissolved in CH_2Cl_2 at lower temperatures ranging from −70°C to −20°C, while benzoylated analog was hardly dissolved under the same conditions. The coupling reactions of bicyclic sialyl dibenzylphosphate donors (**15** and **32**) and glycolipid acceptors (**33** and **34**) were successfully proceeded at low temperatures (−70°C to −60°C), obtaining protected gangliosides at high yield (Fig. 6). Upon selective cleavage of the tether part, the protected GM3, which was obtained by the coupling of **15** and **33**, was converted to C5-amino-modified analogs of GM3, such as *N*-acetyl, *N*-glycolyl a *N*-(^{13}C)acetyl analogs. On the contrary, the sialyllactotetraosyl ceramide (NeuAcLc$_4$Cer) derivative, which was obtained by the coupling of **32** and **34**, was derived into a fluorescent-labeled ganglioside probe via global deprotection and selective introduction of a fluorescent dye at the C9-amino group in the sialic acid residue. The fluorescent NeuAcLc$_4$Cer was used

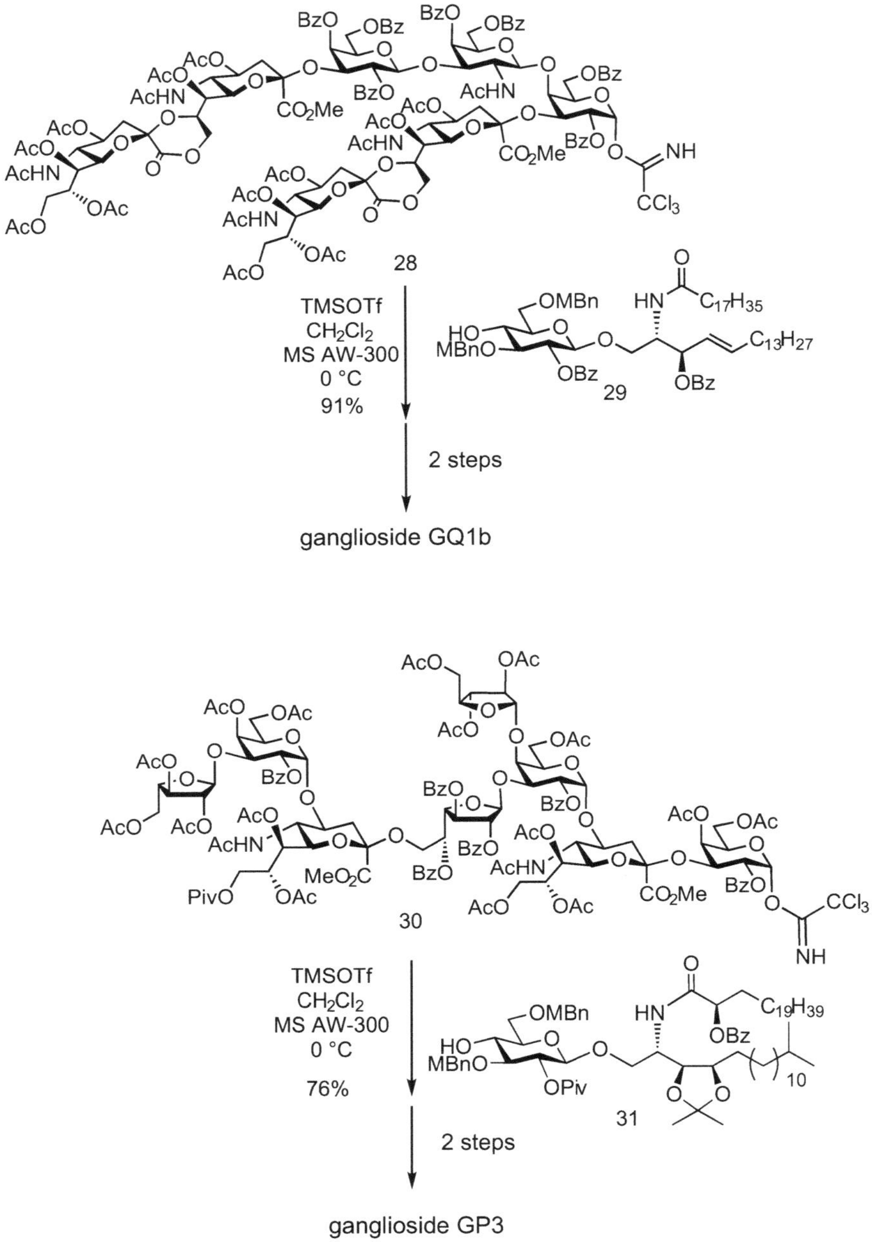

Fig. 5 Construction of glycolipid frameworks based on the Glc-Cer cassette approach

for single-molecule observation of interaction with a GPI-anchored protein (CD59) in the live cell membrane, which showed that Neu-AcLc$_4$Cers were frequently recruited to CD59.

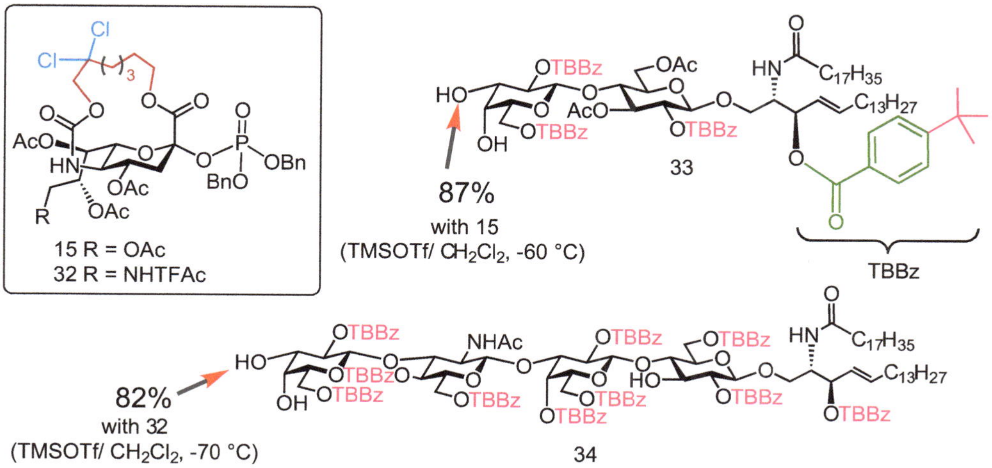

Fig. 6 Construction of glycolipid frameworks based on direct α-sialylation using macrobicyclic donors

2 Materials

All chemicals were of special grades, purchased from commercial suppliers, and used without further purification, unless otherwise noted.

2.1 Preparation of a Bicyclic Sialic Acid Derivative as a Key Intermediate for the Synthesis of Bicyclic Sialyl Donors

Preparation of a Tether Part

1. TLC.
2. Syringe and needle.
3. Pipette.
4. Balloon for purging Ar gas to a flask.
5. Glass (solvent proof) stopcock (two- or three-way).
6. Stopper (glass or rubber).
7. Two-necked round-bottom flask.
8. Magnetic stir bar.
9. Magnetic stirrer.
10. Cooling bath.
11. Rotary evaporator.
12. Separating funnel, funnel, and Erlenmeyer flask.
13. SiO_2.

14. Column chromatotube for silica gel chromatography (for purification).
15. Diaphragm pump (for filtration under reduced pressure).
16. Vacuum pump (for drying).

Introduction of a Tether Part at the C-5 Amino Group of Sialyl Thioglycoside

1. Sialic acid derivative (starting material) prepared using a previously reported procedure [17].
2. TLC.
3. Syringe and needle.
4. Pipette.
5. Balloon for purging Ar gas to a flask.
6. Glass (solvent proof) stopcock (two- or three-way).
7. Round-bottom flask.
8. Two-necked round-bottom flask.
9. Dimroth condenser.
10. Magnetic stir bar with heater.
11. Rotary evaporator.
12. Separating funnel, funnel, and Erlenmeyer flask.
13. SiO_2.
14. Column chromatotube for silica gel chromatography (for purification).
15. Diaphragm pump (for filtration under reduced pressure).
16. Vacuum pump (for drying)

Macrolactonization Reaction

1. TLC.
2. Syringe and needle.
3. Pipette.
4. Balloon for purging Ar gas to a flask.
5. Glass (solvent proof) stopcock (two- or three-way).
6. Two-necked round-bottom flask.
7. Magnetic stir bar with heater.
8. Rotary evaporator.
9. Separating funnel, funnel, and Erlenmeyer flask.
10. SiO_2.
11. Column chromatotube for silica gel chromatography (for purification).
12. Diaphragm pump (for filtration under reduced pressure).
13. Vacuum pump (for drying).

2.2 Direct Sialylation of Oligosaccharyl Cer Acceptor Using a Bicyclic Sialyl Donor

1. Sialyl donor prepared using a previously reported procedure [15].
2. Oligosaccharyl ceramide acceptor prepared using a previously reported procedure [15].
3. TLC.
4. Syringe and needle.
5. Pipette.
6. Balloon for purging Ar gas to a flask.
7. Glass (solvent proof) stopcock (two- or three-way).
8. Stopper (glass or rubber).
9. Cannula.
10. Pear-shaped flask.
11. Two-necked round-bottom flask.
12. Magnetic stir bar.
13. Magnetic stirrer.
14. Cooling bath.
15. Rotary evaporator
16. Separating funnel, funnel, and Erlenmeyer flask.
17. SiO_2.
18. Column chromatotube for silica gel chromatography (for purification).
19. Diaphragm pump (for filtration under reduced pressure).
20. Vacuum pump (for drying).

3 Methods

All procedures were performed in a fume hood, and all reactions were induced under positive Ar pressure, unless otherwise noted.

3.1 Preparation of a Bicyclic Sialic Acid Derivative as a Key Intermediate for the Synthesis of Bicyclic Sialyl Donors

Preparation of the Tether Part

1. Add NaH (4.0 equiv.) and methyl dichloroacetate (4.0 equiv.) in four portions (every 10 min) to a solution of 1,5-dibromopentane **35** (1.0 equiv.) in DMF (300 mM) at 0°C (*see* **Note 1**).

2. Stir for 2 h at 0°C, monitoring the progress of the reaction using TLC (*n*-hexane/EtOAc = 20:1) (*see* **Note 2**).
3. Add satd. aq. NH_4Cl at 0°C.
4. Dilute the mixture with EtOAc and wash with H_2O, satd. aq. $Na_2S_2O_3$, and brine.
5. Dry the organic layer over Na_2SO_4.
6. Filter the mixture and concentrate the filtrate.
7. Purify the residue using silica gel column chromatography with (*n*-hexane/EtOAc = 1:0 to 50:1) as the eluent to obtain the desired product **36** (yield ~32%).
8. Dissolve compound **36** (1.0 equiv.) in THF/MeOH (2:1) (200 mM).
9. Add $NaBH_4$ (3.0 equiv.) to the solution of compound **36** at 0°C.
10. Stir for 26 h at 0°C, monitoring the progress of the reaction using TLC (*n*-hexane/EtOAc = 5:1) (*see* **Note 2**).
11. Add satd. aq. NH_4Cl at 0°C.
12. Dilute the mixture with EtOAc and wash with H_2O and brine.
13. Dry the organic layer over Na_2SO_4.
14. Filter the mixture and concentrate the filtrate.
15. Purify the residue using silica gel column chromatography with (*n*-hexane/EtOAc = 1:0 to 50:1) as the eluent to obtain the desired product **37** (yield ~93%).

Introduction of a Tether Part at the C-5 Amino Group of Sialyl Thioglycoside

1. Add 10 M aq. KOH (15.0 equiv.) to a solution of compound **38** (1.0 equiv.) in EtOH (50 mM) at room temperature.
2. Stir for 48 h under reflux, monitoring the progress of the reaction using TLC ($CHCl_3$/MeOH/H_2O = 5:4:1) (*see* **Note 2**).
3. Neutralize the reaction mixture with HCl (2 M solution in EtOH, 15.0 equiv.) (solution A).
4. Add triethylamine (8.0 equiv.) and di(*N*-succinimidyl)carbonate (DSC; 4.5 equiv.) to a solution of compound **37** (4.5 equiv.) in MeCN (220 mM) at 0°C.
5. Stir for 50 min at room temperature, monitoring the progress of the reaction using TLC (*n*-hexane/EtOAc = 2:1) (*see* **Note 2**). This reaction yields an activated ester **39**.
6. Add the mixture containing **39** to solution A at 0°C.
7. Stir for 2 h at room temperature, monitoring the progress of the reaction using TLC ($CHCl_3$/MeOH/H_2O = 5:2:0.2) (*see* **Note 2**).

8. Dilute the reaction mixture with EtOAc and wash with 2 M HCl, H_2O, satd. aq. $NaHCO_3$, and brine.
9. Dry the organic layer over Na_2SO_4.
10. Filter the organic layer and concentrate the filtrate.
11. Purify the residue using silica gel column chromatography with $CHCl_3$/MeOH (20:1 to 5:1) as the eluent and size-exclusion column chromatography on a Sephadex LH-20 column using $CHCl_3$/MeOH (1:1) as the eluent to obtain the desired product **40**.

Macrolactonization reaction

1. Add Drierite ($CaSO_4$) (0.2 g for 1 mL of THF) and K_2CO_3 (2.0 equiv.) to a solution of compound **40** (1.0 equiv.) in DMF/THF (3:1) (2.0 mM) at room temperature.
2. Stir for 5.5 h at 50°C, monitoring the progress of the reaction using TLC ($CHCl_3$/MeOH = 8:1) (*see* **Note 2**).
3. Filter the reaction mixture through a pad of Celite and wash the pad with EtOAc.
4. Neutralize the combined filtrate and washings with AcOH.
5. Concentrate the neutralized solution.
6. Purify the residue using silica gel column chromatography with ($CHCl_3$/MeOH = 20:1) to obtain the desired product **41**.

3.2 Direct Sialylation of Oligosaccharyl Cer Acceptor Using a Bicyclic Sialyl Donor

1. Mix sialyl donor **32** (4.5 equiv.) and Lc_4Cer acceptor **34** (1.0 equiv.) in a pear-shaped flask.
2. Dissolve the mixed compound in dry toluene and evaporate toluene in vacuo to remove residual water (repeat this step two or three times).
3. Expose the mixture to high vacuum for 3 h.
4. Add CH_2Cl_2 (9 mM) to the mixture and cool down to −70°C (*see* **Note 3**).
5. Add TMSOTf (1.3 equiv. for sialyl donor **32**) at −70°C.
6. Stir for 5 h at −70°C, monitoring the progress of the reaction using TLC (toluene/acetone = 6:1 developed twice or toluene/EtOAc = 2:1) (*see* **Note 2**).
7. Add satd. aq. $NaHCO_3$ to the reaction mixture to quench the reaction and dilute with $CHCl_3$.
8. Wash the mixture with $NaHCO_3$ and brine.
9. Dry the organic layer over Na_2SO_4.
10. Filter the organic layer and concentrate the filtrate.
11. Purify the residue using silica gel column chromatography with *n*-hexane/acetone (5.5:1) as the eluent and size-exclusion column chromatography on a Sephadex LH-20 column

using $CHCl_3$/MeOH (1:1) as the eluent to obtain the desired product **42**.

4 Notes

1. Avoid contact of NaH with water.
2. Compound detection is conducted by either exposing the sample to ultraviolet light (253.6 nm) or soaking it in H_2SO_4 solution (10% in EtOH) or phosphomolybdic acid solution (20% in EtOH), followed by heating.
3. Molecular sieves are not used to avoid the precipitation of the acceptor.

Acknowledgments

This work was supported in part by the Japan Society for the Promotion of Science (JSPS) KAKENHI Grant No. JP22H00359 (H. A.) and JP23K05054 (A. I.); JSPS Core-to-Core Program Grant No. JPJSCCA20200007 (H. A.); Japan Science and Technology Agency (JST) CREST Grant No. JRMJCR18H2 (H. A.).

Conflict of Interest The authors have no conflicts of interest to declare that are relevant to the content of this chapter.

References

1. Smith DC, Lord JM, Roberts LM et al (2004) Glycosphingolipids as toxin receptors. Semin Cell Dev Biol 15:397–408
2. Lopez PHH, Schnaar RL (2009) Gangliosides in cell recognition and membrane protein regulation. Curr Opin Struct Biol 19:549–557
3. Fleming FE, Bohm R, Dang VT et al (2014) Relative roles of GM1 ganglioside, *N*-acetylneuraminic acids, and α2β1 integrin in mediating rotavirus infection. J Virol 88:4558–4571
4. Blaum BS, Frank M, Walker RC et al (2016) Complement factor H and simian virus 40 bind the GM1 ganglioside in distinct conformations. Glycobiology 26:532–539
5. Kawashima N, Yoon S, Itoh JK et al (2009) Tyrosine kinase activity of epidermal growth factor receptor is regulated by GM3 binding through carbohydrate to carbohydrate interactions. J Biol Chem 284:6147–6155
6. Fukuda Y, Fukui T, Hikichi C et al (2015) Neurotropin promotes NGF signaling through interaction of GM1 ganglioside with Trk neurotrophin receptor in PC12 cells. Brain Res 1596:13–21
7. Jung WR, Kim HG, Shin MK et al (2010) The effect of ganglioside GQ1b on the NMDA receptor signaling pathway in H19-7 cells and rat hippocampus. Neuroscience 165:159–167
8. Kaneko M, Yamada K, Miyamoto T et al (2007) Neuritogenic activity of gangliosides from echinoderms and their structure–activity relationship. Chem Pharm Bull 55:462–463
9. Vibhute AM, Komura N, Tanaka H-N et al (2021) Advanced chemical methods for stereoselective sialylation and their applications in sialoglycan syntheses. Chem Rec 21:3194–3223
10. Komura N, Kato K, Udagawa T et al (2019) Constrained sialic acid donors enable selective synthesis of α-glycosides. Science 364:677–680
11. Kraus GA, Hon YS, Thomas PJ et al (1989) Organic synthesis using bridgehead carbocations and bridgehead enones. Chem Rev 89:1591–1598

12. Imamura A, Ando H, Ishida H et al (2009) Ganglioside GQ1b: efficient total synthesis and the expansion to synthetic derivatives to elucidate its biological roles. J Org Chem 74:3009–3023
13. Goto K, Sawa M, Tamai H et al (2016) The total synthesis of starfish ganglioside GP3 bearing a unique sialyl glycan architecture. Chem Eur J 22:8323–8331
14. Takahashi M, Shirasaki J, Komura N et al (2020) Efficient diversification of GM3 gangliosides via late-stage sialylation and dynamic glycan structural studies with ^{19}F solid-state NMR. Org Biomol Chem 18:2902–2913
15. Takahashi M, Komura N, Yoshida Y et al (2022) Development of lacto-series ganglioside fluorescent probe using late-stage sialylation and behavior analysis with single-molecule imaging. RSC Chem Biol 3:868–885
16. Asano S, Tanaka H-N, Imamura A et al (2019) *p-tert*-Butyl groups improve the utility of aromatic protecting groups in carbohydrate synthesis. Org Lett 21:4197–4200
17. Cao S, Meunier SJ, Andersson FO et al (1994) Mild stereoselective syntheses of thioglycosides under PTC conditions and their use as active and latent glycosyl donors. Tetrahedron: Asymmetry 5:2303–2312

Chapter 3

Chemical Synthesis of Fluorescent Analogs of a- and b-Series Gangliosides

Naoko Komura and Hiromune Ando

Abstract

Fluorescent ganglioside analogs recently developed by our group using a fully chemical method have enabled the observation of the behavior, interaction, and lipid raft formation of gangliosides in cell plasma membranes at the single-molecule level. In this chapter, we describe the chemical synthesis of fluorescent analogs of a- and b-series gangliosides previously reported by our group. The procedures for key chemical reactions—glycosylation, global deprotection of gangliosides, and fluorescent labeling of gangliosides—are described in detail.

Keywords Glycolipids, Gangliosides, Lipid rafts, Glycosylations, Fluorescent analogs, Single-molecule imaging

1 Introduction

Gangliosides are glycosphingolipids that contain one or more sialic acid residues [1]. In cell plasma membranes (PMs), gangliosides are involved in various biological processes, including cell differentiation, cell adhesion, viral infection, toxin uptake, and signal transduction [2]. These biological processes are mediated by clusters of gangliosides and other lipid molecules, called "lipid rafts." Fluorescent analogs of a- and b-series gangliosides (GM3, GM2, GM1, GD3, and GQ1b) have been developed using a fully chemical method to clarify ganglioside-related biological processes [3–10]. The functional evaluation of these fluorescent analogs in cell PMs indicates that they exhibit biophysical properties similar to those of their parent gangliosides. Thus, fluorescent ganglioside analogs have facilitated observation of the behavior, interactions, and lipid raft formation of gangliosides in cell PMs by single-molecule imaging [3–10]. Our previous study revealed that the fluorescent dye should be positioned on the outermost glycan moieties (for example, the C9 position

Koichi Furukawa (ed.), *Gangliosides*, Methods in Molecular Biology, vol. 3035,
https://doi.org/10.1007/978-1-0716-5288-6_3,

of the terminal sialic acid) to preserve the properties of the parent gangliosides [3, 4] (Fig. 1).

Thus, to introduce fluorescent dyes in a site-specific manner, we designed amino gangliosides as precursors of fluorescent gangliosides (Fig. 2). The synthesis of a-series ganglioside analogs (GM3, GM2, and GM1), whose C9 position of sialic acid was labeled, started with *N*-Troc Neu-Gal unit **1** [11], which could be synthesized on a large scale. *N*-Troc Neu-Gal unit **1** was converted into C9-NHTFAc Neu-Gal **2**, a common intermediate of a-series analogs, by replacing C9-OH with a trifluoroacetamide (NHTFAc) group. C9-NHTFAc Neu-Gal trichloroacetimidate donor **3** was then subjected to a final glycosidation reaction with Glc-Cer cassette **4** developed by our group [12–14] to obtain a ganglioside GM3 framework. Global deprotection yielded the corresponding amino GM3, which was subsequently treated with the *N*-succinimidyl ester of a fluorescent dye (for example, ATTO594, Fluorescein) to yield fluorescent ganglioside GM3 analogs [3, 4]. In contrast, C9-NHTFAc Neu-Gal **2** was converted into C4-OH Neu-Gal acceptor **6** by the regioselective protection of hydroxyl groups to construct the branched structure from the C4 position of Gal for GM2 and GM1 (Fig. 2).

Next, C4-OH Neu-Gal acceptor **6** was glycosylated using GalN thioglycoside donor **7** for GM2 and Gal-GalN thioglycoside donor **8** for GM1 to yield the corresponding trisaccharide and tetrasaccharide,

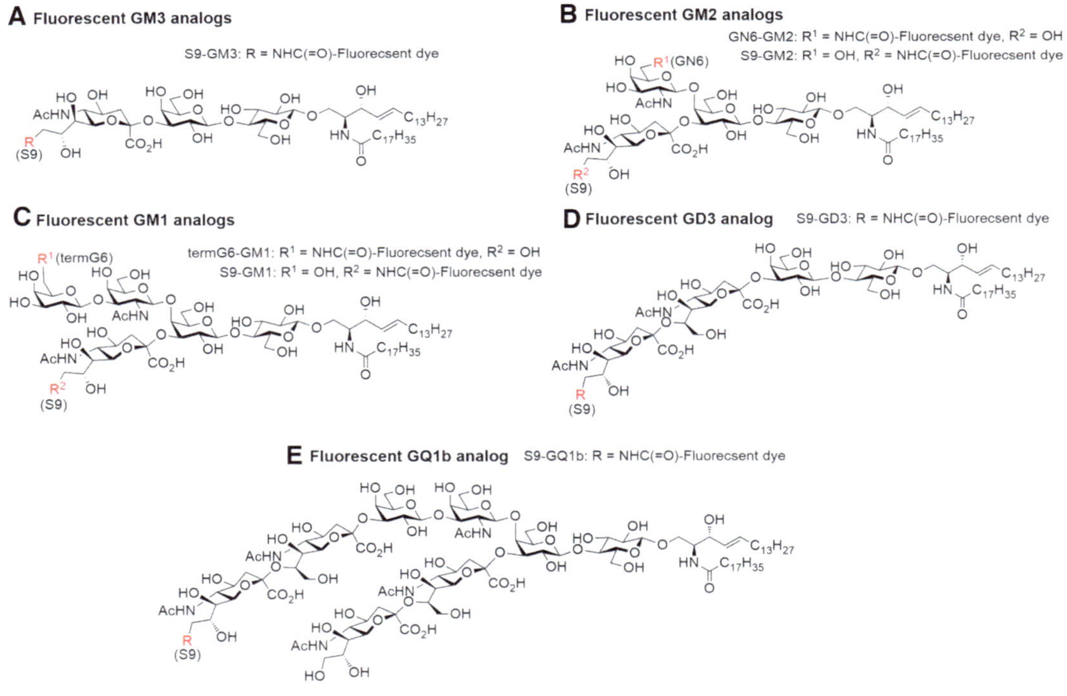

Fig. 1 Fluorescent analogs of a- and b-series gangliosides developed by our group

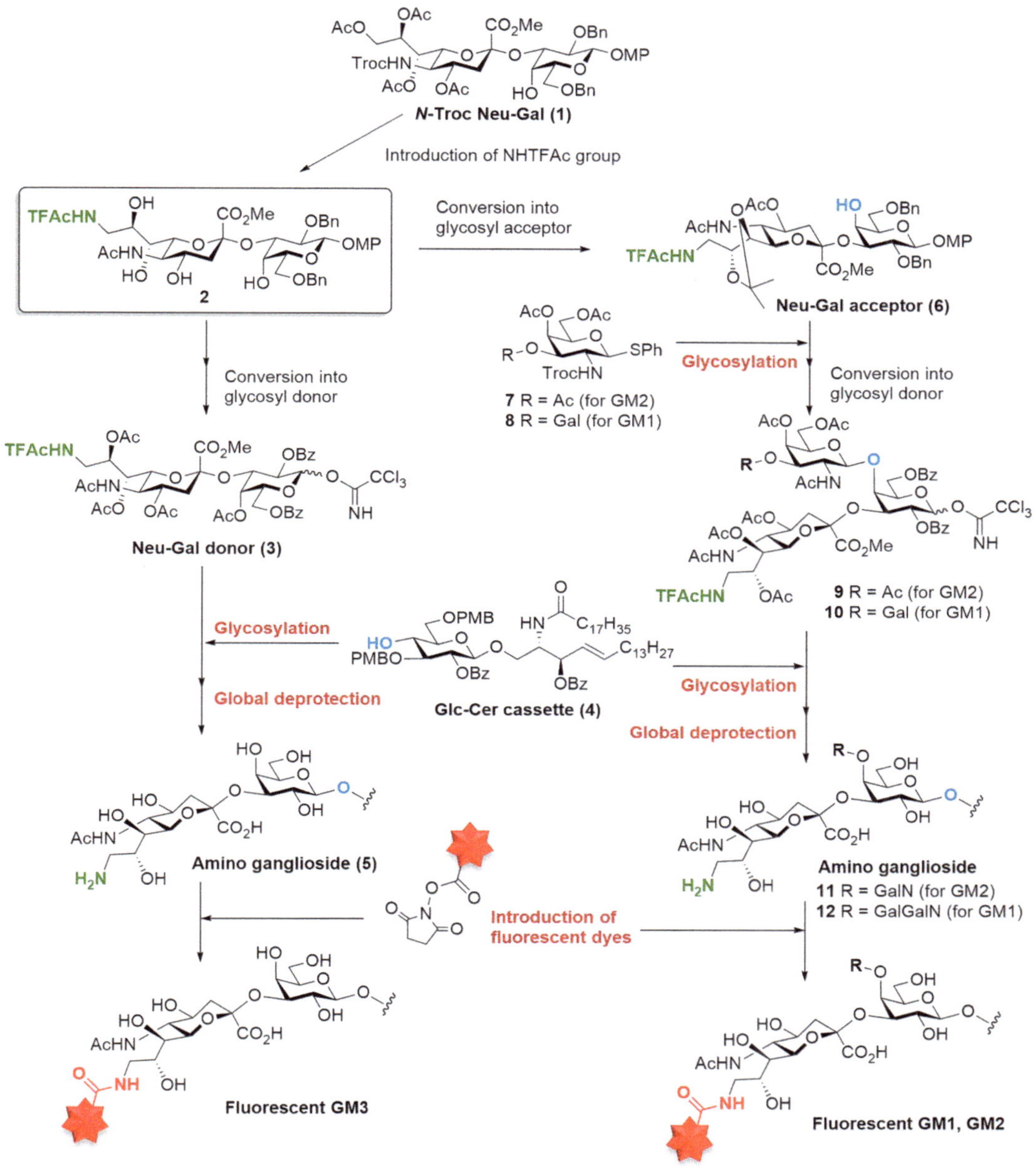

Fig. 2 Outline of the synthesis of a-series ganglioside analogs. *Ac* acetyl, *Bn* benzyl, *Bz* benzoyl. *MP* p-methoxyphenyl, *PMB* p-methoxybenzyl, *TFAc* trifluoroacetyl, *Troc* 2,2,2-trichloroethoxycarbonyl

respectively. The glycan moieties obtained were transformed into trichloroacetimidate donors (**9** and **10**) for the final glycosylation. The final glycosidations of the glycosyl donors (**9** and **10**) with Glc-Cer acceptor **4** were conducted by following the procedure for GM3 to afford the GM2 and GM1 ganglioside frameworks. Global deprotection yielded C9-amino gangliosides, which were subjected to fluorescent dye conjugation reactions to produce the fluorescent GM2 and GM1 analogs.

The synthetic method for a-series analogs can be applied to the synthesis of b-series analogs (GD3 and GQ1b) (Fig. 3) [5].

First, *N*-Troc Neu-Gal unit **1** was transformed into 1,5-lactamized acceptor **13** to enhance the reactivity of C8-OH of the sialic acid in constructing a common sequence of b-series gangliosides:

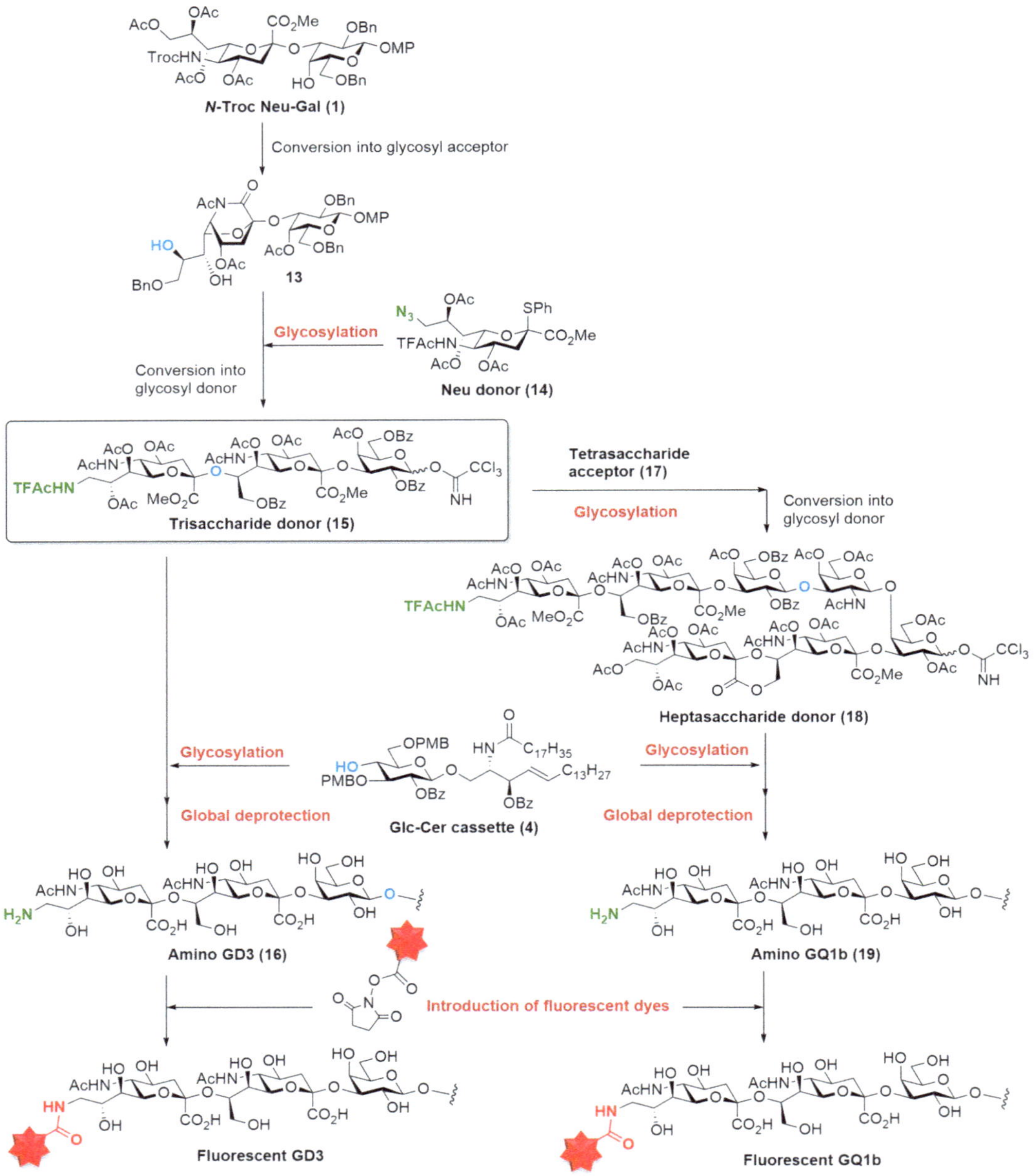

Fig. 3 Outline of the synthesis of b-series ganglioside analogs

Neuα(2,8)Neuα(2,3)Gal [15]. The C8-OH Neu-Gal acceptor **13** was then sialylated by C9-azide thioglycoside donor **14** to afford the Neuα(2,8)Neuα(2,3)Gal trisaccharide unit, which was converted into trichloroacetimidate donor **15**, a key trisaccharide donor of GD3 and GQ1b analogs. Based on the synthetic method of the parent GQ1b previously described by our group [12], trisaccharide donor **15** was subjected to glycosidation with tetrasaccharide acceptor **17** [12] to generate the heptasaccharide for the GQ1b analog. Based on the procedure developed for the a-series gangliosides, trisaccharide donor **15** and heptasaccharide donor **18** were glycosidated with Glc-Cer cassette **4**, yielding ganglioside GD3 and GQ1b frameworks, respectively. Finally, global deprotection and the subsequent fluorescent dye incorporation yielded fluorescent GD3 and GQ1b analogs.

The synthetic methods described above are also applicable to the synthesis of other ganglioside analogs [6–9]. In this chapter, we describe the common chemical reactions involved in the synthesis of fluorescent ganglioside analogs: glycosylation, global deprotection of gangliosides, and fluorescent labeling of gangliosides.

2 Materials

All chemicals are of special grades, purchased from commercial suppliers, and used without further purification, unless otherwise noted.

2.1 Glycosylation Reaction Using Thioglycoside Donors

1. Fully protected thioglycoside donor was synthesized using the reported method [3–5].
2. Glycosyl acceptor was synthesized using the reported method [3–5].
3. Dry toluene.
4. Dry CH_2Cl_2 (for glycosylations other than sialylation).
5. Dry EtCN (for sialylation) is distilled from P_2O_5 to remove basic impurities and stored in CaH_2.
6. Argon gas.
7. Molecular sieves (Powder).
8. *N*-Iodosuccinimide (NIS).
9. Trifluoromethanesulfonic acid (TfOH).
10. Celite®.
11. Saturated aqueous $Na_2S_2O_3$.
12. Saturated aqueous $NaHCO_3$.
13. Saturated aqueous NaCl (Brine).
14. Na_2SO_4.
15. TLC Silica gel 60 F_{254}.
16. Pear-shaped flask.

17. Two-necked round-bottom flask.
18. Cannula.
19. Magnetic stir bar.
20. Drying oven.
21. Magnetic stirrer with heater.
22. Cooling bath.
23. Rotary evaporator.
24. Separating funnel, Funnel, and Erlenmeyer flask.
25. SiO_2.
26. Column chromatotube for silica gel chromatography (for purification).
27. Diaphragm pump (for filtration under reduced pressure).
28. Vacuum pump (for drying).

2.2 Glycosylation Reaction Using Trichloroacetimidate Donors

1. Fully protected trichloroacetimidate donor was synthesized using the reported method [3–5].
2. Glycosyl acceptor was synthesized using the reported method [3–5].
3. Dry toluene.
4. Dry CH_2Cl_2.
5. Argon gas.
6. Molecular sieves (powder).
7. Trimethylsilyl trifluoromethanesulfonate (TMSOTf).
8. Celite®.
9. Saturated aqueous $NaHCO_3$.
10. Saturated aqueous NaCl (Brine).
11. Na_2SO_4.
12. TLC Silica gel 60 F_{254}.
13. Pear-shaped flask.
14. Two-necked round-bottom flask.
15. Cannula.
16. Magnetic stir bar.
17. Drying oven.
18. Magnetic stirrer with heater.
19. Cooling bath.
20. Rotary evaporator.
21. Separating funnel, Funnel, and Erlenmeyer flask.
22. SiO_2.

23. Column chromatotube for silica gel chromatography (for purification).
24. Diaphragm pump (for filtration under reduced pressure).
25. Vacuum pump (for drying).

2.3 Global Deprotection of Gangliosides

1. Protected ganglioside synthesized using the reported method [3–5].
2. Methanol (MeOH).
3. Tetrahydrofuran (THF).
4. $CHCl_3$.
5. 1 M aqueous NaOH.
6. Round-bottom flask.
7. Magnetic stir bar.
8. Magnetic stirrer.
9. TLC Silica gel 60 F_{254}.
10. 5% aqueous $CaCl_2$ (for TLC analysis).
11. SiO_2.
12. Column chromatotube for silica gel chromatography (for purification).
13. Rotary evaporator.
14. Vacuum pump (for drying).

2.4 Fluorescent Labeling

1. Amino ganglioside was synthesized using the reported method [3–5].
2. *N*-Succinimidyl ester of the fluorescent dye: Hydrophilic dyes (for example, ATTO594, Fluorescein) are suitable for retaining the biophysical properties of the parent gangliosides [3, 4].
3. *N*,*N*-Dimethylformamide (DMF).
4. Triethylamine.
5. Distilled water.
6. Pear-shaped flask.
7. Magnetic stir bar.
8. Magnetic stirrer.
9. TLC Silica gel 60 F_{254}.
10. $CHCl_3$.
11. Methanol (MeOH).
12. 5% aqueous $CaCl_2$ (for TLC analysis).
13. Column chromatotube for gel filtration column chromatography (for purification).

14. Sephadex LH-20 (for purification): Swell this in $CHCl_3$/MeOH (1:1).
15. Rotary evaporator.
16. Preparative thin-layer chromatography (PTLC).
17. PTLC-developing chamber.
18. Glass filter Buchner funnel type.
19. Membrane filter.
20. Diaphragm pump (for filtration under reduced pressure).
21. Vacuum pump (for drying).

3 Methods

Perform all the procedures in a fume hood, and induce all reactions under positive argon pressure unless otherwise noted.

3.1 General Method for Glycosylation Reaction Using Thioglycoside Donors

1. Add glycosyl acceptor (1.0 equivalent) and thioglycoside donor (1.0–3.0 equivalent) to a pear-shaped flask, and remove the residual water by azeotroping with dry toluene (*see* **Note 1**).
2. Connect the flask to a high-vacuum line for 3 h and purge argon gas into the flask.
3. Activate molecular sieves in a drying oven at 300°C for 2 h.
4. Add the activated molecular sieves to a two-necked round-bottom flask, and heat the flask at 300°C for 2 h in vacuo.
5. After cooling the flask to ambient temperature, purge argon gas into the flask.
6. Add *N*-iodosuccinimide (NIS) (1.2–4.0 equivalent) to the two-necked round-bottom flask.
7. Add dry CH_2Cl_2 (50–100 mM) (dry EtCN for sialylation, *see* **Note 2**) to the pear-shaped flask in an argon atmosphere, along with the glycosyl acceptor and thioglycoside donor.
8. Transfer the solution to the two-necked flask containing activated molecular sieves through a cannula at ambient temperature.
9. Stir the mixture at −80 to 0 °C for 1 h.
10. Add trifluoromethanesulfonic acid (TfOH) (0.1–0.5 equivalent) to the mixture at −80–0°C and stir the reaction mixture until the thioglycoside donor is completely consumed as the progress of the reaction is monitored using TLC (*see* **Note 3**).
11. Neutralize the reaction mixture with saturated aqueous $NaHCO_3$.

12. Filter the mixture through a pad of Celite®, and wash the pad thoroughly using an organic solvent.
13. Wash the organic layer with saturated aqueous $Na_2S_2O_3$ and saturated aqueous NaCl (brine).
14. Dry the organic layer over Na_2SO_4.
15. Filter the suspension and concentrate the filtrate.
16. Fill a column chromatotube with silica gel swollen in $CHCl_3$.
17. Wash the silica gel with $CHCl_3$.
18. Purify the residue using silica gel column chromatography using $CHCl_3$/MeOH (e.g., 100:1) as the eluent to obtain the desired product.

3.2 General Method for Glycosylation Reaction Using Trichloroacetimidate Donors

1. Add a glycosyl acceptor (1.0 equivalent) and a trichloroacetimidate donor (1.0–3.0 equivalent) to a pear-shaped flask, and remove the residual water by azeotroping with dry toluene (*see* **Note 1**).
2. Connect the flask to a high-vacuum line for 3 h and purge argon gas into the flask.
3. Activate molecular sieves in a drying oven at 300°C for 2 h.
4. Add the activated molecular sieves to a two-necked round-bottom flask, and heat the flask at 300°C for 2 h in vacuo.
5. Add dry CH_2Cl_2 (50–100 mM) to the pear-shaped flask in an argon atmosphere and dissolve the glycosyl acceptor and trichloroacetimidate donor.
6. Transfer the solution to the two-necked flask containing activated molecular sieves through a cannula at ambient temperature.
7. Stir the mixture at −80 to 0°C for 1 h.
8. Add trimethylsilyl trimethylsilyl trifluoromethanesulfonate (TMSOTf) (0.1–0.5 equivalent) to the suspension at −80 to 0°C and stir the reaction mixture until the trichloroacetimidate donor is completely consumed as the progress of the reaction is monitored using TLC (*see* **Notes 3 and 4**).
9. Neutralize the reaction mixture with saturated aqueous $NaHCO_3$.
10. Filter the mixture through a pad of Celite®, and wash the pad thoroughly using an organic solvent.
11. Wash the organic layer with saturated aqueous NaCl (brine).
12. Dry the organic layer over Na_2SO_4.
13. Filter the suspension and concentrate the filtrate.
14. Fill a column chromatotube with silica gel swollen in $CHCl_3$.
15. Wash the silica gel with $CHCl_3$.

16. Purify the residue using silica gel column chromatography using $CHCl_3$/MeOH (e.g., 100:1) as the eluent to obtain the desired product.

3.3 General Method for Global Deprotection Of Gangliosides

1. Add a protected ganglioside to a round-bottom flask, and add a mixed solvent of MeOH/tetrahydrofuran (THF) (1:1) to the flask at ambient temperature.
2. Add 1 M aqueous NaOH (20–50 equivalent) to the solution at ambient temperature.
3. Stir the reaction mixture at ambient temperature until the reaction is completed as the progress of the reaction is monitored using TLC (for example, $CHCl_3$/MeOH/5% aqueous $CaCl_2$ = 5:4:1).
4. Co-evaporate the solution with EtOH (*see* **Note 5**).
5. Fill a column chromatotube with silica gel swollen in $CHCl_3$/MeOH (2:1).
6. Wash the silica gel with $CHCl_3$/MeOH (2:1).
7. Purify the residue via flash column chromatography on silica gel using $CHCl_3$/MeOH/H_2O (e.g., 5:4:1) as the eluent to obtain the corresponding amino ganglioside (*see* **Note 6**).

3.4 General Method for Fluorescent Labeling

1. Add an amino ganglioside (1.0–10.0 mg, 1.0 equivalent) in a pear-shaped flask, and add a mixed solvent of *N*,*N*-dimethylformamide (DMF)/H_2O (10:1) in the flask at ambient temperature.
2. Add the *N*-succinimidyl ester of a fluorescent dye (1.5 equivalent) to the solution at ambient temperature.
3. Add triethylamine (10.0 equivalent) to the solution at ambient temperature.
4. Stir the reaction mixture at ambient temperature until the amino ganglioside is completely consumed as the progress of the reaction is monitored using TLC (for example, $CHCl_3$/MeOH/5% aqueous $CaCl_2$ = 5:4:1) (*see* **Note 3**).
5. Load the reaction mixture directly onto a gel filtration column chromatography on Sephadex LH-20 and purify it using $CHCl_3$/MeOH (1:1) as the eluent to remove the remaining fluorescent reagent and DMF (*see* **Note 7**).
6. Load the residue on the PTLC and develop the PTLC in a developing chamber (e.g., $CHCl_3$/MeOH/H_2O = 5:4:0.5) (Fig. 4A).
7. Scratch the fluorescent color band on the PTLC with a spatula, and load the scratched silica gel flakes onto a glass filter of the Buchner funnel type (Fig. 4B). Extract the fluorescent

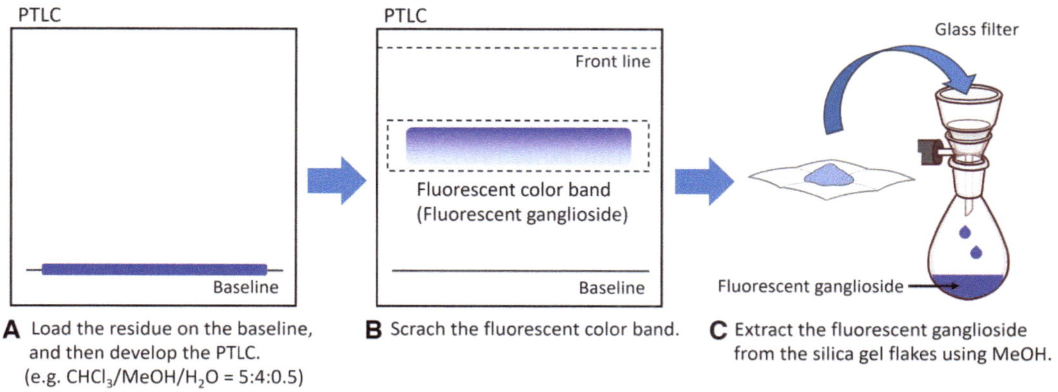

Fig. 4 Purification of fluorescent gangliosides using PTLC

ganglioside from the silica gel flakes using MeOH (Fig. 4C) (*see* **Note 8**).

8. Filter the fluorescent ganglioside solution through a membrane filter to remove any remaining silica gel, and wash the filter with MeOH.
9. Concentrate the solution to obtain the pure fluorescent ganglioside analog.

4 Notes

1. This reaction may produce a hemiacetal as a byproduct. Therefore, all experimental procedures should be performed carefully to prevent the invasion of water into the reaction mixture and prevent side reactions.
2. EtCN is used for stereoselective α-sialylation [16]. EtCN that is obtained commercially contains basic impurities, which may interfere with sialylation. Thus, EtCN should be distilled from P_2O_5 and stored in CaH_2 to eliminate the basic impurities.
3. Compound detection is conducted either by exposing to ultraviolet light (253.6 nm) or by soaking in a H_2SO_4 solution (10% in EtOH) or phosphomolybdic acid solution (20% in EtOH), followed by heating.
4. This reaction may generate *N*-glycosyl trichloroacetamide as possible byproduct.
5. The sialic acid residues of deprotected gangliosides may be cleaved under acidic conditions. Hence, it is preferable not to neutralize the reaction mixture to prevent cleavage.

6. In general, amino gangliosides are soluble in $CHCl_3$/MeOH (1:1). However, they are insoluble in neither $CHCl_3$ nor MeOH.
7. The remaining fluorescent reagent and DMF in the reaction mixture can be removed by gel filtration column chromatography on Sephadex LH-20, and the remaining amino gangliosides can be removed by PTLC.
8. Generally, the fluorescent gangliosides are soluble in MeOH.

Acknowledgments

This work was supported in part by the Japan Society for the Promotion of Science (JSPS) KAKENHI Grant No. JP22H00359 (H. A.), and JP23K05055 (N. K.); JSPS Core-to-Core Program Grant No. JPJSCCA20200007 (H. A.); Japan Science and Technology Agency (JST) CREST Grant No. JRMJCR18H2 (H. A.); JST FOREST Grant No. JPMJFR2004 (N. K.).

Conflict of Interest Statement The authors have no conflicts of interest to declare that are relevant to the content of this chapter.

References

1. Lopez PH, Schnaar RL (2009) Gangliosides in cell recognition and membrane protein regulation. Curr Opin Struct Biol 19:549–557
2. Lingwood D, Simons K (2010) Lipid rafts as a membrane-organizing principle. Science 327:46–50
3. Komura N, Suzuki KGN, Ando H et al (2016) Raft-based interactions of gangliosides with a GPI-anchored receptor. Nat Chem Biol 12:402–410
4. Komura N, Suzuki KGN, Ando H et al (2017) Syntheses of fluorescent gangliosides for the studies of raft domains. Methods Enzymol 597:239–263
5. Konishi M, Komura N, Hirose Y et al (2020) Development of fluorescent ganglioside GD3 and GQ1b analogs for elucidation of raft-associated interactions. J Org Chem 85:15998–16013
6. Koikeda M, Komura N, Tanaka HN et al (2019) Synthesis of ganglioside analogs containing fluorescently labeled GalNAc for single-molecule imaging. J Carbohydr Chem 38:509–527
7. Asano S, Pal R, Tanaka HN, et al (2019) Development of fluorescently labeled SSEA-3, SSEA-4, and Globo-H glycosphingolipids for elucidating molecular interactions in the cell membrane. Int J Mol Sci 20:6187
8. Yamaguchi E, Komura N, Tanaka HN et al (2023) Fluorescent GD2 analog for single-molecule imaging. Glycoconjugate J 40:247–257
9. Takahashi M, Komura N, Yoshida Y et al (2022) Development of lacto-series ganglioside fluorescent probe using late-stage sialylation and behavior analysis with single-molecule imaging. RSC Chem Biol 3:868–885
10. Suzuki KGN, Komura N, Ando H (2023) Recently developed glycosphingolipid probes and their dynamic behavior in cell plasma membranes as revealed by single-molecule imaging. Glycoconjugate J 40:305–314
11. Fuse T, Ando H, Imamura A et al (2006) Synthesis and enzymatic susceptibility of a series of novel GM2 analogs. Glycoconjugate J 23:329–343
12. Imamura A, Ando H, Ishida H et al (2009) Ganglioside GQ1b: efficient total synthesis and the expansion to synthetic derivatives to elucidate its biological roles. J Org Chem 74:3009–3023
13. Tamai H, Ando H, Tanaka HN et al (2011) The total synthesis of the neurogenic ganglioside

LLG-3 isolated from the starfish Linckia laevigata. Angew Chem Int Ed 50:2330–2333
14. Nakashima S, Ando H, Saito R et al (2012) Efficiently synthesizing lacto-ganglio-series gangliosides by using a glucosyl ceramide cassette approach: the total synthesis of ganglioside X2. Chem Asian J 7:1041–1051
15. Ando H, Koike Y, Koizumi S et al (2005) 1, 5-Lactamized sialyl acceptors for various disialoside syntheses: novel method for the synthesis of glycan portions of Hp-s6 and HLG-2 gangliosides. Angew Chem Int Ed 44:6759–6763
16. Kanie O, Kiso M, Hasegawa A (1988) Glycosylation using methylthioglycosides of *N*-acetylneuraminic acid and dimethyl(methylthio)sulfonium triflate. J Carbohydr Chem 7:501–506

Chapter 4

Identification and Evaluation of CMP-Sialic Acids

Di Wu, Takayuki Omoto, Chihiro Sato and Ken Kitajima

Abstract

Cytidine 5′-monophosphate-sialic acid (CMP-Sia) is a donor substrate of all sialyltransferases in both prokaryotes and eukaryotes, and biosynthesized by the CMP-Sia synthetase (CSS or CMAS) using Sia and cytidine 5′-triphosphate (CTP) as the substrates. Since CSSs have unique substrate specificity for Sia species depending on organisms, cells, and organs, and developmental stages, various kinds of CMP-Sia exist in nature. This chapter describes how to quantitatively detect CMP-Sia in the CSS reaction mixture as well as in cells and organs.

Keywords Sialic acid, *N*-acetylneuraminic acid, *N*-glycolylneuraminic acid, Deaminoneuraminic acid, Cytidine 5′-monophosphate-sialic acid, CMP-sia synthetase, High-performance liquid chromatography, DMB labeling, Liquid chromatography–mass spectrometry

1 Introduction

The vertebrate glycosyltransferases (GTs) usually use nucleotide-conjugated monosaccharides as donor substrates, which are high-energy intermediates such as UDP-N-acetylglucosamine (GlcNAc), GDP-mannose (Man), GDP-fucose (Fuc), and CMP-sialic acid (Sia). These nucleotide monosaccharides are usually synthesized in the cytosol and transported into the lumen of the Golgi apparatus, where GTs transfer the monosaccharides onto acceptor substrates, which are often glycans of glycoproteins and glycolipids [1]. CMP-Sia is a donor substrate for all kinds of sialyltransferases (STs), which account for 20 kinds in humans, and is synthesized by the CMP-sialic acid synthetase (CSS or CMAS) [2]. Notably, Sia consists of *N*-acetylneuraminic acid (Neu5Ac), *N*-glycolylneuraminic acid (Neu5Gc), deaminoneuraminic acid (Kdn), and their modified derivatives with O-acetylation, O-methylation, O-sulfation on the 4-, 7-, 8-, and 9-hydroxyl groups [2–4]. CMP-Neu5Ac and CMP-Kdn are de novo synthesized from Neu5Ac and Kdn, respectively [5, 6], although CMP-Neu5Gc is de novo synthesized by conversion from CMP-Neu5Ac in the cytosol through hydroxylation by the CMP-Neu5Ac hydroxylase (CMAH)

Koichi Furukawa (ed.), *Gangliosides*, Methods in Molecular Biology, vol. 3035,
https://doi.org/10.1007/978-1-0716-5288-6_4,

[2]. Interestingly, the CMP-Neu5Ac synthesis occurs in the nucleus [2, 7, 8] as well as in the cytosol [9], and CMP-Neu5Ac is actually detected in the nucleus fraction in mammalian cells [10]. On the other hand, CMP-Kdn is always detected in the cytosol, but not in the nucleus [11, 12]. The modifications of Sia happen on Sia residues after transfer from CMP-Sia onto the glycan termini as in the case with WSCD1- and WSCD2-mediated sulfation reactions [13], while in contrast, 9-O-acetylation occurs on CMP-Sia by CASD1 [14, 15]. It still remains unclarified not only which enzymes are involved in other modifications, but also whether the modifications occur at the CMP-Sia or the Sia residue levels. Therefore, CMP-Sia is usually present in the nucleus or the cytosol of cells as CMP-Neu5Ac, CMP-Neu5Gc, CMP-Kdn, and CMP-O-acetylated Sia, although the last ones have never been isolated due to their instability [14].

CSSs are well conserved in the primary and tertiary structures among various organisms [5, 16–18], and their cDNAs are easily cloned from any organisms, once we obtain the genomic DNA or mRNAs of the organisms or even in the gene database [16, 17, 19]. The existing CMP-Sia species in the organisms can be deduced from the substrate specificity of their recombinant CSSs. This is easy and useful, compared with the direct detection of CMP-Sias in cells and tissues of the organisms. In this chapter, we mainly describe the procedures for the identification and quantification of CMP-Neu5Ac, CMP-Neu5Gc, and CMP-Kdn in the reaction mixture obtained using recombinant CSS proteins, although we also describe a direct detection method of CMP-Sias in cell lysates and tissue homogenates as the note section. **Procedure 1** is a preparation of recombinant CSS using the *Escherichia coli* (*E. coli*) expression system. **Procedure 2** is a preparation of authentic CMP-Sias using the recombinant rainbow trout CSS, because the rainbow trout CSS can equally synthesize CMP-Neu5Ac, CMP-Neu5Gc, and CMP-Kdn, while the mouse CSS can hardly synthesize CMP-Kdn [5, 20]. **Procedure 3** is a determination of the CMP-Sias in the reaction mixture by the anion-exchange high-performance liquid chromatography (HPLC) using authentic CMP-Sias as the standard compounds. **Procedure 4** is to precisely identify Sia species in the CMP-Sia fraction obtained by the anion-exchange HPLC. Since the CMP-Sia fraction especially from crude samples may contain various CMP-Sia species, the CMP-Sia fraction is applied to the fluorometric HPLC analysis to determine the Sia species in the fraction, after acid-hydrolysis and labeling with a fluorescent dye, 1,2-dimethylenedioxybenzene (DMB).

2 Materials

2.1 Preparation of Recombinant CSS or CMAS

1. pET32a plasmid (Novagen, Madison, WI) containing the His-tag in the N-terminus of the insert rainbow trout CSS is described in the previous report [5].

2. *Escherichia coli* strain, BL21(DE3)pLysS.
3. IPTG solution: 100 mM isopropyl-1-D-galactopyranoside (IPTG) solution.
4. Sonication buffer: 50 mM Tris–HCl (pH 8.0) containing 2 mM EDTA, 1 mM phenylmethylsulfonyl fluoride, and 1 μg/ml each of aprotinin, leupeptin, and pepstatin A.
5. Ni^{2+} -NTA agarose (QIAGEN; Valencia, CA)
6. Equilibration buffer: 5 mM imidazole, 0.5 M NaCl, and 20 mM Tris–HCl (pH 8.0)
7. Washing buffer: 30 mM imidazole, 0.5 M NaCl, and 20 mM Tris–HCl (pH 8.0).
8. Elution buffer: 100 mM imidazole,0.5 M NaCl, and 20 mM Tris–HCl (pH 8.0).
9. VIVASPIN® 20 (10 K) ultrafiltration spin column
10. Anti-6His antibody (Cosmo Bio Co., LTD., Japan)
11. An electrophoresis and blotting system for western blotting
12. Western blot detection system
13. Pre-stained Mw marker was obtained from Bio-Rad (Hercules, CA).
14. Bicinchoninic acid (BCA) protein assay kit was purchased from Thermo Scientific Pierce (Rockford, IL).

2.2 Preparation of Standard CMP-Sia

1. Neu5Ac and CMP-Neu5Ac are obtained from Nacalai Tesque, Kyoto, Japan.
2. Neu5Gc is obtained from Tokyo Chemical Industry Co., Tokyo, Japan.
3. Kdn is available in Toronto Research Chemicals Inc. Kdn is also prepared in the laboratory as described previously [21].
4. Alkaline phosphatase from calf intestine (Böhringer Mannheim, Mannheim, Germany)
5. High-performance liquid chromatography (HPLC) system: JASCO HPLC system using a resource Q anion-exchange column (6.5 × 30 mm) and UV detector

2.3 Detection and Identification of CMP-Sia

1. Standard CMP-Sia: Commercially available CMP-Neu5Ac and CMP-Neu5Gc, and CMP-Kdn prepared as described in Sect. 3.2 is used.
2. An integrator connected to UV detector: The peak height and area of the chromatogram are calculated.

2.4 DMB Labeling and the Fluorescent HPLC Analysis

1. DMB solution: 7 mM 1,2-dimethylenedioxybenzene (DMB; Dojindo, Kumamoto, Japan) in 5.0 mM trifluoroacetic acid

containing 1 M 2-mercaptoethanol and 18 mM sodium hydrosulfite.

2. A C18 column, e.g., a CAPCELL PAK C18 (250×4.6 mm i.d.) column (OSAKA SODA, Japan) is connected to an HPLC system. For example, a Jasco LC-900 HPLC system equipped with a Jasco FP-920 fluorescence detector.
3. Liquid chromatography–mass spectrometry (LC-MS) system: A Shimadzu Nexera system and a ZenoTOF 7600 system. The Nexera system controller (SCL-40) equipped with a pump unit (LC-40D), a degasser (DGU-405), and auto-sampler (SIL-40C) is used to apply DMB-Sia using a CAPCELL CORE C18 column (2.7 μm, 2.1 mm i.d. × 150 mm) at a column temperature of 30°C by column oven (CTO-40S). Mobile phase A is 0.1% (v/v) formic acid in water, and mobile phase B is acetonitrile/methanol (50:50, v/v). The flow rate was set at 0.2 ml/min. One ng of the standard sample is injected and eluted with the gradient: 8% B (0–24.0 min), 8%–30% B (24.0–80.0 min), 30%–95% B (80.0–80.1 min), 95% B (80.1–85.0 min), 95%–8% (85.0–85.1 min), 8% B (85.1–90.0 min) in 40 min (*see* **Note 1**). Fluorescence is detected using an RF-20A XS fluorescence detector (excitation and emission, 373 and 448 nm, respectively).

3 Methods

3.1 Preparation of Recombinant Rainbow Trout CSS or CMAS

1. The rainbow trout CSS gene is subcloned into the pET32a plasmid (*see* **Note 2**).
2. *E. coli* strain BL21(DE3) pLysS is transformed with the plasmid.
3. A single colony of transformed cells is isolated and cultured in 1 ml of Luria broth supplemented with ampicillin (LA) at 37°C, overnight.
4. The overnight culture is inoculated to 500 ml LA medium, and cultured at 37°C until the OD_{600} of the culture reach the value between 0.4 and 0.8.
5. IPTG solution is added to the culture at a final concentration of 0.4 mM.
6. After incubation at 15°C for 4 h, the cells are harvested by centrifugation (6000 × *g*, 4°C, 10 min).
7. The cell pellet is suspended in 50 ml sonication buffer and sonically disrupted on ice.

8. The supernatant is collected by centrifugation (21,600 × *g*, 4°C, 15 min).
9. For purification of the recombinant CSS protein, the supernatant is mixed with an equal volume of 10 mM imidazole, 1.0 M NaCl, and 40 mM Tris–HCl (pH 8.0).
10. The mixture is applied to Ni^{2+} chelating column (1.4 × 1.3 cm), which has been equilibrated with the equilibration buffer.
11. The column is washed with the washing buffer.
12. The recombinant CSS is eluted with the elution buffer.
13. The eluted fraction is desalted and three times exchanged to 50 mM Tris–HCl (pH 8.0) by VIVASPIN® 20 (10 K) ultrafiltration spin column to be concentrated to about 1 ml.
14. The purified CSS protein amount is quantified by BCA assay, Coomassie Brilliant Blue (CBB) staining and western blotting using anti-His tag antibody (*see* **Note 3**).

3.2 Preparation of CMP-Sia Using the Recombinant Rainbow Trout CSS

1. The reaction mixture is prepared in 25 μl containing appropriate amount of the recombinant rainbow trout CSS, 1.0 mM Sia (Neu5Ac, Neu5Gc, and Kdn), 2–5.0 mM CTP, 20 mM $MgCl_2$, and 0.1 mM Na_3VO_4 (phosphatase inhibitor) in 100 mM Tris–HCl (*see* **Note 4**).
2. The reaction mixture is incubated at 25°C for 1–3 h (*see* **Note 5**).
3. The reaction is stopped by adding 25 μl of 32 units of alkaline phosphatase from calf intestine in 200 mM NaOH-glycine (pH 10.0) and further incubated at 37°C for 3 h to digest excess CTP.
4. The proteins in the reaction mixture are removed by ethanol precipitation. To the alkaline phosphatase reaction mixture, 100% cold ethanol is added to adjust 70%–90% ethanol and kept at −30°C for 3 h, and centrifuged (16,000 × *g*, 4°C, 30 min). The obtained supernatant is usually applied to an anion-exchange chromatography directly for analytical purpose (see Sect. 3.3); however, for the purpose of purification of CMP-Sia, it is lyophilized and dissolved in 100 μl of 1 mM aqueous ammonium (*see* **Note 6**) and applied to an anion-exchange chromatography and a gel filtration [22].

3.3 Detection and Quantitation of CMP-Sia

1. The supernatant after the ethanol precipitation (*see* **Note 7**) is subjected to a resource Q HPLC to separate CMP-Sia from cytidine and free Sia in the supernatant. The column is eluted at 1.0 ml/min with a linear gradient of 0–0.2 M NaCl in 20 mM Tris–HCl buffer (pH 8.0) in 10 min. The elution profile is monitored by absorbance at 271 nm, and the peak area is integrated for quantitation (Fig. 1). An authentic sample of CMP-Neu5Ac is used for quantitative analysis.

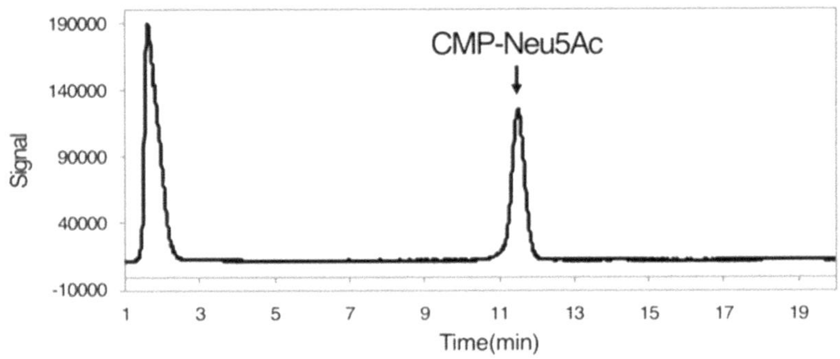

Fig. 1 Anion-exchange chromatography of the reaction mixture of the recombinant medaka CSS. The reaction mixture of Neu5Ac with the rainbow trout CSS was, after ethanol precipitation, applied to a Resource Q HPLC to obtain the CMP-Sia fraction (CMP-Neu5Ac). The column is eluted at 1.0 ml/min with a linear gradient of 0–0.2 M NaCl in 20 mM Tris–HCl buffer (pH 8.0) in 10 min. The elution profile is monitored by absorbance at 271 nm. The bar indicates the CMP-Sia fraction

2. An authentic sample of CMP-Neu5Ac is used for quantitative analysis (*see* **Note 8**), utilizing a calibrated standard curve, and concentrations of CMP-Neu5Ac, CMP-Neu5Gc, and CMP-Kdn in the reaction mixture are calculated using the standard curve.

3.4 DMB Labeling and the Fluorescent HPLC Analysis

1. Since CMP-Neu5Ac, CMP-Neu5Gc, and CMP-Kdn cannot be well separated from each other on an anion-exchange chromatography like a resource Q column, the CMP-Sia fraction is pooled and further analyzed for identifying Sia species in the CMP-Sias using the fluorometric HPLC analysis [23, 24] (*see* **Note 9**).
2. The CMP-Sia fraction is lyophilized and hydrolyzed in 20 μl of 0.1 M trifluoroacetic acid at 80°C for 2 h, and then dried completely by SpeedVac. To the dried sample, 20 μl of 0.01 N TFA solution and 20 μl of the DMB solution are added, and the reaction mixture is incubated at 50°C for 2 h.
3. The reaction mixture in 2- to 20-μl portions is directly injected into a C18 column (250 × 4.6 mm i.d.) and eluted with methanol/acetonitrile/water (7:7:86, v/v/v) [24] (*see* **Note 10**). An HPLC system equipped with a fluorescence detector (excitation, 373 nm; emission, 448 nm), operating isocratically at 1.0 ml/min at a column temperature of 26°C, is used.
4. Authentic Neu5Ac, Neu5Gc, and Kdn (see Sect. 2.2) are used as the standard samples to quantify the amount of these Sias. Each DMB-Sia is separated on the C18 column HPLC, and assigned by the retention time of each authentic Sia peak (Fig. 2), as well as by the mass spectrum of LC–MS of each

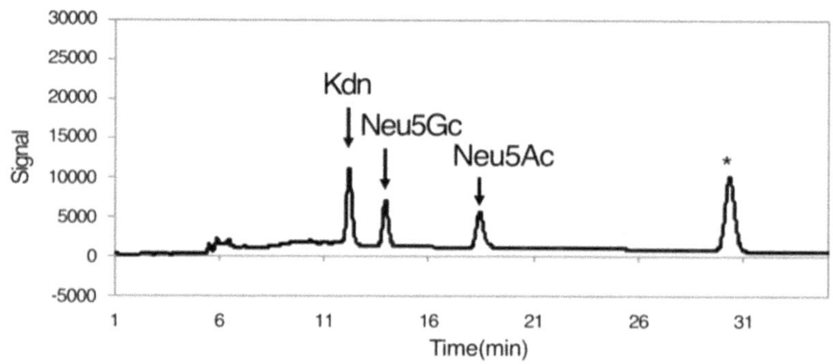

Fig. 2 C18 column chromatography of DMB-Neu5Ac, DMB-Neu5Gc, and DMB-Kdn. An equimolar mixture of authentic Neu5Ac, Neu5Gc, and Kdn was labeled with DMB and subjected to C18 column HPLC. The column was isocratically eluted with methanol/acetonitrile/water (7/7/86, v/v/v), and monitored by a fluorescence detector (excitation, 373 nm; emission, 448 nm). Each peak is assigned as shown in the figure. The reagent peak was marked by *

peak (*see* **Note 11**). A mass spectrum of the DMB-Neu5Ac peak is shown (Fig. 3).

4 Notes

1. When the sample contains more diverged derivatives of CMP-Sia, 90 min but not 40 min is usually adopted for the total elution time.
2. Vertebrate CSSs are soluble proteins without any glycan modifications, which could easily be prepared by *E. coli* system. On the other hand, insect CSSs, which are membrane proteins, could be prepared by *E. coli* system only after the transmembrane domain or the signal peptide sequence in the N-terminus is removed [16].
3. It is important to measure the protein amount to determine the specific activity of the CSS. Bovine serum albumin (BSA) is usually as a standard for the protein amount. If the expression level is high enough to measure it, BCA colorimetric assay in the enzyme solution and CBB staining of a gel after SDS-polyacrylamide gel electrophoresis of the enzyme can be performed. However, if the expression is not high enough for these methods, western blotting of the enzyme solution using anti-His tag antibody can be applied, because the recombinant CSS is designed to have the His tag in the N-terminus.
4. Na_3VO_4 (phosphatase inhibitor) at 0.1 mM is usually added to the reaction mixture to prevent dephosphorylation of

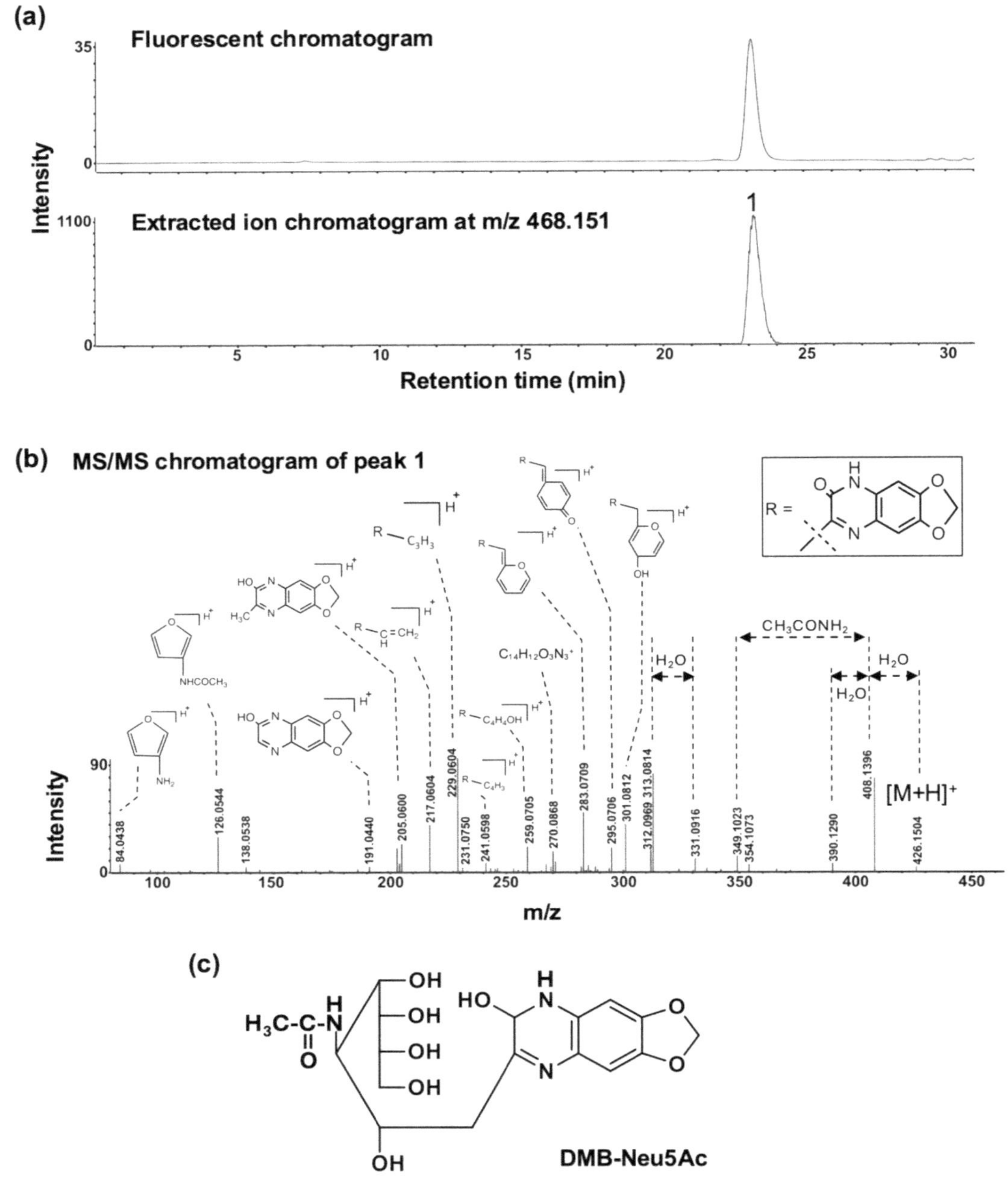

Fig. 3 (**a**) Liquid chromatography of DMB-Neu5Ac. (Upper panel) Fluorescent detection (Ex: 337 nm Em: 448 nm); (Lower panel) Extracted ion chromatogram (XIC) at m/z 468.151. (**b**) LC-MS/MS chromatogram of peak 1. The fragmentation patterns of CID of the DMB-Neu5Ac peak on LC–MS and their deduced structures are shown. (**c**) Structure of DMB-Neu5Ac

CTP. However, if the recombinant CSS is highly pure, we can remove this reagent from the reaction mixture.

5. Reaction temperature varies depending on organism origins as follows: insect CSS, 37°C; fish CSS, 25°C; vertebrate CSS, 37°C [5, 16]. The optimal pH for vertebrate CSSs is usually 8.0 [5]. Reaction time can be variable, depending on the enzyme properties and the amount.
6. CMP-Sia usually comes to the supernatant of ethanol precipitation. After this step, pH of the solution containing CMP-Sia should be kept at alkaline conditions. For this purpose, 1 mM aqueous ammonium is usually used. Particular care should be taken at this point, because the CMP-Kdn is much more fragile than CMP-Neu5Ac [22].
7. When cell lysates or tissue homogenates are used for the determination of the CMP-Sia amount, protein concentrations are often much higher than the recombinant CSS reaction mixture. Acetone precipitation can be applied instead of ethanol precipitation. In this case, cold acetone (−30°C) is added to adjust 80%–95% to keep at −30°C for 3 h or overnight.
8. The molar extinction coefficients of CMP-Neu5Gc and CMP-Kdn are the same as that of CMP-Neu5Ac. CMP-Neu5Ac is used as a standard compound for quantification of CMP-Sia using absorbance at 271 nm, because of its frequent availability.
9. To analyze the Sia level in the homogenates of medaka embryos, the following methods are available [19]. Each medaka fry at 9 dpf (body length about 3.8–4.2 mm) is treated with the lysis buffer (*see* **Note 12**). After centrifugation, 10 μl of the supernatant is subjected to the DMB derivatization (see Sect. 3.4 2). The obtained Sia value indicates the sum of free and bound forms of Sia. If we separate the CMP-Sia fraction from other components, the supernatant can be applied to the Resource Q column chromatography (see Sect. 3.4 1).
10. To obtain a stable baseline of elution profile, the ODS column is recommended to be equilibrated for overnight with the elution solution: methanol/acetonitrile/water (7:9:84, v/v/v).
11. The structure of DMB-Sia is determined by the fragmentation patterns of collision-induced dissociation (CID). For the MS/MS CID fragments, the CAD energy intensity is set at CAD30. The fragment at m/z 168.066 corresponded to the *N*-acetyl group at the C5 position, which is unique to Neu5Ac (Fig. 3). Fragment ions reflecting the differences at the C5 position are observed at m/z 184.060 for Neu5Gc and m/z 127.039 for Kdn (data not shown).
12. Proteinase K treatment is adopted when the genotyping of the embryo is necessary. The tail of medaka fry at 9 dpf is fixed in 40 μl of methanol and subjected to cell lysis in 25 μl

of proteinase K solution (10 mM Tris–HCl, pH 7.5, 10 mM EDTA, and 2 mg/ml proteinase K) at 55°C for 3 h, followed by standing at 95°C for 10 min. For preparing medaka lysate, each medaka fry is subjected to lysis in 25 μl of lysis buffer (PBS containing 1% Triton X-100, 2 mM EDTA, 1 mM phenylmethylsulfonyl fluoride, and 1 μg/ml each of aprotinin, leupeptin, and pepstatin A, followed by centrifugation).

Acknowledgments

This work was supported by Grants-in-Aid for Scientific Research (B) (23K23523) (to KK) and (21H02425 and 23K21291) (to CS) for Young Scientists (21K15040) (to DW) from the Japan Society for the Promotion Science. This work is also supported at least in part by Joint Usage/Research Center Glyco network (J-GlycoNet) and Human Glycome Atlas project (HGA) from the MEXT, Japan.

Conflict of Interest

The authors have no conflicts of interest to declare that are relevant to the content of this chapter.

References

1. Rini JM, Moremen KW, Davis BD et al (2022) Chapter 6. Glycosyltransferases and glycan-processing enzymes. In: Varki A, Cummings R, Esko JD et al (eds) Essentials of glycobiology, 4th edn. Cold Spring Harbor Laboratory Press, Cold Spring harbor (NY)
2. Lewis AL, Chen X, Schnaar RL et al (2022) Chapter 15. Sialic acids and other nonulosonic acids. In Varki A, Cummings R, Esko JD et al (eds) Essentials of glycobiology. 4th edn. Cold Spring Harbor Laboratory Press, Cold Spring Harbor, NY
3. Sato C, Kitajima K (2019) Sialic acids in neurology. Adv Carbohydr Chem Biochem 76:1–64. https://doi.org/10.1016/bs.accb.2018.09.003
4. Kitajima K, Varki N, Sato C (2015) Advanced technologies in sialic acid and sialoglycoconjugate analysis. Top Curr Chem 367:75–103
5. Nakata D, Münster AK, Gerardy-Schahn R et al (2001) Molecular cloning of a unique CMP-sialic acid synthetase that effectively utilizes both deaminoneuraminic acid (KDN) and N-acetylneuraminic acid (Neu5Ac) as substrates. Glycobiology 11:685–692
6. Inoue S, Kitajima K (2006) KDN (deaminated neuraminic acid): dreamful past and exciting future of the newest member of the sialic acid family. Glycoconj J 23:277–290
7. Münster AK, Eckhardt M, Potvin B et al (1998) Mammalian cytidine 5'-monophosphate N-acetylneuraminic acid synthetase: a nuclear protein with evolutionarily conserved structural motifs. Proc Natl Acad Sci U S A 95:9140–9145
8. Tiralongo J, Fujita A, Sato C et al (2007) The rainbow trout CMP-sialic acid synthetase utilises a nuclear localization signal different from that identified in the mouse enzyme. Glycobiology 17:945–954
9. Fujita A, Sato C, Kitajiima K (2007) Identification of the nuclear export signals that regulate the intracellular localization of the mouse CMP-sialic acid synthetase. Biochem Biophys Res Commun 355:174–180
10. Kean EL (1991) Sialic acid activation. Glycobiology 1:441–447
11. Go S, Sato C, Yin J et al (2007) Hypoxia-enhanced expression of free deaminoneuraminic

acid in human cancer cells. Biochem Biophys Res Commun 357:537–542
12. Schaper W, Bentrop J, Ustinova J et al (2012) Identification and biochemical characterization of two functional CMP-sialic acid synthetases in Danio rerio. J Biol Chem 287:13239–13248
13. Ertunc N, Phitak T, Wu D, et al (2022) Sulfation of sialic acid is ubiquitous and essential for vertebrate development. Sci Rep 12:12496
14. Arming S, Wipfler D, Mayr J et al (2011) The human Cas1 protein: a sialic acid-specific O-acetyltransferase?. Glycobiology 21:553–564
15. Baumann AMT, Bakkers MJG, Buettner FFR et al (2015) 9-O-Acetylation of sialic acids is catalysed by CASD1 via a covalent acetyl-enzyme intermediate. Nat Commun 6:7673
16. Wu D, Fujita A, Hamaguchi K et al (2017) Diverse subcellular localizations of the insect CMP-sialic acid synthetases. Glycobiology 27:329–341
17. Guérardel Y, Chang LY, Fujita A et al (2012) Sialome analysis of the cephalochordate Branchiostoma belcheri, a key organism for vertebrate evolution. Glycobiology 22:479–491
18. Oschlies M, Dickmanns A, Haselhorst T et al (2009) A C-terminal phosphatase module conserved in vertebrate CMP-sialic acid synthetases provides a tetramerization interface for the physiologically active enzyme. J Mol Biol 393:83–97
19. Wu D, Arakawa H, Fujita A et al (2021) A point-mutation in the C-domain of CMP-sialic acid synthetase leads to lethality of medaka due to protein insolubility. Sci Rep 11:23211
20. FujitaA SatoC, Münster-Kühnel AK et al (2005) Development of a simple and efficient method for assaying cytidine monophosphate sialic acid synthetase activity using an enzymatic reduced nicotinamide adenine dinucleotide/oxidized nicotinamide adenine dinucleotide converting system. Anal Biochem 337:12–21
21. Angata T, Matsuda T, Kitajima K (1998) Synthesis of neoglycoconjugates containing deaminated neuraminic acid (KDN) using rat liver alpha2,6-sialyltransferase. Glycobiology 8:277–284
22. Terada T, Kitazume S, Kitajima K et al (1993) Synthesis of CMP-deaminoneuraminic acid (CMP-KDN) using the CTP:CMP-3-deoxynonulosonate cytidylyltransferase from rainbow trout testis. Identification and characterization of a CMP-KDN synthetase. J Biol Chem 268:2640–2648
23. Sato C, Inoue S, Matsuda T et al (1999) Fluorescent-assisted detection of oligosialyl units in glycoconjugates. Anal Biochem 266:102–109
24. Sato C, Inoue S, Matsuda T et al (1998) Development of a highly sensitive chemical method for detecting 2,8-linked oligo/polysialic acid residues in glycoproteins blotted on the membrane. Anal Biochem 261:191–197

Chapter 5

Enzyme Activity of Glycolipid Specific Glycosyltransferase Enzymes GM2/GD2 Synthase β1,4-GalNAc Transferase (B4GALNT1)

Robiul Hasan Bhuiyan, Farhana Yesmin, Keiko Furukawa, Yuhsuke Ohmi and Koichi Furukawa

Abstract

Measurement of enzyme activity of glycosyltransferases enables the direct observation of the reaction of substrate specificity, enzyme kinetics, and interaction with other molecules. Cell-membrane fractionations containing glycosyltransferases are collected by nitrogen cavitation, and are utilized for enzyme reaction in cell-free glass tubes. Thin-layer chromatography is used to separate reaction products, and the products are visualized by an imaging analyzer. Here, we describe the whole method in detail and step by step.

Keywords Glycosyltransferase enzymes, GM2/GD2 (B4GALNT1), Enzyme activity, Thin-layer chromatography (TLC), Autoradiography

1 Introduction

Observation of enzyme-mediated reaction of glycosyltransferases became possible since the achievement of successful molecular cloning of glycosyltransferase enzymes by Furukawa's group in 1992. It has become possible to examine substrate specificity, enzyme kinetics, and interaction with other molecules [1, 2]. In this chapter, methods for measuring glycosyltransferase activity of GM2/GD2 synthase (β1,4-GalNAc transferase) (Fig. 1) were described in cell-free enzyme reaction using membrane preparations from cultured cells and tissues. This protocol can be applied not only for GM3 as an acceptor substrate, but also for lactosylceramide and GD3. Previous studies were reported based on these methods [1–3].

Koichi Furukawa (ed.), *Gangliosides*, Methods in Molecular Biology, vol. 3035,
https://doi.org/10.1007/978-1-0716-5288-6_5,

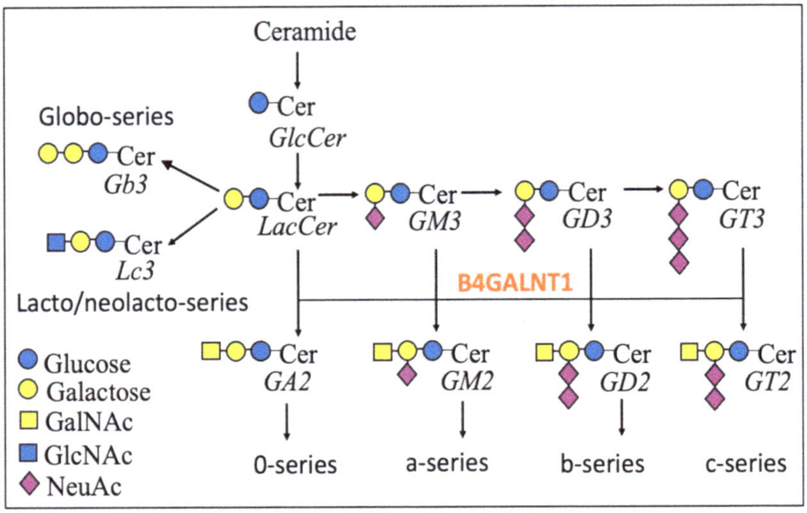

Fig. 1 A synthetic pathway of complex gangliosides. B4GALNT1 is a key enzyme for the synthesis of downstream complex gangliosides. Loss of activity of B4GALNT1 results in the loss of all complex gangliosides [4].

2 Materials

2.1 Reagents

1. GM3 (or GD3) as an acceptor (Sigma-Aldrich, St. Louis, MO, USA).
2. CDP-choline (Sigma-Aldrich, St. Louis, MO, USA).
3. UDP-GalNAc (Sigma-Aldrich, St. Louis, MO, USA).
4. [^{3}H]-UDP-GalNAc [American Radiolabeled Chemicals, Inc. (St. Louis, MO, USA)].
5. Sodium cacodylate (Wako Pure Chemical Industries Ltd., Osaka, Japan), 0.1 M Cacodylate HCl buffer, pH 7.2.
6. Extracts (membrane fractionation) from cells as an enzyme source.
7. Triton X-100 (Sigma-Aldrich).
8. Chloroform.
9. Methanol.
10. Distilled water.
11. $MnCl_2$ (100 mM).
12. Solvent systems for TLC
    a. Chloroform:Methanol: H_2O= 60:35:8 (0.02% $CaCl_2$) (for neutral glycolipids).
    b. Chloroform:Methanol: $CaCl_2$= 55:45:10 (0.02% $CaCl_2$) (for acidic glycolipids).

13. Cultured cell lines
    a. B78 (a subline of B16).
    b. L cells (mouse fibroblast line).

2.2 Apparatus and Instruments

1. N_2 cavitation apparatus (Parr Instrument Co.).
2. Compressed nitrogen gas tank.
3. N_2 evaporator (multiple injectors).
4. Water bath.
5. SepPak C18 cartridge (Waters Corp., Milford, MA).
6. Ultracentrifuge and swing-type bucket (Beckman Coulter, Optima XPN-100).
7. Aluminum-backed Thin-layer chromatography (TLC) plate (HPTLC Silica gel 60 F254, Millipore Corp).
8. Microsyringe, 10 μL (Hamilton).
9. Imaging analyzer.
10. Glass tank for TLC.
11. Glass tubes (up to 10 mL).
12. Plastic syringe (5–10 mL).

3 Methods

3.1 Preparation of Membrane Fraction

1. Cells (10^7) in 1 mM PMSF containing cold PBS are used for N_2 cavitation at 400 psi on ice for 30 min.
2. Transfer membrane fractionations in a plastic tube from the N_2 cavitation apparatus (*see* **Note 1**).
3. Centrifuge at 100 g for 10 min at 4°C, and collect supernatants.
4. Ultracentrifuge the supernatant at 40 K (Beckman, SW 5Ti) for 1 h at 4°C.
5. Resuspend the pellets in 200 μL 0.1 M cacodylate buffer, pH 7.2.

3.2 Enzyme Assay

1. Evaporate the following items in a round-bottom glass tube:
    - UDP-GalNAc (400 μM)
    - GM3 (or GD3) (325 μM)
    - [^{3}H]-UDP-GalNAc (3.5×10^5 dpm)
    - CDP-choline (10 mM)
2. Add 10 mM of $MnCl_2$ and 0.3% Triton X-100 in cacodylate buffer.

3. Sonicate for 10 s.
4. Add membrane proteins (extract) (100 μg).
5. Adjust to make the final volume 50 μL with 0.1 M cacodylate buffer.
6. Incubate at 37°C for 2–3 h with shaking.
7. Add 1 mL of distilled water to stop the reaction.
8. Separate the products by SepPak C18 column.
9. Dry with N_2 stream.
10. Count radioactivity of one-fifth of the products using a scintillation counter (*see* **Note 2**).
11. Apply four-fifths of the products to the TLC plate and separate the products by TLC. Use bovine brain ganglioside mixture (Sigma) for standards.
12. After drying, visualize bands in the plate using autoradiography with an imaging analyzer (*see* **Note 3**). An example image of TLC-autoradiography was shown in Fig. 2.

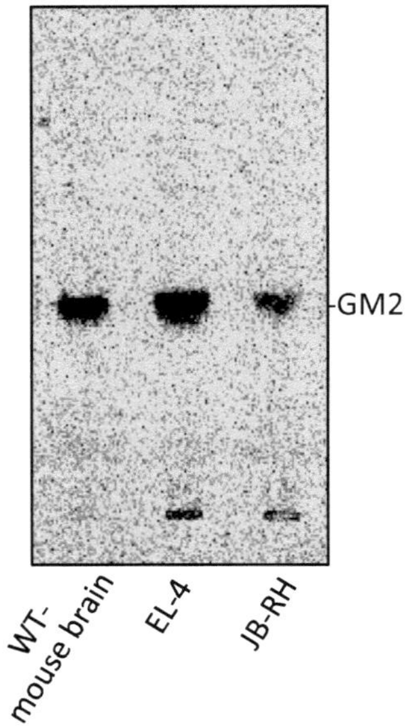

Fig. 2 TLC-autoradiography image. Three reaction products of glycolipid-specific glycosyltransferase enzyme were applied to a TLC plate. In the reactions, GM3 was used as a substrate. Glycosyltransferase GM2/GD2 synthase (B4GALNT1) enzyme was used from wild-type mouse brain, mouse EL-4 cells, and JB-RH cells

4 Notes

1. Carefully transfer the membrane fractionation from the N_2 cavitation apparatus into a 50 mL tube because there is a chance of loss of membrane fractionation due to the N_2 force.
2. Discussion

   The enzyme activity is usually expressed as a unit (pmol/mg/h) and corresponds well with the glycolipid expression on the cell surface [4].
3. Low levels of the enzyme activity should be confirmed using long exposure in autofluorography [4].

Conflict of Interest

The authors have no conflicts of interest to declare that are relevant to the content of this chapter.

References

1. Yamashiro S, Ruan S, Furukawa K et al (1993) Genetic and enzymatic basis for the differential expression of GM2 and GD2 gangliosides in human cancer cell lines. Cancer Res 53:5395–5400
2. Furukawa K, Tsuchida A, Okajima T et al (2009) Glycoconjugate glycosyl-transferases. Glycoconj J 26:987–998
3. Furukawa K, Furukawa K (2013) Enzyme assay of glycolipid glycosyltransferases ~GM2/GD2 synthase (β1,4GalNAc transferase). GlycoPOD. Available at: https://jcggdb.jp/GlycoPOD/protocolShow.action? nodeId=t54
4. Bhuiyan RH, Ohmi Y, Ohkawa Y et al (2019) Loss of enzyme activity of mutated B4GALNT1 gene products found in patients with hereditary spastic paraplegia induces relatively mild neurological disorders: Similarity with phenotypes of B4galnt1 knockout mice. Neuroscience 397:94–106

Chapter 6

Degradation System: Sialidase

Kazuhiro Shiozaki, Koji Yamamoto and Taeko Miyagi

Abstract

Sialidases are crucial enzymes responsible for the desialylation of gangliosides. In vertebrates, four sialidases—Neu1, Neu2, Neu3, and Neu4—exhibit distinct reactivity profiles toward gangliosides. This study aims to elucidate the physiological roles of ganglioside degradation mediated by these four sialidases, while also providing a comprehensive analysis of the methodologies employed to study their activity. By understanding the specific functions and analytical approaches associated with each sialidase, we can gain insight into their contributions to various physiological phenomena.

Keywords Gangliosides, Sialidase, Sialic acid, *N*-acetylneuraminic acid

1 Introduction

Sialidases are enzymes that cleave sialic acids from the non-reducing ends of glycoconjugates, including gangliosides, and are found across a wide range of organisms. In vertebrates, four types of sialidases—Neu1, Neu2, Neu3, and Neu4 have been well characterized. Sialidases are also present in invertebrates, bacteria, and viruses, each serving diverse biological roles. These sialidases differ in cellular localization, substrate specificity, and optimal pH [1]. For instance, Neu1 is primarily localized in lysosomes and exhibits optimal activity at acidic pH, whereas Neu2 is found in the cytosol, functioning best at neutral pH. Neu3 is associated with the plasma membrane and also prefers an acidic pH. Neu4 displays variability in its localization among species; in humans, it resides in lysosomes, mitochondria, and the endoplasmic reticulum (ER), whereas in fish, it can also localize in the nucleus [2–5]. These characteristics reflect the diverse biological roles of sialidases across different species and cellular environments.

Extensive studies have investigated the enzymatic activity of sialidases on gangliosides and the physiological significance of their desialylation. Among the four sialidases in vertebrates, Neu3, which localizes to the plasma membrane, plays a predominant role in ganglioside hydrolysis. Since Neu3 exhibits minimal activity toward the

Koichi Furukawa (ed.), *Gangliosides*, Methods in Molecular Biology, vol. 3035,
https://doi.org/10.1007/978-1-0716-5288-6_6,

desialylation of glycoproteins, it is regarded as a ganglioside-specific sialidase. Recent research has highlighted Neu3's significant involvement in various pathological conditions, particularly in cancer. Neu3 is frequently overexpressed in cancers such as colorectal, renal, prostate, and bladder cancers [6–9]. By hydrolyzing GM3 ganglioside into LacCer, Neu3 enhances EGFR signaling, which activates downstream ERK and AKT pathways, promoting cell proliferation, inhibiting apoptosis, and increasing cell adhesion—key processes in cancer progression [8]. In transgenic (TG) mice and NIH3T-fibroblast overexpressing NEU3, GM3 is degraded, resulting in increased carcinogenesis [10, 11], possibly by the activation of the Wnt signaling pathway by Neu3 [12]. Overexpression of NEU3 increases the cellular radioresistance potential of glioblastoma cells [13]. Conversely, silencing Neu3 with siRNA or using Neu3 inhibitors restores ganglioside patterns, reducing the malignancy of cancer cells, indicating that Neu3 could be a potential therapeutic target in cancer treatment [14, 15].

Indeed, Neu3 has garnered attention for its involvement in diseases beyond cancer. In rheumatoid arthritis, elevated erythropoietin (EPO) levels upregulate NEU3 expression, enhancing inflammatory responses through JAK2/STAT5 pathway activation [16]. The NEU3 inhibitor, Zanamivir, demonstrated anti-inflammatory effects in mouse models of rheumatoid arthritis, suggesting therapeutic potential [17]. Similarly, in GM2 gangliosidosis, NEU3 facilitates an alternative pathway for GM2 degradation. Its deficiency exacerbates GM2 accumulation, leading to neurological damage [18]. In GM1 gangliosidosis, NEU3 converts GM1 to GA1, and its absence intensifies neuroinflammation, with mice demonstrating more efficient conversion than humans [19]. Neu3 has also been implicated in neurodegenerative disorders such as Parkinson's and Alzheimer's diseases, where it contributes to glycocalyx remodeling, leading to neuronal dysfunction [20]. Additionally, the involvement of Neu3 with pulmonary fibrosis has also attracted attention. In patients with pulmonary fibrosis, the expression of DEAD box helicase 3, X-linked (DDX3), is upregulated. Inhibition of DDX3 in fibroblasts suppresses the TGF-β1-induced upregulation of NEU3 expression and alleviates fibrotic symptoms [21]. Extensive desialylation and increased NEU3 expression are observed in human and mouse models of pulmonary fibrosis, and the symptoms are attenuated in NEU3 knockout mice [22]. Neu3 inhibitors have shown efficacy in reducing pulmonary fibrosis in mouse models [23]. The upregulation of NEU3 expression in COVID-19 patients [24], along with the ability of SARS-CoV-2 to bind to the sialic acids of gangliosides [25], suggests that NEU3 may play a role in regulating COVID-19 infection.

Cytosolic Neu2 sialidase is associated with the differentiation of neuronal and muscle cells. Although Neu2 displays broad substrate specificity in vitro, including glycoproteins, gangliosides, and oligosaccharides, its in vivo physiological role in ganglioside degradation remains unclear. In fish, the cytosolic sialidase Neu2 is absent in genome, and instead, an isoform of Neu3 serves as the cytosolic neutral sialidase. Interestingly, in medaka (*Oryzias latipes*), Neu3b hydrolyzes GM2 to increase LacCer levels, which promotes myoblast differentiation into myotube cells [26]. In addition, zebrafish Neu3.2, a cytosolic sialidase hydrolyzing ganglioside, also regulates muscle cell differentiation [27]. This suggests that Neu3b plays a critical role in ganglioside metabolism during muscle cell differentiation in fish.

Like Neu2, Neu4 also demonstrates good substrate specificity for gangliosides and glycoproteins in vitro. There are two splicing variants of human NEU4: the long form (NEU4L), which localizes to the mitochondria, and NEU4S, which is found in the endoplasmic reticulum and other organelles [3]. NEU4S's expression is often decreased in cancers such as colorectal cancer, indicating its potential role in cancer progression [28, 29]. Studies have shown that in neuronal cells undergoing apoptosis through tyrosinase expression [30]. The expression level of NEU4L was dramatically decreased prior to apoptosis, accompanied by the trafficking of ganglioside GD3, a good substrate for NEU4L, to mitochondria, suggesting its involvement in the mitochondrial apoptotic pathway in neuronal cells. Neu4-knockout (Neu4−/−) mice exhibited increased GD1a and decreased GM1 ganglioside levels in the brain, indicating that Neu4 is crucial for ganglioside desialylation [31]. In addition, NEU4 expression is up-regulated in the neurospheres of glioblastoma patients [32]. NEU4. Silencing NEU4 gene in glioblastoma stem cells results in the activation of the glycogen synthase kinase 3β, with the consequent inhibition of Sonic Hedgehog and Wnt/β -catenin signaling pathways, accompanied by the decrease of GD1a, GM1, and GD3.

Neu1 is typically considered to have limited activity toward gangliosides; however, under certain conditions, it can hydrolyze these substrates. Neu1 can translocate to the cell membrane during lysosomal exocytosis [33] and has also been observed in the nucleus in mice [34]. Neu1-deficient zebrafish and mice exhibit abnormal ganglioside accumulation, such as GM1, GM3, LacCer, and Gb4, suggesting a significant, albeit indirect, role in maintaining ganglioside homeostasis [35]. These studies might indicate an indirect effect of Neu1 deficiency on glycosphingolipid metabolism or transport.

As summarized in Table 1, the four sialidases differ in subcellular localization, substrate specificity, and enzymatic properties, therefore involving various pathological significance.

Table 1
Comparison of mammalian sialidases

	Neu1	Neu2	Neu3	Neu4
Major subcellular localization	Lysosomes	Cytosol	Plasma membrane	Lysosomes[a]
Mitochondria[a] and ER				
Good substrates	Oligosaccharides	Oligosaccharides	Gangliosides	Oligosaccharides
	Glycopeptides	Glycoproteins		Glycoproteins
		Gangliosides		Gangliosides
Optimal pH	4.4–4.6	6.0–6.5	4.6–4.8	4.4–4.5
Total amino acids				
Human	415	380	428	496 (484)
Mouse	409	379	418	497 (413)
Chromosomal location				
Human	6p 21.3	2q 37	11q13.5	2q37.3
Mouse	17	1	7	10
Possible function	Lysosomal degradation	Myoblast differentiation	Neural differentiation	Apoptosis
				Adhesion
	Immune function	Neural differentiation	Apoptosis	
			Adhesion	
			Upregulation in various cancers [6–9,12)]	

[a]For the subcellular localization of human NEU4, two different results have been described

2 Materials

2.1 Enzymatic Reaction for Sialidases

1. Buffers: 0.5 M sodium acetate buffer (pH 4.0–5.5) and sodium phosphate buffer (pH 5.5–6.5). Choose the adequate buffer for the sialidase you have interest.
2. Substrates: 4-Methylumbelliferyl N-acetylneuraminic acid (4MU-NeuAc, Nacalai Tesque, Inc., Kyoto, Japan), Gangliosides (mixed gangliosides from bovine brain, Merck, Darmstadt, Germany) or each ganglioside such as GM3, GD3, GM2, GM1a, GD1a, GT1a, GQ1a, etc. (Merck).
3. Triton X-100 (Merck).
4. Bovine serum albumin (Fraction V, Sigma-Aldrich, St. Louis, MO).
5. Water bath.
6. Dye-binding assay (Bio-Rad Laboratories, Hercules, CA).

2.2 Thiobarbituric Acid (TBA) Assay

1. Metaperiodate solution: 0.2 M metaperiodate in 9 M phosphoric acid.
2. Arsenite solution: 10% sodium arsenite in 0.5 M sodium sulfate/0.1 N H_2SO_4.
3. TBA solution: 0.6% TBA in 0.5 M sodium sulfate.
4. Cyclohexanone (Nacalai).
5. Spectrophotometer.

2.3 Fluorescence Method Using 4MU-NeuAc

1. 4-Methylumbelliferone (MU, Nacalai).
2. 0.25 M glycine-NaOH (pH 10.4).
3. Fluorescence spectrophotometer.

2.4 HPLC Method Using Malononitrile

1. Malononitrile (Nacalai).
2. 0.15 M tetraborate buffer (pH 9.5).
3. HPLC system: Single pump, Degasser, Column oven, Autosampler, Fluorescence detector.
4. Methanol.
5. 0.01 M ammonium acetate buffer (pH 9.5).

2.5 HPLC Method Using DMB

1. 1, 2-diamino-4,5 methylene dioxybenzene (DMB dihydrochloride (DOJINDO, Kumamoto, Japan).
2. Acetic acid.
3. *p*-mercaptoethanol (Nacalai).
4. Sodium hydrosulfite (Nacalai).
5. Acetonitrile.

3 Methods

3.1 Sialidase Activity Assay

It is essential to understand the characteristics of each sialidase, including their preparation methods from biological samples and stability. For instance, Neu1 tends to exhibit unstable enzyme activity, making fresh samples preferable. While Neu2 is a soluble protein, both Neu1 and Neu3 are membrane-bound, requiring careful consideration of this difference. Attention must also be given to Neu4, particularly its animal species origin, as solubilization of the nucleus is necessary when assessing the activity of nuclear-localized Neu4. The reaction pH is another critical factor. Neu1, Neu3, and Neu4 exhibit optimal activity under acidic conditions, whereas Neu2 functions best at a neutral to slightly alkaline pH. However, when simulating reactions on cell membranes, even Neu1, Neu2, and Neu3 may be tested at a neutral pH. Reaction temperature for sialidase activity should be maintained at 37°C. Although the native habitat temperatures for fish and bacteria differ significantly from 37°C, their sialidases often exhibit optimal activity at this temperature.

Regarding the substrate for the reaction, using the artificial substrate 4-methylumbelliferyl N-acetylneuraminic acid (4MU-NeuAc) is convenient when measuring sialidase activity in samples like sputum [36]. However, human NEU3 shows relatively low reactivity toward 4MU-NeuAc, unlike Neu1, Neu2, and Neu4. Non-human Neu3 generally exhibits higher reactivity with 4MU-NeuAc. The hydrolyzed product, 4-methylumbelliferone (4MU), can be easily measured using a fluorescence spectrophotometer to determine sialidase activity. When utilizing sialylated glycoconjugates like gangliosides as substrates, sialidase activity can be assessed by measuring the amount of liberated sialic acid using several methods. If the released sialic acid is abundant, the thiobarbituric acid (TBA) method provides a straightforward colorimetric quantification [37]. For higher sensitivity, sialic acid analysis by HPLC using malononitrile [38] or DBA [39] can be employed, but it is essential to note the specific features of each analytical method when performing the analysis.

3.2 Enzymatic Reaction Mixture

1. Place the reaction mixture containing 5 μmol of sodium acetate or phosphate buffer (pH; optimal pH for target sialidase), 10 nmol of substrate (gangliosides for TBA and HPLC method, and 4MU-NeuAc for fluorescence method), 50 μg of Triton X-100, 50 μg of bovine serum albumin, and enzyme (0.25 mg of protein) in 50 μL in a test tube.
2. Incubate for 0.5–2 h at 37°C.
3. Stop the incubation by freezing or immediately determine released sialic acid by TBA, MU fluorescence, or HPLC method.

4. One unit was defined as the amount of enzyme cleaving 1 nmol of sialic acid/h.

   Protein concentrations were determined by dye-binding assay (Bio-Rad Laboratories).

3.3 TBA Method

1. Terminate the reaction (50 μL) by adding 25 μL of metaperiodate solution.
2. Stand for 20 min at room temperature.
3. Add 250 μL of arsenite solution, and vortex the tube until the yellow–brown color disappears.
4. Add 750 μL of TBA solution and vortex and heat in a boiling water bath for 15 min.
5. Place in cold water for 5 min.
6. Add 1 mL of cyclohexanone and vortex and centrifuge at 3000 rpm for 10 min.
7. Measure the absorbance of the upper phase at 549 and 532 nm.
8. If the absorbance ratio A549/A532 for the sample is lower than the standard, calculate the value according to the formula: (0.9 × A549 - 0.3 × A532), to compensate for the interference (*see* **Note 1**).

3.4 Fluorescence Method Using 4MU-NeuAc

1. Terminate the reaction by adding 1.25 mL of 0.25 M glycine-NaOH (pH 10.4).
2. Fluorometrically measure the 4-methylumbelliferylone(4-MU) released (emission 448 nm, excitation 365 nm).
3. Prepare the Standard Curve using 4-MU.

3.5 HPLC Method Using Malononitrile

1. Add 100 μL of malononitrile (0.8%) to the reaction mixture (50 μL).
2. Add 750 μL of tetraborate (0.15 M, pH 9.5).
3. Incubate for 20 min at 80°C.
4. Apply 10–20 μL to a reversed-phase column (C18) and separate with methanol and 0.01 M ammonium acetate buffer (15:85 v/v, pH 5.5). Flow rate is 1 mL/min. Set the column oven at 40°C.
5. Measure fluorescence at 434 nm for emission and 357 nm for excitation (*see* **Note 2**).

3.6 HPLC Method Using DMB

1. Add 200 μL of DMB solution (7 mM DMB dihydrochloride in 1.4 M acetic acid containing 0.75 M *p*-mercaptoethanol and 18 mM sodium hydrosulfite) to the reaction mixture (50 μL).
2. Incubate for 2 h and 30 min at 50°C.

3. Apply 10–20 μL to a reversed-phase column (C18) and separate with acetonitrile- methanol-water (9:7:84, v/v/v). Flow rate is 1 mL/min. Set the column oven at 40°C.
4. Measure fluorescence at 448 nm for emission and 373 nm for excitation (*see* **Note 3**).

3.7 Distribution of Sialidases in the Tissue

Endogenous sialidases can be detected using methods such as Western blotting. Additionally, immunostaining with sialidase antibodies can be employed to observe the distribution of sialidases in tissue samples. However, since sialidases are expressed at low levels and many commercially available antibodies may have uncertain specificity, careful evaluation is required when using them. Recently, it has become possible to visualize the distribution of sialidases within tissues through activity staining. In the case of 4MU-NeuAc, its use for tissue staining has been challenging due to the water solubility of MU. 5-bromo-4-chloroindol-3-yl-α-D- *N*-acetylneuraminic acid (X-Neu5Ac) serves as a substrate for sialidases, and the released X is non-water-soluble, making it suitable for tissue staining [40]. However, due to the relatively low fluorescence intensity of X, there are sensitivity issues in detecting endogenous sialidase activity in samples with low activity, such as those from vertebrates. *N*-Acetyl-2-O-[2-(2-benzothiazolyl)-4-bromophenyl]-α-neuraminic Acid (BTP3-Neu5Ac) is water-soluble and non-fluorescent, but upon hydrolysis by sialidase, it releases BTP3, which exhibits strong fluorescence (excitation wavelength: 374 nm, emission wavelength: 542 nm) [41]. Since BTP3 is water-insoluble, it precipitates in tissues, allowing for staining. BTP3-Neu5Ac is commercially available (supplied by Tokyo Chemical Industry Co., Ltd., Tokyo, Japan). The reactivity of sialidases toward this reagent is considered to be almost identical to that toward 4MU-NeuAc, and nearly all sialidases are expected to react with this reagent.

4 Notes

1. Some interference in TBA methods

   TBA methods have been generally used, but the color reaction is affected by other sugar compounds. With crude homogenates as the enzyme source, the TBA method may be performed after passage through an AG1X-2 mini-column to reduce interference by other sugar compounds.
2. HPLC methods using malononitrile

   This method allows for easy derivatization of sialic acids, enabling rapid analysis of their quantities. However, *N*-acetylneuraminic acid (NANA), *N*-glycolylneuraminic acid (NGNA), and 2-keto-3-deoxy-D-glycero-D-galacto-nononic acid (KDN) are eluted at the same retention time, making it impossible to distinguish them. Using an ion-exclusion chromatography column can separate NANA and NGNA, but it results in a longer analysis time [42]. Additionally, since *N*-acetylgalactosamine and N-acetylglucosamine also react with malononitrile, it is crucial to confirm that their peaks do not overlap with those of the sialic acids.

3. HPLC methods using DMB

DMB can be used as a fluorogenic reagent for sialic acid determination, but the reaction needs a longer time (2.5 h at 50°C in the dark). However, NANA, NGNA, and KDN can be separately quantified by the DMB method.

Conflict of Interest

The authors have no conflicts of interest to declare that are relevant to the content of this chapter.

References

1. Miyagi T, Wada T, Yamaguchi K et al (2008) Human sialidase as a cancer marker. Proteomics 8:3303–3311
2. Seyrantepe V, Landry K, Trudel S et al (2004) Neu4, a novel human lysosomal lumen sialidase, confers normal phenotype to sialidosis and galactosialidosis cells. J Biol Chem 279:37021–37029
3. Yamaguchi K, Hata K, Koseki K et al (2005) Evidence for mitochondrial localization of a novel human sialidase (NEU4). Biochem J 390:85–93
4. Bigi A, Morosi L, Pozzi C et al (2010) Human sialidase NEU4 long and short are extrinsic proteins bound to outer mitochondrial membrane and the endoplasmic reticulum, respectively. Glycobiology 20:148–157
5. Hyodo T, Honda A, Yamate S et al (2023) Elucidation of the mechanism of nuclear localization of Mexican tetra Neu4 via bipartite nuclear localization signal and less conserved regions. Biochimie 212:123–134
6. Kakugawa Y, Wada T, Yamaguchi K et al (2002) Up-regulation of plasma membrane-associated ganglioside sialidase (Neu3) in human colon cancer and its involvement in apoptosis suppression. Proc Natl Acad Sci USA 99:10718–10723
7. Ueno S, Saito S, Wada T et al (2006) Plasma membrane-associated sialidase is up-regulated in renal cell carcinoma and promotes interleukin-6-induced apoptosis suppression and cell motility. J Biol Chem 281:7756–7764
8. Kawamura S, Sato I, Wada T et al (2012) Plasma membrane-associated sialidase (NEU3) regulates progression of prostate cancer to androgen-independent growth through modulation of androgen receptor signaling. Cell Death Differ 19:170–179
9. Tatsuta T, Ito J, Yamamoto K et al (2024) Sialidase NEU3 contributes to the invasiveness of bladder cancer. Biomedicines 12:192
10. Shiozaki K, Yamaguchi K, Sato I et al (2009) Plasma membrane-associated sialidase (NEU3) promotes formation of colonic aberrant crypt foci in azoxymethane-treated transgenic mice. Cancer Sci 100:588–594
11. Yamamoto K, Takahashi K, Shiozaki K et al (2015) Potentiation of epidermal growth factor-mediated oncogenic transformation by sialidase NEU3 leading to src activation. PLoS One 10:e0120578
12. Takahashi K, Hosono M, Sato I et al (2015) Sialidase NEU 3 contributes neoplastic potential on colon cancer cells as a key modulator of gangliosides by regulating Wnt signaling. Int J Cancer 137:1560–1573
13. Orizio F, Triggiani L, Colosini A et al (2019) Overexpression of sialidase NEU3 increases the cellular radioresistance potential of U87MG glioblastoma cells. Biochem Biophys Res Commun 508:31–36

14. Wada T, Hata K, Yamaguchi K et al (2007) A crucial role of plasma membrane-associated sialidase in the survival of human cancer cells. Oncogene 26:2483–2490
15. Yoshinaga A, Kajiya N, Oishi K et al (2016) NEU3 inhibitory effect of naringin suppresses cancer cell growth by attenuation of EGFR signaling through GM3 ganglioside accumulation. Eur J Pharmacol 782:21–29
16. Wu G, Cao B, Zhai H et al (2024) EPO promotes the progression of rheumatoid arthritis by inducing desialylation via increasing the expression of neuraminidase 3. Ann Rheum Dis 83:564–575
17. Sehnert B, Mietz J, Rzepka R et al (2021) Neuraminidase inhibitor zanamivir ameliorates collagen-induced arthritis. Int J Mol Sci 22:1428
18. Deschenes NM, Cheng C, Khanal P et al (2023) Characterization of a phenotypically severe animal model for human AB-variant GM2 gangliosidosis. Front Mol Neurosci 16:1242814
19. Allende ML, Lee YT, Byrnes C et al (2023) Sialidase NEU3 action on GM1 ganglioside is neuroprotective in GM1 gangliosidosis. J Lipid Res 64:100463
20. Delaveris CS, Wang CL, Riley NM et al (2023) Microglia mediate contact-independent neuronal network remodeling via secreted neuraminidase-3 associated with extracellular vesicles. ACS Cent Sci 9:2108–2114
21. Chen W, Pilling D, Gomer RH (2023) The mRNA-binding protein DDX3 mediates TGF-β 1 upregulation of translation and promotes pulmonary fibrosis. JCI Insight 8:e167566
22. Pilling D, Sahlberg K, Karhadkar TR et al (2022) The sialidase NEU3 promotes pulmonary fibrosis in mice. Respir Res 23:215
23. Karhadkar TR, Chen W, Pilling D et al (2022) Inhibitors of the sialidase NEU3 as potential therapeutics for fibrosis. Int J Mol Sci 24:239
24. Wu J, Zhao M, Wei H et al (2022) Neuraminidase inhibitor treatment is associated with decreased mortality in COVID-19 patients: a retrospective analysis. Eur Heart J Cardiovasc Pharmacother 8:392–401
25. Negi G, Sharma A, Chaudhary M et al (2023) SARS-CoV-2 binding to terminal sialic acid of gangliosides embedded in lipid membranes. ACS Infect Dis 9:1346–1361
26. Shiozaki K, Harasaki Y, Fukuda M et al (2016) Positive regulation of myoblast differentiation by medaka Neu3b sialidase through ganglioside desialylation. Biochimie 123:65–72
27. Zizioli D, Codenotti S, Benaglia G et al (2023) Downregulation of zebrafish cytosolic sialidase Neu3.2 affects skeletal muscle development. Int J Mol Sci 24:13578
28. Yamanami H, Shiozaki K, Wada T et al (2007) Down-regulation of sialidase NEU4 may contribute to invasive properties of human colon cancers. Cancer Sci 98:299–307
29. Forcella M, Mozzi A, Stefanini FM et al (2018) Deregulation of sialidases in human normal and tumor tissues. Cancer Biomark 21:591–601
30. Hasegawa T, Sugeno N, Takeda A et al (2007) Role of Neu4L sialidase and its substrate ganglioside GD3 in neuronal apoptosis induced by catechol metabolites. FEBS Lett 581:406–412
31. Pan X, De Aragão CBP, Velasco-Martin JP et al (2017) Neuraminidases 3 and 4 regulate neuronal function by catabolizing brain gangliosides. FASEB J 31:3467–3483
32. Silvestri I, Testa F, Zappasodi R et al (2014) Sialidase NEU4 is involved in glioblastoma stem cell survival. Cell Death Dis 5: e1381
33. Uemura T, Shiozaki K, Yamaguchi K et al (2009) Contribution of sialidase NEU1 to suppression of metastasis of human colon cancer cells through desialylation of integrin β4. Oncogene 28:1218–1229
34. Wang J, Wu G, Miyagi T et al (2009) Sialidase occurs in both membranes of the nuclear envelope and hydrolyzes endogenous GD1a. J Neurochem 111:547–554
35. Ikeda A, Komamizu M, Hayashi A et al (2021) Neu1 deficiency induces abnormal emotional behavior in zebrafish. Sci Rep 11: 13477
36. den Tandt WR, Scharpe S (1984) Methylumbelliferyl-N-acetylneuraminic acid sialidase in human liver. Biochem Med 31:287–293
37. Warren L (1959) The thiobarbituric acid assay of sialic acids. J Biol Chem 234:1971–1975
38. Li K (1992) Determination of sialic acids in human serum by reversed-phase liquid chromatography with fluorimetric detection. J Chromatogr 579:209–213
39. Hara S, Yamaguchi M, Takemori Y et al (1989) Determination of mono-O-acetylatedN-acetylneuraminic acids in human and rat sera by fluorometric high-performance liquid chromatography. Anal Biochem 179:162–166
40. Saito M, Hagita H, Iwabuchi Y et al (2002) Fluorescent cytochemical detection of sialidase activity using 5-bromo-4-chloroindol-3-yl-α-D-N-acetylneuraminic acid as the substrate. Histochem Cell Biol 117:453–458
41. Minami A, Kurebayashi Y, Takahashi T et al (2021) The function of sialidase revealed by sialidase activity imaging probe. Int J Mol Sci 22:3187
42. Honda S, Iwase S, Suzuki S et al (1987) Fluorometric determination of sialic acids using malononitrile in weakly alkaline media and its application to postcolumn labeling in high-performance liquid chromatography. Anal Biochem 160:455–461

Part II

Structure Analysis

Chapter 7

LC–MS-Based Analytical Strategies for Ganglioside Profiling

Kazutaka Ikeda

Abstract

Gangliosides, a class of glycosphingolipids containing sialic acid residues, are abundantly expressed in neuronal membranes and play essential roles in cellular signaling and intercellular communication. Impaired ganglioside metabolism contributes to the pathogenesis of lysosomal storage disorders and neurodegenerative diseases. The structural diversity and functional relevance of gangliosides necessitate analytical methods with high sensitivity, resolution, and specificity. In recent years, liquid chromatography–mass spectrometry (LC–MS) has emerged as a powerful tool for comprehensive ganglioside profiling. This review presents current LC–MS-based strategies for ganglioside analysis, with a particular focus on sample preparation methods, chromatographic separation techniques, and mass spectrometric approaches aimed at enabling accurate structural elucidation and pathway-level insights into ganglioside metabolism.

Keywords Gangliosides, LC–MS, Lipidomics, Mass spectrometry, Glycosphingolipids

1 Introduction

Gangliosides are a subclass of glycosphingolipids consisting of a glycan headgroup containing one or more sialic acid residues linked via glycosidic bonds, and a hydrophobic ceramide tail that anchors the molecule to cellular membranes (Fig. 1). They are particularly enriched on the surface of neuronal cells, where they modulate signaling pathways and mediate cell-cell interactions [1–3]. Dysregulation in ganglioside metabolism is implicated in lysosomal storage disorders such as Tay–Sachs disease [4], and may also contribute to the pathophysiology of Parkinson's disease and Huntington's disease [5, 6].

To elucidate the complex biological roles of gangliosides, highly sensitive and accurate analytical techniques capable of characterizing them at both the class and molecular levels are required. Conventionally, ganglioside profiling was conducted using thin-layer chromatography (TLC) [7]. However, TLC lacks the sensitivity and structural resolution necessary for comprehensive analysis. Consequently, liquid chromatography–mass spectrometry (LC–MS) has gained attention for its superior sensitivity, molecular specificity,

Koichi Furukawa (ed.), *Gangliosides*, Methods in Molecular Biology, vol. 3035,
https://doi.org/10.1007/978-1-0716-5288-6_7,

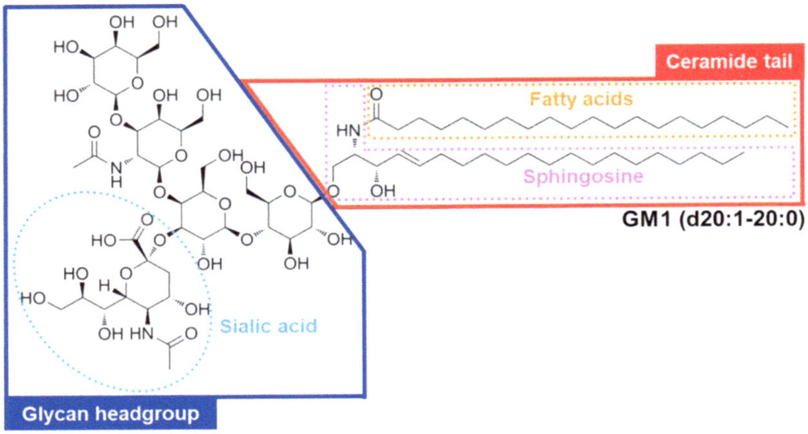

Fig. 1 Ganglioside molecular structure. Gangliosides consist of a sialic acid-containing glycan headgroup and a ceramide tail, which varies in the carbon chain length and degree of unsaturation of the sphingosine and fatty acid components

and throughput [8–10]. In this review, we provide an overview of LC–MS-based analytical strategies for ganglioside characterization.

2 Materials

2.1 Experimental Consumables

1. Safe-lock tube, 1.5 mL.
2. Glass-jacketed vial, 2 mL.
3. Disruption tube, 3 mL.
4. Metal cone, 3 mL.
5. EquiSPLASH.
6. Ganglioside GM1-d_3 (d18:1/18:0-d3).
7. Ganglioside GM2-d_7 (d18:1/18:0-d7).
8. Ganglioside GM3-d_3 (d18:1/18:0-d3).
9. Ganglioside GD3-d_3 (d18:1/18:0-d3).
10. Palmitic-16,16,16-d_3 acid.
11. Stearic-18,18,18-d_3 acid.
12. Inert MS-grade vial, 2 mL.
13. Inert glass insert, 250 μL.
14. ACQUITY UPLC BEH C18 Column (130 Å pore size, 1.7 μm particle size, 2.1 mm I.D. × 50 mm length).

2.2 Instruments and Equipment

1. Tissue homogenizer (Multi-Beads Shocker® MB3000 series).
2. Ultra performance liquid chromatography system (ACQUITY UPLC system, Waters) for untargeted analysis.
3. Ultra performance liquid chromatography system (UltiMate 3000, Thermo Fisher Scientific) for targeted analysis.
4. Mass spectrometer (TripleTOF 6600 system, SCIEX) for untargeted analysis.
5. Mass spectrometer (4000Q TRAP system, SCIEX) for targeted analysis.

3 Methods

Ganglioside sample preparation is a critical step that significantly impacts the accuracy and reproducibility of LC–MS analysis. Various methods have been reported, and the choice often depends on the sample type and analytical objectives [11, 12].

Ganglioside extraction methods are generally categorized into two main types: liquid–liquid extraction (LLE) and solid-phase extraction (SPE) [13, 14]. In LLE, the sample is partitioned into two phases, and the layer containing gangliosides is then collected. Following this, stationary phases such as octadecylsilane (ODS) columns are employed in SPE to specifically enrich gangliosides. This approach effectively reduces other lipid components, minimizing ion suppression and noise in mass spectrometry (MS) analysis, thereby enhancing the accuracy of quantification.

Recent advancements in lipidomics have led to the adoption of mono-phasic extraction (MPE) protocols, which allow simultaneous recovery of a wide range of lipid species, including upstream lipids in ganglioside biosynthesis such as ceramides [15, 16]. This comprehensive extraction approach facilitates a deeper understanding of the metabolic pathways. Representative MPE methods for plasma/serum, cultured cells, and tissue samples are as follows.

3.1 Preparation for Plasma/Serum Samples [17]

1. 20 μL of plasma (or serum) was transferred into a 2 mL glass-jacketed vial (*see* **Note 1**).
2. 200 μL of methanol containing internal standards (adjusted to final concentrations of 5 μM EquiSPLASH, 10 μM deuterated gangliosides, and 10 μM deuterated fatty acids was added and mixed thoroughly. The mixture was incubated on ice for 2 h.
3. 100 μL of chloroform was added, mixed thoroughly, and the sample was incubated on ice for 1 h (*see* **Note 2**).
4. The mixture was centrifuged at 2000 × *g* for 10 min at 4°C, and approximately 200 μL of the upper phase was collected and transferred into an LC vial (*see* **Note 3**).

3.2 Preparation for Cultured Cell Samples [15, 16]

1. A cell pellet (approximately 1×10^6 cells), collected in a 2 mL Eppendorf tube, was resuspended in 500 μL of methanol containing internal standards (adjusted to final concentrations of 5 μM EquiSPLASH, 10 μM deuterated gangliosides, and 10 μM deuterated fatty acids). The suspension was sonicated for 1 min to ensure mixing (*see* **Note 4**), and incubated on ice for 1 h.
2. After vortexing for 1 min, 200 μL of the mixture was transferred to a glass-jacketed vial.
3. 100 μL of chloroform was added to the vial, and the mixture was thoroughly mixed and incubated on ice for 1 h.
4. 20 μL of ultrapure water was added, followed by incubation on ice for 15 min.
5. After vortexing for 1 min, the sample was centrifuged at 2000 × *g* for 10 min at 4°C, and approximately 200 μL of the upper phase was collected and transferred into an LC vial.

3.3 Preparation for Tissue Samples [15, 16]

1. Frozen tissue samples were placed into 3 mL disruption tubes pre-cooled with liquid nitrogen. Metal cones, also pre-cooled with liquid nitrogen, were added to the tubes. The tubes were then set in an aluminum block and further cooled with liquid nitrogen.
2. After cryogenic disruption of the tissues (2500 rpm, 15 s, repeated twice), methanol was added to achieve a consistent weight-to-volume concentration (e.g., 20 mg/mL) (*see* **Note 5**). The suspension was then sonicated for 1 min and vortexed for 1 minu. After removing the metal cones, the samples were incubated on ice for 30 min.
3. After vortexing for 1 min, 100 μL of the methanolic tissue suspension (equivalent to approximately 2 mg of tissue) was transferred into a glass-jacketed vial. Subsequently, 100 μL of methanol containing internal standards (adjusted to final concentrations of 5 μM EquiSPLASH, 10 μM deuterated gangliosides, and 10 μM deuterated fatty acids) was added and mixed. The mixture was incubated on ice for 30 min.
4. 100 μL of chloroform was added, and the mixture was thoroughly mixed and incubated on ice for 1 h.
5. 20 μL of ultrapure water was added, followed by a 15-min incubation on ice. After vortexing for 10 s, the samples were centrifuged at 2000 × *g* for 10 min at 4°C. Approximately 200 μL of the upper phase was collected and transferred into an LC vial (*see* **Note 4**).

3.4 Chromato graphic Separation

GM3 ganglioside has been reported to influence the activity of innate immune receptors such as TLR4, depending on the variations in the fatty acyl chain length of its ceramide moiety [18]. In contrast, structural isomers such as GD1a and GD1b, which differ in the linkage position of their sialic acid residues, exhibit distinct distributions and neuromodulatory functions within neural tissues [19]. Therefore, understanding the structural architecture of gangliosides—particularly the roles of the ceramide moiety and the glycan head group—is crucial for elucidating their specific biological roles.

The separation of gangliosides depends significantly on the type of LC column employed (Fig. 2). When targeting structural information in the ceramide portion, such as sphingoid base types and fatty acyl chain features (e.g., chain length, unsaturation, number and position of hydroxyl groups), reversed-phase chromatography using ODS columns is appropriate [8] (Fig. 3). However, this approach may not sufficiently resolve isomers with different sialic acid linkage positions.

Alternatively, hydrophilic interaction chromatography (HILIC) can achieve better separation of ganglioside isomers such as GD1a and GD1b making it particularly effective for profiling class-specific changes in ganglioside metabolism (Fig. 4). A challenge associated with HILIC-based separation is its complexity, which can compromise the reproducibility of retention time (RT) across different LC batches [20]. Therefore, maintaining consistency and rigorous validation is crucial when employing this technique (*see* **Note 6**).

The Representative Separation Conditions for Reversed-Phase Chromatography for Untargeted Analysis [8]

1. Column Setup
   - Reversed-phase LC separation was performed using an ODS column (e.g., 2.1 mm I.D. × 50 mm length; Waters) maintained at 45°C.
2. Mobile Phase Preparation
   - Solvent A: acetonitrile/methanol/water (1/1/3, v/v/v) containing 1 mM ammonium formate/10 nM EDTA.

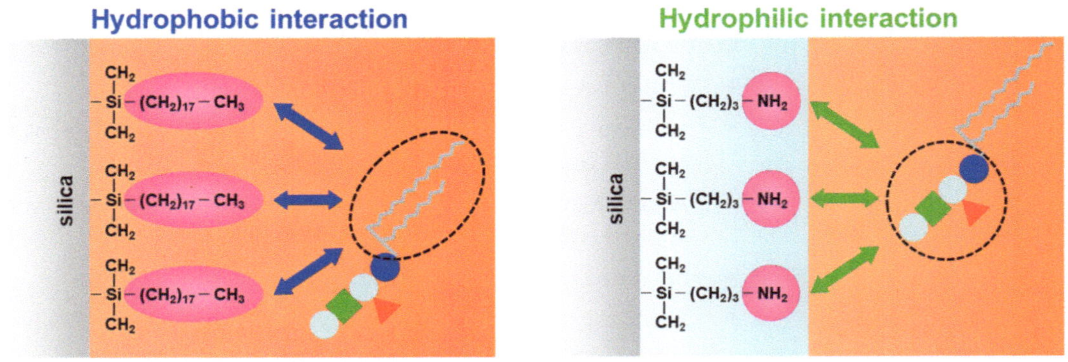

Fig. 2 Typical LC separation strategies for gangliosides. Hydrophobic interaction chromatography using octadecylsilane (ODS) columns enables separation based on the ceramide moiety, while hydrophilic interaction chromatography (HILIC) facilitates separation based on the glycan moiety

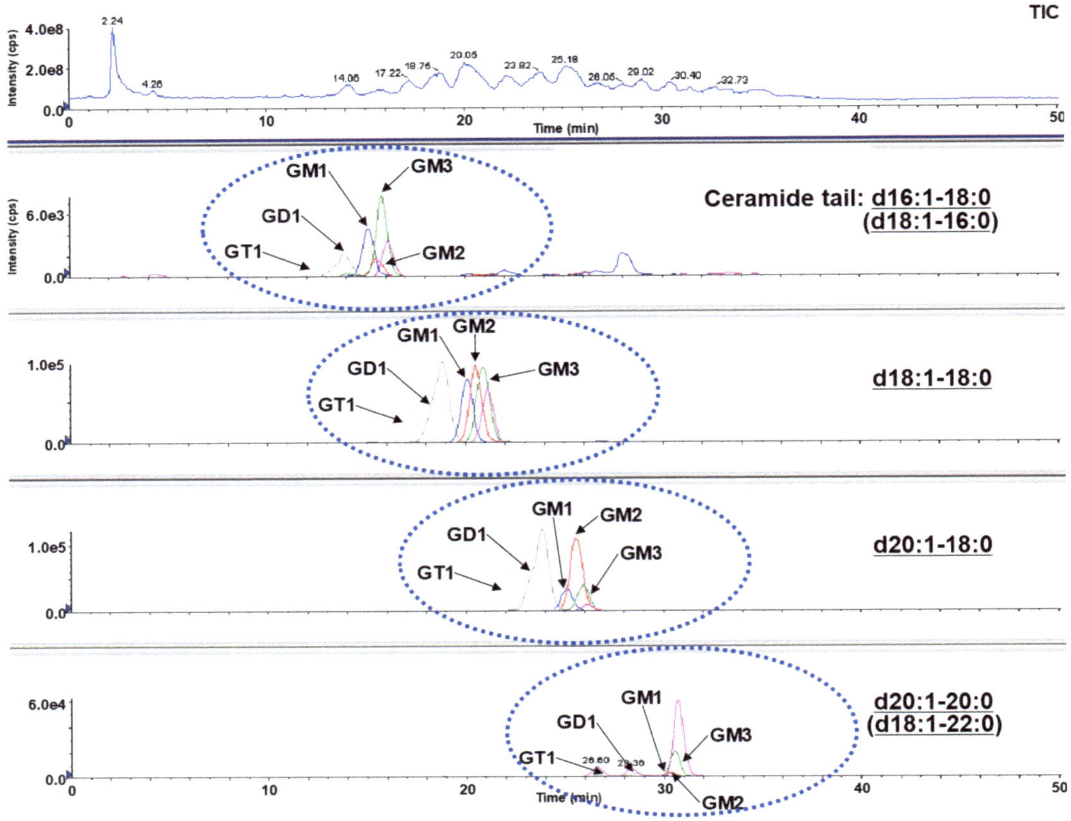

Fig. 3 Separation patterns of gangliosides by hydrophobic interaction. Ganglioside species with shorter ceramide acyl chain lengths and higher degrees of unsaturation tend to elute at earlier retention time (RT), whereas species with structurally similar ceramide moieties exhibit comparable RT. This result is based on MRM analysis of glycosphingolipid mixtures extracted from porcine brain

- Solvent B: isopropanol containing 1 mM ammonium formate/10 nM EDTA.

3. Gradient Program
   - Solvent A was held for 1 min.
   - A linear gradient to Solvent C (A/B = 60/40, v/v) was applied over 4 min.
   - A linear gradient to Solvent D (A/B = 36/64, v/v) was applied over 3 min.
   - Solvent D was held for 4 min.
   - A linear gradient to Solvent E (A/B = 17.5/82.5, v/v) was applied over 1 min.
   - A linear gradient to Solvent F (A/B = 15/85, v/v) was applied over 6 min.

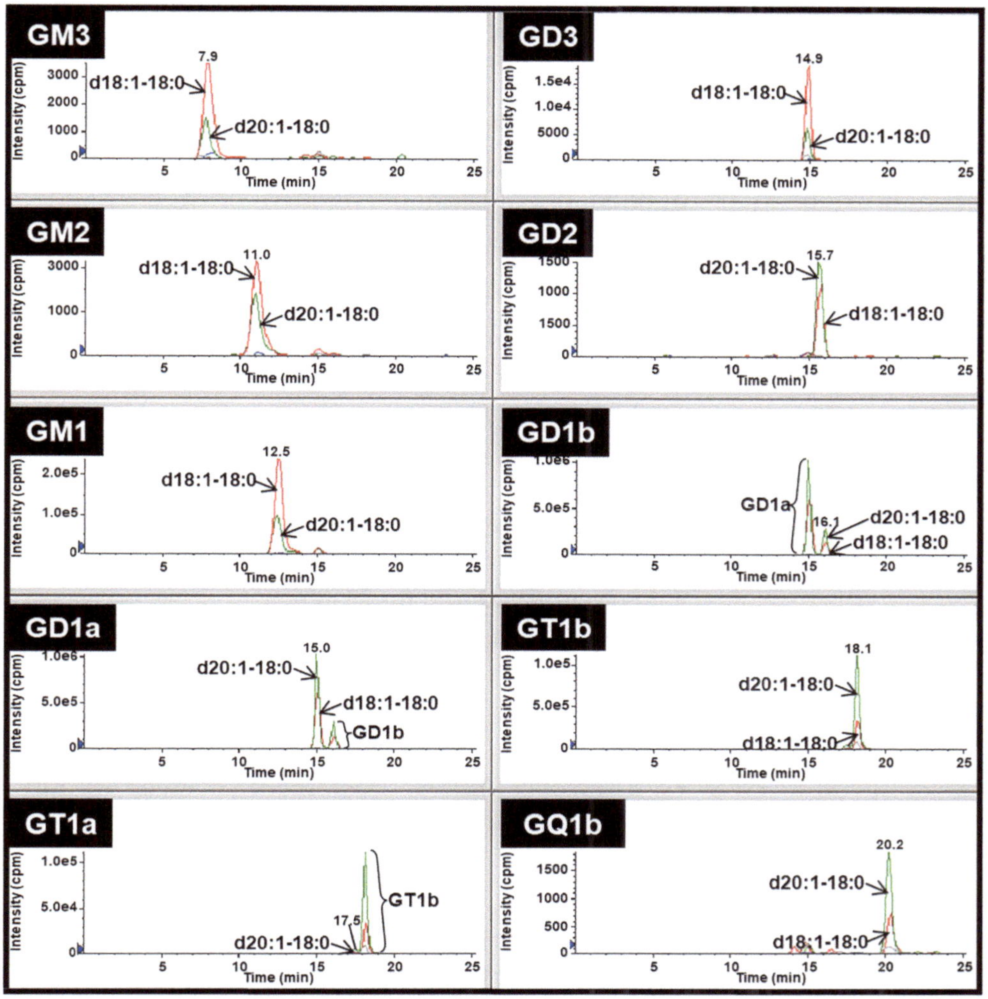

Fig. 4 Separation patterns of gangliosides by hydrophilic interaction. Gangliosides with identical glycan head group structures exhibit similar RT, whereas variations in the ceramide moiety do not substantially affect RT. This result is based on MRM analysis of glycosphingolipid mixtures extracted from porcine brain

- A linear gradient to Solvent G (A/B = 5/95, v/v) was applied over 1 min.
- Re-equilibration was carried out using the initial composition.

4. Flow Rate

- The mobile phase was delivered at a flow rate of 300 μL/min during gradient elution.

The Representative Separation Conditions for Reversed-Phase Chromatography for Targeted Analysis [8]

1. Column Setup
   - Reversed-phase LC separation was performed using an ODS column (e.g., 0.3 mm I.D. × 150 mm length; LC Packings) maintained at 45°C.
2. Mobile Phase Preparation
   - Solvent A: methanol
   - Solvent B: isopropanol
   - Solvent C: water containing 200 mM ammonium formate

   Solvents were mixed according to the gradient program described below.
3. Gradient Program
   - Solvent D (A/B/C = 55/25/20, v/v/v) was held for 5 min.
   - A linear gradient to Solvent E (A/B = 60/40, v/v) was applied over 50 minu.
   - Solvent E was held for 35 min.
   - Re-equilibration was carried out using the initial composition.
4. Flow Rate
   - The mobile phase was delivered at a flow rate of 7.5 μL/min during gradient elution.

The Representative Separation Conditions for Hydrophilic Interaction Chromatography (HILIC) for Targeted Analysis [9]

1. Column Setup
   - HILIC separation was carried out using an NH_2 column (e.g., 0.5 mm I.D. × 150 mm length; GL Sciences) maintained at 40°C.
2. Mobile Phase Preparation
   - Solvent A: acetonitrile/water (83/17, v/v) containing 1 mM ammonium formate
   - Solvent B: acetonitrile/water (50/50, v/v) containing 50 mM ammonium formate
   - Solvents were mixed according to the gradient program described below.
3. Gradient Program
   - Solvent A was held for 5 min.
   - A linear gradient to Solvent C (A/B = 25/75, v/v) was applied over 15 min.
   - A linear gradient to Solvent D (A/B = 10/90, v/v) was applied over 5 min.
   - Solvent D was held for an additional 10 min.

- Re-equilibration was carried out using the initial composition.

4. Flow Rate

- The mobile phase was delivered at a flow rate of 15 μL/min.

3.5 Mass Spectrometric Analysis

Mass spectrometric analysis of gangliosides can be broadly divided into targeted and untargeted approaches. For targeted analysis, triple quadrupole MS operating in multiple reaction monitoring (MRM) mode provides high selectivity and is particularly suitable when analyte concentrations are low or sample amounts are limited [8, 9]. When conducting MRM-based analysis, it is essential to optimize the transitions in advance using authentic standards of the target gangliosides. During the actual measurement, RT consistency should also be verified with the corresponding standards to ensure accurate peak identification.

Alternatively, untargeted analysis using high-resolution MS platforms such as quadrupole time-of-flight (Q-TOF) instruments becomes advantageous, as they allow simultaneous acquisition of qualitative and quantitative data without prior knowledge of target lipids [21–23]. When combined with MPE (see Sect. 3.1, 3.2, and 3.3) and comprehensive separation methods (see Sect. 3.4), untargeted MS—utilizing data-dependent acquisition (DDA)—enables global lipidomic profiling, including the identification of other sphingolipids (Fig. 5), and may lead to the discovery of novel bioactive molecules or metabolic pathways.

However, untargeted workflows generate large and complex datasets. Data processing requires peak detection from full MS

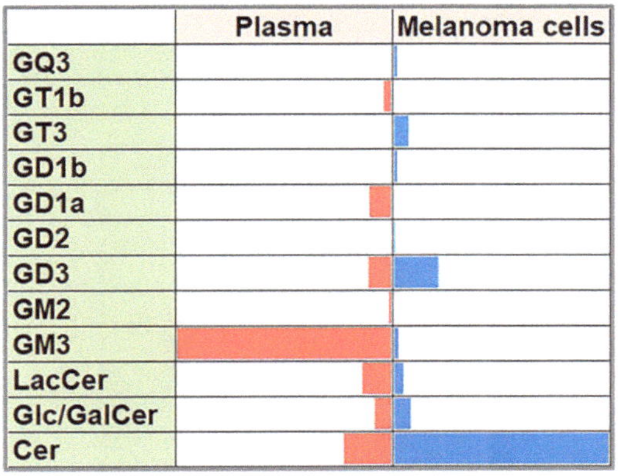

Fig. 5 Distribution patterns of glycolipids in human plasma and melanoma cells. Glycolipids were screened by non-targeted analysis, according to previously published methods [15–17], and the signal intensities of the identified molecules for each lipid class were summed. These results are shown as the relative ratios of each lipid class normalized to the class with the highest total intensity

scans and lipid identification from MS/MS spectra. Therefore, software tools such as LipidSearch and MS-DIAL (https://systemsomicslab.github.io/compms/msdial/main.html) are commonly used to streamline analysis and improve throughput [23, 24]. However, researchers must critically evaluate the reliability of identifications based on MS/MS spectra to minimize the misidentification, ensuring that informatics-based analyses are both accurate and reproducible.

Structural Characterization in Targeted Analysis [8, 9]

In targeted analysis, precursor and product ion pairs for multiple reaction monitoring (MRM) were selected based on MS/MS spectra. The following MRM *m/z* transitions were monitored in the negative-ion mode: from precursor ions ($[M\text{–}H]^-$ or $[M\text{–}2H]^{2-}$) to sialic acid ions (SA^-) or to sialic acid-containing oligosaccharide fragments specific to gangliosides (Fig. 6). For the analysis of regioisomeric gangliosides, fragment ions derived from the terminal regions of the glycan chains were employed to distinguish between isomers.

For example, in the case of GD1 and GT1 isomers, each precursor ion ($[M\text{–}2H]^{2-}$) was analyzed by MS/MS and confirmed using diagnostic carbohydrate-derived fragment ions (highlighted in gray).

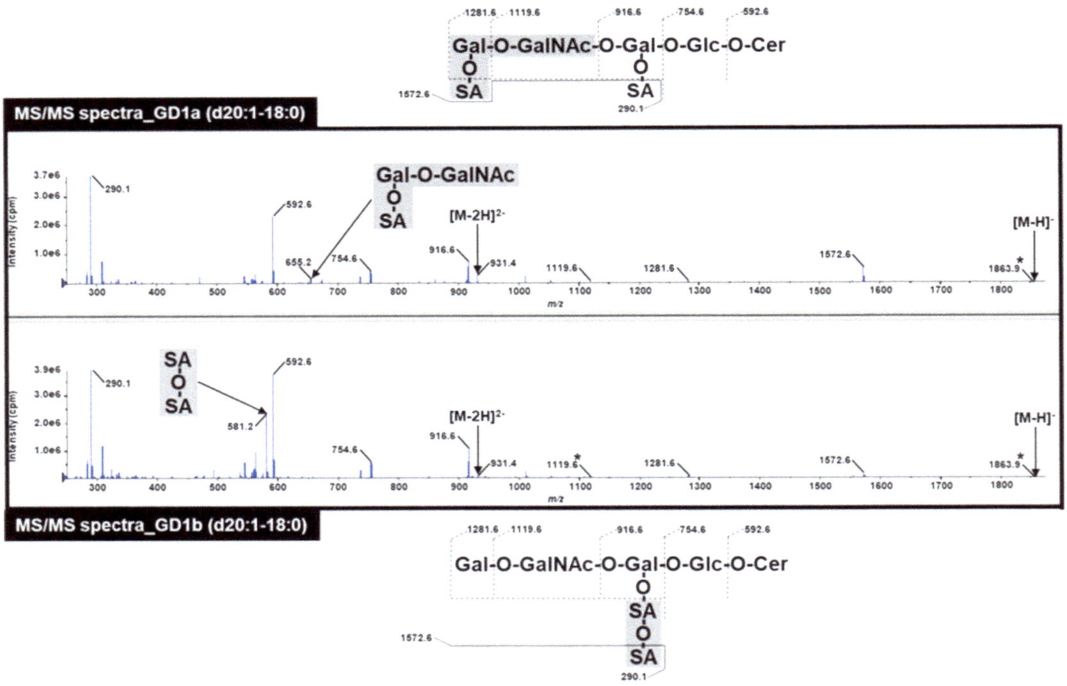

Fig. 6 Identification of ganglioside structural isomers, GD1a,b, and GT1a,b by MS/MS. For identification of these isomers, each precursor ion ($[M\text{–}2H]^{2-}$) of d20:1-18:0 was analyzed by MS/MS and confirmed by each diagnostic fragment ion (indicated by gray color) from the carbohydrate chain; $[SA\text{-}O\text{-}Gal\text{-}O\text{-}GalNAc]^-$ (*m/z* 655.2) for GD1a and GT1b and $[SA\text{-}O\text{-}SA]^-$ (*m/z* 581.2) for GD1b and GT1a,b. Asterisk* denotes undetectable peaks

Specifically, [SA–O–Gal–O–GalNAc]$^-$ (m/z 655.2) was detected in GD1a and GT1b, while [SA–OSA]$^-$ (m/z 581.2) was observed in GD1b and GT1a,b.

In the positive-ion mode, ceramide molecular species were analyzed by detecting long-chain base (LCB) fragment ions characteristic of each sphingoid base: m/z 236 for d16:1 sphingenine, m/z 264 for d18:1 sphingenine, and m/z 292 for d20:1 sphingenine. The combination of ceramide- and LCB-derived MS/MS spectra enabled the identification of ganglioside molecular species.

Structural Characterization in Untargeted Analysis [15, 16, 22, 23]

In untargeted analysis, structural information was obtained from MS/MS spectra acquired automatically using data-dependent acquisition (DDA). It is recommended that structural elucidation be performed by combining information from both ionization modes: ceramide-related fragments in the positive-ion mode and glycan-related fragments in the negative-ion mode.

4 Conclusion

LC–MS-based approaches have significantly advanced the analysis of gangliosides by enabling high-sensitivity, high-resolution profiling of structurally diverse species. The choice of extraction method, chromatographic separation column, and MS acquisition mode must be tailored to the specific analytical goals, whether focused on individual molecular species or comprehensive lipidomic coverage. Targeted analyses using MRM offer high selectivity for the known lipids, while untargeted high-resolution MS approaches expand the scope of discovery and metabolic insight. As data complexity increases, the integration of robust informatics tools and rigorous validation becomes essential. Continued developments in analytical workflows will deepen our understanding of ganglioside biology and its implications in health and disease.

5 Notes

1. Glass-jacketed vials may not be compatible with all centrifuge rotors. It is necessary to verify compatibility with the centrifuge model in advance.
2. When handling solvents containing chloroform, the use of glass pipette tips is recommended over plastic tips, as they help reduce MS noise such as polymer peaks.
3. For the collection of lipid extracts—including those using vials with inserts—containers with inert surface treatment are preferable to minimize lipid adsorption.

4. In some cell types, complete resuspension may be difficult; in such cases, extending the sonication or incubation time may help improve dispersion.
5. When analyzing lipids that tend to adsorb onto metal surfaces (e.g., PA, LPA, S1P), it is recommended to remove the metal cone prior to mixing with methanol and to perform sample weighing after its removal in order to ensure consistent tissue mass.
6. Some lipid molecules are prone to carryover in the analytical system. To minimize data bias resulting from the order of sample injection, it is recommended to randomize the measurement sequence.

Acknowledgments

The author gratefully acknowledges Dr. Koichi Furukawa for his valuable support and guidance throughout this paper.

Conflict of Interest The authors have no conflicts of interest to declare that are relevant to the content of this chapter.

Funding This work was supported by the Kazusa DNA Research Institute Foundation, Japan.

Author Contributions K.I. conceived the study, reviewed the literature, and wrote the manuscript.

References

1. Yu RK, Tsai YT, Ariga T (2012) Functional roles of gangliosides in neurodevelopment: an overview of recent advances. Neurochem Res 37:1230–1244
2. Lopez PHH, Schnaar RL (2009) Gangliosides in cell recognition and membrane protein regulation. Curr Opin Struct Biol 19:549–557
3. Sipione S, Monyror J, Galleguillos D et al (2020) Gangliosides in the Brain: Physiology, Pathophysiology and Therapeutic Applications. Front Neurosci 14:572965
4. Ryckman AE, Brockhausen I, Walia JS (2020) Metabolism of glycosphingolipids and their role in the pathophysiology of lysosomal storage disorders. Int J Mol Sci 21:6881
5. Blennow K, Davidsson P, Wallin A et al (1992) Differences in cerebrospinal fluid gangliosides between "probable Alzheimer's disease" and normal aging. Aging 4:301–306
6. Desplats PA, Denny CA, Kass KE et al (2007) Glycolipid and ganglioside metabolism imbalances in Huntington's disease. Neurobiol Dis 27:265–277
7. Ledeen RW, Yu RK (1982) Gangliosides: structure, isolation, and analysis. Methods Enzymol 83:139–191
8. Ikeda K, Shimizu T, Taguchi R (2008) Targeted analysis of ganglioside and sulfatide molecular species by LC/ESI-MS/MS with theoretically expanded multiple reaction monitoring. J Lipid Res 49:2678–289
9. Ikeda K, Taguchi R (2010) Highly sensitive localization analysis of gangliosides and sulfatides including structural isomers in mouse cerebellum sections by combination of laser microdissection and hydrophilic interaction liquid chromatography/electrospray ionization-mass spectrometry with theoretically expanded multiple reaction monitoring. Rapid Commun Mass Spectrom 24:2957–2965
10. Hořejší K, Kolářová D, Jirásko R et al (2024) Recent advances, challenges, and future

directions in the mass spectrometry analysis of glycosphingolipids in biological samples. Trends Analyt Chem 178:117827
11. Sanni A, Bennett AI, Huang Y et al (2024) An optimized liquid chromatography-mass spectrometry method for ganglioside analysis in cell lines. Cells 13:1640
12. Khoury S, Masson E, Sibille E et al (2020) Rapid sample preparation for ganglioside analysis by liquid chromatography mass spectrometry. J Chromatogr B Analyt Technol Biomed Life Sci 1137:121956
13. Williams MA, McCluer RH (1980) The use of Sep-Pak™ C18 cartridges during the isolation of gangliosides. J Neurochem 35:266–269
14. Arends M, Weber M, Papan C et al (2022) Ganglioside lipidomics of CNS myelination using direct infusion shotgun mass spectrometry. iScience 25:105323
15. Honda A, Nozumi M, Ito Y et al (2023) Very-long-chain fatty acids are crucial to neuronal polarity by providing sphingolipids to lipid rafts. Cell Rep 42:113195
16. Kanno T, Konno R, Sato et al (2024) The integration of metabolic and proteomic data uncovers an augmentation of the sphingolipid biosynthesis pathway during T-cell differentiation. Commun Biol 7:622
17. Nishiumi S, Izumi Y, Hirayama A et al (2022) Comparative evaluation of plasma metabolomic data from multiple laboratories. Metabolites 12:135
18. Inokuchi J-i, Kanoh H (2022) Pathophysiological significance of GM3 ganglioside molecular species with a particular attention to the metabolic syndrome focusing on toll-like receptor 4 binding. Front Mol Biosci 9:918346
19. Vajn K, Viljetić B, Degmečić IV et al (2013) Differential distribution of major brain gangliosides in the adult mouse central nervous system. PLoS One 8:e75720
20. Serafimov K, Knappe C, Li F et al (2024) Solving the retention time repeatability problem of hydrophilic interaction liquid chromatography. J Chromatogr A 1730:465060
21. Almeida R, Pauling JK, Sokol E et al (2015) Comprehensive lipidome analysis by shotgun lipidomics on a hybrid quadrupole-orbitrap-linear ion trap mass spectrometer. J Am Soc Mass Spectrom 26:133–148
22. Tsugawa H, Ikeda K, Arita M (2017) The importance of bioinformatics for connecting data-driven lipidomics and biological insights. Biochim Biophys Acta 1862:762–765
23. Tsugawa H, Ikeda K, Takahashi M et al (2020) A lipidome atlas in MS-DIAL 4. Nat Biotechnol 38:1159–1163
24. Taguchi R, Ishikawa M (2010) Precise and global identification of phospholipid molecular species by an orbitrap mass spectrometer and automated search engine lipid search. J Chromatogr A 1217:4229–4239

Chapter 8

Structure Analysis of Gangliosides: Mass Spectrometry Imaging

Ariful Islam and Mitsutoshi Setou

Abstract

Mass spectrometry imaging (MSI) is a powerful technique that is widely used for the mapping and molecular identification of biomolecules directly from the surface of biological samples. Among these biomolecules, gangliosides are complex glycosphingolipids primarily found in the nervous system. They play crucial roles in cell signaling and neurodevelopment and are responsible for many diseases. Label-free mapping and structural analysis of these gangliosides using MSI offer unique insights into their spatial distribution and concentration in healthy and diseased tissues. MSI techniques such as matrix-assisted laser desorption/ionization (MALDI) and desorption electrospray ionization (DESI) have been widely employed to study gangliosides. These MSI tools enable high-resolution visualization of specific ganglioside species, contributing to the understanding of their roles in human diseases including neurodegenerative diseases, cancer, etc. Continuous improvements in sensitivity and resolution of MSI tools, continues to offer new insights into ganglioside biology, paving the way for novel therapeutic approaches in a wide range of diseases. Here, we have described the mapping and structural analysis of gangliosides using MSI.

Keywords Mass spectrometry imaging, Gangliosides, MALDI, DESI, Neurodegenerative diseases

1 Introduction

Mass spectrometry imaging (MSI) is a newly developed cutting-edge analytical technology that offers spatial information of biomolecules and exogenous analytes such as drugs and their metabolites directly from biological tissues [1, 2]. Nowadays, it is widely used for the detection, structural analysis, identification, quantification, and localization of different biomolecules present in the samples, including lipids, proteins, drugs, and metabolites [3, 4]. Among these molecules, gangliosides are a group of glycosphingolipids containing cell type-specific sialic acid primarily found in the nervous system. They have attracted significant interest of researchers throughout the world due to their important roles in cellular communication, neurodevelopment, and the pathophysiology of several diseases,

Koichi Furukawa (ed.), *Gangliosides*, Methods in Molecular Biology, vol. 3035,
https://doi.org/10.1007/978-1-0716-5288-6_8,

particularly neurological disorders like Alzheimer's and Parkinson's diseases [5, 6].

Gangliosides are the structural components of the membranes of neuronal cells and play essential roles in cellular growth and development, synaptic functions, and modulating signal transduction [7, 8].Their spatial information is crucial to understand their role in the development and function of the brain. Alterations in ganglioside expression have been found in neurodegenerative diseases such as Alzheimer's, Parkinson's, and Huntington's diseases [7, 9]. Detection and visualization of gangliosides in animal tissues applying the powerful tools of MSI helps to unravel their complex biological functions and contributions to disease mechanisms [10, 11]. The capability of MSI tools to detect, identify, and visualize these molecules without the need for radio-labeling or molecular labeling makes it highly advantageous for studying their natural distribution in tissues [12, 13].

Several MSI techniques, including matrix-assisted laser desorption/ionization (MALDI), desorption electrospray ionization (DESI), and secondary ion mass spectrometry (SIMS) have been successfully applied to analyze biomolecules [14, 15] including gangliosides in both healthy and diseased tissues [16–19]. These techniques allow for the high-resolution mapping of ganglioside species, offering valuable insights into their spatial organization within distinct regions of the brain or other organs. For instance, MSI has been used to reveal changes in ganglioside patterns associated with amyloid plaques in Alzheimer's disease, contributing to a better understanding of disease progression and potential therapeutic targets [20, 21].

The application of MSI to ganglioside research has also extended beyond neurodegenerative diseases. Studies have employed MSI to investigate gangliosides in cancer, metabolic disorders, and inflammatory diseases, where altered ganglioside metabolism plays a critical role [18, 20]. As MSI technology continues to advance, with improvements in sensitivity, resolution, and molecular coverage, it holds great promise for expanding our understanding of ganglioside biology in health and disease.

In conclusion, MSI has become an indispensable tool for the study of gangliosides, enabling researchers to explore their distribution and function in unprecedented detail, ultimately advancing the understanding of their roles in both physiological and pathological conditions. Here we have described procedure and related information for the imaging (Fig. 1) [22] and structure analysis (Fig. 2) [23] of gangliosides applying MSI tools.

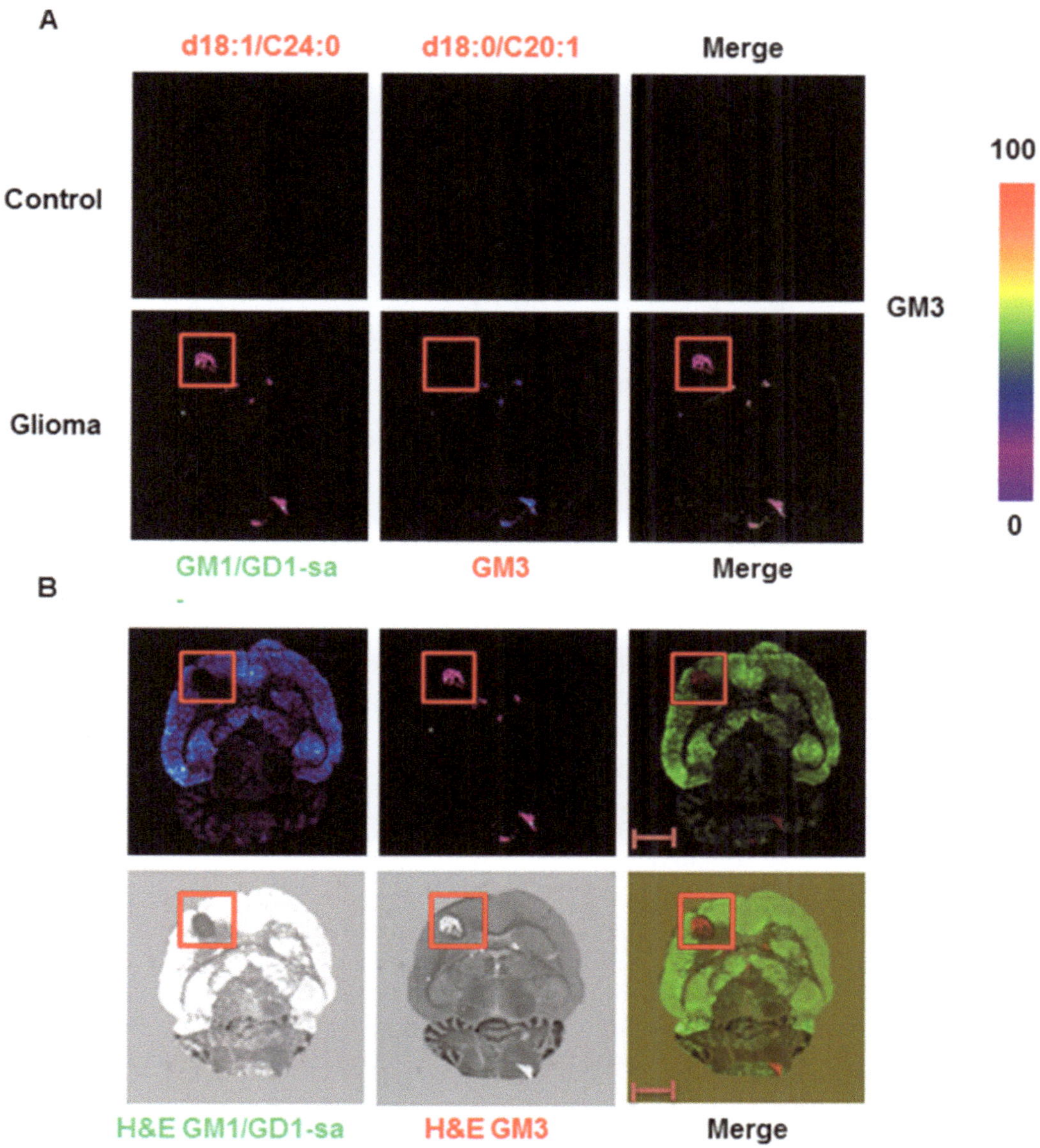

Fig. 1 MALDI-MSI of GM3 in intracranial allografts of rat glioma. (**A**) MSI of GM3 $[M-H]^-$ d18:1/C24:0 (*m/z* 1264) and d18:0/C20:1 (*m/z* 1208) in intracranial allografts of rat glioma. Intensities of the ions are represented in color based on the intensity scale provided. (**B**) Molecular histology distribution of GM1/GD1-sialic acid (d18:1/C18:0) and GM3 (d18:1/C24:0) in intracranial allografts of rat glioma. Red squares indicate the position of the glioma xenograft. Scale bar: 5 mm. GD1-sa = GD1-sialic acid. (Reproduced from [22] with a license approved by PLoS One)

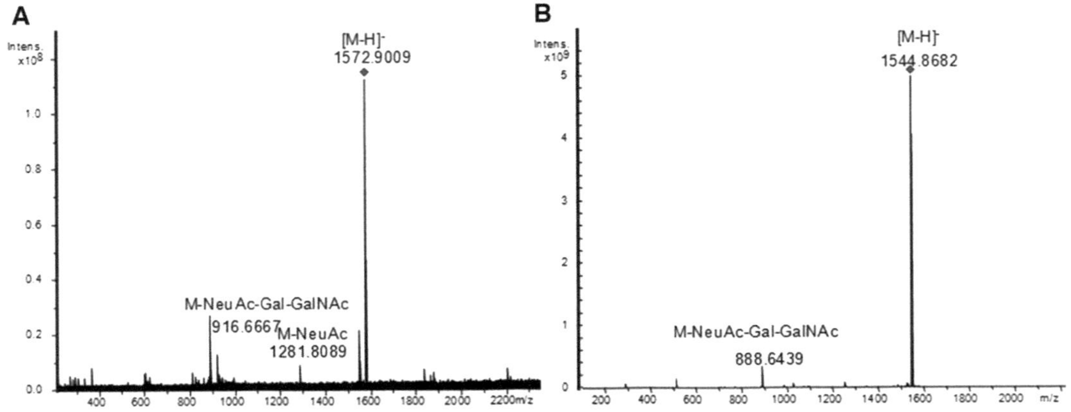

Fig. 2 MALDI MS^2 spectra of ganglioside d18:1/20:0 GM1 (**A**) and d18:1/18:0 GM1 (**B**). (Reproduced from [23] with a license approved by Scientific Reports)

2 Materials

1. Optimal cutting temperature (OCT) compound. This compound can be used for the attachment of sample with the sample holder of cryostat system.
2. Embedding media for the preparation of freeze blocks of small samples, such as mice eye. Different embedding media are available for MSI, including carboxymethyl cellulose (CMC), gelatin, and poly *N*-(hydroxypropyl)-methacrlyamide (HPMA) solutions [24]. Polymeric compounds like OCT are not good for preparing freeze blocks of small samples, which will be subjected to MSI analysis. These compounds can produce several ions containing different numbers of monomers during MSI data acquisition. Ion suppression caused by these ions can be responsible for lower sensitivity of MSI tools. Using monomeric compounds like CMC and HPMS, such a problem could be avoided.
3. Conductive glass slides, such as indium tin-oxide (ITO) coated conductive glass slides for MALDI-MSI and SIMS, and normal glass slides for DESI-MSI. A conductive surface is commonly required for MALDI-MSI data acquisition. It helps to enhance the efficiency of the laser desorption and ionization process by stabilizing the electric field across the sample and ensures that when the laser hits the matrix and analytes, the ions generated can be efficiently drawn towards the mass spectrometer for detection [25]. In SIMS, a conductive slide plays a critical role to reduce sample charging which is important to analyze insulating or non-conductive samples (such as biological tissues). However, several MSI instruments, such as DESI-MSI, do not require such a conductive coating on the slide glass.

4. Matrix compounds for MALDI-MSI. Matrices (singular "matrix") play a crucial role in MALDI-MSI. They are small organic molecules that can absorb the laser energy during MALDI-MSI. This absorbed energy leads to the desorption of both matrix molecules and analytes present in the sample. Matrix molecules transfer some of the absorbed energy to the analyte molecules during desorption process, and help to ionize the analytes present in the sample by protonation or deprotonation while preserving the molecular structure of the analytes (soft nature of the ionization). Several matrices are available for MALDI-MSI of different analytes. Selection of a suitable matrix is crucial for the detection of target analytes with higher efficiency [26]. To detect gangliosides by MALDI-MSI, several matrices have been used previously, such as 2,5-dihydroxyacetophenone (2,5-DHAP) [27], 2,6-dihydroxyacetophenone (DHAP), 1,5-diaminonaphthalene (1,5-DAN) [16], 9-aminoacridine (9-AA) [17], 3-aminoquinoline (3-AQ) [23], and α-cyano-4-hydroxycinnamic acid (α-CHCA) [28]. Both 2,5-DHAP and 2,6-DHA are structurally similar, and they can help to ionize acidic lipids like sphingolipids in negative ion mode [16, 29]. However, 2,5-DHAP has higher sensitivity for sialylated gangliosides and lower background noise compared to 2,6-DHAP. The major drawback of using DHAP is its sublimation under vacuum (low pressure) during MSI data acquisition for longer time studies applying vacuum MALDI-MSI. To overcome this problem of sublimation, application of atmospheric pressure (AP)-MALDI-MSI tool could be a better option. 1,5-DAN is also widely used for the ionization of acidic lipids in negative ion mode in MALDI-MSI. It is also well-suited for the analysis of lipids in positive ion mode. It offers higher sensitivity, as well as lower ion suppression and background noise. In addition to these matrices, 9-AA and 3-AQ are suitable for the detection of gangliosides and other acidic lipids in negative ion mode with better sensitivity and higher signal-to-noise ratio. They efficiently deprotonate acidic lipids and minimize the interference of neutral lipids present in the samples. Both 9-AA and 3-AQ have lower ion suppression effects compared to other matrices, but ion suppression effect of 3-AQ is much lower compared to 9-AA. α-CHCA is a widely used matrix for the detection of smaller metabolites, lipids, and peptides by MALDI-MSI. It can also be used as a matrix for the detection of gangliosides by MALDI-MSI, but it shows higher background noise and lower signal-to-noise ratio at higher mass regions. As a result, it is not suitable for the imaging of intact gangliosides (~1500–2000 Da) by MALDI-MSI.
5. Silica gel particles to store MALDI-MSI and SIMS imaging samples in dry conditions. It prevents moisture absorption by biological samples. Excess moisture can lead to salt adduct

formation and degradation of lipids by hydrolysis. In addition to these, trapped moisture in case of storing MSI samples at −80°C can cause ice crystal formation, which can change tissue morphology. By using silica gel particles, we can maintain a dry environment by absorbing moisture and maintain the quality of MSI samples during storage.

6. Charge-enhancing agents for MSI. They are commonly used to improve the ionization efficiency of large, polar, or acidic molecules like proteins, lipids, etc. They help to increase the generation of ions of the analyte in higher numbers and improve signal sensitivity. Here are some agents commonly used in MSI for charge enhancement: ammonium acetate, ammonium formate, formic acid, acetic acid, trifluoroacetic acid, and acetonitrile.

3 Methods

3.1 Animal Sample Preparation

Collect animal tissues and organs, and freeze them rapidly using dry ice or liquid nitrogen. Before MSI data measurement, prepare 10–20 μm thick section of animal tissues/organs using a cryostat system (*see* **Note 1**), transfer them onto conductive glass slide, and thaw-mount the section using a finger for MALDI-MSI and SIMS imaging data acquisition [3, 16, 27]. For DESI-MSI, normal glass slide is used instead of conductive glass slides. Put slides prepared for MALDI-MSI and SIMS imaging in 50 mL tube containing silica gel particles covered with a thin paper, and store them at −80°C until use (*see* **Note 2**). Silica gel particles are not recommended to store slides at −80°C prepared for DESI-MSI measurement.

3.2 Rinsing the Sections

This step is not mandatory, but it could offer some benefits for better detection of gangliosides by MSI (see **Note 3**). Following are the steps of rinsing the sample sections;

1. Put 30 mL of 150 mM ammonium formate aqueous solution in a 10-cm dish on ice (*see* **Note 4**). Ammonium formate helps to improve the ionization of analytes by forming an ammonium adduct during MSI data acquisition.
2. Carefully immerse the prepared slide (after it has returned to room temperature) in the solution for 15 s.
3. Dry the rinsed slide in a desiccator for 15 min.
4. If required, spot standard gangliosides outside of the section on the glass slide and dry it using a desiccator. It could be helpful for the optimization of MSI parameters to achieve better detection capability, and it could also be used for the structural analysis of detected gangliosides.

3.3 Taking Images

Take images including the sections and several coordinates for setting the location of measurement using a scanner or a microscope. Set the appropriate resolution depending on the required spatial resolution for analysis. In some MSI tools, they have built-in microscopes, such as iMScope and iMScope QT (MALDI-MSI tools developed by Shimadzu, Japan). For such tools, images of slides with samples could be taken after matrix application.

3.4 Matrix Coating for MALDI-MSI

In MALDI-MSI, matrix application is required to assist ionization of analytes present on the sample surface and acquire MSI data. It is crucial to select a suitable matrix and method for matrix application onto the sample sections (sublimation or spray) depending on the purpose of the study or the available type of ion source equipped with the instruments (*see* **Note 5**). Different matrices can be used for the detection of gangliosides by MALDI-MSI according to the previous reports, such as 2,5-dihydroxyacetophenone (2,5-DHAP) [27], 2,6-dihydroxyacetophenone (DHA), 1,5-diaminonaphthalene (1,5-DAN) [16], 9-aminoacridine (9-AA) [17], 3-aminoquinoline (3-AQ) [23], α-cyano-4-hydroxycinnamic acid (α-CHCA) [28], etc. For MSI instrument equipped with an atmospheric pressure MALDI (AP-MALDI) ion source, sublimation of 2,5-DHAP by the following steps could be a good choice;

1. Put 600 mg 2,5-DHAP in the heating box and set it under the prepared sample facing the bottom in a vacuum chamber. Set the distance between the sample and box at 7.5 cm (*see* **Note 6**).
2. Then close the chamber, set the thickness of the matrix (if this option is available), and then start the instrument for matrix sublimation.
3. When the vacuum pressure reaches 50 Pa, start heating the box, up to 180°C.
4. Cool the box until the temperature reaches 100°C and then return the pressure inside the box to atmospheric pressure.
5. Take out the coated sample slide of the chamber.

For MSI tools equipped with a vacuum MALDI ion source, 1,5-DAN (and other matrix, such as 9-AA) can be sprayed on the samples as follows using a solvent sprayer, such as HTX M5 (HTX Technologies, Chapel Hill, USA) [20];

1. Prepare 18 mg/mL 1,5-DAN solution in 70% ACN. 70% ACN helps to increase the solubility of 1,5-DAN and ensure a homogeneous matrix solution. Higher or lower concentration of ACN might decrease the solubility of 1,5-DAN leading to poor matrix distribution.
2. Spray 1,5-DAN solution at a flow rate of 0.1 mL/min, speed of 2250 mm/min, nozzle height of 40 mm, nozzle temperature of 70°C, plate temperature of 30°C, track spacing of 3 mm, and layers of 15.

3.5 MALDI-MSI of Gangliosides

1. Load the sample slide into an instrument for MALDI-MSI. Find the coordinates captured in the MALDI-MSI instrument and set the positions as the corresponding coordinates in the picture taken before applying the matrix.
2. For MSI tools with built-in microscope, load the sample slide into MALDI-MSI tool after applying matrix, take high-resolution image of the slide, and select the area on the slide for MSI data acquisition.
3. Select the ionization mode (+Ve or −Ve) and a target *m/z* range depending on the target. Generally, gangliosides are detected in the negative ion mode, but some gangliosides are also detected in the positive ion mode (*see* **Note 7**).
4. Search appropriate MALDI-MSI parameters to detect target signal for both imaging and structural analysis of gangliosides (tandem mass spectrometry, MS^2 or MS^n) by test shots outside of the target region on the slide, as well as standard gangliosides applied on the slide (*see* **Note 8**). To improve mass accuracy, perform a mass calibration using the dried standard spot.
5. Select a spatial resolution and run the measurement. Depending on the spatial resolution, MSI parameters need to be adjusted and confirm ionization efficiency (*see* **Note 9**).
6. After optimization of MSI parameters, start data acquisition. Data acquisition time could vary depending on the spatial resolution of that measurement, and scanning speeds of MSI tools going to be used for that experiment.

3.6 DESI-MSI of Gangliosides

1. Take an image of that slide containing sections of sample tissues or organs. Generally a high-resolution image is not preferred for DESI-MSI experiment. For DESI-MSI, normal glass slide is preferred, and there is no need to apply matrix [17]. Unlike MALDI-MSI, it is not recommended to dry sample prior to DESI-MSI measurement.
2. Load the photo of the slide into software designed for DESI-MSI tools. Coordinate the slide position and select the area on the slide for DESI-MSI data acquisition.
3. Load the sample slide into a DESI source of DESI-MSI instrument. Adjust the starting point of that slide with the stage position.
4. Select the ion mode and a target *m/z* depending on the purpose.
5. Find out appropriate DESI-MSI parameters to detect target signals for imaging of gangliosides by checking signals from outside of the target region, as well as standard gangliosides applied on the slide. Major parameters are solvent composition, solvent flow rate and stability, ionization mode, source

temperature, sampling cone voltage, pressure of nitrogen gas, capillary voltage, and data accumulation time [30].

6. To improve mass accuracy, perform a mass calibration using the dried standard spot or perform online calibration using the *m/z* value of a known MS peak.
7. For structural analysis, adjust CE or CV for on-tissue tandem mass spectrometry (*see* **Note 8**).
8. Set expected spatial resolution and run DESI-MSI measurement. Depending on the spatial resolution and the area of measurement, data acquisition time will vary.

3.7 SIMS Imaging of Gangliosides

1. Same as MALDI-MSI, dry the sample sections prior to SIMS analysis.
2. If required, apply the matrix after drying the sample. Generally, matrix application is not required for SIMS imaging. However, the primary ion beam used in SIMS causes fragmentation of molecules present on the sample surface, and makes it difficult to detect intact analytes. To reduce fragmentation and enhance the ionization of analytes as intact ions, researchers use different strategies, including the application of a matrix molecule prior to SIMS imaging (*see* **Note 10**).
3. Select a suitable primary ion beam and optimize SIMS parameters for the expected spatial resolution and *m/z* range, such as primary ion beam energy, primary ion current, etc. The acquire SIMS imaging data.
4. For structural analysis, optimize CV or CE for better fragmentation in tandem mass spectrometry, and acquire data [31].

3.8 Data Analysis

1. Use proper software for analyzing MSI data. Different software is available for analyzing the data of different MSI tools.
2. If required, normalize the signal intensity of each pixel by using total ion current (TIC), extracted ion current (XIC), or other methods (*see* **Note 11**).
3. Select the peak(s) of interest with a selecting window.
4. Use MSI data of standard gangliosides for the assignment of gangliosides detected by MSI tools. Information available in different databases can also be helpful the assignment of gangliosides, such as the **Human Metabolome Database** (https://hmdb.ca/), **Metlin** (https://metlin.scripps.edu/landing_page.php?pgcontent=mainPage), and **Lipid Maps** (www.lipidmaps.org).
5. For the structural analysis of gangliosides, use tandem mass spectrometry (MS^2 or MS^n) data acquired by MSI tools. Information of MS^2 or MS^n spectra of targeted peaks detected from the sample could be matched with data acquired from

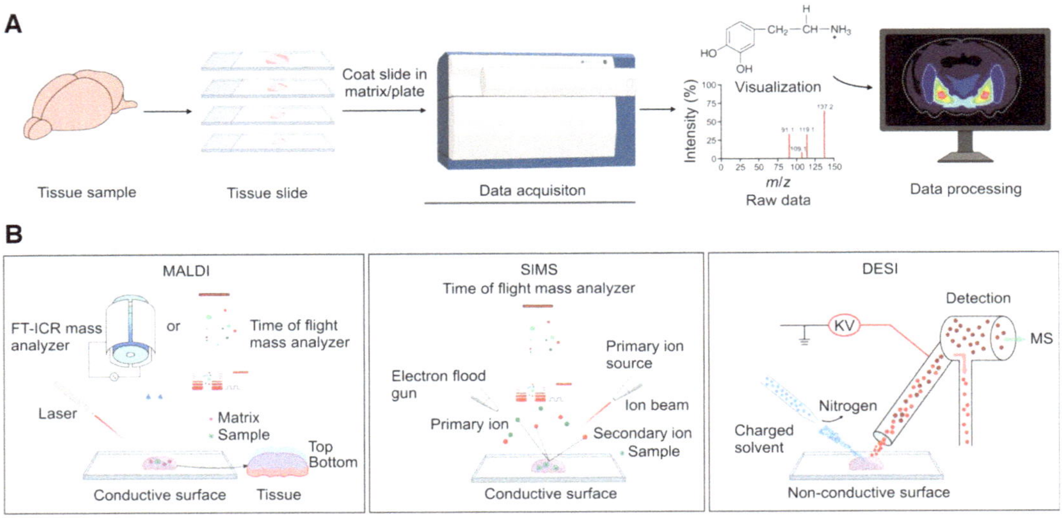

Fig. 3 The diagram of mass spectrometry imaging (MSI) technique. (**A**) General work flow of MSI. (**B**) Schematic illustration of ionizing mechanism in matrix assisted laser desorption ionization (MALDI), secondary ion mass spectroscopy (SIMS), and desorption electrospray ionization (DESI). *m/z*: mass-to-charge ratio; FT-ICR: Fourier transform ion cyclotron resonance; MS: mass spectrometry. (Reproduced from [32] with a license approved by Journal of Pharmaceutical Analysis)

standard gangliosides or information available in the databases and literature.

General work flow of DESI-MSI, MALDI-MSI, and SIMS-MSI is shown in Fig. 3 [32].

4 Notes

1. For MSI measurement, the sample surface should be flat. It directly affects the accuracy, reproducibility, and spatial resolution of MSI data. For these reasons, flat sections of biological samples are prepared using a cryostat system for MSI data acquisition. The thickness of the sections could be different depending on the type of MSI instruments and targeted analytes. The use of thicker sections generally requires a higher amount of matrix.
2. The tube prevents the sections from getting wet when they are moved from −80°C to room temperature. Additionally, silica particles help to keep the sample sections dry. It is important to dry the sample before and SIMS imaging [33] and before applying the matrix for MALDI-MSI [34]. Moisture in the sample can lead uneven distribution and failure to crystallize matrix molecules properly. It can also cause physical

distortions and lead artifacts during matrix application, and can interfere with the ionization process of MALDI.

3. The main purpose of the rinsing of sample sections prior to matrix application is to remove sodium (Na) and potassium (K) from the sample sections. It could provide mass spectra with higher signal intensity by removing Na and K ions, because some gangliosides are detected in multiple ion forms (e.g., $[M-H]^-$, $[M+Na-2H]^-$, $[M+K-2H]^-$). However, this step needs special care to avoid damaging of the sections.
4. Ammonium formate is commonly used as an ionization enhancing agent in MSI as well as in liquid chromatography mass spectrometry (LC-MS). Optimized concentration of these salts can enhance the ionization and sensitivity by forming ammonium adduct during MSI measurement. On the other hand, higher concentration of ammonium formate and other ionization-enhancing agents can cause ion suppression by forming unwanted adducts with matrix ions or other components.
5. Several matrices for analyzing gangliosides by MALDI-MSI. Some matrices easily sublimate and the sublimation causes a change of ionization efficacy of analytes within a run. Such matrix, such as 2,5-DHAP is not suitable for MALDI-MSI tools with a vacuum MALDI source. Additionally, the method for applying the matrix can be different according with the required spatial resolution and sensitivity. Sublimation provides small crystal sizes and is thus suitable for higher spatial resolution (<10 μm). A spray is generally considered to provide higher sensitivity through a liquid extraction step, but there is a risk of delocalization of the target analytes or formation of large-sized crystals. Some researchers have recently used a combination of the two that involves an initial sublimation step followed by a "recrystallization" step using vapor for other molecule.
6. Depending on the available system, parameters for the sublimation of MALDI matrix can be changed. The pressure inside of the chamber, temperature of the heating box, distance between the sample surface and heating box, sublimation time and matrix thickness are the major parameters. The optimal amount of matrix depends on the thickness of the section or the sample type.
7. Most of the gangliosides contain sialic acid side chain. Due to the negative charge on their sialic acid moieties, they are detected in the negative ion mode. Some gangliosides do not contain sialic acid residue on their structure such as asialo-gangliosides. As they do not process negative charge on their structure, they are detected in the positive ion mode.

8. For MALDI-MSI measurement, basic parameters are laser diameter, laser power, *m/z* range, spatial resolution, and mass resolution [35]. Set the laser diameter depending on the required spatial resolution and start test shots with a lower laser power [35]. Change the laser power to get expected signals with higher sensitivity. In case of gangliosides, GM1 (d18:1/18:0) is common in mice brain and can be detected as $[M-H]^-$ ion at *m/z* 1544.9 in the cerebral cortex. Several biomolecules such as gangliosides and phosphatidylcholines are known to be fragmented during ionization in case of using higher laser power. It is recommended to check and adjust laser power to keep fragmentation of analytes as lower as possible before starting data acquisition. Other parameters could also be critical, but they depend on the type of mass spectrometer used for the analysis. For that structural analysis, find out the CE or CV for the better fragmentation using standard gangliosides, and run that experiment with optimized parameters.
9. For higher spatial resolution, laser diameter should be decreased (laser diameter should be same or smaller than the pixel size), but required laser power for smaller laser diameter is lower compared to bigger laser diameter. Therefore, laser power should be adjusted to confirm higher sensitivity and lower tendency of fragmentation of analytes.
10. DESI-MSI and MALDI-MSI tools use soft ionization method suitable for the imaging of intact analytes present in the samples as well as their structural analysis by tandem mass spectrometry. SIMS offers higher spatial resolution compared to MALDI-MSI and DESI-MSI at nm scale, but primary ion beam used in SIMS causes fragmentation of biomolecules and makes it difficult to detect intact larger molecules like gangliosides present in the samples [33]. In order to reduce the fragmentations of analytes present on the sample surface, researcher use several approaches, including the use of cluster ion sources (C_{60}^+, Bi_3^+, Au_3^+, and SF_5^+) [36], matrix application or surface coating with a metal (thin layer of gold or silver) [37].
11. Sensitivity and ionization efficiency of MSI instruments could vary over time. Therefore, normalization is performed to adjust the differences in ionization efficacy and the condition of co-crystallization between measuring points. The most frequently used normalization method is TIC normalization. It divides each signal intensity by the sum of all signals, and helpful in some cases [38]. It can be affected by artifacts depending on the data and setting, such as *m/z* range. Another common normalization method is XIC normalization using a signal from an internal standard (known

compound uniformly applied before matrix) can offer better results [39, 40], but additional optimization is required. In addition to these two normalization methods, several other normalization methods have been developed recently [41, 42]. A careful check before and after the normalization is required to interpret the data accurately.

Conflict of Interest

The authors have no conflicts of interest to declare that are relevant to the content of this chapter.

References

1. Guo X, Wang X, Tian C et al (2023) Development of mass spectrometry imaging techniques and its latest applications. Talanta 264:124721
2. Zhang H, Lu KH, Ebbini M et al (2024) Mass spectrometry imaging for spatially resolved multi-omics molecular mapping. Npj Imaging 2:20
3. Mamun A, Islam A, Eto F et al (2020) Mass spectrometry-based phospholipid imaging: methods and findings. Expert Rev Proteomics 17:843–854
4. Murayama C, Kimura Y, Setou M (2009) Imaging mass spectrometry: principle and application. Biophys Rev 1:131–139
5. Cutillo G, Saariaho A-H, Meri S (2020) Physiology of gangliosides and the role of antiganglioside antibodies in human diseases. Cell Mol Immunol 17:313–322
6. Sandhoff R, Schulze H, Sandhoff K (2018) Ganglioside metabolism in health and disease. Prog Mol Biol Transl Sci 156:1–62
7. Sipione S, Monyror J, Galleguillos D et al (2020) Gangliosides in the brain: physiology, pathophysiology and therapeutic applications. Front Neurosci 14:572965
8. Kumar S, Panda SP (2024) Targeting GM2 ganglioside accumulation in dementia: current therapeutic approaches and future directions. Curr Mol Med 24:1329–1345
9. Ledeen R, Chowdhury S (2022) Gangliosides in neurodegenerative diseases. In: Glycobiology of the nervous system. Springer, pp 391–418
10. Tobias F, Pathmasiri KC, Cologna SM (2019) Mass spectrometry imaging reveals ganglioside and ceramide localization patterns during cerebellar degeneration in the Npc1−/− mouse model. Anal Bioanal Chem 411:5659–5668
11. O'Neill KC, Liapis E, Harris BT et al (2022) Mass spectrometry imaging discriminates glioblastoma tumor cell subpopulations and different microvascular formations based on their lipid profiles. Sci Rep 12: 17069
12. Buchberger AR, DeLaney K, Johnson J et al (2018) Mass spectrometry imaging: a review of emerging advancements and future insights. Anal Chem 90:240
13. Cui X, Geng H, Guo H et al (2024) Visualizing the transdermal delivery of berberine loaded within chitosan microneedles using mass spectrometry imaging. Anal Bioanal Chem 416:6869–6877
14. Aramaki S, Tsuge S, Islam A et al (2023) Lipidomics-based tissue heterogeneity in specimens of luminal breast cancer revealed by clustering analysis of mass spectrometry imaging: a preliminary study. PLoS One 18:e0283155
15. Piehowski PD, Zhu Y, Bramer LM et al (2020) Automated mass spectrometry imaging of over 2000 proteins from tissue sections at 100-μm spatial resolution. Nat Commun 11:1–12
16. Jackson SN, Muller L, Roux A et al (2018) AP-MALDI mass spectrometry imaging of gangliosides using 2, 6-dihydroxyacetophenone. J Am Soc Mass Spectrom 29:1463–1472
17. Škrášková K, Claude E, Jones EA et al (2016) Enhanced capabilities for imaging gangliosides in murine brain with matrix-assisted laser desorption/ionization and desorption electrospray ionization mass spectrometry coupled to ion mobility separation. Methods 104:69–78
18. Neittaanmäki N, Zaar O, Cehajic KS et al (2024) ToF-SIMS imaging reveals changes in tumor cell lipids during metastatic progression of melanoma. Pigment Cell Melanoma Res 37:793–800

19. Wang W, Whitehead S (2020) Imaging mass spectrometry allows for neuroanatomic-specific detection of gangliosides in the healthy and diseased brain. Analyst 145:2473–2481
20. Enzlein T, Lashley T, Sammour DA et al (2024) Integrative single-plaque analysis reveals signature Aβ and lipid profiles in the Alzheimer's brain. Anal Chem 96:9799–9807
21. Ollen-Bittle N, Pejhan S, Pasternak SH et al (2024) Co-registration of MALDI-MSI and histology demonstrates gangliosides co-localize with amyloid beta plaques in Alzheimer's disease. Acta Neuropathol 147:105
22. Ermini L, Morganti E, Post A et al (2017) Imaging mass spectrometry identifies prognostic ganglioside species in rodent intracranial transplants of glioma and medulloblastoma. PLoS One 12:e0176254
23. Zhang Y, Wang J, Liu Ja et al (2016) Combination of ESI and MALDI mass spectrometry for qualitative, semi-quantitative and in situ analysis of gangliosides in brain. Sci Rep 6: 25289
24. Dilmetz BA, Lee YR, Condina MR et al (2021) Novel technical developments in mass spectrometry imaging in 2020: A mini review. Anal Sci Adv 2:225–237
25. Mezger ST, Mingels AM, Bekers O et al (2021) Mass spectrometry spatial-omics on a single conductive slide. Anal Chem 93:2527–2533
26. Leopold J, Prabutzki P, Engel KM et al (2023) A five-year update on matrix compounds for MALDI-MS analysis of lipids. Biomolecules 13:546
27. Sugiyama E, Setou M (2018) Visualization of brain gangliosides using MALDI imaging mass spectrometry. In: Gangliosides: methods and protocols, pp 223–229
28. Suzuki A, Suzuki M, Ito E et al (2018) Mass spectrometry of gangliosides. In: Gangliosides: methods and protocols, pp 207–221
29. Djambazova KV, Gibson-Corley KN, Freiberg JA et al (2024) MALDI TIMS IMS reveals ganglioside molecular diversity within murine S. aureus kidney tissue abscesses. J Am Soc Mass Spectrom 35:1692–1701
30. Islam A, Takeyama E, Nabi MM et al (2022) Stress upregulates 2-arachidonoylglycerol levels in the hypothalamus, midbrain, and hindbrain, and it is sustained by green nut oil supplementation in SAMP8 mice revealed by DESI-MSI. Biochem Biophys Res Commun 609:9–14
31. Agüi-Gonzalez P, Jähne S, Phan NT (2019) SIMS imaging in neurobiology and cell biology. J Anal At Spectrom 34:1355–1368
32. Fangma Y, Liu M, Liao J et al (2023) Dissecting the brain with spatially resolved multi-omics. J Pharm Anal 13:694–710
33. Yoon S, Lee TG (2018) Biological tissue sample preparation for time-of-flight secondary ion mass spectrometry (ToF–SIMS) imaging. Nano Converg 5:1–13
34. Jens JN, Breiner DJ, Phelan VV (2022) Spray-based application of matrix to agar-based microbial samples for reproducible sample adherence in MALDI MSI. J Am Soc Mass Spectrom 33:731–734
35. Islam A, Sakamoto T, Zhai Q et al (2022) Application of AP-MALDI imaging mass microscope for the rapid mapping of imipramine, chloroquine, and their metabolites in the kidney and brain of wild-type mice. Pharmaceuticals 15:1314
36. Passarelli MK, Winograd N (2011) Lipid imaging with time-of-flight secondary ion mass spectrometry (ToF-SIMS). Biochim Biophys Acta-Mol Cell Biol Lipids 1811:976–990
37. Hao-Wen L, Xin H, Yi-Tao L (2018) Metal/matrix enhanced time-of-flight secondary ion mass spectrometry for single cell lipids analysis. Chin J Anal Chem 46:61–66
38. Sugiura Y, Konishi Y, Zaima N et al (2009) Visualization of the cell-selective distribution of PUFA-containing phosphatidylcholines in mouse brain by imaging mass spectrometry [S]. J Lipid Res 50:1776–1788
39. Pirman DA, Kiss A, Heeren RM et al (2013) Identifying tissue-specific signal variation in MALDI mass spectrometric imaging by use of an internal standard. Anal Chem 85:1090–1096
40. Takai N, Tanaka Y, Saji H (2014) Quantification of small molecule drugs in biological tissue sections by imaging mass spectrometry using surrogate tissue-based calibration standards. Mass Spectrom 3:A0025
41. Boskamp T, Casadonte R, Hauberg-Lotte L et al (2021) Cross-normalization of MALDI mass spectrometry imaging data improves site-to-site reproducibility. Anal Chem 93:10584–10592
42. Holbrook JH, Kemper GE, Hummon AB (2024) Quantitative mass spectrometry imaging: therapeutics & biomolecules. Chem Commun 60:2137–2151

Chapter 9

NMR of Gangliosides

Yoshiki Yamaguchi

Abstract

Gangliosides influence numerous physiological functions through cell-surface interactions with exogenous and endogenous molecules. The importance of the glycan moiety of gangliosides has been demonstrated, and studies are underway to elucidate the mechanisms underlying their physiological functions, particularly with regard to structure–activity relationships. Nuclear magnetic resonance (NMR) spectroscopy provides unique insight into the chemical structure, conformation, dynamics, and interactions of biological samples at the atomic level. In this chapter, we outline protocols for NMR determination of the chemical structure of gangliosides, with a particular focus on identification of the sialic acid residues, and for analysis of protein–ganglioside interactions by saturation transfer difference NMR.

Keywords Anomericity, Chemical structure, Glycosidic linkage, Sialic acid, Saturation transfer difference (STD) NMR, Group epitope mapping

1 Introduction

Glycolipids comprising a hydrophobic ceramide (fatty acid amide of sphingosine) are classified as glycosphingolipids. Among various negatively charged glycosphingolipids, gangliosides containing sialic acid exert a range of physiological functions, including in nervous and immune systems, through interactions with exogenous and endogenous molecules [1, 2]. The importance of the glycan moiety of gangliosides has been demonstrated through experimental studies, including the analysis of glycosyltransferase-knockout mice [3]. In light of the growing body of knowledge, research is currently aimed at elucidating the underlying molecular mechanisms of these physiological functions, with a particular focus on structure–activity relationships.

While various methods such as X-ray crystallography, nuclear magnetic resonance (NMR) spectroscopy, and cryo-electron microscopy can be used to study structure–activity relationships, NMR is distinct in its capacity to provide insight into the chemical structure, conformation, dynamics, and interactions of a biological sample at

Koichi Furukawa (ed.), *Gangliosides*, Methods in Molecular Biology, vol. 3035,
https://doi.org/10.1007/978-1-0716-5288-6_9,

the atomic level. With regard to gangliosides, NMR can identify or confirm the structure of the glycan and ceramide parts, and typically provide information on anomericity (α or β) and the glycosidic linkage with minimal ambiguity. NMR can also be used to probe the interactions of gangliosides. For example, saturation transfer difference (STD) NMR, a powerful technique for studying interactions between a macromolecule and small molecules in solution, has been widely used to map group epitopes in the glycan region [4]. It is particularly well-suited for examining weak glycan–protein interactions [5–11], which have dissociation constants within the millimolar range.

In this chapter, we outline NMR protocols for studies of ganglioside structure and function. Given the recent publication of basic NMR methodology for determining the chemical structure of glycolipids [12], we place particular emphasis on identification of the sialic acid residues present in gangliosides. We subsequently outline the analysis of interactions between a protein and a glycan derived from a ganglioside by STD-NMR.

2 Materials

Tetramethylsilane (TMS) was obtained from Sigma-Aldrich (Cat No. T24007), and sodium 2,2-dimethyl-2-silapentane-5-sulfonate (DSS) was from Cambridge Isotope Laboratories, Inc. (Cat No. DLM-32-1). All reagents utilized in NMR experiments must be of analytical grade. It is strongly recommended that the ganglioside sample is free from unnecessary small chemicals and salts, which may interfere with detailed NMR analysis and reduce sensitivity, respectively.

The following materials are required:

1. The ganglioside or ganglioside-derived glycan of interest (0.5–1.0 mg);
2. Deuterium oxide (^2H, 99.96%);
3. Deuterated dimethyl sulfoxide ([$^2H_6$]DMSO) (^2H, 99.96%);
4. Tetramethylsilane $(CH_3)_4Si$ (TMS) or sodium 2,2-dimethyl-2-silapentane-5-sulfonate $(CH_3)_3SiCH_2CH_2CH_2SO_3^-Na^+$ (DSS) (*see* **Note 1**).

3 Methods

3.1 Sample Preparation and One-Dimensional ^{1}H-NMR Measurement

It is imperative that clean plastic disposables, pipettes, and NMR sample tubes are used for sample preparation. Materials that come into contact with the sample can be purified using a plasma cleaner [13].

1. Dissolve the ganglioside sample (0.5–1.0 mg) in 0.60 mL of a mixed solvent comprising [$^2H_6$]DMSO and D_2O (98:2). D_2O is added to exchange -NH and -OH groups to -ND and -OD, respectively. To ensure that all labile protons are exchanged completely, the ganglioside sample should be dissolved in D_2O and lyophilized prior to dissolution in the mixed solvent.
2. Add a small amount of TMS (chemical shift standard for ^1H and ^{13}C) to the solution.
3. Transfer the solution to an NMR sample tube (diameter 5 mm) and secure the cap in place.
4. Set the probe temperature of the NMR magnet to 298–303 K and transfer the NMR sample tube with a spinner turbine into the NMR magnet, a 600 MHz spectrometer (DRX-600, Bruker) equipped with a 5 mm triple resonance inverse (TXI) probe.
5. Allow the sample temperature to reach equilibrium over a period of 15–20 min.
6. Following the completion of shimming and tuning procedures, determine the ^1H 90 pulse length and the power level.
7. Measure the one-dimensional (1D) ^1H-NMR spectrum with high resolution. After Fourier transformation, the digital resolution will be approximately 0.2–0.3 Hz per point.
8. Identify the TMS methyl signal and set it to 0 ppm.
9. Identify the anomeric signals that appear around 4.5–5.5 ppm. Check their coupling constants with H2 protons. Anomeric signals are paramount to the identification of monosaccharide type and anomericity (α or β).
10. Identify the sharp *N*-acetyl methyl signal(s) at approximately 2.0 ppm, which originate from GlcNAc, GalNAc, and Neu5Ac residues. In the case of Neu5Gc, the *N*-glycolyl methylene signal will appear at ~4.1 ppm.
11. Identifyα-Neu5Ac/Neu5Gc H3(equatorial, eq) and H3(axial, ax) signals (~2.7 and ~1.7 ppm, respectively) with characteristic splittings. The chemical shift values of H3(eq) and H3(ax) are useful for the discrimination of α(2-3) or α(2-6) linkages of sialic acid to galactose (Fig. 1). In the case of Neu5Acα (2-3)Gal, the chemical shift of H3(ax) is 1.80 ppm and that of H3(eq) is 2.76 ppm. For Neu5Ac α(2-6)Gal, the chemical shifts of H3(ax) and H3(eq) are 1.72 ppm and 2.67 ppm, respectively [14].

3.2 Saturation Transfer Difference NMR for Mapping the Group Epitope

1. Prepare an NMR sample of ganglioside-derived glycan mixed with protein dissolved in an appropriate buffer in D_2O (*see* **Note 2**). The glycan-to-protein molar ratio should be in the range of 20:1 to 100:1.

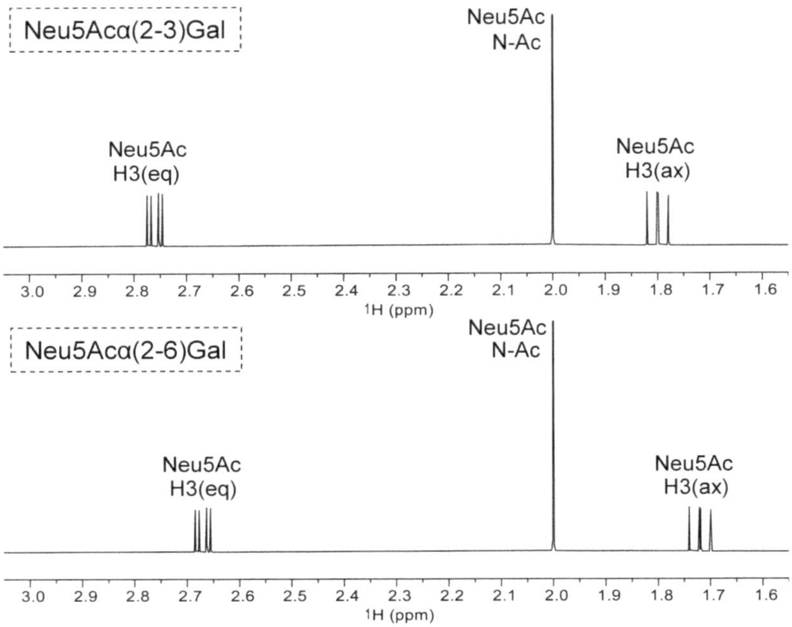

Fig. 1 Simulated ^{1}H-NMR spectra of Neu5Ac residues with particular emphasis on Neu5Ac H3(ax), H3(eq), and *N*-acetyl methyl signals. Upper panel shows Neu5Acα(2-3)Gal; lower panel shows Neu5Acα(2-6)Gal.

2. Add a small amount of DSS (chemical shift standard for ^{1}H and ^{13}C) to the NMR sample.
3. Obtain the 1D NMR spectrum of the sample to confirm that glycan and protein signals are observed (*see* **Note 3**).
4. Identify the DSS methyl signal and set it to 0 ppm.
5. Optimize the parameters for STD-NMR measurement. The parameters that warrant consideration are the probe temperature, ligand-to-protein ratio, and saturation condition (i.e., saturation frequency, power level, and duration) (*see* **Note 4**).
6. Measure STD-NMR under the optimized conditions and confirm that STD signals are observed at various intensities (*see* **Note 5**) (Fig. 2).
7. Read the intensities of NMR signals from the STD-NMR spectra and control NMR spectra (*see* **Note 6**).
8. Calculate the STD-amplification factor (STD-AF) (*see* **Note 7**). Plotting the STD-AF value at varying ligand concentrations enables the dissociation constant to be calculated (Fig. 3, left).
9. Identify the group epitope exhibiting a relatively high STD-AF (Fig. 3, right).

NMR provides a set of observable parameters that can be used to analyze biomolecules from a multitude of perspectives. In general, the amount of gangliosides derived from natural sources is limited.

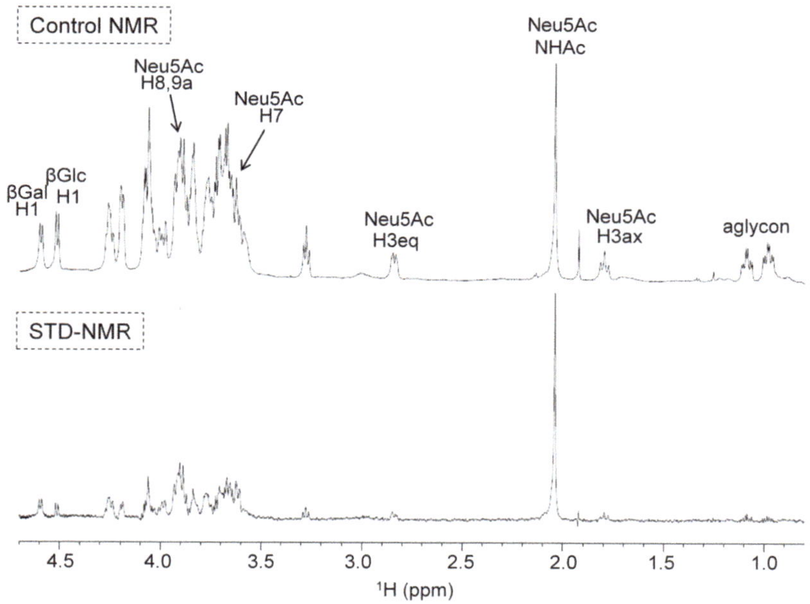

Fig. 2 Control and STD-NMR spectra of starfish ganglioside-derived glycan AG2 in the presence of human Siglec-2 protein [11]. The ligand-to-protein molar ratio was 20:1. NMR spectra were obtained at a ^{1}H observation frequency of 600 MHz and a probe temperature of 5°C. In the STD-NMR experiment, on-resonance irradiation of the protein was performed at 7.51 ppm; off-resonance irradiation was performed at 40 ppm

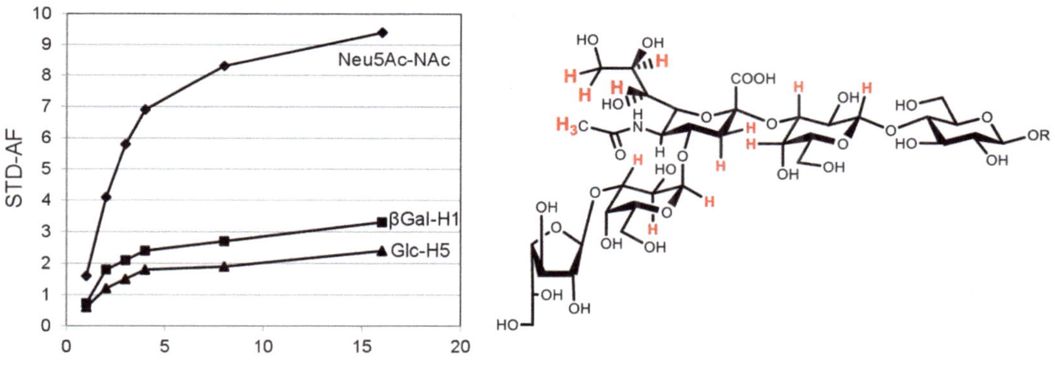

Galfβ(1-3)Galpα(1-4)Neu5Acα(2-3)Galpβ(1-4)Glcpβ-OR

Fig. 3 Plot of STD-AF against ligand concentration (left) and structures of the epitopes of AG2 glycan recognized by human Siglec-2 (right). Protons with a high STD-AF (>2.3) are indicated in red within the structure of AG2 glycan [11]. The dissociation constant is estimated to be 2 mM based on the STD-AF plot

With increases in magnetic field strength and the continuing development of detection probes, the sensitivity and resolution of NMR will improve, enabling its application to the analysis of even smaller sample quantities. In the future, it will be essential to consider the cooperative physiological roles of the glycan and ceramide moieties. Clarifying the role of gangliosides in biological membranes will undoubtedly present a challenge.

4 Notes

1. As chemical shift standards for ^1H and ^{13}C, TMS is used for organic solvents and DSS for water. Use of 3-(trimethylsilyl) propanoic acid $(CH_3)_3CH_2CH_2COOH$ (TSP) as a chemical shift standard is not recommended due to the pH-sensitivity of the methyl signal in TSP. The chemical shift of DSS is less pH-sensitive [15]. It is important to note that internal chemical shift references may interact with peptides [16], proteins [17], and membrane mimics [18], although such instances are exceedingly rare.
2. Anomeric signals frequently overlap with the HDO signal. To ascertain the presence of anomeric signals within a substantial HDO signal, we advise making ^1H-NMR measurements at varying probe temperatures. The chemical shift of the HDO signal is temperature-sensitive, whereas those of anomeric signals are less sensitive to temperature.
3. Use of D_2O is required for STD-NMR; use of ^1H_2O is not advised.
4. This optimization is crucial for obtaining STD-NMR spectra with high sensitivity.
5. It is essential to ensure that the DSS signal is weak in STD-NMR spectra due to signal cancellation.
6. Overlap with signals from non-anomeric protons frequently impedes the analysis. A potential solution is to use a 2D version of STD-NMR (specifically 2D ^1H-^{13}C STD-HSQC) [9]. It is important to note that the sensitivity of 2D ^1H-^{13}C STD-HSQC is relatively low; ^{13}C-labeling of the glycan may enhance the sensitivity of the technique [19].
7. STD-AF is defined as $(I_0 - I_{sat})/I_0 \times$ ligand excess, where I_0 is the signal intensity of the off-resonance spectrum, and I_{sat} is the signal intensity of the on-resonance spectrum [4].

Acknowledgments

The author would like to express gratitude to Dr. Hanashima of Tottori University for his invaluable contributions to the NMR study and chemical synthesis. This study was supported in part by JSPS KAKENHI Grant Number 19H03362.

Conflict of Interest The authors have no conflicts of interest to declare that are relevant to the content of this chapter.

References

1. Schnaar RL (2016) Gangliosides of the vertebrate nervous system. J Mol Biol 428:3325–3336
2. Sonnino S, Prinetti A (2010) Gangliosides as regulators of cell membrane organization and functions. Adv Exp Med Biol 688:165–184
3. Stanley P (2016) What have we learned from glycosyltransferase knockouts in mice? J Mol Biol 428:3166–3182
4. Mayer M, Meyer B (2001) Group epitope mapping by saturation transfer difference NMR to identify segments of a ligand in direct contact with a protein receptor. J Am Chem Soc 123:6108–6117
5. Yamakawa N, Yasuda Y, Yoshimura A et al (2020) Discovery of a new sialic acid binding region that regulates Siglec-7. Sci Rep 10:8647
6. Hanashima S, Götze S, Liu Y et al (2015) Defining the interaction of human soluble lectin ZG16p and mycobacterial phosphatidylinositol mannosides. ChemBioChem 16:1502–1511
7. Hanashima S, Ikeda A, Tanaka H et al (2014) NMR study of short β(1-3)-glucans provides insights into the structure and interaction with Dectin-1. Glycoconj J 31:199–207
8. Yuasa N, Koyama T, Subedi GP et al (2013) Expression and structural characterization of anti-T-antigen single-chain antibodies (scFvs) and analysis of their binding to T-antigen by surface plasmon resonance and NMR spectroscopy. J Biochem 154:521–529
9. Hanashima S, Sato C, Tanaka H et al (2013) NMR study into the mechanism of recognition of the degree of polymerization by oligo/polysialic acid antibodies. Bioorg Med Chem 21:6069–6076
10. Tanaka H, Kawai T, Adachi Y et al (2012) Synthesis of β(1,3) oligoglucans exhibiting a Dectin-1 binding affinity and their biological evaluation. Bioorg Med Chem 20:3898–3914
11. Hanashima S, Sato K, Naito Y et al (2010) Synthesis and binding analysis of unique AG2 pentasaccharide to human Siglec-2 using NMR techniques. Bioorg Med Chem 18:3720–3725
12. Yamaguchi Y (2023) NMR analysis of mammalian glycolipids. Methods Mol Biol 2613:181–188
13. Fellenberg M, Coksezen A, Meyer B (2010) Characterization of picomole amounts of oligosaccharides from glycoproteins by ^{1}H NMR spectroscopy. Angew Chem Int Ed Engl 49:2630–2633
14. Dorland L, Haverkamp J, Viliegenthart JF et al (1978) 360-MHz ^{1}H nuclear-magnetic-resonance spectroscopy of sialyl-oligosaccharides from patients with sialidosis (mucolipidosis I and II). Eur J Biochem 87:323–329
15. Wishart DS, Bigam CG, Yao J et al (1995) ^{1}H, ^{13}C and ^{15}N chemical shift referencing in biomolecular NMR. J Biomol NMR 6:135–140
16. Laurents DV, Gorman PM, Guo M et al (2005) Alzheimer's Aβ40 studied by NMR at low pH reveals that sodium 4,4-dimethyl-4-silapentane-1-sulfonate (DSS) binds and promotes β-ball oligomerization. J Biol Chem 280:3675–3685
17. Shimizu A, Ikeguchi M, Sugai S (1994) Appropriateness of DSS and TSP as internal references for ^{1}H NMR studies of molten globule proteins in aqueous media. J Biomol NMR 4:859–862
18. Morash B, Sarker M, Rainey JK (2018) Concentration-dependent changes to diffusion and chemical shift of internal standard molecules in aqueous and micellar solutions. J Biomol NMR 71:79–89
19. Moure MJ, Gimeno A, Delgado S et al (2021) Selective ^{13}C-labels on repeating glycan oligomers to reveal protein binding epitopes through NMR: polylactosamine binding to galectins. Angew Chem Int Ed Engl 60:18777–18782

Check for updates

Chapter 10

Analysis of Ganglioside Molecular Species Focusing on Ceramide Structures Using Thin-Layer Chromatography–Mass Spectrometry

Noboru Hashimoto, Shizuka Ito and Akihito Yamamoto

Abstract

Thin-layer chromatography is widely used for the analysis of glycolipids and phospholipids. Typically, after separation with various solvents, molecular species are identified by chemical detection or immunostaining. This chapter describes a method for structural analysis of molecules by directly performing mass spectrometry after chromatographic separation. We introduce a classical ganglioside analysis method and a technique for structure analysis of separated lipids using a robotic ion source for direct MS analysis.

Keywords Ganglioside, Ceramide, Hydroxylation, Thin-layer chromatography, Robotic ion source, Mass spectrometry

1 Introduction

Lipids that constitute biological membranes are primarily phospholipids, which contain phosphate ester groups [1]. Glycolipids, on the other hand, consist of a sugar moiety instead of a phosphate group and a ceramide backbone composed mainly of fatty acids and long-chain bases (LCBs) [1, 2]. Glycolipids are involved in various biological phenomena, although their abundance is lower than that of phospholipids [2]. Glycolipids are categorized based on their sugar composition. Various analytical methods such as thin-layer chromatography (TLC), immunostaining, and mass spectrometry (MS) are used for their characterization [3–5]. Importantly, it has been observed that the localization of glycolipids in the cell membrane varies depending on the type of ceramides they contain [6, 7]. Ceramide diversity encompasses variations in acyl chain length, degrees of saturation and unsaturation, as well as hydroxylation at specific carbon positions. Therefore, structural analysis of lipids is essential for understanding their functions. However, available analytical methods are limited. Previous reports include methods such

Koichi Furukawa (ed.), *Gangliosides*, Methods in Molecular Biology, vol. 3035,
https://doi.org/10.1007/978-1-0716-5288-6_10,

as re-extracting and analyzing TLC-separated samples by scraping silica gel [see Chap. 11 for details], as well as analysis using MALDI-TOF MS [3]. This chapter introduces a method for analyzing structural differences in glycosphingolipids by combining TLC and MS. It covers comprehensive lipid structural analysis using direct MS analysis of extracted lipids and detailed structural characterization with a triple quadrupole mass spectrometer [5–7]. This provides a comprehensive approach to characterize structural differences in glycolipids to understand their pathological behaviors. We hope this information will be useful for reference.

2 Materials

All solutions should be prepared using ultrapure water (DW) (deionized and further purified to achieve a resistivity of 18 MΩ·cm at 25°C) and guaranteed-grade (analytical-grade) reagents. Unless otherwise specified, all reagents should be prepared and stored at room temperature. When disposing of waste, follow all applicable waste disposal regulations.

2.1 Thin-Layer Chromatography (TLC)

1. Aluminum HPTLC silica gel 60 plates (size 20 × 20 cm) (Merck Millipore) are used for TLC (*see* **Note 1**).
2. Pencil for writing where to load samples.
3. MeOH, $CHCl_3$, LC/MS-grade solvents for development and extraction. All solvents of MeOH and $CHCl_3$ are mixed at a 2:1 ratio.
4. TLC Resorcinol reagent: Dissolve 0.2 g of resorcinol (Fujifilm WAKO) in 10 ml of water, add 80 ml of concentrated hydrochloric acid and 0.25 ml of 1.0 M CuSO4, and make up to 100 ml with water. When mixing water and concentrated hydrochloric acid, do so slowly and carefully in a draft.
5. Conventional TLC: Glass HPTLC silica gel 60 plate (Merck Millipore)
6. A rectangular TLC developing tank that is larger than the plate size (20 × 20 cm).
7. Hamilton syringe (10 μl capacity) (Hamilton company) for loading samples. The needle is of the beveled-tip type (*see* **Note 2**).
8. TLC-developing solvent, $CHCl_3$: MeOH: 0.2% $CaCl_2$ = 55:45:10.

2.2 TLC-Transfer

1. Primulin (Fujifilm WAKO) solution (0.001% in acetone/water, 80:20) (*see* **Note 3**).
2. Transfer buffer: Isopropanol: 0.2% $CaCl_2$: MeOH = 40: 20: 7.

3. Glass fiber filter (AC-5972, 100×100 mm, Atto).
4. Polytetrafluoroethylene (PTFE) membrane (AC-5973, 100×100 mm, Atto)
5. Clear Blot membrane-P (PVDF) (AE-6665, 85×90 mm, ATTO).
6. TLC thermal blotter (AC-5970, ATTO).

2.3 MS Analysis

1. Analyze TLC-developed samples directly using ESI robotic ion source TriVersa NanoMate LESA™ system (Advion Interchim Scientific) and QTRAP 6500™ (Sciex).
2. Extraction solvent (5 mM ammonium formate (Fisher) in $CHCl_3$: MeOH = 2:1) (*see* **Note 4**).
3. Conductive pipette tip for sample extraction and direct MS analysis (ESI-Chip nozzle, A-Chip HD_A_384, inter diameter = 5.0 μm, Advion).
4. Automatically extract target lipid on the pre-programmed position by the robot with a special tip containing 10 μl of extraction solvent (5 mM ammonium formate in $CHCl_3$: MeOH = 2:1) and then introduced into the ion source.

2.4 TLC-Immunostaining

1. Blocking solution: 5%BSA/PBS.
2. Washing buffer: PBS.
3. Antibody solution: dilute appropriately with 1% BSA/PBS.
4. ABC kit: VECTASTAIN Elite ABC Kit (Vector Laboratories).
5. A peroxidase detection kit: Konica immunostaining HRP-1000 (Konica Minolta).

3 Methods

The experimental procedure is as shown in Fig. 1.

3.1 Transfer the Lipids Separated by TLC onto a PVDF Membrane

1. Prepare glycolipid samples such as extracts by $CHCl_3$ and MeOH or acidic fraction from DEAE Sephadex A50 (GE Healthcare) column (*see* **Note 5**).
2. Dry up a TLC plate, 10 min at 110°C.
3. Place a piece of filter paper of the same height in the tank, add the solvent, and saturate it for at least 10 min.
4. Draw a line 1.5 cm from the bottom edge of the TLC plate using a pencil. Mark sample spots at 1 cm intervals, spacing a 0.5 cm gap between samples.
5. Apply samples onto TLC plate.

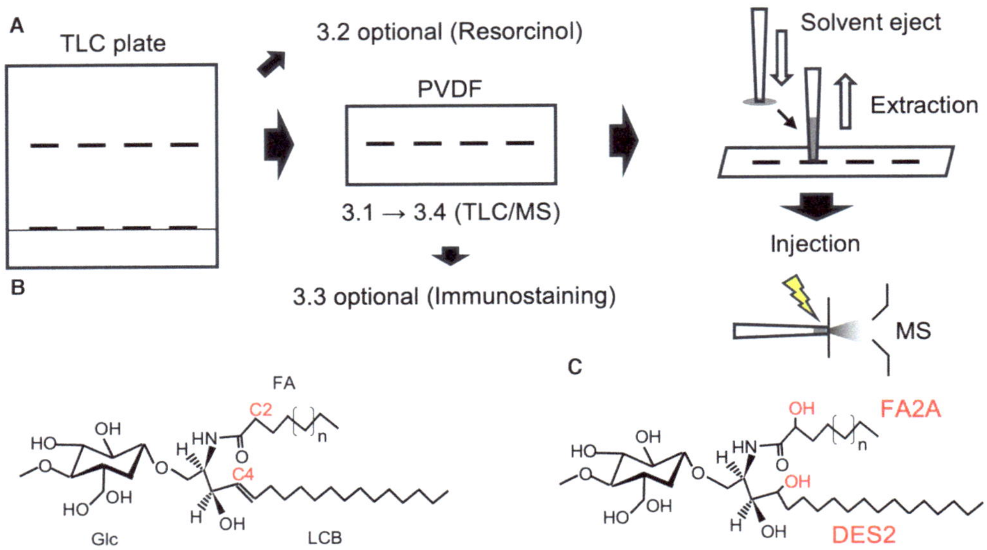

Fig. 1 Experimental Workflow and structure diversity of ceramide moiety. (**a**) The sample is developed by TLC, and the target is transferred onto PVDF. It is then extracted using a $CHCl_3$/MeOH solvent with a robotic ion source and directly subjected to MS analysis. Alternatively, after developing, they may be used for resorcinol spray or for TLC-immunostaining. (**b**) Ubiquitous ceramide structure. (**c**) In DLD-1 colon cancer cell line, ceramide is hydroxylated at the C2 position of fatty acid (FA) and the C4 position of long-chain base (LCB). *Glc* glucose

6. Hamilton syringe is roughed up with $CHCl_3$ and MeOH by inhaling and exhaling at least ten times each time to prevent sample carry-over.
7. Develop the TLC plate with optimal solvent: TLC-developing solvent, $CHCl_3$: MeOH: 0.2% $CaCl_2$ = 55:45:10 (Fig. 2a).
8. Remove the plate from the development chamber when the solvent front reaches about 1 cm from the top edge (*see* **Note 6**).
9. Ensure complete drying (*see* **Note 7**).
10. Spray primulin solution (*see* **Note 3**) onto TLC plate and mark with red pencil under UV light (UV transilluminator such as gel imaging device) (*see* **Note 8**).
11. Cut TLC plate with cutter and prepare the PVDF, PTFE, and glass filter (*see* **Note 9**).
12. Turn on the TLC thermal blotter (ATTO Corp., Tokyo, Japan).
13. Dip the TLC plate into transfer buffer for 45 s.
14. Put PVDF membrane onto TLC plate on glass filter, and subsequently put PTFE membrane onto PVDF. Finally cover by glass filter membranes.

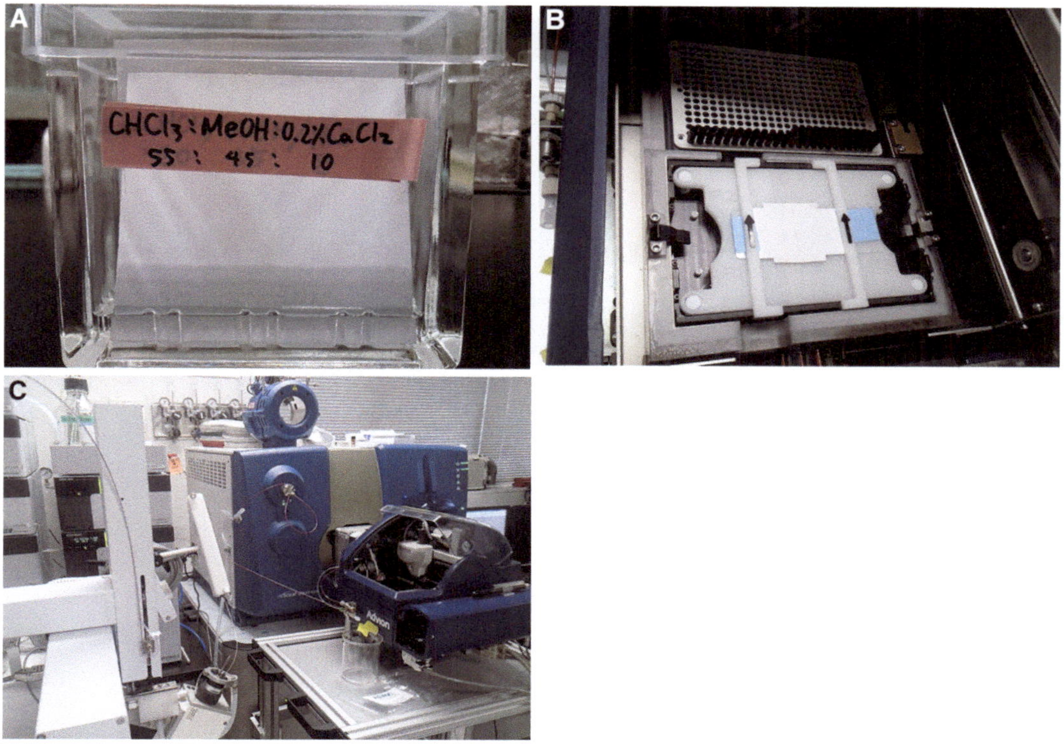

Fig. 2 Experimental Workflow (**a**) Developing TLC. (**b**) A PVDF membrane sample was placed on the sample stage of the robotic ion source TriVersa NanoMate LESA™ system. (**c**) Triple Quadrupole QTRAP 6500™ MS system equipped with LESA

15. After getting the plate between glass filters, put them onto the blotter.
16. Transfer the lipid on TLC plate to PVDF membrane. (heater temperature: 110°C, pressure: 8 (6.4 kg), transfer time: 30 s) (Fig. 2b) (*see* **Note 10**).

3.2 TLC-Staining (Optional)

Preliminary analysis using TLC with markers or standards such as bovine brain gangliosides and chemical detection (e.g., resorcinol) is useful for reliable identification.

1. Apply samples onto a glass plate HPTLC along with ganglioside markers and ganglioside standards (e.g., bovine milk GD3).
2. Develop the samples on the glass HPTLC plate under the same conditions as in Sect. 3.1.7.
3. Spray with resorcinol hydrochloric acid reagent to stain and analyze gangliosides using the resorcinol spray (*see* **Note 11**) (Fig. 3).

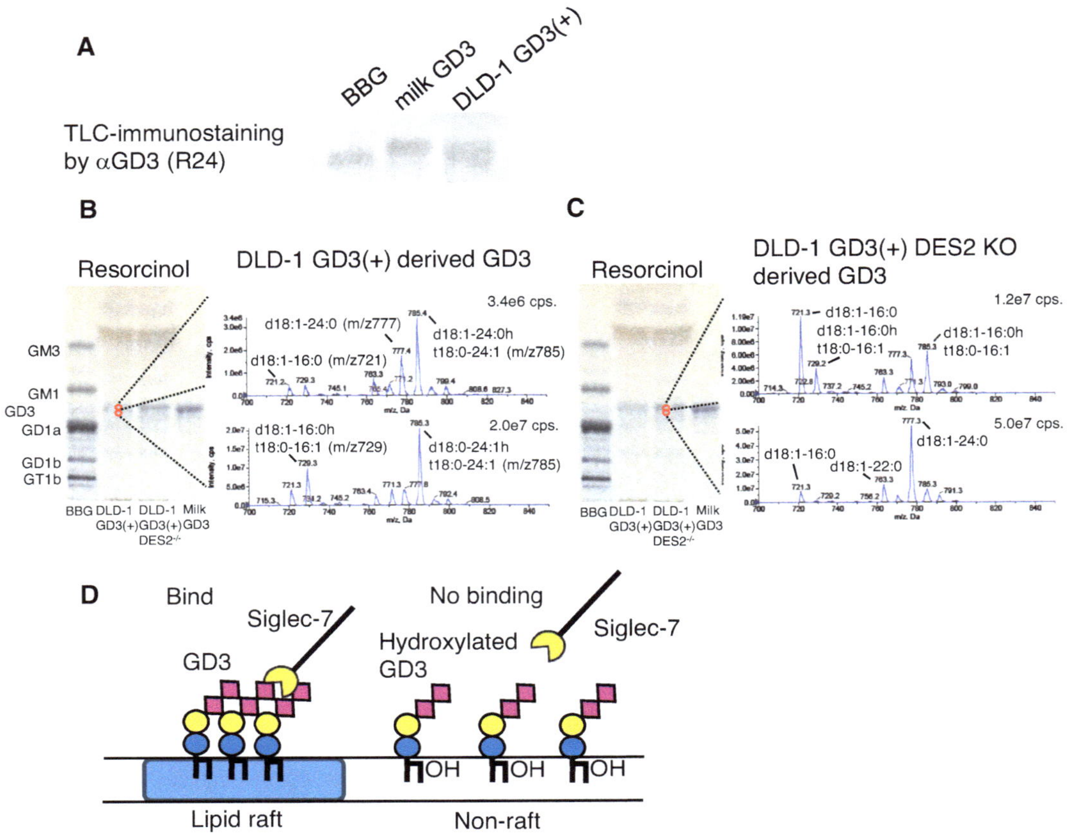

Fig. 3 Example of this experiment. (**a**) TLC-immunostaining with anti-GD3 antibody. (**b** and **c**) Resorcinol staining of TLC and precursor ion scan peaks (Precursor ions of 290.0 (sialic acid)) in negative ion mode. BBG: Bovine brain ganglioside (ganglioside standard). DLD-1 GD3(+) and DLD-1 GD3(+) DES2 KO: Glycolipids extracted from each cell line. Milk GD3: GD3 standard derived from bovine milk. Cps = counts per second. (**b**) DLD-1 GD3(+): GD3 derived from the colon cancer cell line DLD-1, in which GD3 synthase is overexpressed. (**c**) GD3 derived from DES2-KO cells. (**d**) GD3 with a typical ceramide structure localizes in lipid rafts, whereas hydroxylated ceramide-containing GD3 exists outside of this microdomain. Siglec-7 specifically recognizes GD3 localized in lipid rafts (see [7])

4. Cover the developed plate with a glass cover, seal it with adhesive tape, and heat at 110°C for 5–20 min.
5. Stop heating once coloration appears
6. Capture an image.

3.3 TLC-Immunostaining (Optional, see Note 12)

When specific antibodies toward the target molecule are available, immunostaining strongly supports molecular identification. This technique serves as an important complement for validating the results of mass spectrometry.

1. Continue the procedure from Sect. 3.1, Step 16.
2. After transfer lipid from TLC plate to PVDF membrane, block the PVDF membrane in 5% BSA/PBS at 4°C overnight.
3. Incubate with first antibody (e.g., anti-GD3 antibody) for 1 h at room temperature, then wash three times with PBS. Incubate with appropriate secondary antibody (e.g., anti-mouse IgG-biotin) for 1 h at RT, wash three times with PBS, then incubate with ABC reagent for 30 min according to manufacturer's protocol.
4. Detect by a peroxidase detection kit (e.g., Konica Immunostain HRP-1000) for 5–15 min according to manufacturer's protocol.
5. Capture an image (Fig. 3a).

3.4 Direct Extraction and MS Analysis of Lipids Transferred to PVDF Membrane Using Triversa Nanomate LESA

A conductive pipette tip is used to directly extract the sample from the PVDF membrane using a solvent (as shown below). The tip is replaced for each sample, and the solvent is automatically filled. These operations are performed using the TriVersa NanoMate LESA™ system and QTRAP 6500™ (Fig. 2b, c).

1. Set the PVDF membrane, onto which the target region of the TLC-separated sample has been transferred, on the stage of the TriVersa NanoMate LESA system.
2. Use the TriVersa NanoMate LESA system control software to specify the extraction points.
3. Automatically extract target lipids on the pre-programmed position by the robot with a special tip containing 10 μl of extraction solvent ($CHCl_3$: MeOH = 2:1, 5 mM ammonium formate). When the tip is in contact with the PVDF membrane, 10 μl of solvent is dispensed, penetrates the membrane, and is then aspirated. The extracted area is empirically a circular region with a diameter of 1 mm.
4. Transfer the extracted solvent directly to the MS ESI-tip, where the tip is charged and the solvent is sprayed, and then introduced into the ion source.
5. Perform the precursor ion scan of 290 m/z ($[Neu5Ac-H]^-$) in negative ion mode for screening of GD3 molecular species (*see* **Note 13**).
6. Perform the precursor ion scan of 264 m/z ([LCB (d18:1)+ $H]^+$) in positive ion mode for screening of GD3 molecular species.
7. Analyze the detected precursor ion by enhanced MS scan (MS/MS) mode for structure analysis.

8. For Structural analysis, perform an Enhanced Product Ion (EPI) scan for the target *m/z* using a trap fill time of 1 ms in both positive and negative ion modes.
9. Configure the mass spectrometer settings as follows:
   Negative Ion Mode
   - Ion spray voltage: –4500 V
   - Declustering potential: –120 V
   - Collision energy: –40 V

   Positive Ion Mode
   - Ion spray voltage: 5500 V
   - Declustering potential: 45 V
   - Collision energy: 45 V

3.5 Data Analysis

As an example, we present the results of GD3 analysis in the GD3 synthase transfectant and ceramide hydroxylase DES2-knockout (KO) of a colorectal cancer cell line, DLD-1 (Fig. 3).

1. Based on the TLC-resorcinol results, identify the spot with the same migration as the standard. In this example, the ganglioside in the sample is GD3 because of the same migration as bovine milk GD3.
2. Based on the precursor ion scan results obtained by TLC-LESA-MS and the predicted masses of ganglioside species identified in Step 1, estimate the long-chain base and fatty acid components of the ceramide moiety. Saturated ceramides are generally predominant. An example of a case in which a discrepancy from the predicted mass is observed is provided below. TLC-LESA-MS analysis shows that in DLD-1 GD3(+) cells, the major species is *m/z* 753, which is 16 Da higher than representative GD3 molecular species, suggesting the presence of an additional oxygen atom on the lipid moiety (Fig. 3b). On the other hand, the analysis results of delta (4)-desaturase *DES2* knock-out cells show that most of the structures are typical ceramide structures (Fig. 3c).
3. Perform EPI-MS (Enhanced Product Ion, i.e., MS/MS) analysis, if necessary, to obtain a detailed fragmentation pattern of the m/z obtained from precursor scans (*see* **Note 14**).
4. For glycan analysis, perform negative ion mode. For lipid analysis, perform positive ion mode. In negative ion mode, the carboxyl (-COO^-) group of sialic acid is detected, whereas in positive ion mode, the long-chain base (LCB) carries a positive charge (N^+).

5. Establish strains deficient in genes related to lipid modification and identify presumed structural changes. Upon KO of DES2, a ceramide hydroxylase, the *m/z* 777 species becomes predominant, indicating that the major GD3 species shift to a saturated ceramide structure (*see* **Note 15**).

Glycolipids are present in low abundance compared to total lipids, making their detection by LC/MS challenging. This method enables the high-sensitivity analysis of target glycolipids while maintaining the high separation resolution of conventional TLC.

4 Notes

1. The reason for using Aluminium plates is that it is necessary to decide the area to be analyzed in later experiments and cut the plate size accordingly.
2. Needle is a 10 μl needle with a cut surface.
3. Add 200 μl of 0.1% primulin in DW (stock solution stored in the dark) to 20 ml of acetone: DW (4:1).
4. Ammonium formate is ion-pair reagent. Suitable for mass spectrometry (MS) analysis of gangliosides.
5. It is desirable that all samples are desalted.
6. The development time is approximately 30 minutes.
7. At least 30 minutes using a hair dryer.
8. Drying is not necessary. Always use a UV protection shield.
9. Prepare PVDF and PTFE slightly larger than the cut TLC plate (+0.5 cm on all sides: top, bottom, left, and right).
10. Mark the PVDF membrane on the facing side to TLC plate. Use the reverse side for antibody reaction.
11. As an option, it is recommended to check the sample concentration suitable for staining based on the resorcinol spray, and then proceed to mass spectrometry (MS) analysis using equal amounts.
12. By using antibodies, target molecules can be detected selectively and with high sensitivity. Additionally, after separation by TLC, specific molecules can be identified, allowing clear discrimination between similar compounds.
13. In this section, a triple quadrupole mass spectrometer is used. The precursor ion scan is a method in MS/MS that selectively detects precursor ions by generating a specific fragment ion. This approach allows comprehensive screening of precursor ions that produce a specific product ion, making it widely applicable for unknown compound screening and lipid

analysis. Sects. 3.4, Steps 4–8 may vary depending on the mass spectrometer used and should be evaluated accordingly.

14. Using a triple quadrupole mass spectrometer, precursor ions can be selected with the Q1 filter for MS/MS analysis to identify molecular structures. Switching between positive and negative ion modes allows analysis of lipid and glycan structures.

15. The GD3 in colorectal cancer cell lines tends to contain a ceramide structure containing a hydroxyl group, differing from the ubiquitous GD3 ceramide structure. We previously reported that GD3 with a hydroxylated ceramide structure shifts from its native location in the highly hydrophobic lipid raft microdomain of the cell membrane to the outside of the microdomain [7]. Siglec-7, expressed on NK cells and other immune cells, recognizes GD3, which is highly expressed on cancer cells. However, it failed to recognize hydroxylated ceramide-containing GD3 on a colon cancer line. Surprisingly, since it was able to bind to GD3 on DES2 KO cells, suggesting that Siglec-7 recognizes GD3 that forms clusters within lipid rafts (Fig. 3d).

Acknowledgments

This work was supported by Grants-in-aid from the Ministry of Education, Culture, Sports, and Technology of Japan (MEXT, KAKENHI) grant number (26860320 [N.H.], 15K15080 [K.F.], 19K22518 [K.F.], 20K08690 [N.H.], 21K06828 [K.F.], 23K06414 [N.H.]) and JST-CREST grant number JPMJCR17H2 (K.F.).

Conflict of Interest Statement The authors have no conflicts of interest to declare that are relative to the content of this chapter.

References

1. Harayama T, Riezman H (2018) Understanding the diversity of membrane lipid composition. Nat Rev Mol Cell Biol 19:281–296
2. Furukawa K, Ohmi Y, Ohkawa Y et al (2019) New era of research on cancer-associated glycosphingolipids. Cancer Sci 110:1544–1551
3. Fuchs B (2012) Analysis of phospolipids and glycolipids by thin-layer chromatography–matrix-assisted laser desorption and ionization mass spectrometry. J Chromatogr A 1259:62–73
4. Furukawa K, Aixinjueluo W, Kasama T et al (2008) Disruption of GM2/GD2 synthase gene resulted in overt expression of 9-O-acetyl GD3 irrespective of Tis21. J Neurochem 105:1057–1066
5. Ikeda K, Shimizu T, Taguchi R (2008) Targeted analysis of ganglioside and sulfatide molecular species by LC/ESI-MS/MS with theoretically expanded multiple reaction monitoring. J Lipid Res 49:2678–2689
6. Dong Y, Ikeda K, Hamamura K et al (2010) GM1 / GD1b / GA1 synthase expression results in the reduced cancer phenotypes with modulation of composition and raft-localization of gangliosides in a melanoma cell line. Cancer Sci 101:2039–2047
7. Hashimoto N, Ito S, Tsuchida A et al (2019) The ceramide moiety of disialoganglioside (GD3) is essential for GD3 recognition by the sialic acid– binding lectin SIGLEC7 on the cell surface. J Biol Chem 294:10833–10845

Check for updates

Chapter 11

Chemical Identification of Glycosphingolipid Antigens Recognized by Antibodies

Koichi Furukawa, Yuhsuke Ohmi, Orie Tajima and Keiko Furukawa

Abstract

A number of monoclonal antibodies have been generated aiming to detect markers for tissue development and degeneration, cell differentiation, and malignant transformation. Among antigens recognized by these monoclonal antibodies, glycosphingolipids including gangliosides are included. In this chapter, fundamental methods for chemical identification of ganglioside antigens recognized by antibodies are summarized. After extraction of lipids from cells and tissues using organic solvent, neutral and acidic glycosphingolipids are separated with/without pre-elimination of phospholipids. Extracted glycolipids are applied for thin-layer chromatography combined with immunodetection to identify specific antigen component recognized by the antibody. To finally define the antigen structure, mass spectrometry is effectively used. Confirmation with transfection of supposed glycosyltransferase cDNA has also been a supportive way.

Keywords Glycosphingolipid, Monoclonal antibody, Lipid extraction, Thin-layer chromatography, TLC-immunoblotting, Evaporation

1 Introduction

Importance of glycosphingolipids as target antigens recognized by antibodies has been well known in the research fields of developmental biology, cell differentiation, immunobiology, and cancer biology [1]. In particular, since the end of 1970, monoclonal antibodies reactive with known and unknown target antigens with intriguing specificity have been generated, chemical identities of those antigens have been elucidated with various efforts [2]. Glycosphingolipids are unique molecules mainly expressed on the cell membrane embedding the lipid moiety (ceramide) into the outer layer of the lipid bilayers membrane [3]. Generally, they are not water-soluble, making the biochemical analyses uneasy in contrast with protein antigens. Furthermore, GSLs are not directly encoded by genome, but by glycosyltransferase genes and many other factors involved in the synthesis of their lipid portions, making analyses of GSLs more

Koichi Furukawa (ed.), *Gangliosides*, Methods in Molecular Biology, vol. 3035,
https://doi.org/10.1007/978-1-0716-5288-6_11,

complex. In this chapter, main and standard protocols for chemical identification of glycosphingolipid antigens recognized by monoclonal antibodies are summarized.

2 Materials

2.1 Monoclonal Antibodies Generated Toward Known and/or Unknown Antigens with Interesting Specificity

Either mouse (and rat) or human monoclonal antibodies are examined on the identity of recognized antigens as described below, when various supportive data indicate that the targets are glycosphingolipids (GSLs).

Culture supernatants or mouse ascites, and crude or purified antibodies can be used.

2.2 Cells and Tissues

Prepare ~10^7 or more cultured cells or living cells isolated from body tissues are washed twice with PBS (or plain culture medium). Cell pellets are stored in plastic tubes at −80°C until use. When resected animal tissues are used, they are quickly stocked at frozen state in deep freezer after measuring weights. When enough numbers of cells cannot be obtained, utilization of radioisotope labeling can be considered (*see* **Note 1**) [4].

2.3 Culture Medium

Usually, DMEM or RPMI 1640 containing 7.5%–10% fetal calf serum (FCS). When special medium is required, it should be used according to each occasion. When contamination of FCS components must be avoided, non-serum medium, i.e., ITS (insulin, transferrin, and selenium [Corning ITS+ Premix]) should be used for culture medium.

2.4 Organic Solvents for Extraction of Glycosphingolipids (GSLs)

1. Mixture of chloroform:methanol (C:M) (2:1, 1:1, 1:2 sequentially), i.e., about 200 mL for 1 g (or 0.1 mL) of original samples.
2. Round-bottom flasks (50–500 mL).
3. Small glass tubes (4 mL) with screw caps.
4. Magnetic stirrer and stirring bars.

2.5 Desalting

1. Sep-Pak™C_{18} cartridges (Waters).
2. 10 mL syringe and stand.
3. Chloroform and MeOH: C:M (2:1), (1:1).
4. Round-bottom flask (50 mL, 100 mL).

2.6 Ion-Exchange Column Chromatography

1. DEAE Sephadex™ A-50 (Cytiva), 1.0 g for 0.1–0.5 g starting sample.
2. Glass filter and flask for aspiration.

3. Solvent systems for ion-exchange column: solvent A (C:M:DW = 30:60:8) and solvent B (C:M:0.8 M sodium acetate = 30:60:8).
4. 10 mL syringe, glass wool, and stand.
5. Spatula.

2.7 TLC Plates and Running Apparatus

Glass-backed silica gel plates (e.g., HPTLC silica gel plate 20 cm × 20 cm; Merck HX60732741) (*see* **Note 2**). Organic solvents to be used are C:M:0.2% $CaCl_2$ (60:35:8) for neutral fraction and C:M:0.2% $CaCl_2$ (55:45:10) for acidic fraction. (Alternatively, C:M:2.5 N NH_4OH (60:35:8) can be used for the acidic fraction.)

2.8 Detection Reagents

1. **Orcinol spray (for neutral fraction)**: Orcinol solution (0.16 g orcinol in 20 mL DW; add 60 mL H_2SO_4). Stock at 4°C.
2. **Resorcinol spray (for acidic fraction)**: Resorcinol solution (200 mg resorcinol in 10 mL DW; add 250 μL of 0.1 M $CuSO_4$ in 0.250 g/10 mL DW; add 80 mL HCl). Stock at −80 to −20°C with light shielding.

2.9 Reagents for TLC-Immunoblotting

To identify target GSLs on TLC plate, TLC-immunostaining has been used using TLC plate after direct fixation with PIBM. However, blotting of developed GSLs on TLC plate onto PVDF membrane is easier to perform if the blotter is available. So, this thermal blotting (TLC-immunoblotting) is mainly described here [5], and iron-immunoblotting (*see* **Note 3**), classical TLC-imunostaining methods are also attached(*see* **Note 4**).

1. HPTLC Silica gel 60 F254 (25 aluminum sheets, 20 cm × 20 cm plate) (Millipore HX68052248 or MERCK 1.05548).
2. Glass fiber filter (ATTO AC5972).
3. PTFE membrane (100 mm × 100 mm, AC-5973).
4. Clear blot membrane-P (PVDF membrane, AE-6666).
5. TLC-blot solution: isopropanol:0.2% $CaCl_2$:MeOH = 40:20:7.
6. Antibodies: either hybridoma culture supernatant or ascites (crude or purified).
7. Peroxidase-conjugated secondary antibody.
8. 2% BSA in PBS.
9. PBS.
10. Detection kit, e.g., Konica immunostain kit.
11. TLC-thermal blotter (ATTO AC-5970).
12. Electric iron (alternative for thermal blotter).

2.10 Reagents for Preparative TLC

1. TLC plate with no solid fixation, e.g., TLC Silica gel 60 glass plate (Merck 1.05721 or Milipore HX081299).
2. Primulin spray: 0.1 g primulin in 100 mL of DW (0.1% primulin stock solution). For spray, dilute this stock solution ×100 times in acetone: DW (4:1).
3. UV detector.
4. Scrapers.
5. Ion-exchange resin Dowex 50W-X8 (Dow Chemical Company).
6. 1 N NaOH.
7. Glass filter and aspirator.
8. isoProOH:Hex:H_2O (55:31:14) for extraction.
9. N_2-stream evaporator.
10. 0.1 N NaCl.
11. Sep-Pak C_{18} for desalting.
12. Plastic syringes (10 mL or smaller size).

2.11 Identification of the Chemical Structures

TLC-LESA-MS (AB Sciex) [6].

3 Methods

After extraction of lipids from cell pellets or tissues, the acidic fraction, mainly containing gangliosides, is separated andthen applied for TLC to know whole aspect of the composition. The remaining neutral fraction is preserved for just in case needed to be analyzed. To identify the recognized structure(s) by mAbs, perform TLC-Immunoblotting (or TLC-immunostaining). Then, directly isolate the target components by preparative TLC (Fig. 1).

3.1 Extraction of Lipids Containing Glycosphingolipids

1. Take out frozen cell pellets or tissues, measure the weights when needed, and quickly transfer into a glass bottle (100–100 mL).
2. Add C:M (2:1) of 20 times or more of the sample volume, and stir overnight at room temperature (RT).
3. Filter the extract solution through glass wool/funnel into a round-bottom flask, and return the residue to the glass bottle to sequentially extract using C:M (1:1) of more than ten times volume of the original sample. Stir for more than 2 h at RT, then repeat the same procedure. The filtered extracts are evaporated by rotary evaporator warming at 45°C. After filtration, the residue is extracted again using C:M (1:2) of the same volume for more than 2 h. The filtered extracts are evaporated each time and finally combined in a round-bottom flask.

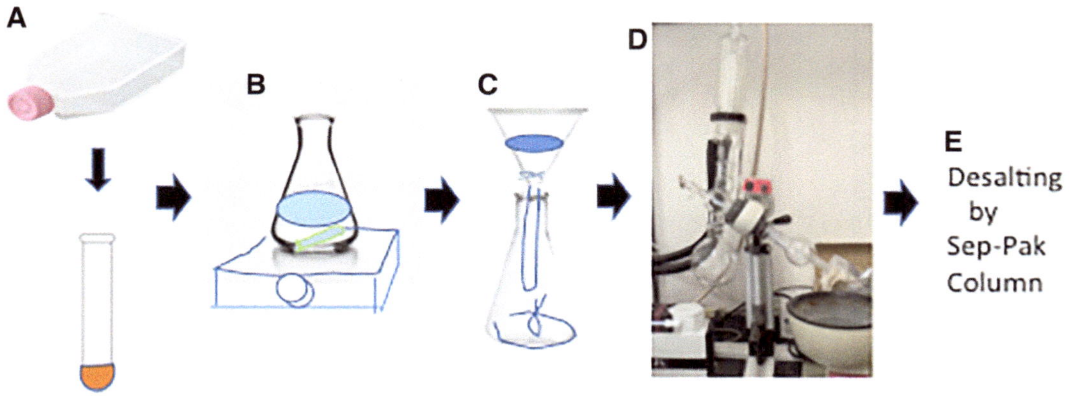

Fig. 1 A scheme of experimental flow for chemical identification of GSLs recognized by antibodies described in this chapter. (**a**) Collection of cell pellets from cultured cells. (**b**) Extraction of lipids using C:M (2:1, 1:1, 1:2, sequentially). (**c**) After each extraction, filter the extracts through glass wool and combine them. (**d**) Evaporate. (**e**) Desalt by Sep-Pak C_{18} cartridge

3.2 Separation of Gangliosides

1. Apply the lipid extracts to mild saponification in 0.1 N NaOH in chloroform/MeOH (1:1) at 37°C for 2 h and then evaporate to dryness [7]. Apply the extracts to ion-exchange chromatography using DEAE-Sephadex column (*see* **Note 4**).

   For separation of phospholipids, use Florisil column after acetylation (*see* **Note 5**). When *O*-acetyl derivatives are important, we need to avoid alkaline exposure for protection of *O*-acetylated gangliosides such as GD3 (*see* **Note 6**) [8].

2. **Desalting**. Prepare Sep-Pak C_{18} column using 10 mL syringe, flow MeOH 10 mL, C:M (2:1) 5 mL, then DW more than 10 mL. Dissolve the extracts in DW about 3 mL by shaking well and pipetting many times. Apply the samples to Sep-Pak C_{18} and push out slowly into the round-bottom flask. Repeat to apply to the column, and then, wash the column by DW. Elute the bound GSLs into a new round-bottom flask using MeOH 2 mL, C:M (2:1) 5 mL, and C:M (1:1) 5 mL, sequentially. Evaporate the combined eluates by rotary evaporator.

3. Separation of neutral and acidic fractions using ion-exchange column chromatography (Fig. 2).

Take DEAE-Sephadex™A50 beads into glass filter, add the solvent B containing 0.8 M sodium acetate. After mixing, leave for 30 min at RT. Aspirate water, add the solvent A and stir well, then aspirate. Repeat the same procedure five times, and cover with parafilm. After preparation of column consisting of 10 mL syringe and glass wool, pack the treated DEAE-Sephadex beads into the syringe column, and wash by ~8 mL of neutral solvent five times. After dissolving the sample in 2 mL of the solvent A, apply it into the

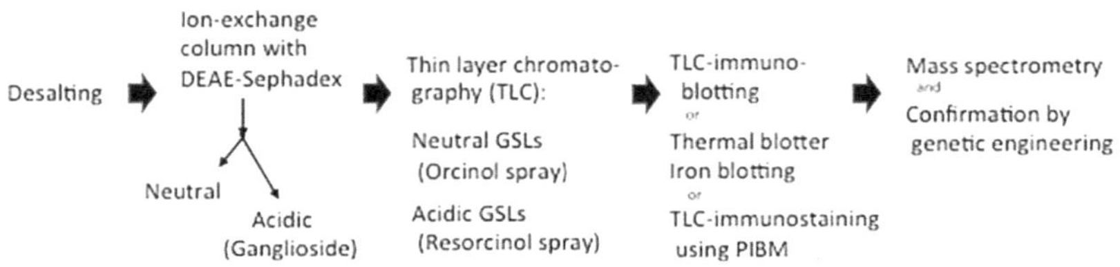

Fig. 2 Experimental flow 2. Separation of neutral and acidic fractions using ion-exchange column chromatography after lipid extraction, then identify the GSLs recognized by antibodies

column, and receive the flow-through fraction into the original flask and apply again. The flow-through fraction is then received in a new round-bottom flask as neutral fraction. Add solvent A ~8 mL ×3 times to completely collect neutral fraction into the same flask. Changing to a new round-bottom flask, 2 mL of solvent B is added to receive eluates as acidic fraction. Then, 8 mL of solvent B is added three times more, collecting acidic fraction in one flask. Then, separated neutral and acidic fractions are evaporated by rotary evaporator completely. Desalt the acidic fraction by Sep-Pak C_{18} column as explained in Sect. 3.2.2, and dry. Finally, transfer the contents into small glass tubes in ~1 mL C:M (1:1) and dry with N_2 stream at 45°C. Dissolve the dried contents in C:M (1:1) with half ~ same volume of the original sample mass.

3.3 TLC and Detection of Bands (see Fig. 2)

Prepare the TLC solvent C:M:0.2% $CaCl_2$ (55:45:10) and glass tank with filter paper inside. After leaving for at least 30 min, TLC plate is put quietly in the tank and develop until the solvent reaches near the plate top. For detection of bands, resorcinol spray is performed for acidic fraction. As for neutral fraction, use TLC solvent C:M:0.2% $CaCl_2$ (60:35:8) and orcinol spray for the detection of bands. After spray by resorcinol, the plate is sealed by glass plate and taping, then heated for 10 min at about 120°C.

3.4 TLC-Immunoblotting by Mainly ATTO Thermal Blotter (Fig. 3)

Basically, detection of target glycolipid on TLC plate has been done by TLC-immunostaining using polyisobutyl methacrylate (*see* **Note 4**). Here, thermal TLC-immunoblotting method is described [5].

Thermal TLC-Immunoblotting

1. Heat TLC plate (HPTLC Silica gel 60 F254 (25 aluminum sheets) after cutting at 10 cm × 10 cm size at 110°C for at least 10 min.
2. Apply samples at 1 cm level from the bottom with 1–2 mm space. Use appropriate amounts, about 0.1–0.5 μg/lane when known structures. Then, dry under dryer for 10 min without heating.

3. Prepare TLC tank with filter paper on one side in inner space by solvent C:M:0.2 % $CaCl_2$ (55:45:10) with 0.5–0.8 cm depth. Leave for at least 30 min.
4. Perform TLC by soaking the TLC plate slowly, leave till the top reach upper edge.
5. Dry at RT for about 10 min, then primulin spray in the draft, and observe bands under UV lamp, mark bands of marker sample and main components.
6. Prepare following materials: glass fiber filter (ATTO AC5972), PTFE membrane (100 mm × 100 mm, AC-5973), clear blot membrane-P (PVDF membrane, AE-6666), TLC-blot solvent: isopropanol:0.2% $CaCl_2$: MeOH = 40:20:7.
7. Soak the membrane in TLC-blot solvent for 40 s.
8. Set TLC-blotter after switching on at 180°C for 30 s with pressure 8.
9. Layer prepared membranes as follows from the bottom: glass filter/TLC plate/membrane-P/PTEF membrane/glass filter (see Fig. 3), then heat-blot for 30 s at 180°C with pressure 8 (ATTO TLC-blotter).
   Alternatively, use iron (*see* **Note 3**).
10. Take out membrane and mark by pencil when needed, and soak the membrane in blocking solution (2% BSA in PBS) for 2 h at RT or O/N at 4°C.
11. After brief washing the membrane with PBS, apply first antibody and incubate for 1 h at RT (or O/N at 4°C). After washing with PBS 5 min, three times, apply peroxidase-conjugated secondary antibody and incubate for 1 h at RT.

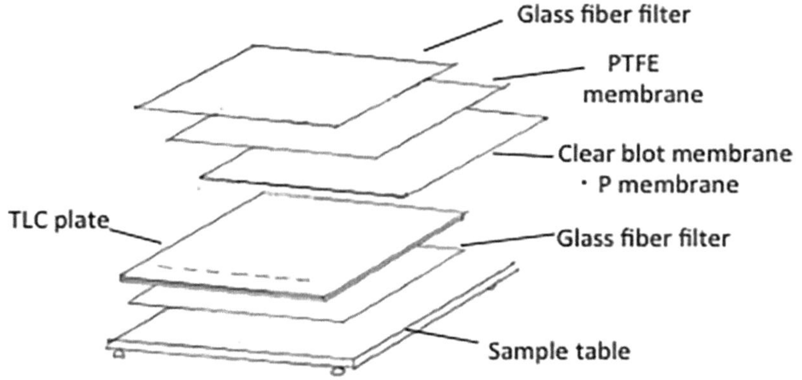

Fig. 3 TLC-immunoblotting using a TLC-thermal blotter. Using a thermal blotter, layer following components on the sample table from the bottom: Glass fiber filter, TLC plate, clean blot membrane (P membrane), PTFE membrane, and glass fiber filter, then blot GSLs at 180°C for 30 s

12. After washing with PBS three times, detect stained bands using current detection kit, e.g., Konica immunostain kit.

When TLC-blotter is not available, iron-blotting method (Fig. 4) or immunostaining with PIBM fixation can be used instead (*see* **Notes 4 and 6**).

3.5 Preparative TLC and Separation of Bands [9]

When the GSL bands are clearly shown in TLC-immunostaining, the bands can be directly isolated from TLC silica gels followed by purification from silica powder.

1. After thin layer chromatography is done, dry the plate in air. Do not dry completely.
2. Primulin spray is done on the TLC plate, and mark stained bands under UV quickly.
3. Scrape each band and put into glass tube after determination of the weight of gel.
4. Add a half-weight Dowex 50W-X8 (Na^+-form*).
5. Extract with isoProOH:Hex:H_2O (55:31:14) using 2 mL/0.1 g silica gel.
   Sonicate for a few minutes in water, then stir with magnetic stirrer.
6. Take the supernatant after centrifugation at 2000–3000 rpm for 15 min.
7. Repeat step 6, and combine the extracts and evaporate by N_2 stream.

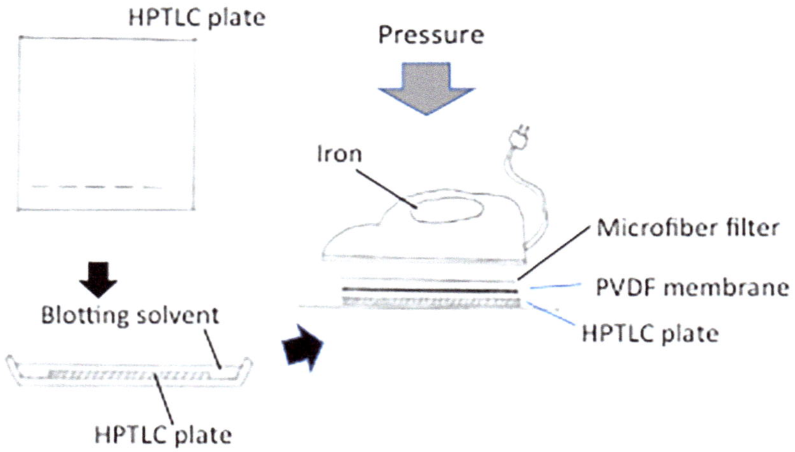

Fig. 4 TLC-iron-blotting method. Instead of TLC-thermal blotter, electric iron can be used for GLC blotting from TLC plate to PVDF membrane. After soaking the TLC plate in blotting solvent (isopropanol:0.2% $CaCl_2$:MeOH = 40:20:7), iron-blotting will be performed by layering HPTLC plate, PVDF membrane, and microfiber filter. Push iron and heat at 180°C for 30 s

8. Suspend the residue in 0.1 N NaCl at 300 μg glycolipids/mL with sonication.
9. Centrifuge at 2500 rpm for 15 min.
10. Supernatants are collected by repeating the same procedure 3 times.
11. Desalt by Sep-Pak C18 column.
12. Confirm separated components on TLC by primulin spray (or resorcinol spray)

*To convert H^+-form Dowex 50W-X8 to Na^+-form, add 2–3 g resin in H_2O on glass filter, wash and add 1 N NaOH four to five volume, and wash with H_2O four to five times.

3.6 Biochemical Characterization of Separated Bands on TLC (TLC-LESA-MS)

Directly analyze the structure of each band by MS from TLC-PVDF membrane-blotted sample using ESI robotic ion source TriVersa NanoMate LESA™system and QTRAP 6500™ (AB Sciex) (*see* **Note 7**) [6].

3.7 Confirm Defined Structures by Genetic Engineering (see Note 8) [10]

According to the identified ganglioside structures as mentioned above, we can confirm the supposed chemical structures based on the genetic engineering of ganglioside profiles on some appropriate cell lines (Fig. 5). As we have reported genetic engineering of glycosphingolipids in cultured cell lines [10–12], supposed structures can be reconstituted and detected by antibodies on the cell surface. This process strongly supports the results of TLC-immunoblotting/MS analysis. One more meaning is to confirm the reactivity of the defined GSLs with monoclonal antibodies in both membrane-attached form and cell surface-expressed condition. Sometimes, these are not completely corresponding, leaving further and fine analysis [6].

4 Notes

1. When the number of cells are too low, radioisotope-labeling of GSLs can be used by [^{3}H]-glucosamine or [^{3}H]-galactose metabolic labeling [4]. Although the experimental size is very small, final detection step is high. Users need to be familiar with radioisotopic procedures. In addition, enzymatic labeling of glycolipids such as globotetraosylceramide (Gb4) is also promising [13].
2. **Various TLC plate:** (10 cm × 10 cm or 20 cm × 10 cm) plate can be also used depending on the sample number and/or required resolution.
3. **Iron-immunoblotting:** Using iron heat blotting, sufficient transfer of GSLs from TLC to PVDF membrane can be achieved. Preparation needed for iron-blotting is almost same

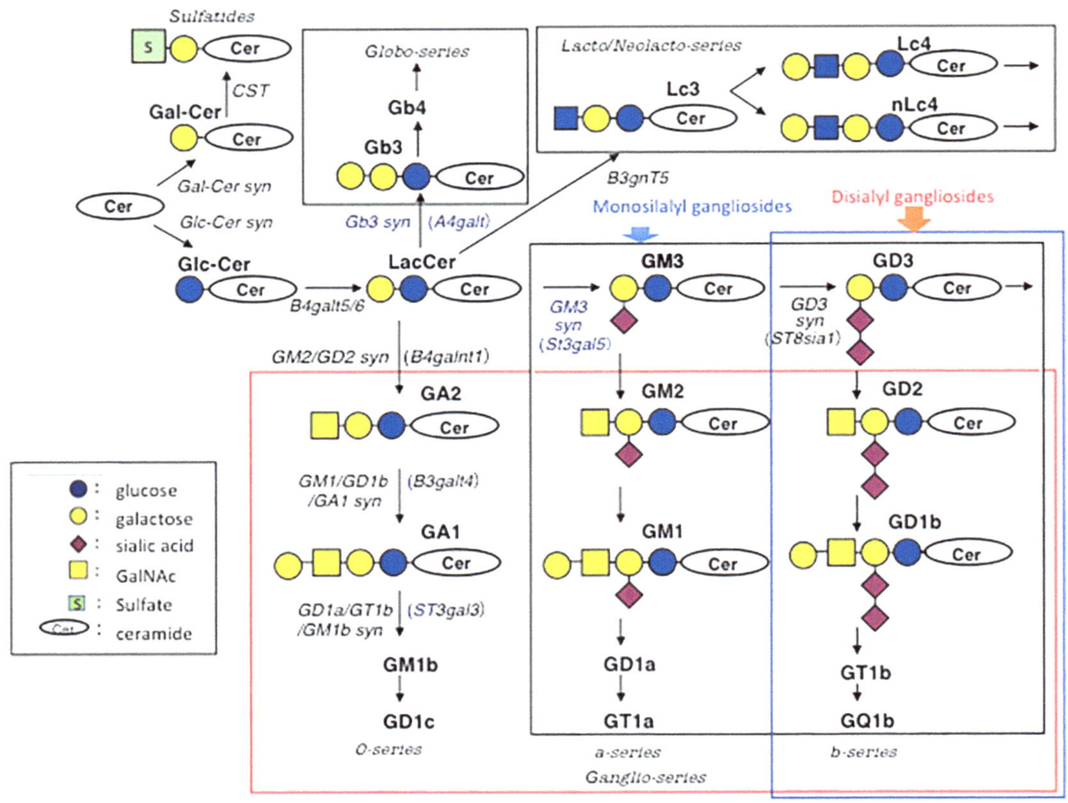

Fig. 5 Synthetic pathway of gangliosides and major GSLs. Depending on this scheme, genetic engineering of expression profiles in some cell lines (or animals) can be performed. Key enzymes involved in the synthesis of gangliosides are shown. Pathways for synthesis of sulfatides, globo-series, and lacto/neolacto-series GSLs are also shown, although they are not described in this chapter

as TLC-blotting as mentioned above: glass fiber filter, PTFE membrane (100 mm × 100 mm), PVDF membrane, TLC-blot solvent: isopropanol:0.2% $CaCl_2$:MeOH = 40:20:7. Soak the membrane in TLC-blot solvent for 40 s.

Set iron at 180°C in advance, then layer prepared membranes as follows from the bottom: glass filter/TLC plate/PVDF membrane/PTEF membrane/glass filter (see Fig. 2), then heat-blot for 30 s at 180°C with pressure by pressing iron.

Immediately after taking out membrane, mark the reverse side by pencil on which antibody should be applied. Block the membrane using 2% BSA in PBS at RT for 1 h. Then, start antibody staining as described above.

4. **TLC-immunostaining:** Instead of thermal blotter-mediated membrane blotting, alternative way using PIBM without thermal blotting is still useful (TLC-immunostaining) [14–16].

Prepare 0.4% polyisobutylmethacrylate solution (PIBM): 2.5% PIBM in $CHCl_3$ solution dilute to 0.4% with hexane, 2% BSA in PBS, PBS, monoclonal antibody, HRP-secondary antibody, and detection system such as Konica immunostain kit. Run TLC using aluminum-backed silica gel F254 sheet (Merck) in C:M:2.5 N NH_4OH (60:35:8). After drying, the sheet was dipped for 1 min in n-hexane saturated with polyisobutylmethacrylate (Polyscience Inc), then sprayed with 2% BSA/PBS (solution A). After being immersed in solution A for 2–3 h, add monoclonal antibody and incubate for 3 h at RT. Thereafter, add HRP-secondary antibody after washing, and let it bind as described above. Finally, detect bands using current detection kit, e.g., Konica immunostain kit.

5. **Florisil column after acetylation**. To keep phospholipids, Florisil column after acetylation should be used, i.e.,

   Extract lipids by C:M (2:1, 1:1, 1:2 sequentially), combine the extracts in round-bottom flask, and evaporate. To acetylate the extracts, dry completely in vacuo over P_2O_5 for 3 h. Add pyridine ~3 mL and acetic anhydride (2 mL) and allow to react in stoppered flask for 18 h at RT. Stop reaction with MeOH (50 mL), and transfer solution to large round-bottom flask and add 450 mL toluene. Evaporate to dryness by rotary evaporator.

   Separation of glycolipids and phospholipids. Prepare Florisil column with 29 g of Florisil either dry or as a slurry in hexane: dichloroethane (DCE) (1:4). If using dry column, wash column with hexane:DCE (1:4) before using. Dissolve sample in hexane: DCE (1:4), 1– 2 mL and apply to column. Elute column as follows:

   A. hexane:DCE (1:4) ~200 mL
   B. DCE ~200 mL
   C. DCE:acetone (1:1) ~200 mL (glycolipid fraction)
   D. DCE:MeOH:water (2:8:1) ~200 mL (phospholipid fraction)

   For isolation of GSLs, evaporate fraction C by rotary evaporator, transfer sample with C:M (1:1) to round-bottom flask (100 mL), and re-evaporate.

   Deacetylation: Dissolve sample in C:M (2:1) 2 mL, and add 0.5% Na methoxide in MeOH 1 mL (prepared by dissolving 0.5 g of sodium metal in MeOH 100 mL). Allow to react in stoppered flask for 90 min. Add ethyl acetate (30 mL) and evaporate to dryness.

   After dissolving residue in water (1–2 mL), desalt using Sep-Pak C_{18} cartridge. Then, go to separation into neutral and acidic fractions with DEAE-Sephadex column.

6. Need to avoid alkaline exposure for protection of *O*-acetylated derivatives such as 9-*O*-acetyl GD3 from degradation. Alkaline exposure removed one of major gangliosides, 9-*O*-acetyl GD3 in the acidic fraction of brain GSLs of B4galnt1 KO mice [8].
7. Using PVDF membrane-blotting samples, each component can be applied to MS with QTRAP 6500™ (AB Sciex) [6].
8. Using cell lines with similar ganglioside profiles and expression vectors of glycosyltransferase cDNAs [10]. Basically, cell line expressing only GM3 is easy to use as a prototype. Even cell line expressing no GM3 is also convenient to start from early stage. As for cDNA expression vectors, GM2/GD2 synthase [17], GD3 synthase [18], GM3 synthase [19], and GM1/GD1b synthase [20] genes are needed. Species of cells and cDNAs are not strict, i.e., murine or human origin materials are equally useful.

Acknowledgments

This work was supported by grants-in-aid from the Ministry of Education, Culture, Sports, Science and Technology (MEXT) of Japan (19K22518, 19K07393, 21K26828) and JST-CREST (JPMJCR17H2). This study was also supported by the Interdisciplinary Joint Research Program of the J-GlycoNet cooperative network, which was accredited by the MEXT. We also thank N. Hashimoto, Y. Ohkawa, K. Hamamura, R. H. Bhuiyan, F. Yesmin, K. Yuji Kondo, Kaneko, M. Kambe, T. Ito, Y. Imao, and all students in our laboratory. We also thank M. Kojima for excellent assistance.

Conflict of Interest The authors have no conflicts of interest to declare that are relevant to the content of this chapter.

References

1. Yu RK, Tsai YT, Ariga T (2012) Functional roles of gangliosides in neurodevelopment: an overview of recent advances. Neurochem Res 37:1230–1244
2. Furukawa K, Lloyd KO (1990) Gangliosides in melanoma. In: Ferrone S (ed) Human melanoma: from basic research to clinical application. Springer-Verlag, New York, pp 15–30.
3. Hakomori S (2002) Glycosylation defining cancer malignancy: new wine in an old bottle. Proc Natl Acad Sci USA 99:10231–10233
4. Furukawa K, Arita Y, Satomi N et al (1990) Tumor necrosis factor enhances GD3 ganglioside expression in cultured human melanocytes. Arch Biochem Biophys 281:70–75
5. Ohkawa Y, Momota H, Kato A et al (2015) Ganglioside GD3 enhances invasiveness of gliomas by forming a complex with platelet-derived growth factor receptor α and Yes kinase. J Biol Chem 290:16043–16058
6. Hashimoto N, Ito S, Tsuchida A et al (2019) The ceramide moiety of disialoganglioside (GD3) is essential for GD3 recognition by the sialic acid-binding lectin SIGLEC7 on the cell surface. J Biol Chem 294:10833–10845
7. Tajima O, Fujita Y, Ohmi Y et al (2023) Ganglioside GM3 prevents high fat diet-induced hepatosteatosis via attenuated insulin signaling pathway. PLoS One 18:e0281414
8. Furukawa K, Aixinjueluo W, Kasama T et al (2008) Disruption of GM2/GD2 synthase gene resulted in overt expression of 9-O-acetyl GD3 irrespective of Tis21. J Neurochem 105:1057–1066

9. Bhuiyan RH, Kondo Y, Yamaguchi T et al (2016) Expression analysis of 0-series gangliosides in human cancer cell lines with monoclonal antibodies generated using knockout mice of ganglioside synthase genes. Glycobiology 26:984–998
10. Ohmi Y, Kambe M, Ohkawa Y et al (2018) Differential roles of gangliosides in malignant properties of melanomas. PLoS One 13:e0206881
11. Tsuchida A, Senda M, Ito A et al (2018) Roles of GalNAc-disialyl lactotetraosyl antigens in renal cancer cells. Sci Rep 8:7017
12. Furukawa K, Ohmi Y, Hamamura K et al (2022) Signaling domains of cancer-associated glycolipids. Glycoconj J 39:145–155
13. Kondo Y, Ikeda K, Tokuda N et al (2013) TLR4-MD-2 complex is negatively regulated by an endogenous ligand, globotetraosyl-ceramide. Proc Natl Acad Sci USA 110:4714–4719
14. Magnani JL, Smith DF, Ginsburg V (1980) Detection of gangliosides that bind cholera toxin: direct binding of 125I-labeled toxin to thin-layer chromatograms. Anal Biochem 109:399–402
15. Young JrWW, Johnson HS, Tamura Y et al (1983) Characterization of monoclonal antibodies specific for the Lewis a human blood group determinant. J Biol Chem 258:4890–4894
16. Taki T, Ishikawa D, Handa S et al (1995) Direct mass spectrometric analysis of glycosphingolipid transferred to a polyvinylidene difluoride membrane by thin-layer chromatography blotting. Anal Biochem 225:24–27
17. Nagata Y, Yamashiro S, Yodoi J et al (1992) Expression cloning of beta 1,4 N-acetylgalactosaminyltransferase cDNAs that determine the expression of GM2 and GD2 gangliosides. J Biol Chem 267:12082–12089. (corrected in 1994)
18. Haraguchi M, Yamashiro S, Yamamoto A et al (1994) Isolation of GD3 synthase gene by expression cloning of GM3 alpha-2,8-sialyltransferase cDNA using anti-GD2 monoclonal antibody. Proc Natl Acad Sci USA 91:10455–10459
19. Fukumoto S, Miyazaki H, Goto G et al (1999) Expression cloning of mouse cDNA of CMP-NeuAc:Lactosylceramide alpha2,3-sialyltransferase, an enzyme that initiates the synthesis of gangliosides. J Biol Chem 274:9271–9276
20. Miyazaki H, Fukumoto S, Okada M et al (1997) Expression cloning of rat cDNA encoding UDP-galactose:GD2 beta1,3-galactosyltransferase that determines the expression of GD1b/GM1/GA1. J Biol Chem 272:24794–24799

Part III

Interacting Molecules and Functions

Chapter 12

Elucidation of Ganglioside Microdomains Using Proximity Labeling

Norihiro Kotani, Arisa Miyagawa-Yamaguchi and Koichi Honke

Abstract

It has long been known that lipid microdomain structures (also called "lipid rafts") are formed in the cell membrane, and gangliosides are concentrated in these domains. Many functional molecules (membrane proteins, receptors, and their associated signaling molecules, etc.) are concentrated in these domains along with gangliosides, and it has been pointed out that these domains play an important role in signal transduction, for instance, from outside the cell. Interestingly, it has been pointed out that each microdomain contains different types of gangliosides and functional molecules, and it is highly likely that the functions of the microdomain structures are different from each other. However, it has been difficult to specifically analyze individual microdomains, for instance, such as the microdomain composed of GM1, whereas the methods for analyzing all microdomains existed in same cell have been established. Here, we explain a method for analyzing "individual" microdomains by each ganglioside using a proximity labeling technique.

Keywords Proximity labeling, Proximity proteomics, Lipid raft, Enzyme-mediated activation of radical sources, Tyramide, Arylazide, Plasma membrane, Horseradish peroxidase, Lipid–protein interactions

1 Introduction

The cell membrane is not a uniform lipid bilayer but forms an ordered structure in which specific molecules are locally associated. These regions are collectively called "lipid microdomains (or lipid raft)" [1, 2]. Lipid microdomains are dynamic structures in which specific functional molecules repeatedly associate and disperse on a millisecond basis and function as platforms for important biological phenomena such as signal transduction and intracellular protein transport [3, 4]. For example, molecules involved in signal transduction and endocytosis proximal to a cell growth factor receptor temporarily form a functional cluster. A lot of studies have revealed that microdomains containing gangliosides also contribute significantly to signal transduction mechanisms (therefore, from here, lipid microdomains

Koichi Furukawa (ed.), *Gangliosides*, Methods in Molecular Biology, vol. 3035,
https://doi.org/10.1007/978-1-0716-5288-6_12,

are referred to as "ganglioside microdomains"). In the unstimulated state, ganglioside microdomains are small, less than a few tens of nanometers (containing several to several tens of protein molecules), but once stimulated, they are known to form large molecular clusters of several hundred nanometers to 1 μm (Fig. 1).

Ganglioside microdomains are often prepared by homogenizing cell membranes using a neutral detergent at low temperature and applying to a sucrose density gradient ultracentrifugation, and then collecting the low-density fraction containing detergent-resistant membrane (DRM) [5]. However, it has been pointed out by some researchers that DRMs do not reflect physiological conditions and are artifacts caused by detergents [6]. In addition, since heterogeneous and diverse ganglioside microdomains are collected together, it should not be assumed that the molecules collected in DRMs are associated in the same functional cluster (Fig. 2).

Therefore, when it comes to determining what molecular interactions cause a certain biological phenomenon, or to confirm whether molecular interactions observed in experiments occur in living cells, it is expected that DRM analysis alone may not be sufficient. In contrast to this, some technologies for observing interactions in real time include fluorescence resonance energy transfer (FRET) and single-molecule tracking. However, these experiments cannot be performed unless the interacting molecules are known in advance. Observations by immunoelectron microscopy are also highly reliable, but this cannot be done unless the associated molecule is known prior to the experiment.

We have previously developed one of novel protein labeling technologies called "Proximity labeling." Our technology is based on radical reaction using the arylazide, typically used as a cross-linking reagent, or tyramide compound resulting in converting to aryl nitrene or phenoxy radicals (active form) by horseradish peroxidase (HRP) (Fig. 3). This is now referred to as an "enzyme-mediated activation of radical sources" (EMARS) reaction [7–9].

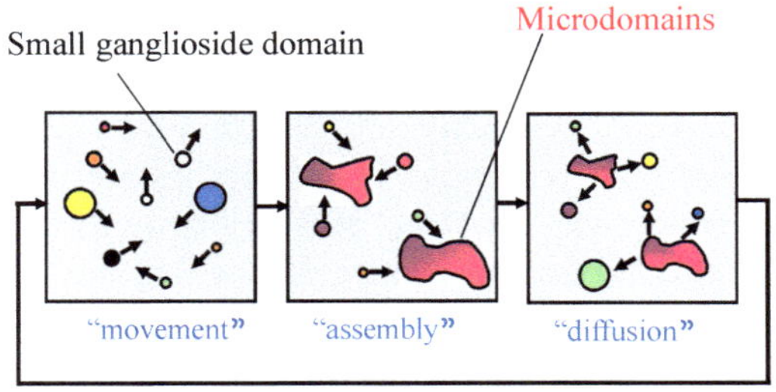

Fig. 1 Summary of ganglioside microdomain cycle on the cell surface. The model displays how ganglioside microdomains are formed in the membrane. Small ganglioside domains are constantly moving around on the cell membrane ("movement"), and some molecules containing protein complexes, resulting in biological functions ("assembly"). Complex formation is reversible, and each domain diffuses and disperses again

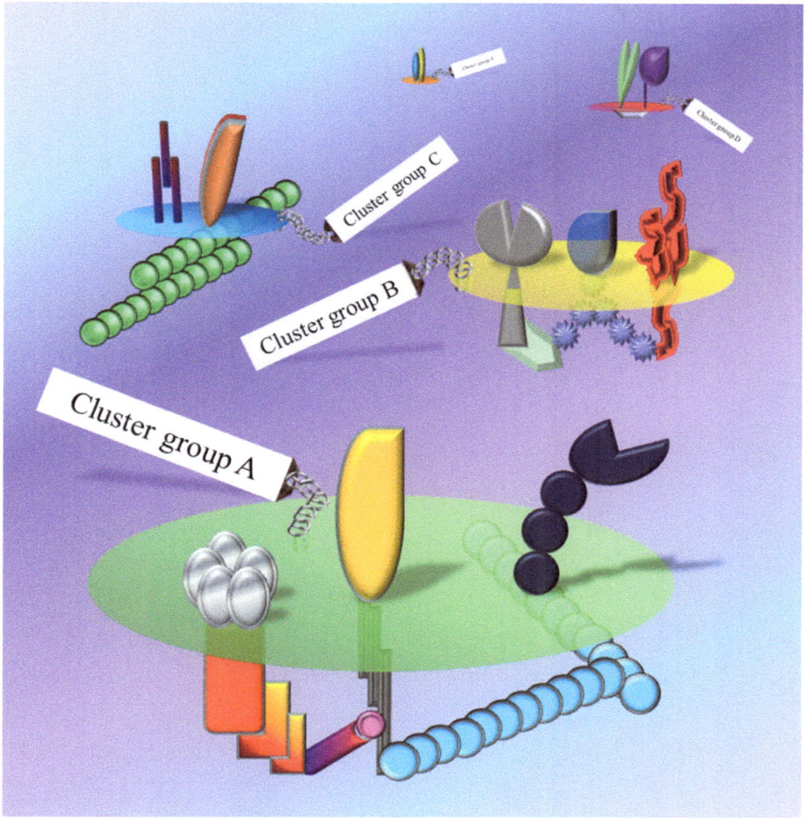

Fig. 2 Heterogeneous and diverse ganglioside microdomains. Each ganglioside microdomain contains different membrane proteins. Therefore, if each cluster is collected and analyzed together under DRM, important information regarding differences in clustered molecules cannot be observed

Because radical is typically a short lifespan, aryl nitrene or phenoxy radicals only exist in very close proximity to HRP. Therefore, if target molecule expressed in the cell surface is treated with an HRP-conjugated recognizing molecule (in the case of ganglioside, ganglioside-recognizing antibodies or lectins) and then with an aryl azide or tyramide compound, proximal molecules which are associated with ganglioside can be specifically labeled with covalently bonded tags (Fig. 3). In EMARS reaction, proximal proteins within 300 nm (aryl azide reagent) or 20 nm (tyramide reagent) from HRP were labeled [8, 9]. This is suitable for identifying molecules that interact in respective ganglioside microdomains, suggesting that the EMARS reaction is an important technique for resolving some problems in conventional experiments. We can also identify EMARS-labeled proteins by proteome [10]. In conjunction with these, effective analysis is possible; we have carried out various experiments using our EMARS method [11–19]. The EMARS method using GPI-anchored

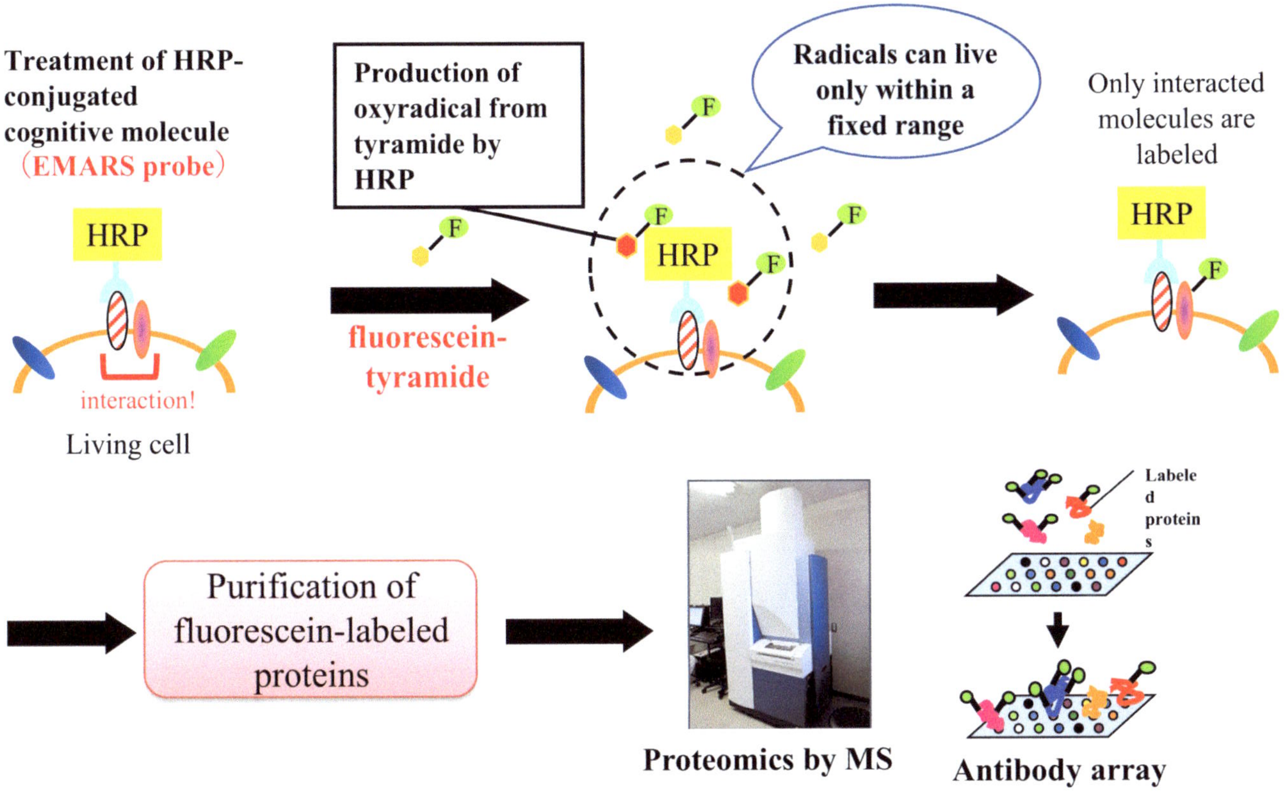

Fig. 3 EMARS, aproximity labeling strategy. The protocols for proximity labeling and proteomes with EMARS reagent (Tyramide-fluorescein). The HRP-conjugated antibody recognizing a given molecule you want to analyze is treated, followed by a tyramide-fluorescein treatment. The reaction proceeds under physiological conditions, and the fluorescein group attaches to proximity proteins within approximately 10 min. They are purified using Anti-fluorescein antibody-conjugated Sepharose and identified via an antibody array or MS-based proteome

HRP fusion proteins is able to segregate individual microdomains associated with a particular GPI-anchored protein [11, 12].

In this chapter, we introduce experimental procedures for "proximity labeling for microdomain analysis" in the cell surface (cell membrane) ganglioside. The first section introduces membrane proximity labeling for ganglioside GM1 and microdomain assay on the cell surface using EMARS, and the second introduces mass spectrometry-based identification of labeled proteins by EMARS for proteomics analysis of these labeling proteins.

2 Materials

2.1 EMARS Probe Preparation

1. Peroxidase labeling kit SH (Dojindo).
2. Peroxidase labeling kit NH2 (Dojindo).
3. HRP-conjugated anti-ganglioside antibody: Label with HRP using Peroxidase labeling kit SH or NH2.
4. Fluorescein (FITC)-conjugated anti-HRP antibody: Transfer an appropriate amount of the anti-HRP antibody (20–100 μ g) to a new 1.5 mL microtube. Add 5 μL of 1 mg/mL FITC solution (0.5 M bicarbonate buffer; pH 9.0). Incubate at room temperature for 1 h. For removing excess FITC, transfer total solutions to a PBS-equibilized 1.2 mL Sephadex G-50 column. Centrifuge at 1500 × *g* for 1 min to spin down the solution. Collect the path-through fraction containing the fluorescein-labeled antibodies. Store at 4°C.

2.2 EMARS Reaction

1. Culture cells: HeLa S3 cells or appropriate cells for your experiment.
2. Cell culture medium: Add 5% or 10% Fetal Bovine Serum and antibiotics.
3. Phosphate Buffered Saline (PBS).
4. HRP-conjugated Cholera Toxin B Subunit (LIST Biological Lab).
5. Anti-ganglioside antibody: Label with HRP using Peroxidase labeling kit SH or NH2.
6. EMARS reagent (Tyramide-fluorescein): Dissolve 100 mg NHS-fluorescein with 5 mL Dimethylacetamide (DMAA). Dissolve 100 mg tyramine with 5 mL DMAA. Mix the NHS-fluorescein and tyramide solutions (total 10 mL). Incubate the mixture at 25°C overnight with rotation. Apply the 5 mL AG50W-X8 cation exchange resin to remove excess tyramide. Apply the aliquot of the pass-through fraction to thin-layer chromatography (TLC) to investigate whether the excess

tyramide is removed. Perform TLC (chloroform: methanol: water = 65:25:4) to confirm the removal of excess tyramine. After confirming the removal of tyramine, dispense the mixture solution and store at −80°C.

7. Hydrogen peroxide (35% H_2O_2).
8. 100 mM Tris–HCl (pH 7.5): Weigh 12.1 g Tris–HCl and prepare a 1 L solution.
   Add protease inhibitor cocktail (optional).
9. Disposable plastic syringe equipped with a 21–26 G needle.
10. Dounce homogenizer.
11. NP-40 lysis buffer: Mix with 20 mM Tris–HCl (pH 7.4), 150 mM NaCl, 5 mM ethylenediaminetetraacetic acid (EDTA), 1% NP-40, 10% glycerol, and a protease inhibitor cocktail (optional).
12. Chloroform/methanol: Mix Chloroform and methanol (2:1 v/v).

2.3 Purification of EMARS Products

1. 100 mM Tris–HCl (pH 7.5): Weigh 12.1 g Tris–HCl and prepare a 1 L solution.
2. Chloroform/methanol: Mix Chloroform and methanol (2:1 v/v).
3. 50% Ethanol solution.
4. Sodium dodecyl-sulfate (SDS) lysis solution: Mix 1% SDS with 100 mM Tris–HCl (pH 7.4).
5. NP-40 lysis buffer.
6. Anti-fluorescein antibody-conjugated Sepharose: When analyzing the labeled EMARS products by proteome analysis, the purification and concentration of these products are required. Be sure to bring NHS-Activated Sepharose Fast Flow (Cytiva) to room temperature before use. Add 5 mL of coupling buffer (0.2 M $NaHCO_3$ and 0.5 M NaCl (pH 8.3)) to a 15 mL tube. Add NHS-Activated Sepharose (appropriate to make 2 mL of beads (net)). Centrifuge at 800 × *g* for 5 min. Remove the supernatant. Wash with 5 mL coupling buffer and discard the supernatant (repeated four times). Add 5 mL coupling buffer and anti-fluorescein antibody (1 mg). Rotate the tube at room temperature overnight. Centrifuge at 800 × *g* for 5 min. Remove the supernatant. Add 5 mL blocking buffer (0.5 M ethanolamine and 0.5 M NaCl (pH 8.3)). Mix and then remove the supernatant. Add 5 mL blocking buffer and rotate the tube at room temperature for 1 h. Centrifuge at 800 × *g* for 5 min and remove the supernatant. Wash with 5 mL PBS (repeat three times). Add PBS to make 10 mL total (a 20% slurry). Stock slurry at 4°C.

7. 50 mM NH_4HCO_3 buffer: Dissolve 0.4 g NH_4HCO_3 in 100 mL water.

2.4 Peptide Sample Preparation and Proteome Analysis

1. EMAERS products elution reagent: 1:1 v/v mixture of Solution B in the MPEX PTS reagent kit (GL Science) and SDS lysis solution (pH 7.4).
2. 10 mM DTT/50 mM NH_4HCO_3 solution: Dissolve 1.54 mg dithiothreitol in 1 mL 50 mM NH_4HCO_3 buffer.
3. 100 mM Iodoacetamide/50 mM NH_4HCO_3 solution: Dissolve 1.84 mg Iodoacetamide in 1 00 μL 50 mM NH_4HCO_3 buffer.
4. SDS-eliminant solution (ATTO).
5. 20 mM $CaCl_2$ solution.
6. Trypsin solution: Dissolve 1 μg Trypsin Gold MS grade (Promega) in 1 μL 50 mM acetic acid.
7. Ethyl acetate containing trifluoroacetic acid (TFA): Mix 365 μ L ethyl acetate with 3.5 μL TFA.
8. 5% Acetonitrile (ACN)/0.1% TFA solution: 50 μL ACN, 950 μL water, 1 μL TFA.
9. 2% Acetonitrile (ACN)/0.1% TFA solution: 20 μL ACN, 980 μL water, 1 μL TFA.
10. GL-Tip SDB (GL Science).
11. nano LC-ESI-MS/MS: LTQ Orbitrap XL equipped with Proteome Discoverer ver. 2.4 or higher (Thermo Fisher Scientific).

3 Methods

3.1 Preparing "EMARS Probe" Recognizing Gangliosides

EMARS reactions require "EMARS probe" which aims to conjugate HRP to the target ganglioside. This is usually an HRP-labeled antibody or recognizing proteins (e.g., lectin) for target gangliosides to be analyzed. It can be purchased and used if it is commercially available. In this case, it is unnecessary to prepare a special EMARS probe in advance. In this chapter, we use commercially available HRP-conjugated cholera Toxin B subunit (recognizing ganglioside GM1) and HRP-conjugated anti-ganglioside antibodies (Abs) as EMARS probes. Here, we also introduce how to prepare HRP-conjugated antibody for the latter HRP-conjugated anti-ganglioside Abs (*see* **Notes 1 and 2**).

1. Transfer an appropriate amount of an anti-ganglioside antibody (not exceeding 200 μg) to a filter tube unit (included in the peroxidase labeling kit SH or NH_2).
2. Centrifuge at 8000 × *g* for 10 min.
3. Add 100 μL reducing reagent (peroxidase labeling kit SH) then incubate at 37°C for 30 min if use the peroxidase labeling kit SH.
4. Add 100 μL of solution B (included in peroxidase labeling kit SH), then centrifuge at 8000 × *g* for 10 min if using the peroxidase labeling kit SH.
5. Add 200 μL of solution B, respectively, then centrifuge at 8000 × *g* for 10 min if using the peroxidase labeling kit SH.
6. Add SH- or NH_2- Reactive peroxidase solution (included in the peroxidase labeling kit SH or NH_2).
7. Incubate at 37°C for 2 h.
8. Add 100 μL of solution A (included in the peroxidase labeling kit SH) or Washing Buffer (included in the peroxidase labeling kit NH_2).
9. Centrifuge at 8000 × *g* for 10 min.
10. Add the appropriate storage buffer (included in the peroxidase labeling kit SH or NH_2) and recover the conjugates into a new 1.5 mL microtube.
11. Store at 4°C.
12. To confirm the EMARS probe, treat the prepared EMARS probe, HRP-conjugated anti-ganglioside antibody, with appropriate culture cells that express the antigen ganglioside, and then Fluorescein (FITC)-conjugated anti-HRP antibody.
13. Observe the binding of prepared EMARS probes using flow cytometry or fluorescence microscopy.

3.2 EMARS Reaction

EMARS probe and EMARS reagent are ready; you can perform the EMARS reaction. Typically, we perform cell-based EMARS reaction targeting the cell membrane ganglioside, but in some cases, it can also be applied to living organisms and materials, such as tissue sections or extracellular vesicles. This section introduces the proximity labeling for the GM1 microdomain-resident proteins (proximal proteins around GM1) by EMARS using CTxB and anti-ganglioside Ab probes.

1. Prepare (primary) cultured cells (approximately 80% of the confluent culture is recommended in a 10 cm plastic dish). In this section, we use cervical epithelioid carcinoma (HeLaS3) cells for GM1 analysis (*see* **Notes 3 and 4**).
2. Wash the cells once with PBS.

3. Add HRP-conjugated Cholera Toxin B Subunit or HRP-conjugated anti-ganglioside antibody dissolved in an appropriate buffer (usually in PBS, 1–2 μg/mL; *see* **Note 5**).
4. Incubate at a suitable temperature (4°C, 25°C, or 37°C) for 20 min (*see* **Note 6**).
5. Wash the cells three times with PBS.
6. Add 0.1 mM of EMARS reagent with H_2O_2 (final concentration: 0.00004% to 0.00008%) in PBS (*see* **Note 7**).
7. Incubate at a suitable temperature (4°C, 25°C, or 37°C) for 10 min.
8. Wash the cells three times with PBS.
9. Add 300 μL of 100 mM Tris–HCl (pH 7.5) with a protease inhibitor cocktail (optional).
10. Harvest the cells using a scraper.
11. Transfer the harvested cells to a microtube.
12. Homogenize the cells using a disposable plastic syringe equipped with a 21–26 G needle (10–15 strokes) or a Dounce homogenizer.
13. Transfer the lysate to a new microtube. If you need to remove the nuclei, centrifuge at 800 × *g* for 5 min at 4°C, then transfer the supernatant to new tubes.
14. Centrifuge at 20,000 × *g* for 15 min at 4°C to collect the microsome fraction containing the membrane.
15. Discard the supernatant and wash the pellet (microsome fraction) once with 500 μL of 100 mM Tris–HCl (pH 7.5).
16. Solubilize the pellet with the NP-40 lysis buffer or add a Chloroform/methanol for mass spectrometry analysis (*see* **Note 8**).
17. Apply the samples to several subsequent experiments (sodium dodecyl-sulfate polyacrylamide gel electrophoresis (SDS-PAGE), western blot, immunoprecipitation, or antibody array; *see* **Note 9**) (Fig. 4).

3.3 Purification of EMARS Products

1. Collect membrane pellets (microsome fraction) after EMARS (see Sect. 3.1).
2. Wash the pellet once with 500 μL of 100 mM Tris–HCl (pH 7.5) and discard the supernatant.
3. Add 500 μL of chloroform/methanol to the microsome pellet and mix gently.
4. Add 500 μL of distilled water and mix gently.
5. Centrifuge at 20,000 × *g* for 2 min at room temperature.

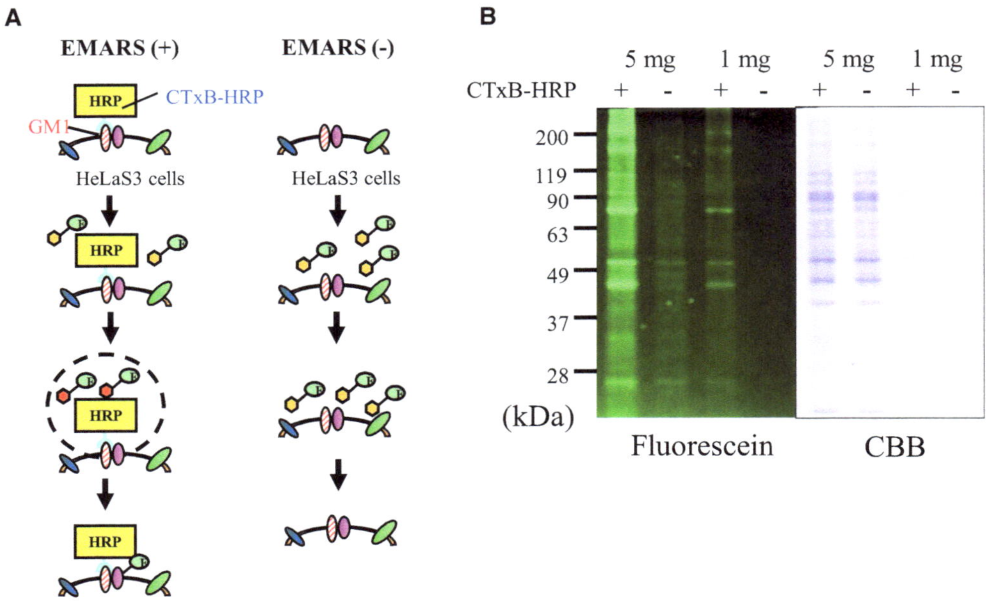

Fig. 4 SDS-PAGE analysis of proximal proteins around GM1 **in HeLaS3 cells**. (**A**) The scheme of EMARS-based membrane proximity labeling using HRP-conjugated Cholera Toxin B Subunit. HeLaS3 cells were treated with HRP-conjugated Cholera Toxin B Subunit and EMARS reagent. The EMARS reaction does not occur in untreated HeLaS3 cells. (**B**) The EMARS products from HRP-conjugated Cholera Toxin B Subunit -treated (+) and untreated (−) cell lysates were prepared and subjected to 10% SDS-PAGE at total protein concentrations of 1 μg and 5 μg, respectively. The proteins labeled with fluorescein were detected with a fluorescent imager (left column), and Coomassie brilliant blue (CBB) staining was performed for the loading control (right column)

6. Remove the supernatant containing the chloroform/methanol solution and distilled water (you can see pellets in the intermediate layer).
7. Wash the pellets with 50% Ethanol solution twice.
8. Evaporate the pellets for 15 min to completely remove the solution.
9. Add 100 μL of SDS lysis solution and heat at 100°C for 10 min.
10. Centrifuge at 20,000 × *g* for 15 min at room temperature.
11. Transfer the supernatant to a new tube.
12. Add 400 μL of the NP-40 lysis buffer and mix.
13. Add 100 μL of Anti-fluorescein antibody-conjugated Sepharose (*see* **Note 10**).
14. Rotate the tube at 4°C overnight
15. Centrifuge at 800 × *g* for 5 min at 4°C and discard the supernatant.

Table 1

Candidate proximal proteins in GM1 microdomains. Proximal proteins labeled with fluorescein by the enzyme-mediated activation of radical sources (EMARS) reaction were subjected to nano LC–ESI–MS/MS. The raw data obtained from the MS system were subsequently analyzed using Proteome Discoverer

Accession no.	Name
P14384|CBPM_HUMAN	Carboxypeptidase M
Q6YHK3|CD109_HUMAN	CD109 antigen
Q5ZPR3|CD276_HUMAN	CD276 antigen
P16070|CD44_HUMAN	CD44 antigen
P13987|CD59_HUMAN	CD59 glycoprotein
Q6UVK1|CSPG4_HUMAN	Chondroitin sulfate proteoglycan 4
P08174|DAF_HUMAN	Complement decay-accelerating factor
Q14126|DSG2_HUMAN	Desmoglein-2
Q14118|DAG1_HUMAN	Dystroglycan
P17813|EGLN_HUMAN	Endoglin
Q9UNN8|EPCR_HUMAN	Endothelial protein C receptor
P15328|FOLR1_HUMAN	Folate receptor alpha
P17931|LEG3_HUMAN	Galectin-3
P04406|G3P_HUMAN	GAPDH
P04899|GNAI2_HUMAN	G(i), alpha-2 subunit
P01891|1A68_HUMAN	HLA class I, A-68 alpha chain
P01591|IGJ_HUMAN	Immunoglobulin J chain
P08648|ITA5_HUMAN	Integrin alpha-5
P05556|ITB1_HUMAN	Integrin beta-1
O75145|LIPA3_HUMAN	Liprin-alpha-3

16. Wash the Sepharose resin five times with the NP-40 lysis buffer, twice with 0.5 M NaCl in PBS, and with 50 mM NH_4HCO_3 buffer.

3.4 Peptide Sample Preparation and Proteome Analysis

This section describes how to identify EMARS products (fluorescein-labeled proteins) via proteomics using mass spectrometry. Purified labeled proteins are eluted from resins followed by trypsinization. The peptides are subjected to a shotgun proteome using mass spectrometry. Here, we introduce a sample preparation and analysis method using electrospray ionization mass spectrometry (nano LC-ESI-MS/MS: LTQ Orbitrap) (Table 1).

1. Add 40 μL of EMAERS products elution reagent (*see* **Note 11**) to the washed Sepharose resin and mix.
2. Heat at 100°C for 10 min (mix after 5 min and continue heating for another 5 min).
3. Transfer the supernatant to a new tube and add 20 μL of 10 mM DTT/50 mM NH_4HCO_3 solution.
4. Incubate at 50°C for 30 min.
5. Add 20 μL of 100 mM Iodoacetamide/50 mM NH_4HCO_3 solution.
6. Incubate at 37°C for 1 h in the dark.
7. Add 5 μL SDS-eliminant solution (*see* **Note 12**).
8. Incubate on ice for 30 min.
9. Centrifuge at 13,000 × *g* for 10 min at 4°C.
10. Transfer the supernatant to a new tube and add 270 μL of 50 mM NH_4HCO_3.
11. Add 2 μL of 20 mM $CaCl_2$ and Trypsin solution (1 μg/μL in 50 mM acetic acid).
12. Incubate at 37°C overnight (*see* **Note 13**).
13. Add 350 μL of Ethyl acetate containing TFA to the trypsinized sample solutions and mix gently.
14. Centrifuge at 20,000 × *g* for 1 min at room temperature.
15. Discard the upper layer.
16. Repeat Steps 13–15 five times (*see* **Note 14**).
17. Evaporate the sample for 2 h to completely remove the ethyl acetate.
18. Add 200 μL of 5% ACN/0.1% TFA solution and mix gently.
19. Desalt the sample using a GL-Tip SDB (*see* **Note 15**).
20. Evaporate the eluate from the desalting cartridge for 10–15 min to remove the acetonitrile (*see* **Note 16**).
21. Dissolve peptides with the appropriate volume of 2% ACN/0.1% TFA solution.
22. Apply shotgun proteomics to the sample using nano LC-ESI-MS/MS: LTQ Orbitrap XL (*see* **Note 17**).
23. Analyze the data using Proteome Discoverer ver. 2.4 or higher.

4 Notes

1. We first recommend here the protocol for making a monovalent probe (HRP-conjugated monovalent antibody) using the **peroxidase labeling kit SH**. This is because the monovalent

probe does not have the artificial cross-linking effect between target molecules, preventing unexpected changes in cell surface protein-protein interactions. However, if the cross-linking effect is not a problem, EMARS can be performed by HRP-conjugated native antibody (not monovalent antibody) produced by the **peroxidase labeling kit NH_2**. Otherwise, we are also able to treat the primary native antibody followed by HRP-labeled secondary antibody as you would for a typical Western blot analysis. In that case, preparing the EMARS probe in advance is unnecessary.

2. Since making the monovalent probe requires partially treating the antibody with a reducing reagent, the antibody may sometimes lose its binding ability and may not work. Take care when selecting antibodies.
3. The attached cells and free cells are both EMARS applicable because EMARS can be performed in both a cultured dish and a microtube.
4. The sample cell number is arbitrary. Approximately 1×10^5–1×10^7 cells per microtube are suitable for the free cell EMARS method.
5. A pilot study is needed to determine the appropriate concentration of the EMARS probe in every experiment. If you have non-specific EMARS probe binding, the EMARS probe buffer containing a small amount of a protein component (e.g., bovine serum albumin) can be used to avoid non-specific binding to the cell surface. There is, however, a possibility that the protein components will affect the mass spectrometry analysis.
6. EMARS reactions can be performed at several temperatures. We typically used 25°C (room temperature). In some cases, incubation at 4°C induces cell detachment and heavy non-specific labeling via unknown causes.
7. Tyramide-based EMARS reagent has been typically used for EMARS reaction compared with arylazide reagent because the tyramide reagent reduces nonspecific labeling more than arylazide. Excess reagents cause cell damage and non-specific reactions. A preferable concentration of each reagent, especially H_2O_2, is needed. By the way, tyramide-biotin and -fluorescein are commercially available at https://www.rndsystems.com/products/fluorescein-tyramide_6456 https://www.sigmaaldrich.com/JP/en/product/sigma/sml2135.
8. The final products of EMARS can be applied to several subsequent experiments. Therefore, you should select the solvent and volume of the final buffer according to subsequent experiments.

9. Both direct detection with a fluorescent imager and a western blot with an anti-fluorescein antibody can be used to identify fluorescein-labeled proteins.
10. Anti-fluorescein antibody-conjugated Sepharose can be used if there is a commercially available one. The anti-fluorescein antibody can also be purchased from Southern Biotech or Rockland. Pay attention to the binding capacity when using an anti-fluorescein antibody from other companies. It should be noted that many commercially available anti-fluorescein antibodies have low binding capacities, making them unsuitable for purifying the fluorescein-labeled proteins described in this section.
11. A detergent, such as SDS, should be used when eluting the fluorescein-labeled protein from the washed Sepharose resin. However, SDS should be removed before trypsinization. In addition, since this analysis often contains membrane proteins, a strong detergent is recommended for the elution process. A specific detergent called MPEX, which can be removed easily by extraction using an organic solvent, is used together with SDS to adapt to these difficult conditions.
12. "SDS-eliminat" is a reagent that can appropriately precipitate SDS and easily remove it. Other SDS removal methods, including a similar reagent, can be used.
13. In this section, proteins are digested by typical tryptic digestion, but other enzymes are also available.
14. It should be noted that a single treatment with ethyl acetate leaves MPEX and adversely affects the mass spectrometry analysis. Multiple extraction steps are required.
15. Other desalting cartridges can be used in the desalting step. For detailed usage of each cartridge, follow the instructions for each cartridge.
16. If the sample is completely dried up at the evaporation stage, some peptides may be adsorbed on the tube surface and cannot be eluted.
17. We finally analyzed the sample by shotgun proteomics, but other methods, such as in-gel digestion, can also be used.

Acknowledgments

We thank Kochi University's experimental training equipment facility and the Saitama Medical University Biomedical Research Center for providing general technical assistance. This work was supported by Grants-in-aid for Scientific Research in Japan (No. JP19790060,

JP22790071, JP24590082, JP15K07941, No. JP18K06663, and JP21K06562 to N.K.), Mizutani Foundation Research Grants (to N.K.), Japan Science and Technology grants (to N.K.), The Vehicle Racing Commemorative Foundation research grants (to N.K.), TERUMO Life Science Foundation research grants (to N.K.), and Takeda Science Foundation (to N.K.).

Conflict of Interest The authors have no conflicts of interest to declare that are relevant to the content of this chapter.

References

1. Lingwood D, Simons K (2010) Lipid rafts as a membrane-organizing principle. Science 327:46–50
2. Vereb G, Szöllosi J, Matkó J et al. (2003) Dynamic, yet structured: the cell membrane three decades after the Singer-Nicolson model. Proc Natl Acad Sci U S A 100:8053–8058
3. Brown DA, London E (1998) Functions of lipid rafts in biological membranes. Annu Rev Cell Dev Biol 14:111–136
4. Simons K, Toomre D (2000) Lipid rafts and signal transduction. Nat Rev Mol Cell Biol 1:31–39
5. Brown DA, Rose JK (1992) Sorting of GPI-anchored proteins to glycolipid-enriched membrane subdomains during transport to the apical cell surface. Cell 68:533–544
6. Heerklotz H (2002) Triton promotes domain formation in lipid raft mixtures. Biophys J 83:2693–2701
7. Kotani N, Gu J, Isaji T et al. (2008) Biochemical visualization of cell surface molecular clustering in living cells. Proc Natl Acad Sci U S A 105:7405–7409
8. Honke K, Kotani N (2012) Identification of cell-surface molecular interactions under living conditions by using the enzyme-mediated activation of radical sources (EMARS) method. Sensors (Basel) 12:16037–16045
9. Honke K, Kotani N (2011) The enzyme-mediated activation of radical source reaction: A new approach to identify partners of a given molecule in membrane microdomains. J Neurochem 116:690–695
10. Jiang S, Kotani N, Ohnishi T et al. (2012) A proteomics approach to the cell-surface interactome using the enzyme-mediated activation of radical sources reaction. Proteomics 12:54–62
11. Miyagawa-Yamaguchi A, Kotani N, Honke K (2014) Expressed glycosylphosphatidylinositol-anchored horseradish peroxidase identifies co-clustering molecules in individual lipid raft domains. PLoS One 9:e93054
12. Miyagawa-Yamaguchi A, Kotani N, Honke K (2015) Each GPI-anchored protein species forms a specific lipid raft depending on its GPI attachment signal. Glycoconj J 32:531–540
13. Kotani N, Ishiura Y, Yamashita R et al. (2012) Fibroblast growth factor receptor 3 (FGFR3) associated with the CD20 antigen regulates the rituximab-induced proliferation inhibition in B-cell lymphoma cells. J Biol Chem 287:37109–37118
14. Yamashita R, Kotani N, Ishiura Y et al. (2011) Spatiotemporally-regulated interaction between β1 integrin and ErbB4 that is involved in fibronectin-dependent cell migration. J Biochem 149:347–355
15. Ishiura Y, Kotani N, Yamashita R et al. (2010) Anomalous expression of Thy1 (CD90) in B-cell lymphoma cells and proliferation inhibition by anti-Thy1 antibody treatment. Biochem Biophys Res Commun 396:329–334
16. Hashimoto N, Hamamura K, Kotani N et al. (2012) Proteomic analysis of ganglioside-associated membrane molecules: substantial basis for molecular clustering. Proteomics 12:3154–3163
17. Kaneko K, Ohkawa Y, Hashimoto N et al. (2016) Neogenin, defined as a GD3-associated molecule by enzyme-mediated activation of radical sources, confers malignant properties via intracytoplasmic domain in melanoma cells. J Biol Chem 291:16630–16643
18. Esaki N, Ohkawa Y, Hashimoto N et al. (2018) ASC amino acid transporter 2, defined by enzyme-mediated activation of radical sources, enhances malignancy of GD2-positive small-cell lung cancer. Cancer Sci 109:141–153
19. Kotani N, Nakano T, Kuwahara R (2022) Host cell membrane proteins located near SARS-CoV-2 spike protein attachment sites are identified using proximity labeling and proteomic analysis. J Biol Chem 298:102500

Check for updates

Chapter 13

Analysis of Intracellular Trafficking of Cholera Toxin B Subunits Depending on Ganglioside GM1 in Tissue-Resident Macrophages

Izumi Sasaki, Shiori Kaji, Naoko Wakaki-Nishiyama and Tsuneyasu Kaisho

Abstract

Cholera toxin (CT) is a bacterial exotoxin secreted by *Vibrio cholerae*. CT is comprised of one toxic A subunit (CTA) and five nontoxic B subunits (CTB). CTB is required for internalization of CT through binding with a membrane glycolipid ganglioside GM1. Previously, we clarified that CTB could induce production of proinflammatory cytokine interleukin-1β (IL-1β) from murine resident peritoneal macrophages (RPMs) in synergy with lipopolysaccharide (LPS) through a ganglioside GM1-dependent manner. Then, CTB was incorporated into endoplasmic reticulum (ER) depending on GM1 and activate ER stress sensor, inositol-requiring enzyme 1 alpha (IRE1α). IRE1α was required for both ER stress responses and IL-1β production in CTB- plus LPS-stimulated RPMs. These findings indicate that ganglioside GM1 functions not only for cell-surface receptor for CTB but also for cross talk with the ER stress sensor, IRE1α, in RPMs. Here, we describe the protocols for analysis of intracellular trafficking of CTB depending on ganglioside GM1 in RPMs.

Keywords Ganglioside GM1, Resident peritoneal macrophages, Cholera toxin

1 Introduction

Cholera toxin (CT) is a bacterial exotoxin secreted by *Vibrio cholerae* and causes an acute diarrheal disease, well known as cholera. CT is a protein complex which is called holotoxin comprised of one A subunit (CTA) and five B subunits (CTB). CTB binds to ganglioside GM1, which is a membrane glycolipid. Then, CT was internalized. CTA induces elevation of intracellular cAMP levels and leads to diarrhea. CT is also well known as a strong immune adjuvant that can induce generation of antigen-specific IgA- or IgG-producing plasma cells and various types of T cell responses [1–4]. CTB also functions as an immune adjuvant such as interleukin-6 production [5]. However, it

Koichi Furukawa (ed.), *Gangliosides*, Methods in Molecular Biology, vol. 3035,
https://doi.org/10.1007/978-1-0716-5288-6_13,

remains unclear how CTB function as immune adjuvant and whether CTB activate tissue-resident macrophages.

Resident peritoneal macrophages (RPMs) are evolutionally conserved between invertebrate sea urchins and mammals such as mouse, pig, and human [6, 7]. RPMs are critically involved not only in tissue repair [8] but also in the IgA production from peritoneal B-1 cells [9]. However, little is known about the molecular mechanisms of innate immune responses such as proinflammatory cytokine production in RPMs.

We have found that murine RPMs could response to CTB and produce large amounts of interleukin-1β (IL-1β) in the presence of lipopolysaccharides (LPS) [10]. Analysis of beta-1,4-N-acetylgaracrosaminyltransferase 1 (B4galnt1)-deficient mice, which lack ganglioside GM1, showed abolishment of CTB-induced IL-1β production from LPS-primed RPMs. These findings indicate that a ganglioside GM1 is essential for CTB-induced IL-1β production from LPS-primed RPMs [10]. Then, it is important to understand how CTB moves in RPMs. Analysis of *in vitro* cell lines showed that CTB is internalized through binding with ganglioside GM1 and translocated from Golgi apparatus to the endoplasmic reticulum (ER) [11–15]. The traffic is termed as retrograde transport. However, the intracellular trafficking of CTB in tissue-resident macrophages remains unclear. We have developed an assay to investigate the intracellular localization of CTB and found that CTB was incorporated into the ER in a ganglioside GM1-dependent manner in RPMs [16]. The accumulation of CTB with ganglioside GM1 in RPMs at ER was essential for both an ER stress sensor, inositol-requiring enzyme 1 alpha (IRE1α)-mediated unfolded protein responses and inflammasome-mediated IL-1β production [16]. This is the first report showing that the ganglioside GM1 is critical not only for CTB sensor but also for an ER stress sensor–mediated inflammasome activation in tissue-resident macrophages.

Here, we show an analysis of intracellular trafficking of CTB depending on ganglioside GM1 in RPMs.

2 Materials

2.1 Mice

C57BL/6 mice.

2.2 Reagents

1. Phosphate-buffered saline (PBS).
2. Magnetic-activated cell sorting (MACS) buffer: PBS supplemented with 1% fetal calf serum (FCS), 5 mM ethylenediaminetetraacetic acid (EDTA), and 0.01% sodium azide (NaN_3).
3. Antibodies: anti-CD16/32, biotinylated anti-F4/80, anti-IRE1α, anti-KDEL, normal rabbit IgG, Alexa546–labeled

anti-rabbit IgG, Alexa647–labeled anti-mouse IgG; Alexa647–labeled anti-ganglioside GM1.

4. Streptavidin MicroBeads.
5. Culture medium (CM): Roswell Park Memorial Institute (RPMI) 1640 medium supplemented with 10% fetal calf serum (FCS), 100 U/mL penicillin G, 100 μg/mL streptomycin, and 109 nM 2-mercaptoethanol.
6. Cell-Tak solution: 0.1 M sodium hydrogen carbonate ($NaHCO_3$) supplemented with 20 μg/mL Cell-Tak and 16 μM sodium hydroxide (NaOH).
7. Fluorescein isothiocyanate (FITC)-conjugated CTB.
8. ER-Tracker Red.
9. 4% paraformaldehyde in PBS.
10. Normal goat serum.
11. DAPI solution: VECTASHIELD Antifade Mounting Medium with DAPI diluted fivefold with MACS Buffer.

2.3 Equipment

1. Forceps and surgery scissors.
2. 3 mL disposable Droppers (130025, Ina Optika).
3. 15 mL tubes.
4. 40 μm meshes.
5. 20 G needles.
6. LS columns (130-042-401, Miltenyi Biotec).
7. CELLview cell culture dish, 35 × 10 mm, four-separated glass-bottom plate (627975, Greiner Bio-One International).
8. Cell-Tak–treated CELLview cell culture dish: Add 400 μL of Cell-Tak solution per well in CELLview cell culture dish and incubate them at 37°C for 20 min. Aspirate supernatants and add 500 μL of sterile water per well. Repeat this step. Before addition of cells, aspirate the supernatant completely (*see* **Note 1**).

3 Methods

3.1 Preparation of Resident Peritoneal Exudate Cells

1. Sacrifice a C57BL/6 mouse and cut the outer skin of the back using forceps and surgery scissors.
2. Inject 10 mL of ice-cold PBS into the peritoneal cavity using a 10 mL syringe attached with a 20 G needle.
3. Collect the PBS as much as possible and transfer to a 15 mL tube kept on ice.

4. Cut the peritoneal inner skin using forceps and surgery scissors and collect the remaining PBS in the peritoneal cavity using a 3 mL disposable dropper (*see* **Note 2**).
5. Centrifuge the collected cells at 2000 rpm for 5 min at 4°C. Aspirate the supernatants and resuspend them in 2×10^6 cells/mL in CM. Then, the collected cells can be used as resident peritoneal exudate cells (RPECs). Usually, $1–2 \times 10^6$ RPECs can be obtained from one mouse.

3.2 Preparation of RPMs

1. Centrifuge RPECs at 2000 rpm for 5 min at 4°C. Aspirate the supernatants and resuspend the cells in 1×10^7 cells/mL in CM.
2. Add anti-CD16/32 antibody (final concentration (final conc.) 5 µg/mL) and leave them for 5 min on ice to block the unspecific binding of antibodies to Fc receptors.
3. Add biotinylated anti-F4/80 antibody (final conc. 1 µg/mL) and leave them for 20 min on ice.
4. Add 10 mL of MACS buffer and centrifuge them at 2000 rpm for 5 min at 4°C. Aspirate the supernatants and resuspend the cells in 1×10^7 cells/mL in MACS buffer.
5. Add one-fifth volume of strepravidine microbeads and leave them for 20 min on ice.
6. Add 10 mL of MACS buffer and centrifuge them at 2000 rpm for 5 min at 4°C. Aspirate the supernatants, resuspend the cells in 1 mL of MACS buffer, and magnetically sort them with LS column according to the manufacturer's instructions. Microbead-conjugated cells (F4/80 positive cells) can be used as RPMs. Usually, $3–6 \times 10^5$ RPMs can be obtained from one mouse.

3.3 Evaluation of RPMs Purity by Flow Cytometry Analysis

1. Centrifuge about 1×10^5 RPECs or RPMs in a 1.5 mL Eppendorf tube at 8000 rpm for 1 min at 4°C. Aspirate the supernatants.
2. Add 20 µL of MACS buffer with anti-CD16/32 antibody (final conc. 5 µg/mL) and leave them for 5 min on ice.
3. Add 20 µL of MACS buffer with biotinylated anti-F4/80 antibody (final conc. 1 µg/mL) and leave them for 20 min on ice.
4. Add 1 mL of MACS buffer and centrifuge cells at 8000 rpm for 1 min at 4°C. Aspirate the supernatants.
5. Add 20 µL of MACS buffer with PerCP-Cy5.5–conjugated streptavidin (final conc. 2 µg/mL) and V450-conjugated anti-CD11b antibody (final conc. 2 µg/mL) and leave for 20 min on ice with protection from light.
6. Add 1 mL of MACS buffer and centrifuge cells at 8000 rpm for 1 min at 4°C. Aspirate the supernatants and resuspend the cells in 0.5 mL of MACS buffer.
7. Analyze the cells using a flow cytometer (see Fig. 1 and [10, 17]).

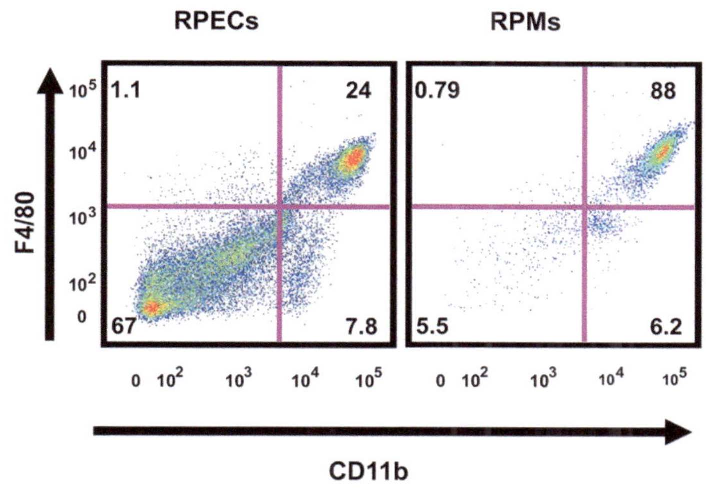

Fig. 1 Preparation of RPECs and RPMs. FACS profiles of those cells were shown. Numbers in pseudocolor plots indicate the percentages of gated cells. RPECs contain F4/80$^+$ CD11b$^+$ cells (RPMs) and F4/80$^-$ cells

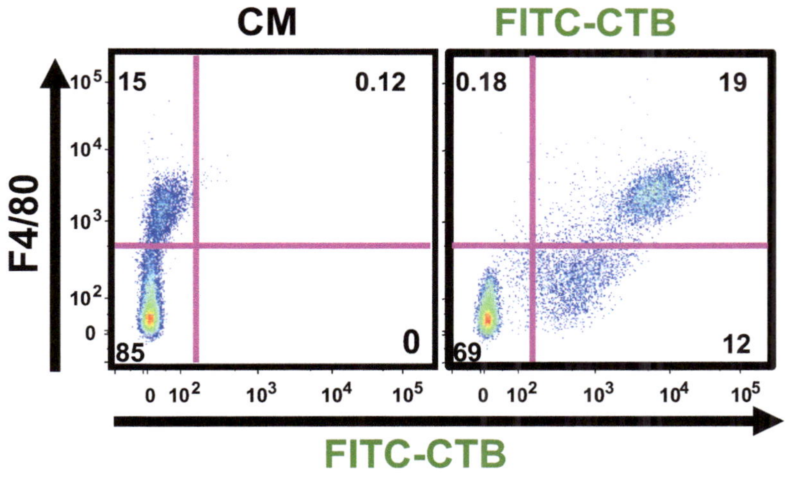

Fig. 2 CTB uptake in RPECs. Flow cytometry analysis of FITC-labeled CTB uptake in RPECs. Numbers in pseudocolor plots indicate the percentages of gated cells. The frequency and intensity of FITC-CTB in F4/80$^+$ cells (RPMs) were significantly higher compared with them in F4/80$^-$ cells (non-RPMs)

3.4 Evaluation of CTB Uptake in RPECs by Flow Cytometry Analysis

1. Start at Sect. 3.1, step 5.
2. Seed 2×10^5 of RPECs in a 1.5 mL Eppendorf tube.
3. Add CM or FITC-labeled CTB (final conc. 1 μg/mL) and culture the cells for 12 h.
4. Continue to the protocol in Sect. 3.3 (see Fig. 2).

3.5 Evaluation of Intracellular Traffic of CTB in RPMs by Immunofluorescence Analysis

1. Centrifuge RPMs at 2000 rpm for 5 min at 4°C. Aspirate the supernatants and resuspend them in 2 × 10^6 cells/mL in CM.
2. Seed 4 × 10^5 of RPMs per well in a Cell-Tak–treated CELLview cell culture dish.
3. Add CM or FITC-labeled CTB (final conc. 1 μg/mL) and culture the cells for 12 h.
4. Centrifuge culture dish at 2000 rpm for 2 min at 4°C. Aspirate the supernatants and carefully add 500 μL of ice-cold PBS per well (*see* **Note 3**). Repeat this step.
5. Centrifuge the culture dish at 2000 rpm for 2 min at 4°C and aspirate the supernatants. Then, add 500 μL of 4% paraformaldehyde in PBS per well and fix for 10 min at room temperature (RT).
6. Centrifuge the dish at 2000 rpm for 2 min at 4°C. Aspirate the supernatants and carefully add 500 μL of ice-cold PBS per well. Repeat this step.
7. Centrifuge the dish at 2000 rpm for 2 min at 4°C and aspirate the supernatants. To block the unspecific binding of antibodies to dish, add 500 μL of 5% of normal goat serum in MACS buffer per well and leave them for 30 min at 4°C.
8. Centrifuge the dish at 2000 rpm for 2 min at 4°C. Aspirate the supernatants and carefully add 500 μL of ice-cold PBS per well. Repeat this step.
9. Centrifuge the dish at 2000 rpm for 2 min at 4°C and aspirate the supernatants. Then, add anti-IRE1α antibody (final conc. 1.45 μg/mL) or normal rabbit IgG (fin conc. 1.45 μg/ml) as the control and leave them for 30 min at 4°C.
10. Centrifuge the dish at 2000 rpm for 2 min at 4°C. Aspirate the supernatants and carefully add 500 μL of ice-cold PBS per well. Repeat this step.
11. Centrifuge the dish at 2000 rpm for 2 min at 4°C and aspirate the supernatants. Then, add 500 μL of Alexa546–labeled anti-rabbit IgG antibody (final conc. 1 μg/mL) per well and leave them for 30 min at 4°C with shedding from light.
12. Centrifuge the dish at 2000 rpm for 2 min at 4°C. Aspirate the supernatants and carefully add 500 μL of ice-cold PBS per well. Repeat this step.
13. Centrifuge the dish at 2000 rpm for 2 min at 4°C and aspirate the supernatants. To stain the nuclei, add 500 μL of DAPI solution per well and leave them for 30 min at 4°C with shedding from light.
14. Centrifuge the dish at 2000 rpm for 2 min at 4°C. Aspirate the supernatants and carefully add 500 μL of ice-cold PBS per well. Repeat this step.

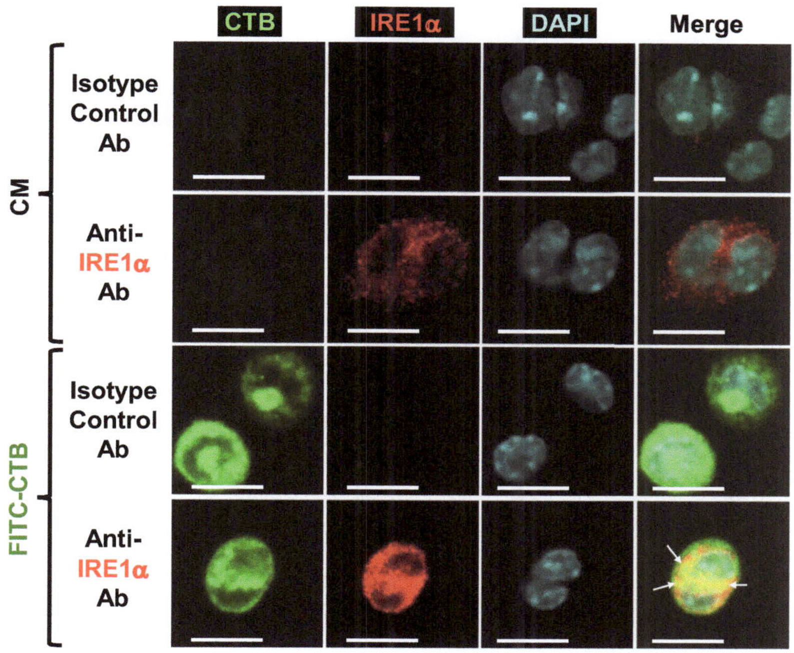

Fig. 3 Intracellular trafficking of CTB in RPMs. Microscopic images of subcellular localization of CTB in RPMs. Green, red, and sky-blue coloring represents CTB internalization, IRE1α, and DAPI staining, respectively. Scale bars indicated by white lines represent 10 μm. Arrows indicate co-localization of CTB and IRE1α

15. Centrifuge the dish at 2000 rpm for 2 min at 4°C and aspirate the supernatants. Then, add 1 mL of MACS buffer per well and analyze the cells using FV10i confocal microscopy (see Fig. 3 and [16]).

3.6 Evaluation of CTB Accumulation in RPMs at ER by Immunofluorescence Analysis

1. Start at Sect. 3.5, step 9. Centrifuge the dish at 2000 rpm for 2 min at 4°C and aspirate the supernatants. Then, add anti-IRE1α antibody (final conc. 1.45 μg/mL) and anti-KDEL antibody (final conc. 10 μg/mL) to stain the ER and leave them for 30 min at 4°C.
2. Centrifuge the dish at 2000 rpm for 2 min at 4°C. Aspirate the supernatants and carefully add 500 μL of ice-cold PBS per well. Repeat this step.
3. Centrifuge the dish at 2000 rpm for 2 min at 4°C and aspirate the supernatants. Then, add 500 μL of antibody mix containing Alexa546–labeled anti-rabbit IgG antibody (final conc. 1 μg/mL) and Alexa647–labeled anti-mouse IgG (final conc. 1 μg/mL) per well and leave them for 30 min at 4°C with shedding from light.
4. Continue to the protocol in Sect. 3.5, steps 12–15 (see Fig. 4 and [16]).

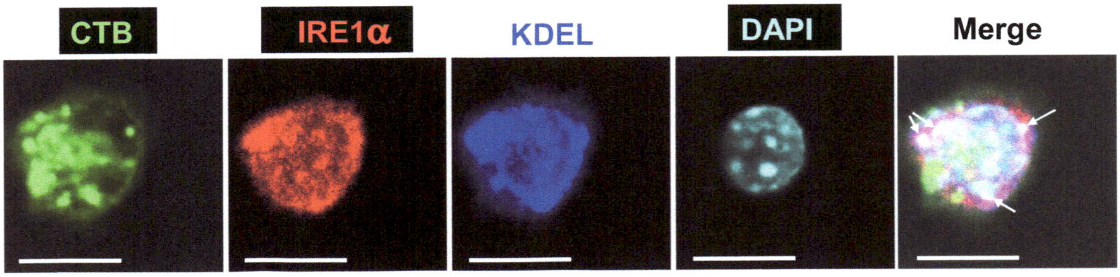

Fig. 4 CTB accumulation in RPMs at the ER. Microscopic images of subcellular localization of CTB in RPMs. Green, red, blue, and sky-blue coloring represent CTB internalization, IRE1α, KDEL, and DAPI staining, respectively. Scale bars indicated by white lines represent 10 μm. Arrows indicate co-localization of CTB, KDEL, and IRE1α

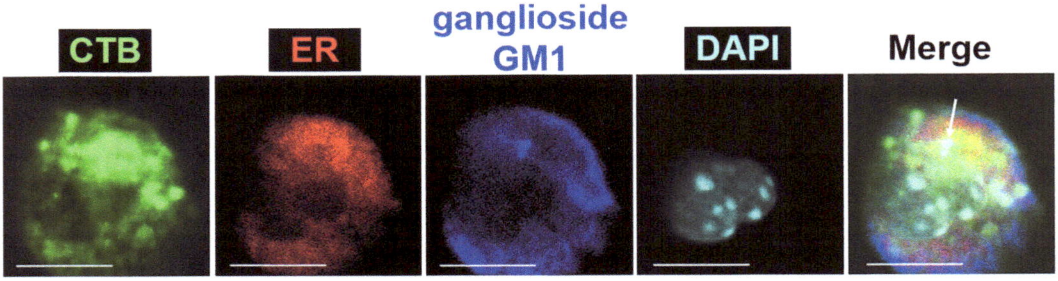

Fig. 5 CTB accumulation in RPMs at the ER with ganglioside GM1. Microscopic images of FITC-CTB in RPMs. Green, red, blue, and sky-blue coloring represent CTB internalization, ER, ganglioside GM1, and DAPI staining, respectively. Scale bars indicated by white lines represent 10 μm. Arrow indicates co-localization of CTB and ganglioside GM1 at the ER

3.7 Evaluation of CTB Accumulation in RPMs at the ER with Ganglioside GM1 by Immunofluorescence Analysis

1. Start at Sect. 3.5, step 3. Add CM or FITC-labeled CTB (final conc. 1 μg/mL) and culture the cells for 12 h. Before collecting culture dish, add the ER-Tracker Red (final conc. 1 μM) and incubate for 30 min.
2. Continue to the protocol in Sect. 3.5, steps 4–8.
3. Centrifuge the dish at 2000 rpm for 2 min at 4°C and aspirate the supernatants. Then, add 500 μL of Alexa647–labeled anti-ganglioside GM1 (final conc. 10 μg/mL) per well and leave them for 30 min at 4°C with shedding from light.
4. Continue to the protocol in Sect. 3.5, steps 12–15 (see Fig. 5 and [16]).

4 Notes

1. Remaining water could impair cell viability. We recommend you to aspirate completely.
2. Please be careful not to injure any organs. This step has a risk of bleeding.
3. RPMs are easy to peel off from dish although treatment of enhancement the adhesion. We recommend add PBS carefully.

Acknowledgments

This work was supported by Grant-in-Aid for Transformative Research Areas (JP22H05182 and JP22H05187 to T. Kaisho and JP22H05187 to I. Sasaki), Grant-in-Aid for Scientific Research (B) (JP20H03505 to T. Kaisho), Grant-in-Aid for Scientific Research (C) (JP22K07006 and JP26K09993 to I. Sasaki), Takeda Science Foundation (to T. Kaisho and I. Sasaki), GSK Japan Research Grant 2021 (to I. Sasaki), and Kowa Life Science Foundation (to I. Sasaki).

Conflict of Interest The authors have no conflicts of interest to declare that are relevant to the content of this chapter.

References

1. Xu-Amano J, Kiyono H, Jackson RJ et al (1993) Helper T cell subsets for immunoglobulin A responses: oral immunization with tetanus toxoid and cholera toxin as adjuvant selectively induces Th2 cells in mucosa associated tissues. J Exp Med 178:1309–1320
2. Marinaro M, Staats HF, Hiroi T et al (1995) Mucosal adjuvant effect of cholera toxin in mice results from induction of T helper 2 (Th2) cells and IL-4. J Immunol 155:4621–4629
3. Wakabayashi A, Nakagawa Y, Shimizu M et al (2008) Suppression of an already established tumor growing through activated mucosal CTLs induced by oral administration of tumor antigen with cholera toxin. J Immunol 180:4000–4010
4. Datta SK, Sabet M, Nguyen KP et al (2010) Mucosal adjuvant activity of cholera toxin requires Th17 cells and protects against inhalation anthrax. Proc Natl Acad Sci U S A 107:10638–10643
5. Nagatomo D, Taniai M, Ariyasu H et al (2015) Cholesteryl pullulan encapsulated TNF-alpha nanoparticles are an effective mucosal vaccine adjuvant against influenza virus. BioMed Res Int 2015:471468
6. Salm L, Shim R, Noskovicova N et al (2023) Gata6(+) large peritoneal macrophages: an evolutionarily conserved sentinel and effector system for infection and injury. Trends Immunol 44:129–145
7. Deniset JF, Belke D, Lee WY et al (2019) Gata6(+) pericardial cavity macrophages relocate to the injured heart and prevent cardiac fibrosis. Immunity 51:131–40.e5
8. Wang J, Kubes P (2016) A reservoir of mature cavity macrophages that can rapidly invade visceral organs to affect tissue repair. Cell 165:668–678
9. Okabe Y, Medzhitov R (2014) Tissue-specific signals control reversible program of localization and functional polarization of macrophages. Cell 157:832–844
10. Orimo T, Sasaki I, Hemmi H et al (2019) Cholera toxin B induces interleukin-1β production from resident peritoneal macrophages through the pyrin inflammasome as well as the NLRP3 inflammasome. Int Immunol 31:657–668

11. Lencer WI, Tsai B (2003) The intracellular voyage of cholera toxin: going retro. Trends Biochem Sci 28:639–645
12. Chinnapen DJ, Chinnapen H, Saslowsky D et al (2007) Rafting with cholera toxin: endocytosis and trafficking from plasma membrane to ER. FEMS Microbiol Lett 266:129–137
13. Wernick NL, Chinnapen DJ, Cho JA et al (2010) Cholera toxin: an intracellular journey into the cytosol by way of the endoplasmic reticulum. Toxins (Basel) 2:310–325
14. Uchida Y, Hasegawa J, Chinnapen D et al (2011) Intracellular phosphatidylserine is essential for retrograde membrane traffic through endosomes. Proc Natl Acad Sci USA 108:15846–15851
15. Luong P, Li Q, Chen PF et al (2020) A quantitative single-cell assay for retrograde membrane traffic enables rapid detection of defects in cellular organization. Mol Biol Cell 31:511–519
16. Sasaki I, Fukuda-Ohta Y, Nakai C et al (2024) A stress sensor, IRE1α, is required for bacterial-exotoxin-induced interleukin-1β production in tissue-resident macrophages. Cell Rep 43:113981
17. Sasaki I, Kaisho T (2022) Preparation and inflammasome activation of murine bone marrow-derived and resident peritoneal macrophages. Methods Mol Biol 2427:95–104

Check for updates

Chapter 14

Electrostatic Interaction Analysis Between GM3 and Insulin Receptor

Kazuya Kabayama

Abstract

The interaction between ganglioside GM3 and the insulin receptor suppresses receptor function and contributes to the development of type 2 diabetes, as demonstrated by experiments in living cells. However, due to the complexity of the cell membrane, detailed analysis of the GM3–insulin receptor interaction has been challenging. Therefore, a model system was constructed to analyze these interactions by synthesizing the transmembrane domain of fluorescently labeled insulin receptors and incorporating it into liposomes. In this chapter, the lipid composition of liposomes is investigated, demonstrating that 3D imaging can detect subtle differences in peptide localization. This model system will be a useful tool for detailed analysis of electrostatic interactions.

Keywords GM3, Insulin Receptor, Phase Separated Liposome, Electrostatic Interaction

1 Introduction

The cell membrane is a structure that separates the interior and exterior of cells; it comprises a lipid bilayer and membrane proteins. Membrane lipids, the components of the lipid bilayer, are divided into glycerolipids and sphingolipids with the exception of cholesterol, and gangliosides are classified as sphingolipids.

Gangliosides exist in the outer leaflet of the plasma membrane and consist of sphingosine and glycan with one or more sialic acid molecules possessing a negative charge. They have various functions on the cell surface via their characteristic sugar chains. Furthermore, numerous gangliosides are reported to be receptors of viruses and toxins, e.g., GM1 for cholera toxin [1–3]. In addition, it is reported that they have the ability to regulate membrane proteins, such as epidermal growth factor receptor [4, 5] and vascular endothelial growth factor receptor [6, 7].

Insulin receptor (IR) is one of the transmembrane proteins regulated by ganglioside. It belongs to the receptor tyrosine kinases

Koichi Furukawa (ed.), *Gangliosides*, Methods in Molecular Biology, vol. 3035,
https://doi.org/10.1007/978-1-0716-5288-6_14,

and is involved in various types of metabolism, such as protein synthesis in muscle, glycogen synthesis in liver, and anorexigenic signal promotion in neurons [8]. Among these roles, the most well-known is in glucose uptake [9] because its malfunction causes high blood glucose levels, leading to type 2 diabetes.

In type 2 diabetes caused by obesity, it is thought that the IR does not sufficiently respond to insulin, which is referred to as insulin resistance, despite insulin being secreted [10]. Previous studies [11] reported that TNFα, a type of cytokine increased by obesity, causes the overexpression of ganglioside GM3 in adipocytes, thereby inhibiting insulin signaling. As a cause of the insulin resistance induced by overexpressing GM3, Kabayama et al. focused on the negative charge of GM3 and the positive charge just above the transmembrane domain of the IR and established a hypothesis that GM3 and IR interact with each other electrostatically (Fig. 1A). An experiment using live cells supported this hypothesis [12]. However, the interaction that actually works has not been elucidated because of the complexity of the cell membrane.

In this chapter, a simple model system in which the IR and cell membrane are replaced by transmembrane peptides and artificial lipid bilayers, respectively, is constructed to clarify this interaction, and an analytical method using this system is presented.

2 Materials

All solvents for peptide synthesis, amino acids and reagents, 4-chloro-7-nitrobenzofurazan (NBD-Cl), and all lipids are commercially available.

2.1 Peptide Synthesis

1. H-Ile-Tyr(OBut)-Leu-Phe-Leu-Arg(Pbf)-Lys(Boc)-Arg(Pbf)-NH-resin.
2. Piperidine/1-methyl-2-pyrrolidone (NMP).
3. Boc-Ser-OH.
4. N,N′-Diisopropylcarbodiimide (DIC).
5. 1-Hydroxybenzotriazole (HOBt).
6. N,N-Dimethylformamide (DMF).
7. Fmoc-Gly-OH.
8. 1-[Bis(dimethylamino)methylene]-1H-1,2,3-triazolo[4,5-b] pyridinium 3-oxid hexafluorophosphate (HATU).
9. N,N′-Diisopropylethylamine (DIEA).
10. Fmoc-Ile-OH.
11. 2-(1H-Benzotriazole-1-yl)-1,1,3,3-tetramethyluronium hexafluorophosphate (HBTU).
12. Boc-Ser(Fmoc-Ile-Gly)-Ile-Tyr(OBut)-Leu-Phe-Leu-Arg(Pbf)-Lys(Boc)-Arg(Pbf)-NH-resin.

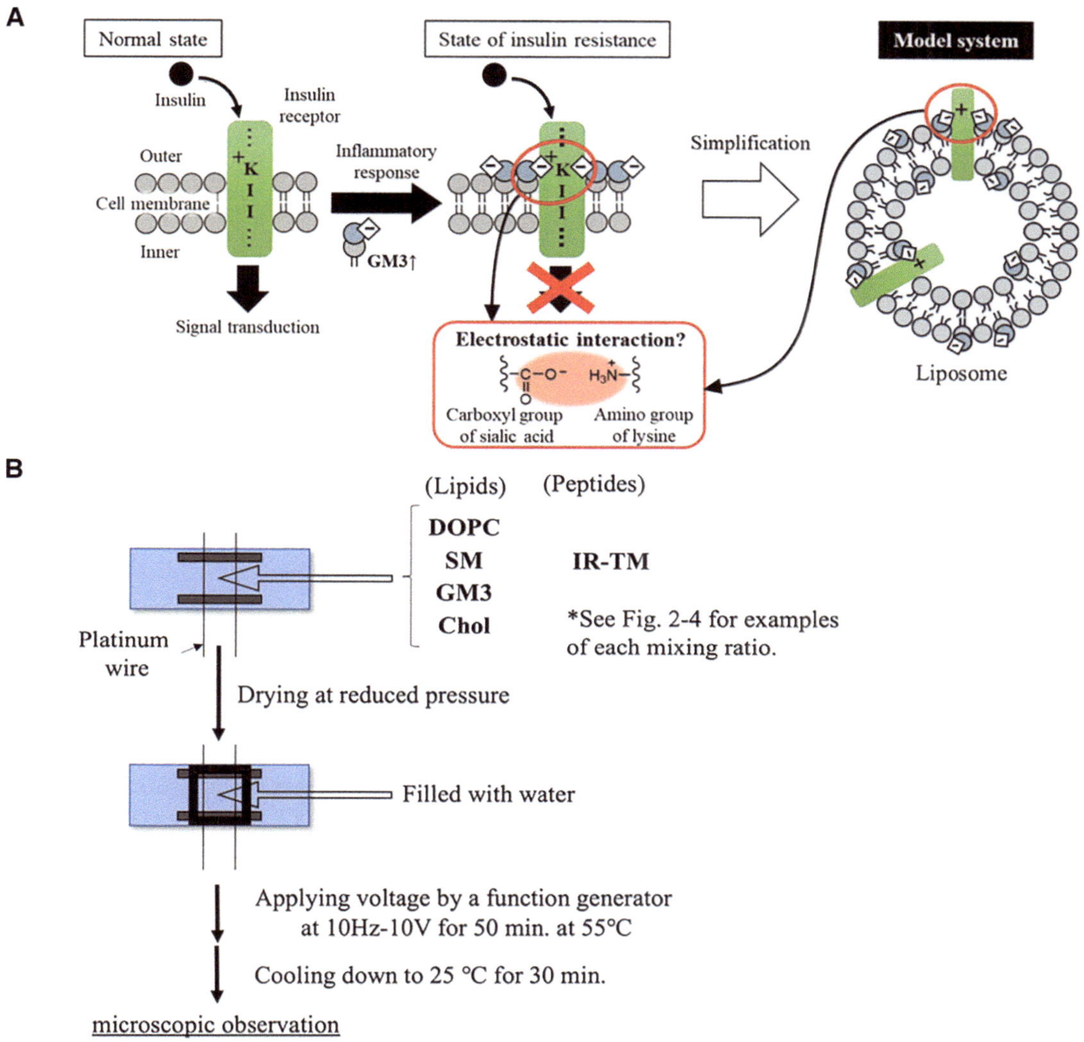

Fig. 1 (**A**) The electrostatic interaction between the insulin receptor and GM3 is thought to inhibit the receptor signaling. To visualize this interaction, it is useful to have a model system that can analyze it. For the synthesis of peptides, refer to Sect. 3.1. (**B**) Method for creating liposomes on a glass slide: By mixing lipids with peptides that have transmembrane domains, peptide liposomes can be created as described in Sect. 3.2

13. TFA cocktail (TFA/triisopropylsilane/H_2O/1,2-ethanedithiol = 92.5/2.5/2.5/2.5).
14. NH_4HCO_3 aq. (sat.).
15. Hydrochloric acid (HCl).

2.2 Confocal Microscopy

1. NBD-conjugated transmembrane peptide of insulin receptor (NBD-IR TM).
2. 2-(4,4-Difluoro-5,7-dimethyl-4-bora-3a,4a,diaza-s-indacene-3-dodecanoyl)-1-hexadecanoyl-sn-glycero-3-phosphocholine (Bodipy-PC).

3. ATTO488-Sphingomyelin (ATTO488-SM).
4. N-(Rhodamine 101 sulfonyl)-1,2-hexadecanoyl-sn-4-phosphoethanolamine (Texas Red-DHPE).

3 Methods

3.1 Peptide Synthesis

1. The transmembrane (TM) peptide of IR is able to be synthesized using Fmoc-based solid-phase and O-acyl isopeptide methods [13].
2. Rink amide resin (0.26 g, 0.1 mmol) is set in an automated microwave peptide synthesizer to carry out microwave-assisted solid-phase peptide synthesis, which provides H-Ile-Tyr(OBut)-Leu-Phe-Leu-Arg(Pbf)-Lys(Boc)-Arg(Pbf)-NH-resin. Half of the resin is treated with 20% NMP for 1 min with vortex mixing.
3. Rink amide resin (0.26 g, 0.1 mmol) is treated with the same reagent for another 5 min.
4. A solution of Boc-Ser-OH (41 mg, 0.2 mmol), DIC (42 μL, 0.24 mmol), and HOBt (32 mg, 0.24 mmol) in DMF is added to the resin and mixed for 1 h at 25°C.
5. After washing with NMP five times, a mixture of Fmoc-Gly-OH (74 mg, 0.25 mmol), HATU (91 mg, 0.24 mmol), and DIEA (87 μL, 0.5 mmol) is added to the resin and vortexed for 20 min at 50°C. This reaction is repeated twice.
6. The resin is subjected to general Fmoc removal conditions, followed by a subsequent reaction with Fmoc-Ile-OH (88 mg, 0.25 mmol), HBTU (83 mg, 0.22 mmol), and DIEA (87 μL, 0.5 mmol) to give Boc-Ser(Fmoc-Ile-Gly)-Ile-Tyr(OBut)-Leu-Phe-Leu-Arg(Pbf)-Lys(Boc)-Arg(Pbf)-NH-resin.
7. In the same way, the remaining amino acids are coupled to the resin. For NBD condensation, NBD-Cl (50 mg, 0.25 mmol) in NMP (0.5 mL), DIEA (87 μL, 0.5 mmol), and the resin are mixed and vortexed together for 15 min at 50°C.
8. Double coupling is performed as follows: part of the resin (50 mg, 8.6 μmol) is treated with a TFA cocktail (1 mL) for 2 h at 25°C.
9. The mixture is then concentrated by an N_2 stream and precipitated with ether.
10. The precipitate is washed three times with ether and then dried in vacuo.
11. The crude product is purified by reversed-phase high-performance liquid chromatography (RP-HPLC), which is carried out on a reverse-phase column.

12. The purified peptide is exposed to NH_4HCO_3 aq. (sat.) for 30 min at 25°C and quantitatively migrated into a native sequence.
13. Electrospray ionization (ESI) mass is found as m/z 1234.7, calculated for $[M + 3H]^{3+}$; 1234.5. Amino acid composition is determined using a commercially available amino acid analyzer after hydrolysis with 6 M HCl at 180°C for 25 min in an evacuated sealed tube.
14. The synthesis of NBD-IR TM is achieved by the method described above.

3.2 Preparation of Giant Unilamellar Vesicles (GUVs) (Fig. 1B)

1. GUVs are prepared by electroformation, a technique in which an electric field is applied to induce self-assembly of lipid membranes to form liposomes (artificial lipid bilayer vesicles) [14, 15].
2. A sample solution containing lipids and 0.1 mol% fluorescent lipids in chloroform–methanol is spread on a cover slip (24 × 60 mm, 0.12–0.17 mm thickness) and dried overnight in vacuo.
3. The cover slip is sealed with a square-shaped rubber sheet to form a chamber and filled with ultrapure water. Two platinum wires are laid between the glass slide and the chamber (*see* **Note 1**).
4. These platinum wires are then connected to a function generator, and an AC field (10 Hz, peak-to-peak voltage (V_{pp}): 10 V) is applied at 55°C for 1 h (*see* **Note 2**).
5. GUVs are gradually cooled down to 25°C and equilibrated for 30 min (*see* **Note 3**).

3.3 Confocal Microscopy

1. Confocal laser-scanning fluorescence microscopy as shown in Figs. 2, 3, and 4 is performed on commercially available microscope with a 60× oil immersion objective (numerical aperture (NA) 1.40 or higher recommended) (*see* **Note 4**).

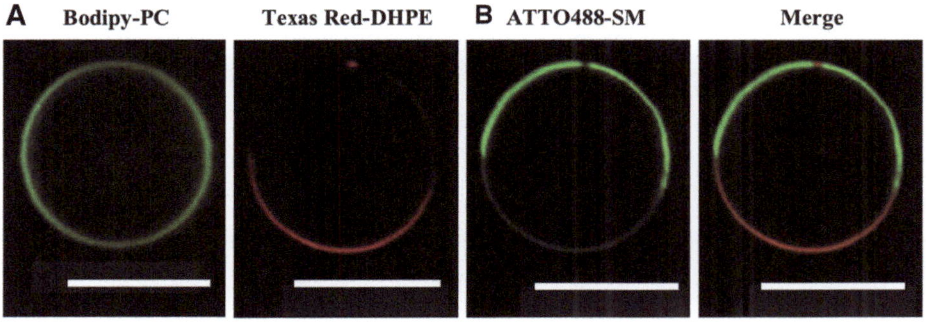

Fig. 2 Sphingomyelin-stabilized phase separation. (**A**) GUV (DOPC/GM3/cholesterol = 2/2/1 with 0.1 mol% Bodipy-PC) did not show phase separation. Scale bar represents 5 μm. (**B**) Sphingomyelin-containing GUV (DOPC/SM/GM3/cholesterol = 2/1/1/1 with 0.1 mol% Texas Red-DHPE and ATTO488-SM) showed phase separation. Scale bar represents 10 μm. For the observation conditions, refer to Sect. 3.3

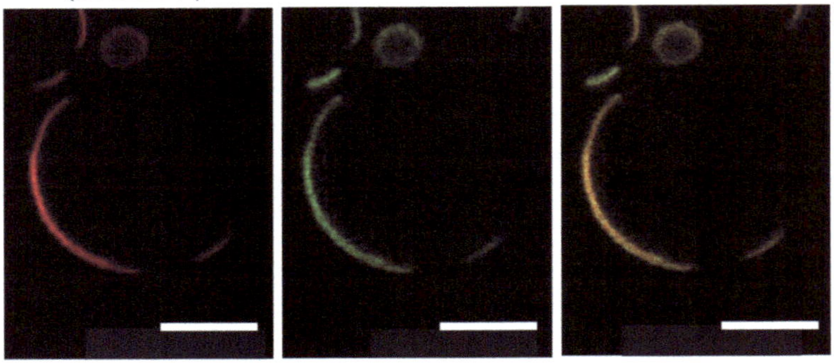

Fig. 3 NBD-IR TM was introduced to GUV. GUV was made at the ratio of DOPC/SM/GM3/cholesterol/IR TM = 33/28/5/33/0.4 with 0.1 mol% Texas Red-DHPE. Scale bar represents 10 μm. For the observation conditions, refer to Sect. 3.3

2. NBD-IR TM, Bodipy-PC, and ATTO488-SM are excited at 473 nm, and Texas Red-DHPE is excited at 559 nm by each excitation laser. The emission signals are detected as bands at 485–545 nm and 570–670 nm, respectively.

4 Notes

1. The commercially available 3 mm thick rubber sheet will be attached using a commercially available polymer adhesive for rubber.
2. Although large unilamellar vesicles (LUVs) are formed by vibration, some small unilamellar vesicles (SUVs) or multilamellar vesicles (MLVs) are also slightly formed.
3. The liposomes about the same size as cells are well formed in the vicinity of the platinum wire.
4. The 3D imaging of phase-separated liposomes in Fig. 4 can be made closer to the actual spherical shape as the scanning speed increases.

Conflict of Interest

The authors have no conflicts of interest to declare that are relevant to the content of this chapter.

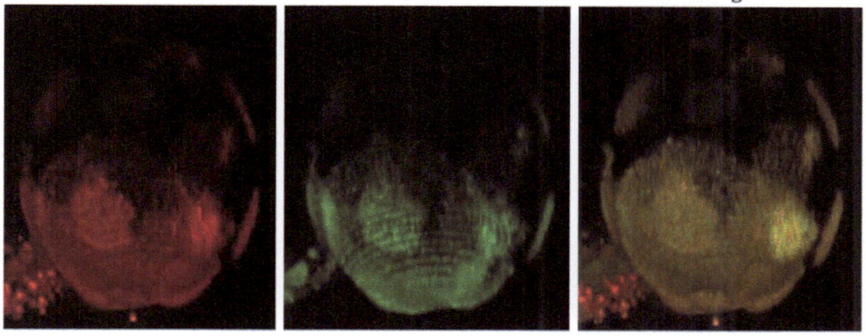

Fig. 4 3D imaging of GUV. GUV was made at the ratio of POPC/SM/GM3/cholesterol/IR TM = 40/25/15/20/0.4 with 0.1 mol% Texas Red-DHPE. For the observation conditions, refer to Sect. 3.3

References

1. Jobling MG, Holmes RK (2002) Mutational analysis of ganglioside GM(1)-binding ability, pentamer formation, and epitopes of cholera toxin B (CTB) subunits and CTB/heat-labile enterotoxin B subunit chimeras. Infect Immun 70:1260–1271
2. Heyningen SV (1974) Cholera toxin: interaction of subunits with ganglioside GM1. Science 183:656–657
3. Russo D, Parashuraman S, D'Angelo G (2016) Glycosphingolipid-protein interaction in signal transduction. Int J Mol Sci 17:1732
4. Bremer EG, chlessinger J, Hakomori S (1986) Ganglioside-mediated modulation of cell growth. Specific effects of GM3 on tyrosine phosphorylation of the epidermal growth factor receptor. J Biol Chem 261:2434–2440
5. Coskun U, Grzybek M, Drechsel D et al (2011) Regulation of human EGF receptor by lipids. Proc Natl Acad Sci USA 108:9044–9048
6. Mukherjee P, Faber AC, Shelton LM et al (2008) Thematic review series: sphingolipids. Ganglioside GM3 suppresses the proangiogenic effects of vascular endothelial growth factor and ganglioside GD1a. J Lipid Res 49:929–938
7. Lopez PH, Schnaar RL (2009) Gangliosides in cell recognition and membrane protein regulation. Curr Opin Struct Biol 19:549–557
8. Haeusler RA, McGraw TE, Accili D (2018) Biochemical and cellular properties of insulin receptor signalling. Nat Rev Mol Cell Biol 19:31–44
9. Boucher JJ, Kleinridders A, Kahn CR (2014) Insulin receptor signaling in normal and insulin-resistant states. Cold Spring Harb Perspect Biol 6:a009191
10. Stephens JM, Lee J, Pilch PF (1997) Tumor necrosis factor-alpha-induced insulin resistance in 3T3-L1 adipocytes is accompanied by a loss of insulin receptor substrate-1 and GLUT4 expression without a loss of insulin receptor-mediated signal transduction. J Biol Chem 272:971–976
11. Tagami S, Inokuchi J, Kabayama K et al (2002) Ganglioside GM3 participates in the pathological conditions of insulin resistance. J Biol Chem 277:3085–3092
12. Kabayama K, Sato T, Saito K et al (2007) Dissociation of the insulin receptor and caveolin-1 complex by ganglioside GM3 in the state of insulin resistance. Proc Natl Acad Sci USA 104:13678–13683
13. Sohma Y, Sasaki M, Hayashi Y et al (2004) Novel and efficient synthesis of difficult sequence-containing peptides through O–N intramolecular acyl migration reaction of O-acyl isopeptides. Chem Commun 1:124–125
14. Angelova MI, Dimitrov DS (1986) Liposome electroformation. Faraday Discuss 81:303–311
15. Nakano M, Hanashima S, Hara T et al (2021) FRET detects lateral interaction between transmembrane domain of EGF receptor and ganglioside GM3 in lipid bilayers. Biochim Biophys Acta Biomembr 1863:183623

Check for updates

Chapter 15

Methods for Investigating Siglec–Ganglioside Interactions

Edward N. Schmidt and Matthew S. Macauley

Abstract

Siglecs are a family of cell surface receptors generally found on immune cells, which are important in many cellular processes such as cell-adhesion, cell-signaling, and internalization of extracellular particles. While all these smaller functions play a part in larger and different physiological roles, one aspect they share is that they are initiated by the binding of a Siglec to its ligand. Siglec ligands are cell-surface carbohydrates appended to proteins and lipids that terminate with the monosaccharide sialic acid. The functions of a Siglec are directly linked to the binding of its ligands. There has been a great effort by many research teams over the past three decades to describe Siglec ligands to better understand their roles in health and disease. The systematic description of Siglec ligands has revealed that Siglecs are important in a range of physiological processes, including but not limited to allergies, antibody production, cancer progression, and reproduction. However, despite these efforts, the description of Siglec ligands remains incomplete. Historically Siglec-glycoprotein interactions have been better described compared to Siglec-glycolipid interactions; this is likely due to the combination of the relative difficulty in studying Siglec-glycolipid interactions and the dearth of tools available to study these interactions. Gangliosides are the major class of sialylated glycolipids in mammals. In this chapter, three protocols used to investigate Siglec-ganglioside interactions are described; these methods can be applied to further our understanding of gangliosides as Siglec ligands and the physiological and pathophysiological roles that they mediate.

Keywords Siglec, Ganglioside, Liposome, Flow cytometry, Avidity

1 Introduction

Cell surface carbohydrates or glycans are positioned to be the first point of contact between two cells, and these glycans contribute to cellular communication [1]. There are two major classes of glycans: glycoproteins and glycolipids. However, glycans are only half of the conversation; after all, what good is talking if you cannot hear. Lectins or glycan-binding proteins are the other half of the conversation [2]. In oversimplified terms, glycans can be thought of as *words* and lectins can be thought of as *ears*. Many mammalian glycans terminate with the monosaccharide sialic acid and gangliosides are the most abundant glycolipid carriers of sialic acid in adult mammals [3]. Sialic acid containing glycans serve many biochemical purposes,

Koichi Furukawa (ed.), *Gangliosides*, Methods in Molecular Biology, vol. 3035,
https://doi.org/10.1007/978-1-0716-5288-6_15,

one of which is to help the immune system recognize *self*-cells. [4]. Immune cells engage with these sialylated glycans using a family of lectins known as the Siglecs (**si**alic-acid-binding immuno**g**lobulin-type **lec**tins) and in humans, there are fifteen Siglecs [5–8]. Siglec-ganglioside interactions have not been described to the same degree as Siglec-glycoprotein interactions; however, from what is relatively little known about Siglec-ganglioside interactions, they have been found to be very important in human physiology and pathophysiology [3]. The importance of Siglec-ganglioside interactions in physiological processes such as viral infection, cancer, and neurological development motivates a more complete description of gangliosides as Siglec ligands [9–11]. Given the ubiquitous nature of ganglioside expression and the recent advancements in the tools to study Siglec-ganglioside interactions, the physiological processes mediated by Siglecs and gangliosides described above are likely just the tip of the iceberg and Siglec-ganglioside interactions are expected to be important in more physiological roles.

1.1 Sialic-Acid-binding Immunoglobulin-type Lectins

Siglecs (Siglecs are sialic-acid-binding immunoglobulin-type lectins) are a family of cell surface receptor generally expressed by immune cells. At the N-terminus of a Siglec is the carbohydrate binding domain (CRD, also known as V-set domain). Within the CRD there is an arginine residue which is the key residue for binding sialylated glycans such as gangliosides. However, compared to other protein-ligand interactions, Siglec-sialoside interactions are relatively weak with dissociation constants (K_d) in the range of 0.1-1 mM [12–16]. To overcome this weak intrinsic affinity, Siglecs harness a phenomenon referred to as avidity. Avidity is a complicated concept; however, it is often likened to Velcro® wherein many small interactions add up to a very strong interaction. With respect to Siglecs and glycans, many copies of the Siglec engage with many copies of the ligand simultaneously [1]. Inspired by nature, nearly all approaches used to describe Siglec ligands also use very multivalent systems [14–19]. For more information regarding Siglecs consult the following reviews [5, 8].

1.2 Biological Siglec-Ganglioside Interactions

Gangliosides are a family of glycosphingolipids that share the carbohydrate back bone β-Gal*p*-(1 → 3)-β-Gal*p*NAc-(1 → 4)-β-Gal*p*-(1 → 4)-β-Glc*p* and are anchored into the lipid bilayer by ceramide [20]. Sialic acid can be linked at the terminal Gal (α(2→3)), the GalNAc (α(2→6)), or the internal Gal (α(2→3)) [21]. Moreover, either or both of the (α(2→3)) linked sialic acids can be further elongated with α(2→8) linked sialic acid. Gangliosides are most abundant on the outer leaflet of the cell membrane and generally compose 0.1–5 mol% of the lipids in the bilayer depending on the cell type [22–25]. Historically, glycoproteins have been better described as Siglec ligands compared to glycolipids for a number of practical reasons, however what is known about Siglec-gangliosides interactions was summarized in a recent review [26]. One of these reasons is that there were few approaches to study Siglec-ganglioside interactions. In the recent past, Siglec-ganglioside interactions have been the subject of many studies [13, 15, 16, 27, 28]. This has led to the discovery of many biological processes mediated by Siglec-ganglioside interactions. One of the first discoveries of a biological interaction between Siglecs and gangliosides was Siglec-4 (AKA MAG) and gangliosides GD1a/GT1b in neuronal health [11]. Following this discovery, it was observed that many enveloped viruses, such as HIV and later Ebola and SARS-CoV-2 use gangliosides acquired in the budding process to exploit Siglec-1 to contribute to their virulence [9, 29–31]. Many studies have also suggested that gangliosides in extracellular vesicles mediate binding to Siglecs, including Siglec-1, -6, and -7 [16, 32, 33]. However, given that gangliosides are found in all cell types in some capacity, and that only a few *bonafide* biologically relevant Siglec-ganglioside interactions have been described, it is likely that there are many more just waiting to be found.

1.3 Traditional Assays

One of the pioneering approaches used in the 1990s to study gangliosides as Siglec ligands was a cell-adhesion assay [34–36]. In this assay, gangliosides were adsorbed to a microplate, and then a cell expressing a Siglec would be added to the well of the microplate, and binding would be quantified using LDH activity following a wash and cell lysis (Fig. 1a). In the early 2000s, this approach was replaced with the enzyme-linked immunoassay (ELISA, Fig. 1b) [37]. Similar to the cell-adhesion assay, gangliosides were adsorbed to a microplate; however, instead of using Siglec-expressing cells, a soluble version of a Siglec known as a Siglec-Fc was used [18]. For more information regarding Siglec-Fcs please see the following references [18, 38]. Siglec-Fcs which bound gangliosides on the plate would be detected using an antibody that was conjugated to an enzyme such as horseradish peroxidase (HRP) and the binding could be quantified colorimetrically. Using this approach, other Siglecs such as Siglec-5 and Siglec-7 were identified as ganglioside binders [37]. This style of assay is also used recently to study Siglec-ganglioside interactions [15, 16, 39, 40].

1.4 Next Generation Assays

As a better understanding of gangliosides as Siglec ligands was taking shape, there was a switch in assay style towards approaches that use cells expressing full-length Siglecs instead of soluble versions of the Siglecs, and the ligands were presented from nanoparticles (Fig. 1c) [9, 15, 16, 27, 28]. This style of approach led to the discovery that viruses use gangliosides to exploit Siglecs [9]. While cell assays are excellent, working with cells comes with their own challenges. For instance, cells can express more than one Siglec making it difficult to resolve which Siglec is interacting with the ganglioside. Moreover, the cells themselves will have a glycocalyx, which could block the Siglec

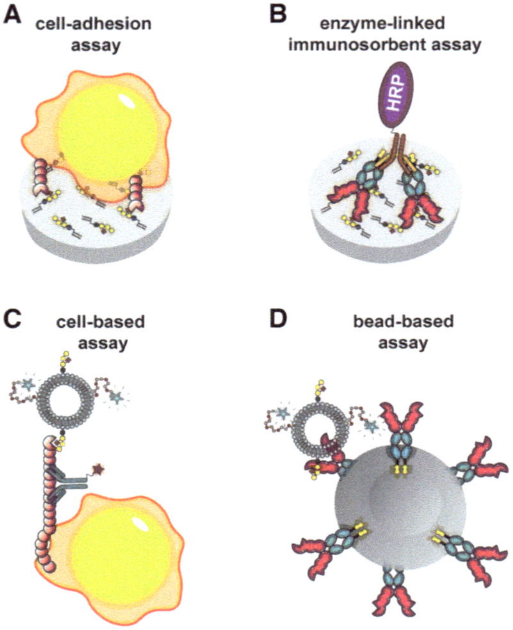

Fig. 1 Assays to probe Siglec-ganglioside interactions. Approaches can be divided into two categories: plate-based approaches (**a** and **b**) and bilayer-based approaches (**c** and **d**). The cell-adhesion assay was the first approach used to systematically probe Siglec-ganglioside interactions, and the majority of work that used this approach was performed in the late 1990s and early 2000s. The enzyme-linked immunosorbent assay (ELISA) became more popular in the early 2000s and is still used in 2024. Approaches where the ganglioside was presented to the Siglecs from a bilayer became more common in the 2010s and continues to be used in 2024

binding to the ganglioside [16]. While competition from *cis* glycans can be overcome many ways (e.g., sialidase [16], pharmacological inhibitors [41], genetically [17, 42], etc.) it can be cumbersome to chemically/enzymatically treat the cells before each experiment or develop a knock out cell line. Lastly, while not as problematic as the other issues, cell-based assays often report on a combination of liposome binding and internalization due to the endocytic nature of Siglecs [7]. To address these issues, a bead assay can be used. In this approach, magnetic beads (approximately 1 μm in diameter) coated with a protein such as streptavidin or protein A/G are used [15, 16]. These beads are then complexed with a Siglec-Fc to generate a Siglec-coated bead and are ready to be used to measure the interaction between a Siglec and ganglioside (Fig. 1d).

1.5 Liposomes

Many nanoparticles have been used to probe Siglec-ganglioside interactions, including liposomes [15, 18], phages [43], gold nanoparticles [44], or labeled protein probes [45]; however, liposomes are an ideal platform to study Siglec-ganglioside. While liposomes are by no means a perfect re-creation of a biological membrane, liposomes present the ganglioside from a bilayer instead of being adsorbed to a surface, a step in the right direction towards a genuine biological membrane. Indeed, ligand presentation has been found to influence the ability of a lectin to engage with a glycan [16, 46]. Many different liposome formulations (or recipes) exist, and each occupies a different niche. Most liposome formulations consist of a phospholipid (e.g., DSPC) and a sterol (e.g., cholesterol) for structural integrity as well as a PEGylated lipid (e.g., PEG_{45}-DSPE) to prevent non-specific interactions and/or improve circulatory half-life [47]. Regardless of the formulations, liposomes are generally described with respect to the concentration of the lipids rather than the number of particles. For instance, if a liposome concentration is reported as 1 mM, this means that the total concentration of lipids is 1 mM. Each component of the liposome is described as a mol%. The liposome formulation we recommend for studying Siglec-ganglioside interactions is 58.5 mol% DSPC, 38% cholesterol, 3% ganglioside, and 0.5% PEG_{45}-DSPE. Following this, in 1 mM liposomes according to the formulation described above, there is 0.585 mM DSPC, 0.38 mM cholesterol, 0.03 mM ganglioside, and 0.05 mM PEG_{45}-DSPE. To easily measure a binding interaction between a Siglec and a ganglioside liposome, the liposomes are often fluorescently labeled [15, 16].

2 Materials

2.1 Siglec-Fc

1. Siglec-Fcs can be purchased from retailers such as *R & D Systems* or can be produced as described elsewhere [18].

2.2 Buffers

1. Many different buffers are required for the assays described in this Chapter. A description of each buffer can be found in Table 2.

2.3 ELISA

1. We recommend using Strep-Tactin conjugated to HRP, which can be purchased from *Fisher Scientific* (Cat. No. NC9523094) as the enzymatic reporter.
2. We recommend using the TMB substrate kit from *Sera Care* (Cat. No. 5120-0047) as the colourmetric substrate.
3. 1 M phosphoric acid is also required to quench the reaction.

2.4 Liposome Production

1. While there are many approaches to produce liposomes, we recommend using the thin film extrusion method, which requires a device such as the *Avanti Mini Extruder®*.
2. Using this method will also require filter supports (also known as pre-filters) and filters. We recommend using *Cytiva* filter supports (Cat. No. 230300) and for 800 nm and 100 nm filters (Cat. No. 10417304 and 800309, respectfully).

2.5 Structural Lipid Solutions

1. Dry lipids (DSPC, cholesterol, PEG_{45}-DSPE) should be stored according to the manufacture's recommendations. When ready to make liposomes, stock solutions should be prepared according to Table 1 using chloroform (*see* **Note 1**).

2.6 Functional Lipids

1. In this context, functional lipids refer to gangliosides and fluorescent lipids. As little amounts of each functional lipid are required for each liposome, it is recommended to make stock solutions of these lipids in DMSO and store them at −80°C.
2. For functional lipids, 4 mg/mL is our recommended concentration for stock solutions; however, this can be adjusted as needed.

2.7 Fluorescent Lipids

1. For any liposome binding assay, a fluorescent reporter can be added to the liposomes making it easy to quantify binding.
   a. One way is to make a fluorescent lipid using an NHS-activated fluorophore (*Thermo Fisher Scientific* Cat. No. A20006) and an amine-terminated lipid such as NH_2-$[PEG]_{45}$-DSPE (*Avanti* Cat. No. 880128 P) in DMSO under basic conditions.
   b. Alternatively, lipids like this can be purchased (*Avanti* Cat. No. 880154 C).

Table 1
Recommended lipid stock solution concentrations for making 1 mL of 1 mM liposomes

Lipid	Cat. No. (*Avanti*)	[Stock Solution]
DSPC	850365 P	4 mg/mL
Cholesterol	700100 P	2 mg/mL
PEG-DSPE	880128 P	0.1 mg/mL

Table 2
Buffer formulations required for each assay to study Siglec-ganglioside interactions

Assay	Buffers required
Cell assay	*Assay Buffer:* PBS pH 7.4, 1% (g/mL) BSA *Flow Buffer:* PBS pH 7.4, 1% (g/mL) BSA, 500 μM EDTA
Bead assay	*Blocking Buffer:* PBS pH 7.4, 5% (g/mL) BSA *Assay Buffer:* PBS pH 7.4, 1% (g/mL) BSA.
ELISA	*Ganglioside Solution:* 50 μM in 100% EtOH *Blocking Buffer:* PBS pH 7.4, 5% (g/mL) BSA *Wash Buffer:* PBS pH 7.4 *Quenching Buffer:* 1 M H_3PO_4 *Developing solution:* 1:1 (V:V) TMB solution part A and B

c. Another option is to use a lipophilic dye such as DiD, which can also be purchased (*Thermo Fisher Scientific* Cat. No. D7757).

2.8 Cell Lines

1. Many cell lines, such as monocyte-derived U937 or B cell-derived Daudi cell lines, naturally express Siglecs and can be used for liposome binding experiments [17, 48]. However, cell lines that express Siglecs often express multiple Siglecs (e.g., U937 cells express Siglec-3, -5, and -7) making it difficult to resolve which Siglec is engaging with the probe [17].
2. To overcome this, a cell line that only expresses one Siglec can be developed [16, 17].
   a. This can be accomplished in two ways: expression of a Siglec from a cell line that does not express Siglecs (e.g., Chinese Hamster Ovary (CHO) cells, Human Embryonic Kidney (HEK) 293T, etc.) or genetically knocking out Siglecs in a cell line [17]. CHO and HEK293T are excellent choices as they have both been engineered with the Flp-In™ System, which enables the facile addition of a gene, such as a full-length Siglec, to the genome of a cell.
3. Many transfection and transduction protocols have been reported elsewhere [14, 19, 49, 50].

2.9 Bead Assay

1. For the bead assay, we recommend using *Pierce™ Magnetic Beads*. These can be coated with either streptavidin or protein A/G, which can be purchased from retailers including *Thermo Fisher Scientific* (Cat. No. 88816, 88847).
2. Streptavidin beads are recommended if primary human samples are being used as the presence of protein A/G can dramatically increase the amount of glycan-independent binding. However, if only liposomes are being used, then either streptavidin or protein A/G beads will work.
3. Regardless of which bead is being used, a 96-well plate magnet is needed. These are sold by many retailers such as *New England Biolabs* (Cat. No. S1511S)

3 Methods

3.1 Siglec ELISA

1. See Fig. 2 for an overview of the assay.
2. 50 μM ganglioside stock solutions are prepared using 100% EtOH.
3. Add 50 μL of stock solution to each well of the microplate as desired (add 50 μL of 100% EtOH to wells as a vehicle control.

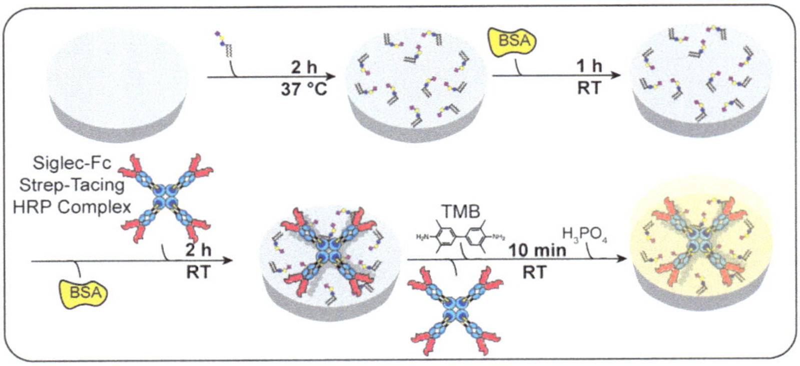

Fig. 2 Schematic description of the ELISA

4. Allow ethanol to evaporate overnight at room temperature or 2 h at 37°C.
5. Block all wells with *Blocking Buffer* and let rest at room temperature for 1 h.
6. Remove blocking solution by inverting and flicking the microplate into a waste reservoir and wash the plate by submerging the microplate in PBS. Remove the PBS by inverting the microplate. Repeat for a total of five times.
7. Add 50 μL of precomplexed-Strep-Tactin-Siglec-Fc solution to each well and incubate at room temperature for 2 h.
8. Remove both parts of the developing solution from 4°C storage and allow to warm to room temperature.
9. Wash the microplate as described in step 5.
10. Combine the developing solution and 75 μL transfer to each well. Allow blue color to develop for approximately 10 min.
11. Quench reaction with 75 μL of *Quenching Buffer*.
12. Quantify absorbance at 450 nm.

3.2 Liposome Preparation (Thin Film)

1. Prepare structural lipid stock solutions using chloroform according to Table 2.
2. Decide on the appropriate formulation for the experiment and make a table such as Table 3 using the equation below, which uses DSPC as example.

a. $$V_{DSPC} = \frac{C_{Lipo} * x * MW_{DSPC} * V_{Lipo}}{C_{DSPC}}$$

b. $$V_{DSPC} = \frac{\left(1 \times 10^{-3} \frac{mol}{L}\right) * \left(\frac{58.5}{100}\right) * 790 \frac{g}{mol} * 1 \times 10^{-3} L}{4 \frac{g}{L}}$$

c. $$V_{DSPC} = 1.15 \times 10^{-4} L = 115\ \mu L$$

Table 3
Representative Liposome formulation table to make 1 mL of 1 mM 3 mol% GM1 liposomes

	Phospholipid (DSPC)	Sterol (cholesterol)	PEG_{45}-DSPE (PEGylated lipid)	Ganglioside (GM1)	Fluorescent Lipid (AF647-PEG_{45}-DSPE)
Molecular Weight (g/mol)	790	387	2900	1546	4060
χ (mol%)	58.5	38.0	0.4	3.0	0.1
Amount(μ mol)	0.59	0.38	4.00E-03	0.03	1.00E-03
Mass (mg)	0.46	0.15	0.01	0.05	4.00E-03
Volume (μL)	115.5	73.5	116.0	11.60	4.0

3. Transfer the appropriate volume of each structural lipid chloroform solution to a glass test tube.
4. Remove the chloroform from the lipids using a gentle flow of N_2 gas to yield a lipid thin film.
5. Add 100 μL of DMSO to each test tube and then add the appropriate volume of each functional lipid from their respective DMSO stock to the glass test tube.
6. Place the test tube at −80°C for 30 min and then transfer to a lyophilizer to remove the DMSO. This will likely only take a few hours but can be left overnight.
7. The next day, remove the sample from the lyophilizer and store at −20 °C until ready to extrude.
8. More information regarding liposome production can be found elsewhere [51].

3.3 Liposome Preparation (extrusion)

1. Remove the thin films from the freezer and allow them to warm to room temperature.
2. Add 1 mL of PBS (or other desired buffer) and cover/wrap the glass test tubes with parafilm.
3. Using a bath sonicator, sonicate the lipid/PBS solutions in a cycle of 1 min on, 5 min rest, for a total of five cycles.
4. While the liposome solutions are being sonicated, wash all components (except the filters and the filter supports) of the *Avanti Mini Extruder®* and two 1 mL *Hamilton syringes* with Milli-Q water and then methanol (Fig. 3a). After washing leave to dry (*see* **Note 2**).
5. Once Dry, assemble the *Avanti Mini Extruder®* as described in Fig. 3b with an 800 nm filter. To make it easier to get the

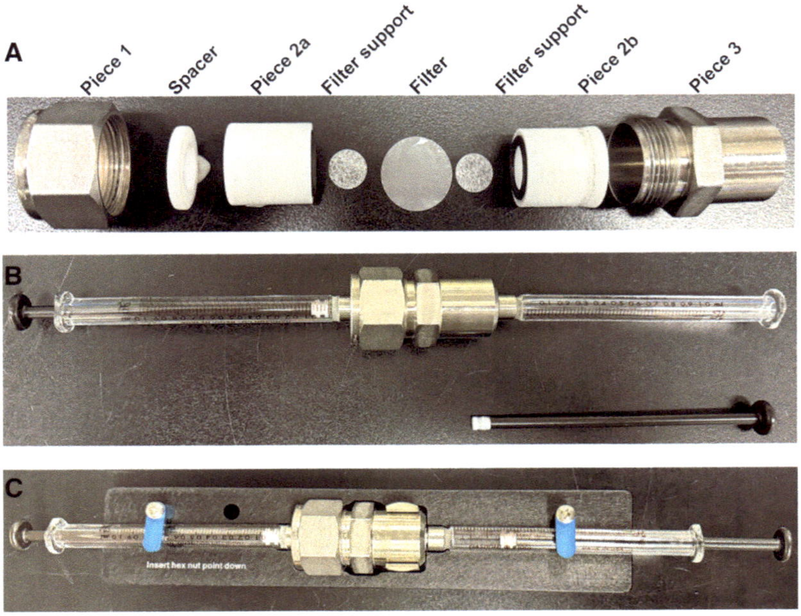

Fig. 3 Assembly of an *Avanti Mini Extruder®*. (**a**) Disassembled *Avanti mini Extruder®*. (**b**) Assembled *Avanti mini Extruder®* with two 1 mL *Hamilton syringes*. Complete *Avanti mini Extruder®* with *Hamilton* syringes in the extruding block

filter supports and filter to stay on pieces 2a and b, a drop of PBS can be placed on the pieces.

6. Transfer 1 mL of the liposome solution to the 1 mL *Hamilton syringe* which does not have the plunger in. Once all the solutions are in the syringe, insert the plunger and place the extruder into the extruder block such that it resembles Fig. 3c. To make extrusion physically easier, the extruder block can be warmed to 60°C prior to extrusion.
7. Push the plunger so that the liposome solution travels through the extruder. Pass the solution through the extruder for a total of twenty times.
8. Gently remove the syringe containing the liposomes and transfer the liposomes to a new container.
9. Disassemble the extruder and replace the 800 nm filter with a 100 nm filter and reassemble the extruder.
10. Repeat steps 6–8.
11. Disassemble the extruder and wash all components with Milli-Q water and then methanol.
12. Liposomes can then be characterized by dynamic light scattering.
13. Liposomes should be stored at 4°C in the dark if they have a fluorescent label.

3.4 Cell Assay

1. See Fig. 4 for an overview of the assay.
2. Plate 1×10^5–2×10^5 cells per well into a 96-well U-bottom 96-well plate and centrifuge at $330 \times g$ for 5 min.
3. Prepare liposome solution by diluting liposome stock solution to 50 μM with *Assay Buffer*.
4. Remove the supernatant by inverting the microplate and resuspend cells in liposome solution and incubate at 37°C for 30 min. Following incubation, centrifuge cells at $330 \times g$ for 5 min.
5. Remove the supernatant by inverting the microplate and resuspend cells in 200 μL of *Flow Buffer* and centrifuge cells at 330 $\times g$ for 5 min.
6. Prepare antibody solution by diluting the anti-Siglec antibody in 1:500 (V:V) in *Flow buffer*.
7. Remove the supernatant by inverting the microplate and resuspend cells in antibody solution (50 μL/well) and incubate on ice (4°C) for 30 min. Following incubation, centrifuge the cells at $330 \times g$ for 5 min.
8. Wash the cells as described in Step 4.
9. Resuspend the cells in 300 μL of *Flow Buffer* and proceed to analyze the cells by flow cytometry (Fig. 5).

3.5 Bead Assay

1. See Fig. 6 for an overview of the assay.
2. Prepare *Blocking Buffer* and add 0.33 μL/replicate to a 1.5 mL microcentrifuge tube (e.g., if 30 replicates are needed add 10 μL of solution).

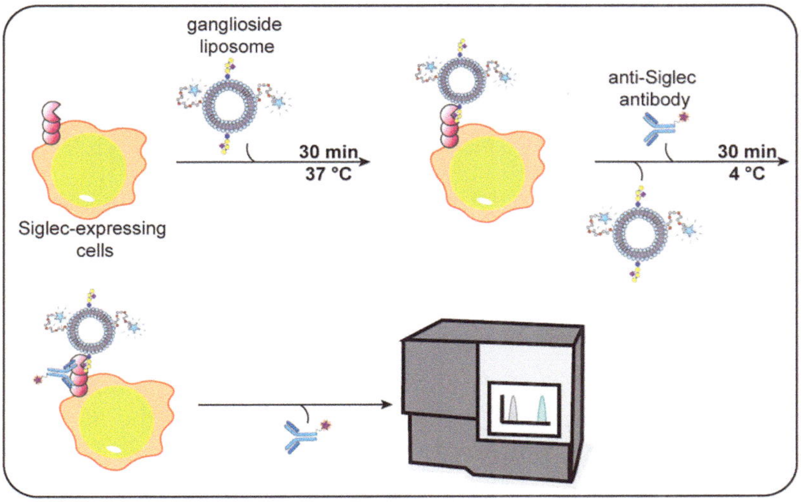

Fig. 4 Schematic description of the cell assay

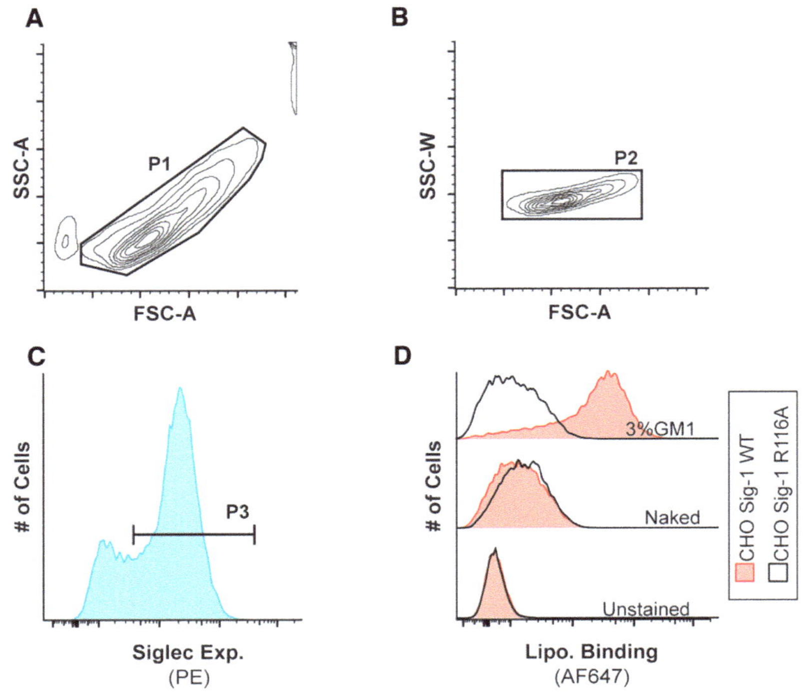

Fig. 5 Gating strategy for the cell assay. (**a**) Side scatter area (SSC-A) *vs* forward scatter area (FSC-A). P1 is gated on the center population. (**b**) Side scatter width (SSC-W) *vs* forward scatter area (FSC-A), where P2 is used to gate out doublets. (**c**) Histogram of Siglec-1$^+$ cells where P3 is used to gate on Siglec-1+ cells. (**d**) Histogram AF647 signal, which represents liposome binding to the Siglec-1$^+$ cells

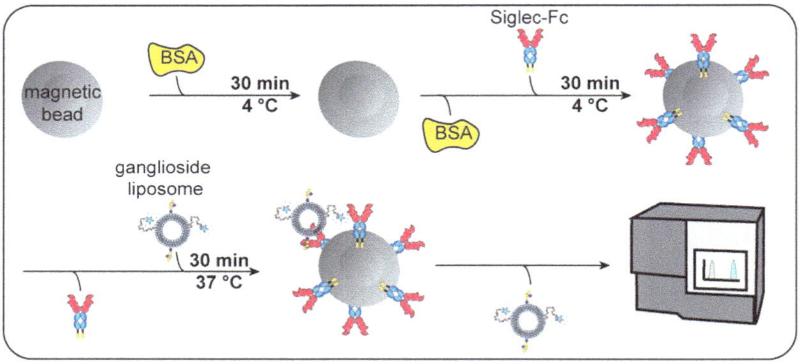

Fig. 6 Schematic description of the bead assay

3. Vortex *Pierce™ Streptavidin Magnetic Beads* for 30 s and add an equal volume of bead solution to the *Blocking Buffer* in the 1.5 mL microcentrifuge tube (e.g., if 30 replicates are needed add 10 μL of *Blocking Buffer* to the 10 μL of bead solution so that the total volume in the 1.5 mL microcentrifuge tube is now 20 μL-*see* **Note 3**). Block the beads on ice (4°C) for 30 min.

4. Add 80 μL of 25 μg/mL Siglec-Fc solution to the blocked beads and let rest on ice (4°C) for 30 min.
5. Prepare liposome solution by diluting liposome stock solution to 50 μM *Assay Buffer*.
6. Resuspend the Siglec-Fc-Beads in enough *Assay Buffer* solution to transfer the beads into a microplate. (e.g., If 920 μL of solution *Assay Buffer* is added to the Siglec-Fc beads, the total volume would then be 1000 μL, 30 μL of this solution can be added to the wells of the microplate for 30 replicates).
7. Place the microplate containing the Siglec-Fc-Beads on the magnet and wait for 5 min. After 5 min, small brown clumps should be visible with the naked eye.
8. GENTLY remove the solution with a multi-channel pipette and remove the microplate from the magnet.
9. Resuspend the Siglec-Fc-Beads in 50 μL of 50 μM liposome solution and incubate the beads with the liposomes at 37°C for 30 min.
10. Following the incubation, add 100 μL of *Assay Buffer* and return the microplate to the magnet.
11. Remove the solution and then wash the beads with Assay Solution as described in Step 7.
12. Repeat Steps 9 and 10.
13. Resuspend the Siglec-Fc-Beads in 300 μL of *Flow Buffer* and analyze by flow cytometry as described in Fig. 7.

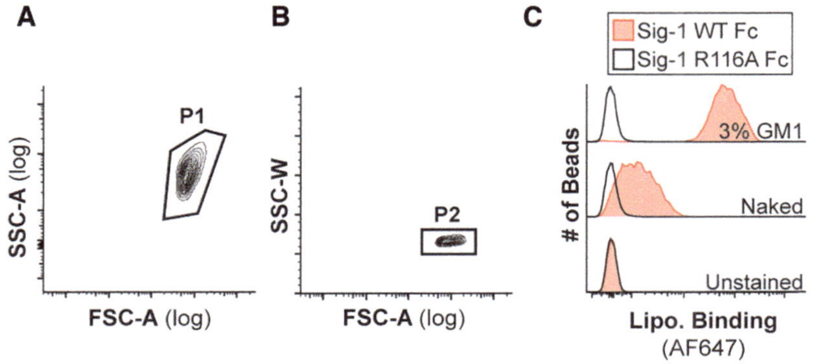

Fig. 7 Gating strategy for the bead assay. (**a**) Side scatter area (SSC-A) *vs* forward scatter area (FSC-A), where P1 is gated on the center population to isolate beads from debris. (**b**) Side scatter width (SSC-W) *vs* forward scatter area (FSC-A), where P2 is used to gate out doublets. Notice that both axes for panels (**a**) and (**b**) are logarithmic instead of linear. (**c**) Histogram AF647 signal, which represents liposome binding to the Siglec-1$^+$ beads

4 Notes

1. These concentrations are recommended when making 1 mL of 1 mM liposomes. However, if a higher concentration of liposome is required, multiple each concentration by the increase factor (e.g., for 3 mM liposomes the new concentrations would be 12 mg/mL, 6 mg/mL, and 0.3 mg/mL for DSPC, cholesterol, and PEG_{45}-DSPE, respectively).
2. *Pieces 2a*, *b* and the *Hamilton syringes* can take longer than the rest of the extruder. To decrease the drying time, the hole in pieces 2a and b can be dried using a 3 mL generic syringe to blow any remaining methanol out of the pieces. It is also recommended to push air through the *Hamilton syringes* to remove any remaining methanol.
3. The amount of bead solution and concentration of Siglec-Fc can be adjusted proportional when scaling up or down the assay. Moreover, if results are not as expected, the beads can be stained with anti-Siglec antibodies to confirm that the Siglec-Fcs have formed a complex with the beads.

Conflict of Interest

The authors have no conflicts of interest to declare that are relevant to the content of this chapter.

References

1. Varki A (2017) Biological roles of glycans. Glycobiology 27:3–49
2. Ambrosi M, Cameron NR, Davis BG (2005) Lectins: tools for the molecular understanding of the glycocode. Org Biomol Chem 3:1593–1608
3. Sipione S, Monyror J, Galleguillos D et al (2020) Gangliosides in the Brain: physiology, pathophysiology and therapeutic applications. Front Neurosci 14:572965
4. Jame-Chenarboo Z, Gray TE, Macauley MS (2024) Advances in understanding and exploiting Siglec-glycan interactions. Curr Opin Chem Biol 80:102454
5. Duan S, Paulson JC (2020) Siglecs as immune cell checkpoints in disease. Annu Rev Immunol 38:365–395
6. Macauley MS, Crocker PR, Paulson JC (2014) Siglec-mediated regulation of immune cell function in disease. Nat Rev Immunol 14:653–666
7. Crocker PR, Paulson JC, Varki A (2007) Siglecs and their roles in the immune system. Nat Rev Immunol 7:255–266
8. Lin SY, Schmidt EN, Takahashi-Yamashiro K et al (2024) Roles for Siglec-glycan interactions in regulating immune cells. Semin Immunol 77:101925
9. Izquierdo-Useros N, Lorizate M, Puertas MC et al (2012) Siglec-1 is a novel dendritic cell receptor that mediates HIV-1 trans-infection through recognition of viral membrane gangliosides. PLoS Biol 10:e1001448
10. Theruvath J, Menard M, Smith BAH et al (2022) Anti-GD2 synergizes with CD47 blockade to mediate tumor eradication. Nat Med 28:333–344
11. Sun J, Shaper NL, Itonori S et al (2004) Myelin-associated glycoprotein (Siglec-4) expression is progressively and selectively decreased in the brains of mice lacking complex gangliosides. Glycobiology 14:851–857
12. Bui DT, Favell J, Kitova EN et al (2023) Absolute affinities from quantitative shotgun glycomics using concentration-independent (COIN) native mass spectrometry. ACS Cent Sci 9:1374–1387

13. Favell JW, Bui DT, Li J et al (2024) Elusive protein-glycosphingolipid interactions revealed by membrane anchor-assisted native mass spectrometry. J Am Chem Soc 146:21700–21709
14. Rodrigues E, Jung J, Park H et al (2020) A versatile soluble siglec scaffold for sensitive and quantitative detection of glycan ligands. Nat Commun 11: 5091
15. Schmidt EN, Guo XY, Bui DT et al (2024) Dissecting the abilities of murine Siglecs to interact with gangliosides. J Biol Chem 300:107482
16. Schmidt EN, Lamprinaki D, McCord KA et al (2023) Siglec-6 mediates the uptake of extracellular vesicles through a noncanonical glycolipid binding pocket. Nat Commun 14: 2327
17. McCord KA, Wang C, Anhalt M et al (2024) Dissecting the ability of siglecs to antagonize fcgamma receptors. ACS Cent Sci 10:315–330
18. Schmidt EN, Jung J, Macauley MS (2023) Flow cytometry-based detection of siglec ligands. carbohydrate-protein interactions. In: Methods in molecular biology. Springer, pp 181–193
19. Jung J, Enterina JR, Bui DT et al (2021) Carbohydrate sulfation as a mechanism for fine-tuning siglec ligands. ACS Chem Biol 16:2673–2689
20. Kolter T (2012) Ganglioside biochemistry. ISRN Biochem 2012:506160
21. Yu RK, Tsai YT, Ariga T et al (2011) Structures, biosynthesis, and functions of gangliosides–an overview. J Oleo Sci 60:537–544
22. Prinetti A, Chigorno V, Prioni S et al (2001) Changes in the lipid turnover, composition, and organization, as sphingolipid-enriched membrane domains, in rat cerebellar granule cells developing in vitro. J Biol Chem 276:21136–21145
23. Manuelidis L, Yu RK, Manuelidis EE (1977) Ganglioside content and pattern in human gliomas in culture. Correlation of morphological changes with altered gangliosides. Acta Neuropathol 38:129–135
24. Chigorno V, Tettamanti G, Sonnino S (1996) Metabolic processing of gangliosides by normal and Salla human fibroblasts in culture. A study performed by administering radioactive GM3 ganglioside. J Biol Chem 271:21738–21744
25. Hansson HA, Holmgren J, Svennerholm L (1977) Ultrastructural localization of cell membrane GM1 ganglioside by cholera toxin. Proc Natl Acad Sci U S A 74:3782–3786
26. Schnaar RL (2023) Gangliosides as siglec ligands. Glycoconj J 40:159–167
27. Affandi AJ, Grabowska J, Olesek K et al (2020) Selective tumor antigen vaccine delivery to human CD169(+) antigen-presenting cells using ganglioside-liposomes. Proc Natl Acad Sci U S A 117:27528–27539
28. Grabowska J, Affandi AJ, van Dinther D et al (2021) Liposome induction of CD8(+) T cell responses depends on CD169(+) macrophages and Batf3-dependent dendritic cells and is enhanced by GM3 inclusion. J Control Release 331:309–320
29. Gutierrez-Martinez E, Benet Garrabe S, Mateos N et al (2023) Actin-regulated Siglec-1 nanoclustering influences HIV-1 capture and virus-containing compartment formation in dendritic cells. eLife 12:e78836
30. Perez-Zsolt D, Erkizia I, Pino M et al (2019) Anti-siglec-1 antibodies block Ebola viral uptake and decrease cytoplasmic viral entry. Nat Microbiol 4:1558–1570
31. Perez-Zsolt D, Munoz-Basagoiti J, Rodon J et al (2021) SARS-CoV-2 interaction with siglec-1 mediates trans-infection by dendritic cells. Cell Mol Immunol 18:2676–2678
32. Saunderson SC, Dunn AC, Crocker PR et al (2014) CD169 mediates the capture of exosomes in spleen and lymph node. Blood 123:208–216
33. Shenoy GN, Loyall J, Berenson CS et al (2018) Sialic acid-dependent inhibition of T cells by exosomal ganglioside GD3 in ovarian tumor microenvironments. J Immunol 201:3750–3758
34. Collins BE, Ito H, Sawada N et al (1999) Enhanced binding of the neural siglecs, myelin-associated glycoprotein and Schwann cell myelin protein, to Chol-1 (alpha-series) gangliosides and novel sulfated Chol-1 analogs. J Biol Chem 274:37637–37643
35. Kelm S, Pelz A, Schauer R et al (1994) Sialoadhesin, myelin-associated glycoprotein and CD22 define a new family of sialic acid-dependent adhesion molecules of the immunoglobulin superfamily. Curr Biol 4:965–972
36. Yang LJ, Zeller CB, Shaper NL et al (1996) Gangliosides are neuronal ligands for myelin-associated glycoprotein. Proc Natl Acad Sci U S A 93:814–818
37. Rapoport E, Mikhalyov I, Zhang J et al (2003) Ganglioside binding pattern of CD33-related siglecs. Bioorg Med Chem Lett 13:675–678
38. Chang LY, Low PY, Sridharan D et al (2020) Preparation of recombinant Siglecs and identification of their ligands. Methods Mol Biol 2132:85–98
39. Yamakawa N, Yasuda Y, Yoshimura A et al (2020) Discovery of a new sialic acid binding region that regulates Siglec-7. Sci Rep 10: 8647
40. Yoshimura A, Hatanaka R, Tanaka H et al (2021) The conserved arginine residue in all siglecs is essential for Siglec-7 binding to sialic acid. Biochem Biophys Res Commun 534:1069–1075
41. Kumawat D, Gray TE, Garnier CR et al (2024) A kinetic trapping approach for facile access

to 3F(ax)Neu5Ac and a photo-cross-linkable sialyltransferase probe. J Am Chem Soc 146:28630–28634
42. Jung J, Schmidt EN, Chang HC et al (2025) Understanding the glycosylation pathways involved in the biosynthesis of the sulfated glycan ligands for siglecs. ACS Chem Biol 20(2): 386–400
43. Sojitra M, Schmidt EN, Lima GM et al (2024) Measuring carbohydrate recognition profile of lectins on live cells using liquid glycan array (LiGA). Nat Protoc 20:989–1019
44. Xu F, Reiser M, Yu X et al (2016) Lipid-mediated targeting with membrane-wrapped nanoparticles in the presence of corona formation. ACS Nano 10:1189–1200
45. Tiemeyer M, Yasuda Y, Schnaar RL (1989) Ganglioside-specific binding protein on rat brain membranes. J Biol Chem 264:1671–1681
46. Han L, Nguyen L, Schmidt EN et al (2022) How choice of model membrane affects protein-glycosphingolipid interactions: insights from native mass spectrometry. Anal Chem 94:16042–16049
47. Andra VVSNL, Pammi SVN, Bhatraju LVKP et al (2022) Correction to: a comprehensive review on novel liposomal methodologies, commercial formulations, clinical trials and patents. Bionanoscience 12:660
48. Madge PD, Maggioni A, Pascolutti M et al (2016) Structural characterisation of high affinity Siglec-2 (CD22) ligands in complex with whole Burkitt's lymphoma (BL) Daudi cells by NMR spectroscopy. Sci Rep 6: 36012
49. Ye J, Kober V, Tellers M et al (2009) High-level protein expression in scalable CHO transient transfection. Biotechnol Bioeng 103:542–551
50. Elegheert J, Behiels E, Bishop B et al (2018) Lentiviral transduction of mammalian cells for fast, scalable and high-level production of soluble and membrane proteins. Nat Protoc 13:2991–3017
51. Bednar KJ, Hardy L, Smeekens J et al (2018) Antigenic liposomes for generation of disease-specific antibodies. J Vis Exp (140):e58285. https://doi.org/10.3791/58285

Chapter 16

Expression of Mono- and Dimeric Ganglioside-Binding Lectins and Their Cellular Functions

Kenichi Kamata, Marco Gerdol, Imtiaj Hasan, Sultana Rajia, S.M. Abe Kawsar, Tatsuya Kawasaki, Masao Yamada, Yukiko Ogawa, Yasuhiro Ozeki, Hideaki Fujita and Yuki Fujii

Abstract

In the 2020s, a novel R-type lectin family with β-trefoil folding was found in Mytilidae (phylum Mollusca). SeviL, as described in this chapter, is the lectin discovered from *Mytilisepta virgata*, which is habituated in the intertidal community in the northwestern Pacific Ocean. SeviL is bound to the glycan moiety of gangliosides such as asialo-GM1 (GA1: Galβ1-3GalNAcβ1-4Galβ1-4Glc), GM1b (Neu5Acα2-3Galβ1-3GalNAcβ1-4Galβ1-4Glc), and SSEA-4 (Galβ1-3GalNAcβ1-3Galα1-4Galβ1-4Glcβ1), leading to apoptosis of cancer cells and M1 macrophage polarization. We describe how to prepare dimeric and monomeric SeviLs, which are valuable in studying the function of gangliosides in terms of glycan recognition. It also introduces how to apply lectins to ganglioside studies on concentration-dependent growth inhibition and cell differentiation.

Keywords Gangliosides, Apoptosis, Differentiation, MAPK Family, Mytilidae, SeviL, R-type Lectin, Signal Transduction

1 Introduction

Mussels, an important resource as food, have adapted to live in the world's oceans. Research on their genomes and transcripts has progressed [1], elucidating the mechanisms of immunity and environmental resistance in detail. Furthermore, many lectins that bind to the glycans of glycosphingolipids have been discovered [2], which is helpful for ganglioside research.

SeviL is a β-galactoside-binding lectin found in the mantle and gill of the Pacific mussel *Mytilisepta virgata*. It was purified from the crude extraction of the mussel homogenates to be found a dimeric form with non-covalently linked 15,000 Da subunits. The primary structure of the lectin had been identified by combining the amino acid sequencing by mass spectrometry of purified lectin and

Koichi Furukawa (ed.), *Gangliosides*, Methods in Molecular Biology, vol. 3035,
https://doi.org/10.1007/978-1-0716-5288-6_16,

transcriptome of these mussels [3, 4]. The entire primary structure of the lectin was R-type lectin consisted of 129 amino acids having triple tandem repeating structure with subdomains 1, 2, and 3 (Fig. 1A). Subdomain-1, the N-terminus side of SeviL, had a characteristic motif of Q-X-W (Fig. 1A, black square), which is the feature of R-type lectin.

The R-type lectin family covers many organisms, from bacteria to humans [5]. In the case of mussels, the orthologue of the mRNA corresponding to SeviL is transcribed in shared in the family (*Mytilus galloprovincialis*, *M. edulis*, *M. californianus*, *M. trossulus*, *Perumytilus purpuratus*, and *Lithophaga lithophaga*) [3], indicating that this lectin family is especially rich in the family Mytilidae (https://unilectin.unige.ch/trefoil/) [6].

The functions of R-type lectins are believed to be many, such as toxins, immune molecules, and enzyme-promoters [5]. They have characteristic and different glycan-binding properties. SeviL strongly binds to asialoGM1 (GA1: Galβ1-3GalNAcβ1-4Galβ1-4Glc) and GM1b (Neu5Acα2-3Galβ1-3GalNAcβ1-4Galβ1-4Glc), and moderately binds to SSEA-4 (Galβ1-3GalNAcβ1-3Galα1-4Galβ1-4Glcβ 1), via divalent cations such as Ca^{++} [3].

Crystallography identified the 3D structure of SeviL as fold β-trefoil [7]. Furthermore, simulation analysis predicted that substituting these two amino acids (12Q, 126 F) into hydrophilic (12 R, 126 K) in the natural (dimeric) SeviL could design a monomeric molecule [7, 8] in bacterial recombination (Fig. 3A).

Recombinant dimer-SeviL dose-dependently negatively regulates the cell growth of human cervical cancer HeLa cells that express asialo-GM1 (GA1) [3]. In addition, recombinant dimer-SeviL activated

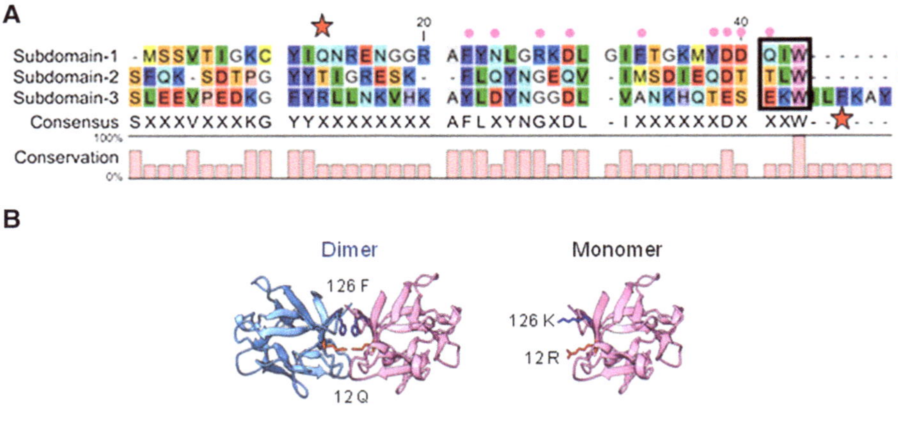

Fig. 1 Structures of SeviL. (**A**) Amino acid sequence alignment of the internal tandem-repeat subdomains of SeviL. The consensus at the bottom summarizes the residues shared by the three domains. The polypeptide sequence alignment is analyzed using the MUSCLE program [13]. (**B**) 3-D Prediction structure of a dimer (natural) and monomer-SeviLs. Dimer form keeping 12Q (blue) and 126 F (red) and monomer form substituting to R and K, respectively. (Molecular graphics showing the 3D structures of SeviL were generated by ChimeraX [14])

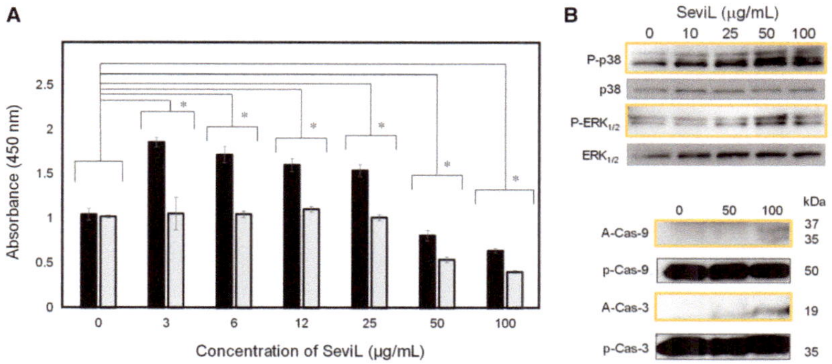

Fig. 2 Cellular effects of SeviL. (**A**) The black and light gray bars indicate the mouse macrophages (RAW264.7) and human ovarian cancer (HeLa) cells. Cells were treated with SeviL at various concentrations (0–100 μg/mL) for 24 h, and a WST-8 assay determined cell viability. Error bars: S.E. (n = 3). *P* values (**P* < 0.05). (**B**) Activations of ERK, p38, caspase-3, and caspase-9 in HeLa cells by the treatment of SeviL (0–100 μg/mL). Phosphorylation levels of $ERK_{1/2}$ and p38 are P- $ERK_{1/2}$ and P-p38 (orange boxes) are activated forms of $ERK_{1/2}$ and p38, respectively. Activated levels of caspase-3 and -9 are A-caspase-3 and A-caspase-9 (orange boxes) derived from proto (p)-caspase-3 and p-caspase-9, respectively [3]

the signal transduction of MAP Kinase in the cells. On the other hand, monomer-SeviL did not regulate the cell growth or activation of cell signaling [8]. These results suggest that the multivalency of the lectin binding site of glycans is necessary to control cell proliferation through SeviL-ganglioside binding.

Furthermore, adding dimeric SeviL (3–10 nM) activated the macrophage cells RAW264.7, which induced polarization into the M1 stage with biochemical and morphological changes [8].

These characteristic lectin properties will be valuable molecules for studying the functions and biological activities of gangliosides. This chapter introduces an approach to studying gangliosides based on mussel lectins.

2 Materials

2.1 Expression and Purification of Dimer and Monomer SeviL

1. BL21(DE3) Competent cells (bacteria).
2. pet28a(+) plasmid vector.
3. LB agar plate containing kanamycin plate: kanamycin 0.05%, tryptone 1%, yeast extracts 0.5%, sodium chloride 1%, agar 1.5%, water 95.9% (w/v).
4. Ni-Sepharose™ 6 Fast Flow.
5. HiTrap Q HP anion exchange chromatography column.
6. HiLoad 16/600 Superdex 200 size exclusion chromatography column.
7. BCA (bicinchoninic acid) Protein Assay Kit.

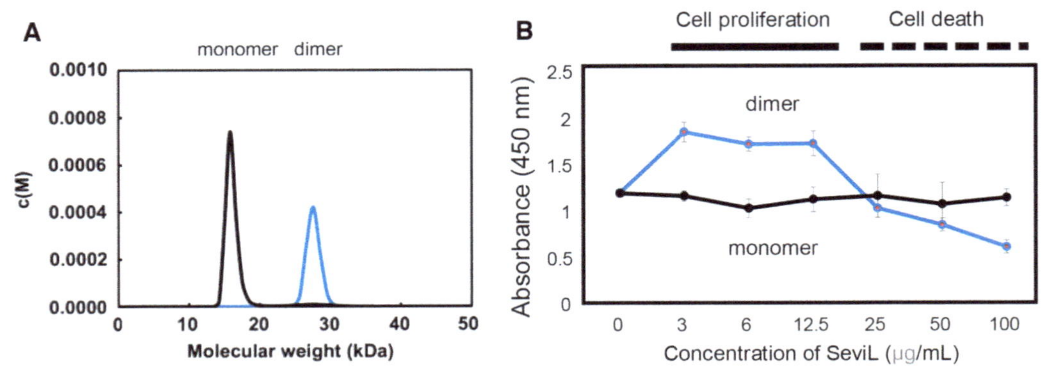

Fig. 3 Design of dimeric and monomeric SeviL and their macrophage proliferation effects. (**A**) Molecular weight estimation by sedimentation velocity experiment. Dimer-SeviL (blue) and monomer-SeviL (black), which were designed by mutation (Q12R/F126K), were confirmed in each form by using analytical ultracentrifugation. (**B**) Comparison of the cellular effect RAW264.7 between dimer-SeviL versus monomer-SeviL [8]. X-axis shows the concentration of monomer or dimer SeviL to the cells. Y-axis shows the cell viability of macrophage proliferation according to the absorbance obtained from WST-8 assay. Each point represents an average of triplicate measurements

8. ÄKTAprime plus is used as a Chromatography System.
9. Analytical ultracentrifugation was carried out using an Optima XL-I analytical ultracentrifuge using an An-50 Ti rotor.

2.2 Cell Culture and Cell Growth Assay

1. Cell lines HeLa (human cervical cancer) and RAW264.7 (mouse macrophage) are obtained from the American Type Culture Collection.
2. Cell culture medium: RPMI 1640 with 10% heat-inactivated fetal calf serum, 2 mM glutamine, 100 U/mL penicillin, and 100 μg/mL streptomycin.
3. Trypsin: 0.5 w/v% Trypsin-5.3 mmol/L EDTA·4Na Solution with Phenol Red.
4. $CaCl_2$.
5. Millex Filer 0.22 μm.
6. 100-mm culture dishes.
7. 15 mL polypropylene conical tubes.
8. Hemocytometer.
9. 96-well flat-bottomed plastic plates.
10. Cell Counting Kit-8: WST (water-soluble tetrazolium salt)-8 in 1-metholy-5-methylphenazinium methylsulfate solution.
11. Microplate spectrophotometer: Multiskan Microplate Spectrophotometer.

2.3 Antibodies and Western Blotting

1. Anti-phospho-$ERK_{1/2}$ antibody.
2. Anti-$ERK_{1/2}$ antibody.
3. Anti-phospho-p38 antibody.
4. Anti-p38 antibody.
5. Anti-caspase-3.
6. Anti-caspase-9.
7. RIPA buffer: 50 mM Tris–HCl (pH 7.5), 150 mM NaCl, 0.1% SDS, 1% Triton X-100, 1% sodium deoxycholate containing 1× protease inhibitor cocktail.
8. PVDF (polyvinylidene difluoride) Membrane P-plus.
9. Semi-dry blotting system.
10. EzFastBlot: Transfer reagent for PVDF membrane.
11. EzBlock Chemi: Blocking reagent for PVDF membrane.
12. PBS-T: PBS containing 0.1% (w/v) Triton X-100.
13. 1% (w/v) BSA (bovine serum albumin) in PBS.
14. ECL (enhanced chemical-luminescence) Western Blotting Substrate.
15. EzWest Blue.
16. Amersham Imager 680: luminescence image analyzer.

3 Methods

3.1 Expression and Purification of Dimer and Monomer SeviL

1. The codon-optimized dimer-SeviL and monomer-SeviL (12Q⇢R, 126F⇢K) genes are cloned into the multiple cloning sites of the pET28 vector using the *NdeI* and *BamHI* restriction sites. Nucleotide sequence data of SeviL (Dimer form) have been deposited in the GeneBank database under the accession number MK434191.
2. These vectors direct the expression of SeviLs carrying a thrombin-cleavable hexahistidine tag (NH_3^+-HHHHHH-) sequence at the N-terminus.
3. Following cleavage of the His-tag with thrombin, the purified protein begins with the NH_3^+-GSHM- sequence.
4. Transform BL21(DE3) cells with the plasmids and select transformed cells on an LB agar plate containing kanamycin (50 μg/mL).
5. Pick kanamycin-resistant colonies and inoculate them in an LB medium containing 20 μg/mL kanamycin. Shake the culture and incubate it at 37°C.

6. When the culture reaches an O.D.600 of approximately 0.6, induce SeviL expression by adding IPTG to a final concentration of 0.5 mM, and continue growth for 3 hours at 37°C.
7. Harvest the cells by centrifugation and resuspend the cell pellet in 100 mM Tris-HCl, pH 8.0, 150 mM NaCl. Sonicate the cell suspension on ice.
8. Centrifuge the sonicated lysate at 15,000 × *g* for 45 min at 4°C.
9. Apply the supernatant to a Ni-Sepharose™ 6 Fast Flow resin (5 mL) and wash with 50 mM Tris-HCl, pH 8.0, 150 mM NaCl (wash buffer).
10. Elute the protein from the column with 50 mM Tris-HCl, pH 8.0, 150 mM NaCl, and 300 mM imidazole.
11. Collect the protein-containing fractions and digest them with thrombin overnight at 18°C while dialyzing against the wash buffer.
12. Reload the protein onto the Ni-Sepharose™ 6 Fast Flow resin (5 mL) and collect the flow-through.
13. Dialyze the protein against 20 mM Tris-HCl, pH 7.4, and 20 mM NaCl before applying it to a Q-Sepharose column.
14. Apply to a Q-Sepharose column and elute with a gradient from 20 mM to 1000 mM NaCl.
15. Concentrate the protein fraction to 5 mL and apply it to a size-exclusion column equilibrated with 20 mM Tris-HCl, pH 7.4, and 150 mM NaCl.
16. Measure the purified lectin concentration using the BCA Protein Assay Kit [9, 10].
17. Dimer and monomer forms are confirmed by analytical ultracentrifuge (*see* **Note 1**).

3.2 Cell Growth Quantification for SeviL Treated Cells

1. Grow the culture cells (HeLa and RAW264.7) to more than 1×10^6 in a 100 mm plastic plate.
2. Aspirate the cell medium and wash it with PBS.
3. After aspirating the PBS, add 2 mL of trypsin to the cells and incubate for 5 min at 37°C to detach the cells.
4. Pipette the cells and add the 8 mL of medium to cancel the activity of trypsin.
5. Place the cell suspension in a 15 mL polypropylene conical tube and centrifuge at 200 × *g* at room temperature.
6. Discard the supernatant and resuspend the cells with 5 mL of medium.
7. Count the cell number with a hemocytometer to adjust their density to 1×10^4 in 80 μL/well.

8. Seed 1 × 10^4 cells (80 μL) with several concentrations (0–100 μg/mL) of SeviL (10 μL) on each well in the 96-well flat-bottomed plastic plate (*see* **Note 2**).
9. Add 10 μL of the calcium-containing TBS to each well in the 96-well flat-bottomed plastic plate (*see* **Note 3**).
10. Incubate the plate at 37°C in the incubator with 5% CO_2 for 24 h.
11. Add WST-8 in 1-metholy-5-methylphenazinium methylsulfate solution (10 μL) of Cell Counting Kit-8 in each well to quantify the ratio of live cells affected by SeviLand incubated with 5% CO_2 for 4 h at 37°C [11] (*see* **Note 4**).
12. Measure each well's absorbance at 450 nm with a microplate spectrophotometer (Figs. 2 and 3B).

3.3 Lectin-Dependent Signal Transduction

1. Grow HeLa cells more than 1 × 10^6 in 100 mm plastic plates.
2. Wash cells as in Sect. 3.2, Steps 5 and 6.
3. Count the cell number with a hemocytometer to adjust their density to 5 × 10^5 in 2.7 mL/well.
4. Seed 5 × 10^5 (2.7 mL) with several concentrations (0–100 μ g/mL) of SeviL (300 μL) in each well of the six-well culture plate (*see* **Note 5**).
5. Incubate the plate at 37°C in the 5% CO_2 incubator for 24 h.
6. Collect the cell suspension in each well to 15 mL polypropylene conical tube and centrifuge at 200 × *g* at room temperature (*see* **Note 6**).
7. Remove the medium and add the PBS (1 mL) to wash the cells.
8. Centrifuge at 200 × *g* at room temperature and repeat the washing procedure twice.
9. After the removal of PBS, add the cell lysis (RIPA) buffer (100 μL) and incubate for 15 min on ice.
10. Extract the cell component by pipetting several times.
11. Apply the cell extract (20 μg) to each well of SDS-PAGE.
12. Separate proteins of the crude cell extraction by electrophoresis and transfer them onto a PVDF membrane.
13. After masking the membrane with blocking reagent EzBlock Chemi for 1 h, incubate it overnight with primary antibodies (1/1000 dilution with 1% BSA in PBS) to detect the activated kinase or caspase, respectively.
14. Wash the PVDF membrane with PBS-T for 5 min and repeat this procedure for three times.

15. Incubate the secondary antibody (1/1000 dilution with 1% BSA in PBS) of each, which was conjugated with horseradish peroxidase, for 2 h at room temperature.
16. Enhance the signal of peroxidase by using ELC Western Blotting Substrate and detect the signal with a luminescence image analyzer, such as Amersham Imager 680 (*see* **Note 7**) (Fig. 2B).

4 Notes

1. If an analytical ultracentrifugation system is unavailable for the lab, a gel-permeation column connecting the HPLC system [12] can be used to determine the relative molecular mass.
2. The purified lectin, dissolved in TBS, should be centrifuged and filtered with a sterile disposable 0.22 μm pore filter membrane (non-absorption type).
3. The calcium ion is necessary to compose the multimer to activate the SeviL against cancer cells. $CaCl_2$ was contained in TBS (20 mM Tris–HCl, 150 mM NaCl) at 200 mM in final concentration. $CaCl_2$-containing TBS was used after filter sterilization.
4. The details of the WST-8 assay are mentioned in the instruction manual provided by the manufacturer.
5. In the case of adhesive cells, seed cells at 50% confluence in a 6-well culture plate, in order to achieve almost complete confluence on the day of treatment.
6. Incubate the secondary antibody (1/1000 dilution with 1% BSA in PBS) of each, which was conjugated with horseradish peroxidase, for 2 h at room temperature.
7. The immunocomplex can also perform qualitative detection by staining the membrane with the peroxidase substrate, such as EzWest Blue.

Acknowledgments

This work was supported by the Assisted Joint Research Program (Exploration type) FY2024 E-26 of the J-GlycoNet cooperative network, accredited by the Ministry of Education, Culture, Sports, Science and Technology, MEXT, Japan, as a Joint Usage/ Research Center. Kenichi Kamata, Yuki Fujii, Yukiko Ogawa, Tatsuya Kawasaki, and Yasuhiro Ozeki are supported by a research grant (24K08716, 24K15170, 23K10950, 23K06190, and 19K06239) from JSPS. Marco Gerdol is supported by the Interconnected

Nord-Est Innovation Ecosystem (iNEST) and received funding from the European Union Next-GenerationEU (PIANO NAZIONALE DI RIPRESA E RESILIENZA (PNRR)—MISSIONE 4 COMPONENTE 2, INVESTIMENTO 1.5—D.D. 1058 23/06/2022, ECS00000043. Yasuhiro Ozeki is supported by the Yokohama Trial Grant for Research and Development YT2024-1050 from the City of Yokohama.

Conflict of interest The authors have no conflicts of interest to declare that are relevant to the content of this chapter.

References

1. Gerdol M, Moreira R, Cruz F et al (2020) Massive gene presence-absence variation shapes an open pan-genome in the Mediterranean mussel. Genome Biol 21:275
2. Gerdol M, Nerelli DE, Martelossi N et al (2023) Taxonomic distribution and molecular evolution of mytilectins. Mar Drugs 21:614
3. Fujii Y, Gerdol M, Kamata K et al (2020) A GM1b/asialo-GM1 oligosaccharide-binding R-type lectin from purplish bifurcate mussels *Mytilisepta virgata* and its effect on MAP kinases. FEBS J 287:2612–2630
4. Gerdol M, Fujii Y, Hasan I et al (2017) The purplish bifurcate mussel *Mytilisepta virgata* gene expression atlas reveals a remarkable tissue functional specialization. BMC Genomics 18:590
5. Cummings RD, Schnaar RL, Ozeki Y (2022) Chapter 31. R-type lectins. In: Varki A, Cummings RD, Esko JD et al (ed) Essentials of glycobiology, 4th edn. Cold Spring Harbor Laboratory Press.
6. TrefLec - A database of predicted β-trefoil lectins (2017). https://unilectin.unige.ch/trefoil/
7. Kamata K, Mizutani K, Takahashi K et al (2020) The structure of SeviL, a GM1b/asialo-GM1 binding R-type lectin from the mussel *Mytilisepta virgata*. Sci Rep 10:22102
8. Fujii Y, Kamata K, Gerdol M et al (2024) Multifunctional cell regulation activities of the mussel lectin SeviL: induction of macrophage polarization toward the M1 functional phenotype. Mar Drugs 22:269
9. Smith PK, Krohn RI, Hermanson GT et al (1985) Measurement of protein using bicinchoninic acid. Anal Biochem 150:76–85
10. Wiechelman KJ, Braun RD, Fitzpatrick JD (1988) Investigation of the bicinchoninic acid protein assay: identification of the groups responsible for color formation. Anal Biochem 175:231–237
11. Ukeda H, Kawana D, Maeda S et al (1999) Spectrophotometric assay for superoxide dismutase based on the reduction of highly water-soluble tetrazolium salts by xanthine-xanthine oxidase. Biosci Biotechnol Biochem 63:485–488
12. Hayashi H, Kamata K, Gerdol M (2024) Novel galectins purified from the sponge Chondrilla australiensis: unique structural features and cytotoxic effects on colorectal cancer cells mediated by TF-antigen binding. Mar Drugs 22:400
13. Edgar RC (2004) MUSCLE: multiple sequence alignment with high accuracy and high throughput. Nucleic Acids Res 32:1792–1797
14. Goddard TD, Huang CC, Meng EC et al (2018) UCSF ChimeraX: meeting modern challenges in visualization and analysis. Protein Sci 27:14–25

Check for updates

Chapter 17

Visualization of Ganglioside–Bacteria Interaction Using a Frozen Tissue Overlay Assay with GFP-Labeled Bacteria and Ganglioside-Specific Antibodies

Shinichi Chisada, Kohei Shimizu, Naoyuki Matsunaga, Nozomu Okino and Makoto Ito

Abstract

This chapter describes a tissue overlay assay for visualizing ganglioside–bacteria interaction on epithelial cells. Bacteria are labeled with green fluorescent protein (GFP) and incubated with frozen tissue sections pretreated with skim milk to minimize nonspecific binding. Staining actin with Alexa Fluor 568-conjugated phalloidin visualizes the epithelial cell surface. Fluorescence microscopy allows observation of bacterial binding to epithelial cells. To determine whether gangliosides mediate bacterial attachment, ganglioside-specific antibodies are used to block bacterial binding. Additionally, sialidase treatment, which removes sialic acid residues from gangliosides, confirms the specificity of the ganglioside–bacteria interaction.

Keywords Ganglioside, Glycosphingolipid, Frozen tissue overlay assay, Bacteria, GFP

1 Introduction

Gangliosides are located on the cell surface of vertebrates, with sugar chains extending outward and lipid moieties (ceramides) embedded in the plasma membrane [1]. Along with cholesterol, gangliosides contribute to the formation of microdomains on the cell surface, playing a crucial role in cell-cell interactions and communication [2]. Additionally, gangliosides, as well as sialic acid-free neutral glycosphingolipids, are exploited by bacteria for invasion or adhesion to epithelial cells [3, 4]. The TLC-overlay assay, described in the next chapter, helps identify gangliosides that serve as bacterial binding sites [5, 6], [7, 8]. This chapter describes the tissue overlay assay to confirm whether candidate gangliosides function as bacterial receptors in vivo [9]. An overview of the method is shown in Fig. 1. The assay consists of eight steps: (1) fluorescent labeling of bacteria, (2) preparation of frozen tissue sections, (3) blocking of

Koichi Furukawa (ed.), *Gangliosides*, Methods in Molecular Biology, vol. 3035,
https://doi.org/10.1007/978-1-0716-5288-6_17,

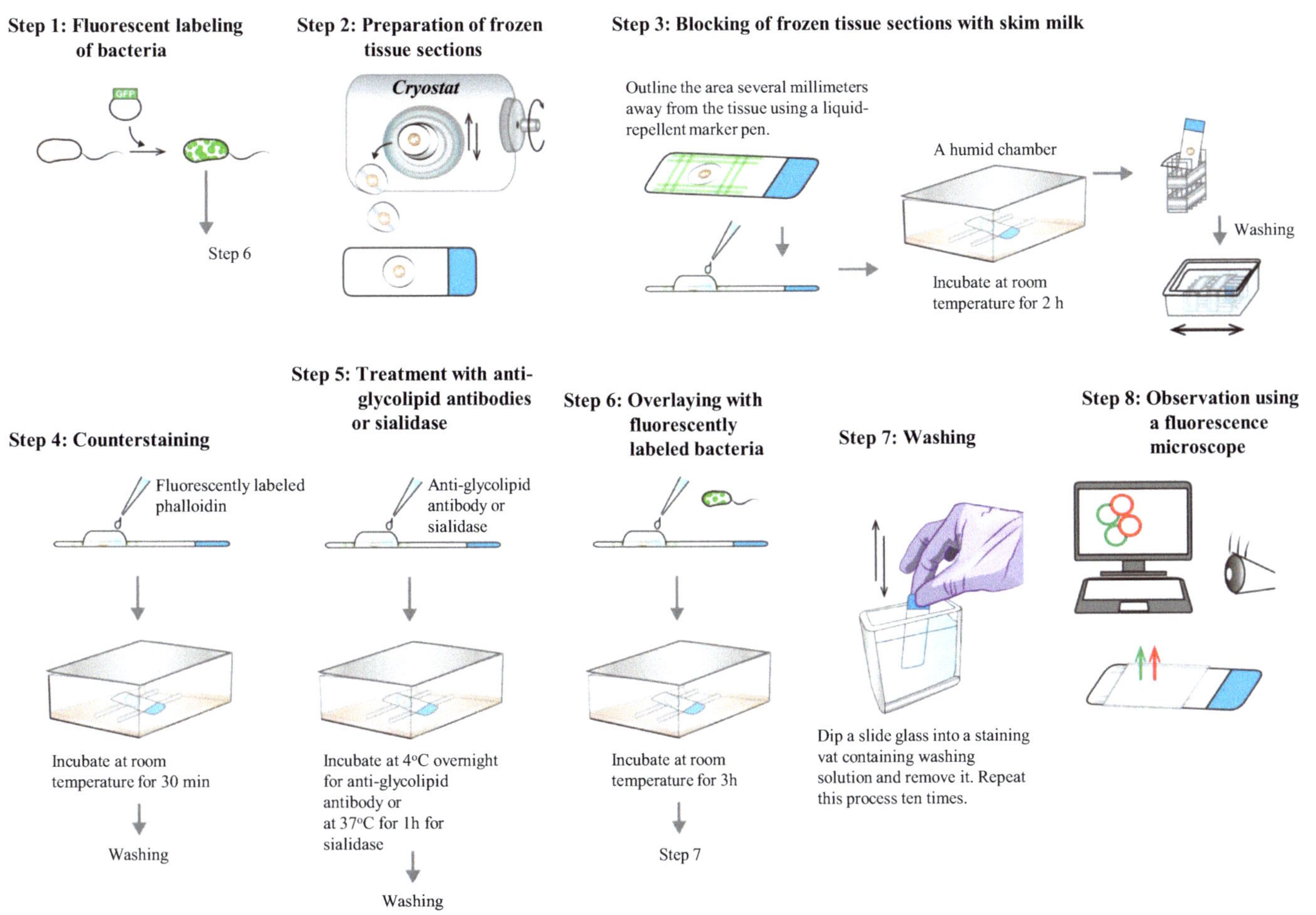

Fig. 1 Schematic diagram showing an outline of the tissue overlay assay to evaluate bacterial ganglioside receptors in biological specimens. The assay consists of eight steps, each corresponding to Sects. 3.1–3.8

tissue sections with skim milk, (4) counterstaining with Alexa Fluor 568-conjugated phalloidin, (5) treatment with anti-ganglioside/neutral glycosphingolipid antibodies, and specific glycosidases such as sialidase, (6) overlay with fluorescently labeled bacteria, (7) washing, and (8) fluorescence microscopy observation. To confirm that gangliosides are involved in bacterial and endothelial cell adhesion, step 5 is required.

2 Materials

Prepare all solutions using deionized water and analytical-grade reagents. Prepare and store all reagents in the refrigerator (unless indicated otherwise). Diligently follow all waste disposal regulations when disposing of waste materials.

2.1 GFP-Labeled Bacteria

Escherichia coli strain CC118 pir carrying the conjugative helper plasmid pEVS104 and *gfp*-containing plasmid pKV111 [10].

2.2 Counterstaining

1. Alexa Fluor 568-conjugated phalloidin (counterstaining F-actin filaments).
2. FITC-conjugated phalloidin (counterstaining F-actin filaments).

2.3 Antibodies and Glycosidase

1. Anti-GM4 polyclonal antibody (anti-ganglioside antibody, Matreya LLC, USA).
2. Anti-GalCer monoclonal antibody (anti-neutral glycosphingolipid antibody, AMR-20).
3. Neuraminidase (sialidase).

2.4 Blocking to Prevent Nonspecific Binding

Skim milk (Sigma-Aldrich, USA).

2.5 Mounting Sections (Slices) of a Sample onto Slides

1. Fluoromount (mounting solution, Cosmo-Bio, Japan).
2. Optimal cutting temperature (OCT) compound (used to embed tissue samples prior to frozen sectioning on a microtome-cryostat, Sakura Finetechnical, Japan).

3 Methods

Carry out all procedures at room temperature otherwise specified.

3.1 Fluorescent Labeling of the Bacteria

1. *Escherichia coli* strain CC118λpir carrying the conjugative helper plasmid pEVS104 and *gfp*-containing plasmid pKV111 was used to transfer the *gfp* gene to *Vibrio* recipients (*see* **Note 1**) [10].

2. Labeling bacteria with GFP has been reported elsewhere [11, 12]. Recently, ATCC has been providing authentic bacteria (*Escherichia coli, Pseudomonas aeruginosa, Shigella flexneri, Salmonella enterica*, etc.) harboring GFP reporter genes, which can be used to analyze host-pathogen interactions (https://www.atcc.org/microbe-products/bacteriology-and-archaea/reporter-labeled-bacteria#t=productTab&numberOfResults=24).

3.2 Preparation of Frozen Tissue Sections

1. Add 60 to 70% volume of 4% paraformaldehyde in PBS in a 15- or 50-mL tube, and add the surgically removed tissues (the tissue should be less than 8 mm long).
2. Shake overnight at 4°C.
3. Exchange the fixative in the tube with PBS and shake at room temperature for 10 min. Repeat this process three times.
4. Exchange PBS in the tube with 15% sucrose in PBS containing 0.02% NaN_3, and then incubate at 4°C for 1 h.
5. Transfer into 30% sucrose in PBS containing 0.02% NaN_3 and incubate at 4°C overnight.
6. Embed the tissues in OCT compound and freeze at −30°C for 1 h to make an OCT block with embedded tissue (and then keep the block at −80°C until use).
7. Serially slice the OCT block into 10-μm-thick slices using a Cryostat (Leica CM1900, Leica Microsystems, Nussloch, Germany) and put the cut section on the aminopropyl silane-coated slide glass (and then keep the slide glass at −80°C until use).

3.3 Blocking of Frozen Tissue Sections with Skim Milk

1. Prepare a humid chamber large enough to accommodate the slide glass (*see* **Note 2**).
2. Place the slide glass with tissue sections at room temperature, allow it to dry, and remove condensation water (*see* **Note 3**).
3. On the slide glass, mark the outline of the area several millimeters away from the tissue using a liquid-repellent marker pen (*see* **Note 4**), commonly used in the staining process. This will help to maintain the condition in which the staining solution to be used later is more likely to cover the tissue due to the surface tension.
4. Place the slide glass in the humid chamber.
5. Inside the area outlined with a liquid-repellent marker pen, carefully pour the blocking solution (*see* **Note 5**) without spilling over the outline.
6. Close the chamber lid to prevent evaporation and let it stand at room temperature for 2 h.

7. Place the slide glass in a dyeing basket, and then immerse the basket in a vat type 1 containing washing solution (*see* **Note 6**), with shaking for 5 min.
8. Replace the solution and repeat this process five times.

3.4 Counterstaining

1. Wipe off moisture outside the area marked with the liquid-repellent marker, and place the slide glass in the humid chamber. Dilute the reagent for counterstaining of tissues (*see* **Note 7**) with the dilution solution (*see* **Note 8**).
2. Carefully pour the counterstain reagent diluent into the area outlined with the indelible marker without spilling over the outline. Close the chamber lid to prevent evaporation and let it stand at room temperature for 30 min.
3. Wash with washing solution three times.

3.5 Treatment of Tissue Sections with Anti-Glycolipid Antibodies and Glycosidases

Treatment with Anti-GM4 and Anti-GalCer Antibodies

1. Wipe off moisture outside the area marked with the liquid-repellent marker, and place the slide glass in the humid chamber. Dilute the anti-glycosphingolipid antibody (*see* **Note 9**) with the dilution solution (*see* **Note 8**).
2. Carefully dispense anti-glycosphingolipid antibody diluent into the area outlined with an indelible marker without spilling over the outline. In parallel, prepare control sections using the same procedure without addition of anti-glycosphingolipid antibodies. Close the chamber lid to prevent evaporation and let it stand at 4°C overnight.
3. Wash the sample with the washing solution five times and put a cover glass on it.
4. Observe the specimen under a confocal laser-scanning microscope (Fig. 2).
5. Count the bacteria before and after treatment with anti-glycosphingolipid antibodies (Fig. 3).

Treatment with Sialidase

1. Wipe off moisture outside the area marked with the liquid-repellent marker, and place the slide glass in the humid chamber.
2. Dilute sialidase (*see* **Note 10**) with the dilution solution (*see* **Note 8**).
3. Carefully pour the sialidase diluent into the area outlined with the indelible marker without spilling over the outline. In parallel, prepare control sections using the same procedure without addition of sialidase.

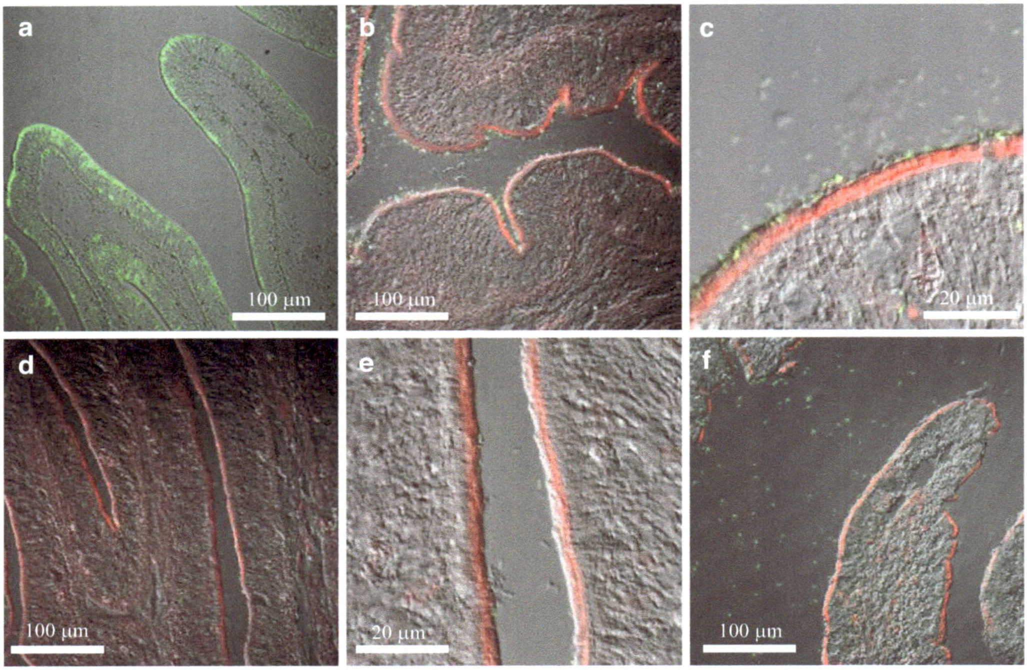

Fig. 2 Tissue overlay assay showing adhesion of GFP-labeled *Vibrio harveyi* to intestinal villi of the red sea bream with and without treatment with anti-GM4 antibody or sialidase. (**a**) Staining with anti-GM4 antibody (green). Immunostaining of intestinal epithelial cells of the red sea bream with polyclonal antibody against GM4 (Matreya LCC). The secondary antibody is Alexa Fluor 488 goat anti-rabbit IgG (H + L) (green, Thermo Fisher Scientific, USA). (**b**) and (**c**) Tissue overlaid with GFP-labeled *V. harveyi*. (b) Fluorescence-labeled phalloidin (red) with GFP-labeled vibrios (green). (**c**) Magnified view of panel (**b**). (**d**) and (**e**) Tissue overlaid with GFP-labeled *V. harveyi* after treatment with anti-GM4 antibody. (**d**) Fluorescence-labeled phalloidin (red) and GFP-labeled vibrios (green). (**e**) Magnified view of panel (d). (**f**) Tissue overlaid with GFP-labeled *V. harveyi* after treatment with sialidase. (**f**) Fluorescence-labeled phalloidin (red) and GFP-labeled vibrios (green). (**d**)–(**f**) Intestinal villi were stained with Alexa Fluor 568 phalloidin (red). (Adapted from [9] with modifications)

4. Close the chamber lid to prevent evaporation and keep it at 37°C for 1 h.
5. Wash with washing solution five times.

3.6 Overlay with Fluorescently Labeled Bacteria

1. Wipe off moisture outside the area marked with the liquid-proof marker and place the slide glass in the humid chamber.
2. Add 100 μL of fluorescently labeled bacteria (*see* **Note 11**) to the area outlined with a liquid-repellent marker pen and incubate for 3 h at room temperature.

3.7 Washing

1. Wash and apply a cover glass with a mouting medium.
2. Pinch the rim of the slide glass with your fingers or cornet tweezers, dip it into a Type 2 Staining Vat containing Wash Solution (*see* **Note 6**), and remove it. Repeat ten times.

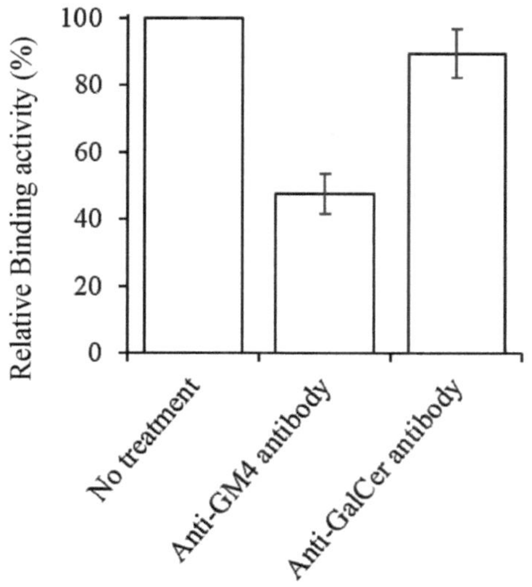

Fig. 3 Inhibition of adherence of GFP-labeled *Vibrio harveyi* to the intestinal villi of red sea bream by anti-GM4 or anti-GalCer antibodies. The tissue overlay assay was performed using GFP-labeled *V. harveyi* and frozen sections of the intestinal tract of red seabream before and after treatment with anti-GM4 antibody or anti-GalCer antibody. The number of bacteria adhering to the intestinal villi (10 μm thick) was counted using a confocal laser scanning microscope (Digital Eclipse, C1, Nikon, Japan) at 600× magnification. Relative values were calculated by setting the value from the tissue overlay assay without anti-glycosphingolipid antibodies (no treatment) to 100. Data were averaged from eleven fields and presented with standard errors. (Adapted from [9] with modifications)

3. Wipe away any moisture outside the area marked with the permanent marker.

3.8 Observation Using a Fluorescent Microscope

1. A coverslip with a mounting medium (Fluoromount, Cosmo-Bio, Japan) is applied to observe the fluorescence specimen.
2. Observe the specimen under a confocal laser-scanning microscope.
3. Count the bacteria bound to epithelial cells if necessary.

4 Notes

1. *Vibrio harveyi* was labeled with GFP by the conjugative transfer of a plasmid containing the *gfp* gene from *gfp*-donor *E. coli* (*E. coli* CC118λ*pir* strain carrying pEVS104 [a conjugal helper, *tra trb* Kn^r] and pKV111 [*gfp* Cp^r]) [10]. *Gfp* donor *E. coli* was pre-cultured in LB broth containing chloramphenicol (Cp, 20 μg/mL) and kanamycin (Kn, 40 μg/mL) at

37°C overnight, and *V. harveyi* (recipient) was pre-cultured in liquid PY medium A (1% polypeptone, 0.2% yeast extract, 0.05% $MgSO_4 \cdot 7H_2O$, 75% filtered seawater, and 25% distilled water, pH 7.2) at 28°C overnight. The 100 μL aliquots of each culture were mixed, centrifuged, washed, and resuspended in liquid PY medium A. This mixture was dropped onto an agar plate containing the same composition of PY medium A and incubated at 20°C for 1–2 days. The resulting bacterial colonies were suspended in 700 μL of liquid PY medium A and plated on TCBS agar (a selective medium for isolation of *Vibrio* species, Nissui Seiyaku, Tokyo, Japan) containing Cp (20 μg/mL). The plate was incubated at 20°C, and GFP-expressing clones were identified under a fluorescence microscope (MZFLIII, Leica, Germany). Purification of GFP-expressing clones was achieved by multiple rounds of streak plating on TCBS agar at 28°C.

2. Place several water-soaked Kimtowls in a plastic case large enough to hold several slide glasses. Place two rods (made of glass or plastic) on top of the Kimtowls to provide a platform for placing the slides.
3. Typically, slides containing frozen tissue sections are stored at a temperature below 0°C. Placing these slides at room temperature will cause condensation, which may interfere with subsequent procedures.
4. As shown in step 3 of Fig. 1.
5. The blocking solution is 2.5% skim milk in PBS, but each experiment is needed to optimize the reagent and conditions.
6. The wash solution is a 1:10 dilution of the blocking solution (0.25% skim milk in PBS), but each experiment is required to optimize the reagent and conditions.
7. The tissue counterstain reagent is Alexa Fluor 568-conjugated phalloidin (200 mU/mL) for counterstaining F-actin filaments of intestinal villi, but it is necessary to select the reagent based on the tissue being observed.
8. The diluent is Can Get Signal Immunostain Immunoreactions Enhancer solution (TOYOBO Inc., Japan).
9. Anti-GM4 antibody is a polyclonal antibody used at 1:100 dilution. The anti-GalCer antibody is a mouse monoclonal antibody used at 1:200 dilution.
10. *Clostridium perfringens* neuraminidase (EC3.2.1.18) was used at a final concentration of 70 mU/mL.
11. The number of *V. harveyi* in the 0.25% skim milk in PBS solution is 4.0 ×108 cells/mL, but it is necessary in each experiment to optimize the conditions.

Acknowledgments

We thank H. Kamada and K. Fujioka at Kyushu University for their technical assistance. We also thank Dr. T. Sawabe at Hokkaido University for providing the *gfp*-donor *E. coli*.

Conflict of Interest This study was supported by JSPS KAKENHI (grant number JP06J09743, SC) The funder had no role in study design, data collection and analysis, decision to publish, or preparation of the manuscript.

References

1. Sonnino S, Mauri L, Chigorno V et al (2007) Gangliosides as components of lipid membrane domains. Glycobiology 17:1R–13R
2. Regina Todeschini A, Hakomori SI (2008) Hakomori, "Functional role of glycosphingolipids and gangliosides in control of cell adhesion, motility, and growth, through glycosynaptic microdomains. Biochim Biophys Acta 1780:421–433
3. Karlsson KA (1989) Animal glycosphingolipids as membrane attachment sites for bacteria. Annu Rev Biochem 58:309–350
4. Ito M, Chisada S, Matsunaga N et al (2023) Vibrio-binding gangliosides in fish intestinal tracts. Glycoconj J 40:315–322
5. Matsunaga N, Chisada S, Fujioka H et al (2011) Glycosphingolipid receptors for pathogenic vibrios in intestines of mariculture fish. Fisheries Science 77:583–590
6. Chisada S, Horibata Y, Hama Y et al (2005) The glycosphingolipid receptor for *Vibrio trachuri* in the red sea bream intestine is a GM4 ganglioside which contains 2-hydroxy fatty acids. Biochem Biophys Res Commun 333:367–373
7. Chisada S, Matsunaga N, Ito M (2022) Assay of the binding of bacteria to glycosphingolipids. Glycoscience Protocols (GlycoPODv2) https://www.ncbi.nlm.nih.gov/books/NBK594065/ Accessed 09 Sep 2024
8. Tsukamoto K, Kohda T, Mukamoto M et al (2005) Binding of Clostridium botulinum type C and D neurotoxins to ganglioside and phospholipid. Novel insights into the receptor for clostridial neurotoxins. J Biol Chem 280:35164–35171
9. Chisada S, Shimizu K, Kamada H et al (2013) Vibrios adhere to epithelial cells in the intestinal tract of red sea bream, pagrus major, utilizing GM4 as an attachment site. FEMS Microbiol Lett 341:18–26
10. Sawabe T, Fukui Y, Stabb EV (2006) Simple conjugation and outgrowth procedures for tagging vibrios with GFP, and factors affecting the stable expression of the *gfp* tag. Lett Appl Microbiol 43:514–522
11. Ma L, Zhang G, Doyle MP (2011) Green fluorescent protein labeling of Listeria, Salmonella, and Escherichia coli O157:H7 for safety-related studies. PLoS One 6:e18083
12. Geva-Zatorsky N, Alvarez D, Hudak JE et al (2015) In vivo imaging and tracking of host–microbiota interactions via metabolic labeling of gut anaerobic bacteria. Nat Med 21:1091–1100

Check for updates

Chapter 18

Identification of Gangliosides as Bacterial Attachment Sites Using a TLC-Overlay Assay with Radioisotope-Labeled Bacteria

Shinichi Chisada, Naoyuki Matsunaga, Nozomu Okino and Makoto Ito

Abstract

This chapter describes a thin-layer chromatography (TLC) overlay assay for identifying gangliosides to which bacteria bind. In this method, gangliosides are first applied to a silica gel TLC plate and separated using an appropriate developing solvent. The TLC plate is then treated with polyisobutylmethacrylate to prevent nonspecific adsorption. Subsequently, the treated TLC plate is incubated with a PBS suspension of radioisotope-labeled bacteria for an appropriate duration. After incubation, unbound bacteria are washed away, and the TLC plate is analyzed using an imaging analyzer. The specificity of a ganglioside for binding to a particular bacterium is determined by quantifying the amount of radioisotope-labeled bacteria bound to the ganglioside on the TLC plate.

Keywords Ganglioside, Glycosphingolipid, Thin-layer chromatography (TLC) overlay assay, Bacteria, Autoradiography

1 Introduction

Gangliosides, glycosphingolipids containing sialic acid, are found in the epithelial cells of vertebrate skin, oral cavity, and gastrointestinal tract. Both gangliosides and neutral glycosphingolipids regulate cell adhesion, motility, and growth through glycolipid-enriched microdomains [1]. They also act as attachment sites for resident and pathogenic bacteria [2]. Resident bacteria attach to gangliosides (neutral glycosphingolipids) to colonize epithelial surfaces, whereas pathogenic bacteria use them as initial receptors for invading host cells [3–7]. Bacterial adhesion is mainly facilitated by specific bacterial adhesion proteins that interact with host glycans, including gangliosides and neutral glycosphingolipids [8].

The TLC overlay assay has been widely used to identify gangliosides and neutral glycosphingolipids that interact with bacteria [4–6], as well as bacterial toxins [7] and adhesins [8, 9]. This chapter

Koichi Furukawa (ed.), *Gangliosides*, Methods in Molecular Biology, vol. 3035,
https://doi.org/10.1007/978-1-0716-5288-6_18,

describes a TLC overlay assay using radiolabeled bacteria to identify gangliosides and neutral glycosphingolipids as bacterial attachment sites (receptors) [4–6, 10].

Figure 1 provides an overview of the method, which consists of six steps: (1) preparation of ganglioside-loaded TLC plates, (2) precoating of the TLC plates with polyisobutylmethacrylate, (3) metabolic labeling of bacteria with [^{35}S]-methionine, (4) incubation of the TLC plates with [^{35}S]-labeled bacteria, (5) detection of gangliosides bound to bacteria on TLC using an imaging analyzer, and (6) comparison of the ganglioside-stained TLC plate with autoradiography to identify gangliosides recognized by bacteria.

Since TLC plates are coated with silica gel, which does not mimic the cell surface, polyisobutylmethacrylate treatment (shown in Sect. 3.2) is used to prevent non-specific adsorption while exposing the sugar moieties of gangliosides on the TLC surface [11]. Combined with bacterial overlay, this treatment better simulates the interaction between bacteria and cell-surface gangliosides.

2 Materials

Prepare all solutions using deionized water and analytical-grade reagents. Prepare and store all reagents in the refrigerator (unless indicated otherwise). Diligently follow all waste disposal regulations when disposing of waste materials.

2.1 Thin-Layer Chromatography (TLC) of Gangliosides

1. Gangliosides are purified from appropriate animal tissues or purchased from commercial sources.
2. TLC is performed using TLC plate (plastic TLC plates precoated with Silica Gel 60 (MACHEREY-NAGEL GmbH & Co. KG, Germany) and developing solvents: chloroform/methanol/0.02% $CaCl_2$ (5:4:1, v/v/v) for gangliosides, and chloroform/methanol/water (60:35:8, v/v/v) for neutral glycosphingolipids.
3. For staining reagents, resorcinol-HCl reagent and orcinol-H_2SO_4 reagent are used for gangliosides, and for both gangliosides and neutral glycosphingolipids, respectively.
4. TLC plates are pre-treated with polyisobutylmethacrylate (FUJIFILM Wako Pure Chemical, Osaka, Japan, or TCI Chemical Co., Tokyo, Japan).

2.2 Binding of Bacteria to TLC Plates

1. Phosphate-buffered saline (PBS): 1.5 mM of KH_2PO_4, 14 mM of Na_2HPO_4, and 2.7 mM of KCl, pH 7.2, containing 0.85% NaCl (for marine bacteria, 0.85% NaCl is replaced with 2%–3% NaCl).
2. Bacteria are maintained as a slant culture.

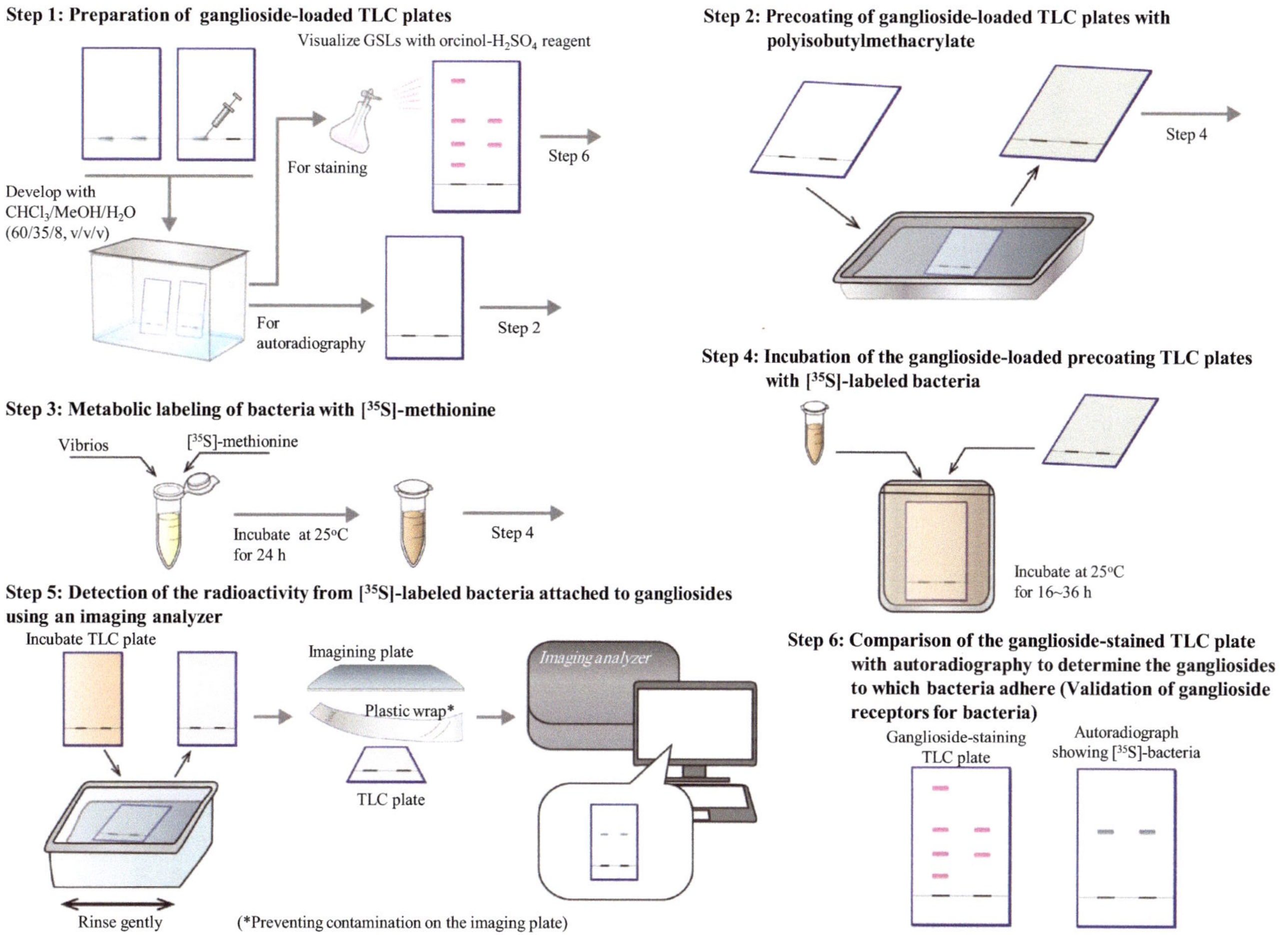

Fig. 1 Schematic diagram showing an outline of the thin-layer chromatography overlay assay for detecting bacterial ganglioside receptors. The assay consists of six steps, each corresponding to Sects. 3.1–3.6

3. Liquid medium suitable for the growth of bacteria used, e.g., *Vibrio*, PY medium A (1% polypeptone, 0.2% yeast extract, 0.05% $MgSO_47H_2O$, 75% filtrated seawater, 25% distilled water, pH 7.2–7.4); *Pseudomonas*, PY medium B (0.5% polypeptide, 0.1% yeast extract, 0.5% NaCl, pH 7.2); *Flavobacterium*, TY medium (1% bacto-tryptone, 0.1% yeast extract, 0.2% NaCl, pH 7.2).
4. [^{35}S]-methionine (L-Methionine, [^{35}S], 500 mCi (>1000 Ci/mmol) in a solution of 50 mM L-lysine, pH 7.4, containing 10 mM 2-mercaptoethanol, In vitro Translation Grade, MP Biomedicals, CA, USA).
5. Hybridization bag (Cosmo Bio Co., LTD., Tokyo, Japan).

3 Methods

Carry out all procedures at room temperature otherwise specified.

3.1 Preparation of Ganglioside-Loaded TLC Plates

1. Apply gangliosides (10–30 nmol dissolved in chloroform/methanol, 1/1 or 2/1, *see* **Note 1**) to 1.5 cm from the edge of a TLC plate with a width of 5 mm using a microsyringe. Prepare two of the same TLC plates. Alternatively, gangliosides are spotted on TLC as dots (dot-TLC overlay assay). For dot-TLC assay, Steps 2 and 3 are omitted.
2. Develop two TLC plates using an appropriate developing solvent (*see* **Note 2**) and remove TLC from the chamber when the solvent reaches 1–2 cm below the top of each plate.
3. Dry the plates completely with warm air using a hair dryer.
4. One TLC plate is used to visualize the gangliosides and neutral glycosphingolipids with a suitable dyeing reagent (*see* **Note 3**), while another TLC plate is used for the TLC overlay assay using radioisotope-labeled bacteria, as described in Sect. 3.4.

3.2 Precoating of Ganglioside-Loaded TLC Plates with Polyisobutylmethacrylate

1. Put diethyl ether/*n*-hexane (1/5, v/v) containing 0.5% (w/v) polyisobutylmethacrylate in a stainless-steel tray with a depth of 5 mm.
2. Put the GSL-loading TLC plate in the solvent from the edge quickly and immerse it for 30 s.
3. Use tweezers with flat tips to pick up the TLC plate from the tray and dry it completely at room temperature.

3.3 Metabolic Labeling of Bacteria with [^{35}S]-Methionine

1. Culture bacteria in 1 mL of suitable liquid medium by the late logarithmic growth phase or early stationary phase under appropriate conditions for bacteria used (*see* **Note 4**).
2. Add 50 μL of the bacterial suspension to 5 mL of a sterile fresh medium containing [^{35}S]-methionine (2.0 mCi/mL) in a 15-mL disposable test tube (*see* **Note 5**).

3. Incubate the bacterial suspension at 25–30°C for an appropriate period (12–24 h) (*see* **Notes 6 and 7**).

3.4 Incubation of the Ganglioside-Loaded Precoating TLC Plate with [^{35}S]-Labeled Bacteria

1. Collect [^{35}S]-labeled bacterial cells in a centrifugation tube as a pellet by centrifugation (6000 × *g* for 5 min).
2. Add PBS (*see* **Note 8**) to the pellet and prepare a bacterial suspension (2.0–3.0 × 10^{9-11} cells/mL, 2.0–5.0 × 10^5 cpm/mL).
3. Incubate the ganglioside-loaded precoating TLC plate with a bacterial suspension in a plastic hybridization bag for 12–24 h (*see* **Note 9**) at 25°C with gentle shaking.

3.5 Detection of the Radioactivity from [^{35}S]-Labeled Bacteria Attached to Gangliosides Using an Imaging Analyzer

1. Put PBS (*see* **Note 8**) in a stainless steel tray with a depth of 5–10 mm.
2. Place the TLC plate in the buffer carefully and shake it gently for 30 s (*see* **Note 10**).
3. Change the buffer to a new one and rinse the plate twice gently.

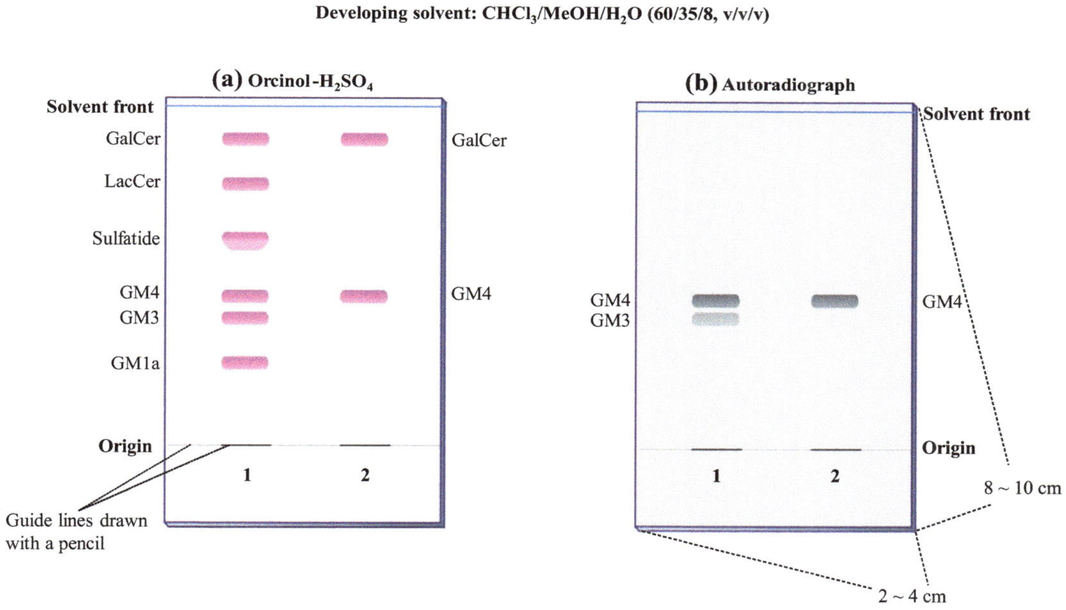

Fig. 2 Adhesion of [^{35}S]-labeled *Vibrio anguillarum* to various gangliosides and neutral glycosphingolipids. (**a**) Thin-layer chromatography (TLC) showing gangliosides visualized with orcinol-H_2SO_4 reagent. Various gangliosides were developed on a TLC plate with chloroform/methanol/water (60/35/8, v/v/v) and visualized with orcinol-H_2SO_4 reagent. (**b**) TLC autoradiograph showing gangliosides attached to [^{35}S]-labeled *V. anguillarum* NBRC12710. Gangliosides were developed on a TLC plate under the same conditions in panel (**a**). In both panels, lane *1* contained authentic glycosphingolipids, which were GalCer, LacCer, Sulfatide, GM4, GM3, and GM1a. Lane *2* contained GalCer and GM4, which were extracted from the red sea bream intestine. Each GSL was 30 nmol. Therefore, *V. anguillarum* binds to GM3 and GM4 but not GalCer, LacCer, sulfatide, or GM1a, indicating that the ganglioside receptors for *V. anguillarum* are GM3 and GM4. (Adapted from [4] with modifications)

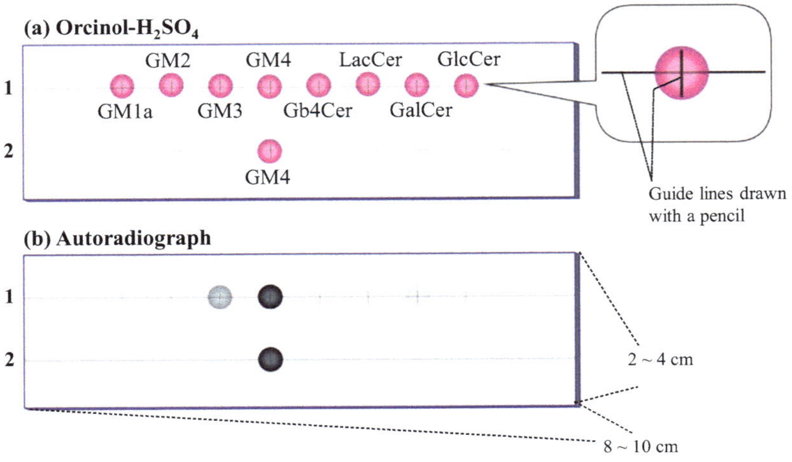

Fig. 3 Dot-thin-layer chromatography overlay assay showing the binding of [^{35}S]-labeled *Vibrio harveyi* to various gangliosides and neutral glycosphingolipids. (**a**) TLC showing the glycosphingolipids visualized with orcinol-H_2SO_4 reagent. (**b**) Autoradiograph after incubation of TLC, which was loaded with glycosphingolipids in the same manner in panel (a), with [^{35}S]-labeled *V. harveyi*. The bacteria were isolated from the intestine of the red sea bream. Each GSL was 60 nmol. This result shows that *V. harveyi* specifically binds to GM4 but not GM1a, GM2, Gb4Cer, GlcCer, or LacCer. Weak adhesions of *V. harveyi* to GM3 and GalCer are also observed. (Adapted from [5] with modifications)

4. Cover the TLC plate with plastic wrap after absorbing the excess buffer with filter paper or similar material.
5. Expose the wrapped TLC plate to an imaging plate for 10 h.
6. Using an imaging analyzer FLA-5000 (Fuji Film Co., Japan), detect and quantify the radioactivity from [^{35}S]-labeled bacteria attached to gangliosides and neutral glycosphingolipids.

3.6 Validation of Ganglioside Receptors for Bacteria

1. Compare the ganglioside and neutral glycosphingolipid-visualizing TLC plate and radioautogram showing gangliosides and neutral glycosphingolipids adhering to [^{35}S]-labeled bacteria, as shown in Fig. 2.
2. When purified gangliosides and neutral glycosphingolipids are available, the TLC overlay assay could be performed using TLC without being developed by organic solvents (dot-TLC overlay assay), as shown in Fig. 3.
3. The relative binding activity of bacteria to gangliosides is calculated using an imaging analyzer FLA-5000 (Fuji Film Co., Japan), as shown in Fig. 4.

4 Notes

Notes 1–4, 7, and 8 show the conditions for the TLC overlay assay shown in Fig. 2.

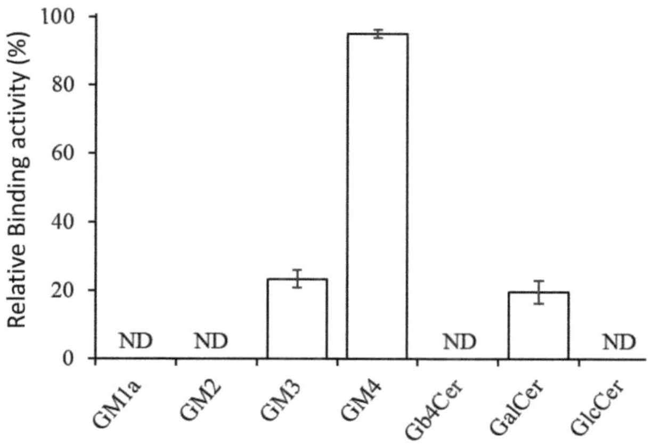

Fig. 4 Quantification of binding of [^{35}S]-labeled *V. harveyi* to gangliosides and neutral glycosphingolipids by dot thin-layer chromatography overlay assay. The binding of [^{35}S]-labeled *V. harveyi* to gangliosides and neutral glycosphingolipids on TLC was quantified using an imaging analyzer. Each value is expressed as a relative percentage of the value for *V. harveyi* binding to GM4 as 100%. Data were averaged from three different experiments and presented with standard errors. ND indicates that the value is below the limit of detection. (Adapted from [5] with modifications)

1. Each 30 nmol of ganglioside or neutral glycosphingolipids is applied to TLC.
2. The developing solvent is chloroform/methanol/water (60:35:8, v/v/v).
3. Orcinol-H_2SO_4 reagent is used for the dyeing reagent.
4. Bacteria are cultured at 25°C in PY medium A (1% polypeptone, 0.2% yeast extract, 0.05% $MgSO_47H_2O$, 75% filtrated seawater, and 25% distilled water, pH 7.2) for 16 h.
5. Cultures are performed using several test tubes in parallel to obtain bacteria ($2.0–3.0 \times 10^{9-11}$ cells/mL). The radioactivity increases with the number of bacteria; thus, preliminary experiments are needed to determine suitable numbers of bacteria for TLC-overlay assay.
6. The incubation time is also related to the intensity of the radioactivity signals. It is usually 12–24 h, but it is recommended to investigate the optimum reaction time to obtain adequate signals.
7. For the experiment in Fig. 2, incubation is performed at 25°C for 16 h.
8. PBS is usually used, but in this case (Fig. 2), phosphate buffer containing 3% NaCl is used because the bacteria tested are of marine origin (genus *Vibrio*).

9. The incubation time is also related to the intensity of the radioactivity signals. It is usually 12–24 h, but it is recommended to investigate the optimum reaction time to obtain adequate signals.
10. Wash the plates one by one to prevent overlapping with each other.

Acknowledgments

We thank K. Shimizu, H. Kamada, and K. Fujioka at Kyushu University for their technical assistance.

Conflict of Interest This study was supported by JSPS KAKENHI (grant number JP06J09743, SC). The funder had no role in study design, data collection and analysis, decision to publish, or preparation of the manuscript.

References

1. Regina Todeschini A, Hakomori SI (2008) Functional role of glycosphingolipids and gangliosides in control of cell adhesion, motility, and growth, through glycosynaptic microdomains. Biochim Biophys Acta 1780:421–433
2. Karlsson KA (1989) Animal glycosphingolipids as membrane attachment sites for bacteria. Annu Rev Biochem 58:309–350
3. Ito M, Chisada S, Matsunaga N et al. (2023) Vibrio-binding gangliosides in fish intestinal tracts. Glycoconj J 40:315–322
4. Matsunaga N, Chisada S, Fujioka H et al. (2011) Glycosphingolipid receptors for pathogenic vibrios in intestines of mariculture fish. Fisheries Science 77:583–590
5. Chisada S, Horibata Y, Hama Y et al. (2005) The glycosphingolipid receptor for Vibrio trachuri in the red sea bream intestine is a GM4 ganglioside which contains 2-hydroxy fatty acids. Biochem Biophys Res Commun 333:367–373
6. Chisada S, Shimizu K, Kamada H et al. (2013) Vibrios adhere to epithelial cells in the intestinal tract of red sea bream, Pagrus major, utilizing GM4 as an attachment site. FEMS Microbiol Lett 341:18–26
7. Tsukamoto K, Kohda T, Mukamoto M et al. (2005) Binding of Clostridium botulinum Type C and D neurotoxins to ganglioside and phospholipid: novel insights into the receptor for clostridial neurotoxins. J Biol Chem 280:35164–35171
8. Lingwood CA (1992) Bacterial adhesins/glycolipid receptors. Curr Opin Struct Biol 2:693–700
9. Payne D, O'Reilly M, Williamson D (1993) The K88 fimbrial adhesin of enterotoxigenic Escherichia coli binds to beta 1-linked galactosyl residues in glycosphingolipids. Infect Immun 61:3673–3677
10. Chisada S, Matsunaga N, Ito M (2022) Assay of the binding of bacteria to glycosphingolipids. In: Glycoscience protocols (GlycoPODv2). https://www.ncbi.nlm.nih.gov/books/NBK594065/ Accessed 9 Sep 2024
11. Yiu SC, Lingwood CA (1992) Polyisobutylmethacrylate modifies glycolipid binding specificity of verotoxin 1 in thin-layer chromatogram overlay procedures. Anal Biochem 202:188–192

Chapter 19

Analyzing Molecular Complexes Containing Gangliosides via Co-Immunoprecipitation

Yuki Ohkawa

Abstract

Gangliosides are expressed on the cell surface, particularly in cholesterol-enriched microdomains that are referred to as lipid rafts. These are considered platforms for cell signaling because they contain dense gatherings of growth factor receptors and their intracellular signaling molecules. Gangliosides are believed to play important roles in cell-signal transduction due to their restricted localization in lipid rafts. A method for the detection of protein-ganglioside complexes would be indispensable in understanding the molecular function of gangliosides. This chapter introduces a basic immunoprecipitation method that can be used to analyze complexes that contain gangliosides. The process begins by performing immunoprecipitation against membrane proteins, which is followed by using thin-layer chromatography (TLC) combined with immunoblotting to analyze the gangliosides.

Keywords Ganglioside, Lipid raft, Growth factor receptor, Immunoprecipitation (IP), Thin-layer chromatography (TLC), Immunoblotting

1 Introduction

Gangliosides are sialic acid-containing glycosphingolipids consisting of a ceramide and a sugar chain moiety. They are considered unique molecules having both a hydrophobic lipid portion and hydrophilic carbohydrate moieties within a single molecule [1]. The simplest form of a ganglioside is the GM3, and addition reactions of monosaccharides such as sialic acid, N-acetylgalactosamine (GalNAc), and galactose sequentially produce longer glycan-containing gangliosides. The biosynthesis of gangliosides is tightly regulated in both tissues and cells, and, therefore, the expression patterns of gangliosides are considered to be unique to their respective organs [2, 3].

In the central nervous systems of mammals, GM1, GD1a, GD1b, and GT1b gangliosides are abundantly expressed, whereas the expression of GD3 remains undetectable [4, 5]. On the other hand, a strong expression of GD3 can be observed in gliomas, which are tumors of the central nervous system that derive mainly from either glial cells

Koichi Furukawa (ed.), *Gangliosides*, Methods in Molecular Biology, vol. 3035,
https://doi.org/10.1007/978-1-0716-5288-6_19,

or glial precursor cells [6]. In our previous study using a genetically engineered mouse model, glioma tissues strongly expressed GD3, which is the case in humans as well [7]. To understand the molecular functions of GD3, we used proximity labeling in an attempt to analyze the molecular association with GD3 in glioma cells. Dr. Kotani and co-workers developed a nomenclature that is referred to as enzyme-mediated activation of radical sources (EMARS), which is one of the methods for proximity labeling. Using this method, the molecules that fall within the 200-300 nm diameter of a target are labeled with fluorescein isothiocyanate (FITC) dye [8]. Following the EMARS reaction, the labeled molecules can then be purified with anti-FITC antibody and identified via mass spectrometry analysis. This method and co-immunoprecipitation have allowed us to demonstrate how GD3 binds directly to the platelet-derived growth factor receptor α (PDGFRα) and additionally reported that GD3 enhances the association of PDGFRα with Yes kinase in the lipid raft to transduce malignant cell signals [9]. A key portion of this study was to demonstrate the presence of ganglioside-containing complexes via co-immunoprecipitation.

Based on our experience, determining the binding partners has been an important factor in understanding the function of gangliosides in both cancer research and biological research. Herein, we describe the details of using co-immunoprecipitation to analyze molecular complexes containing gangliosides. We expect this information to be helpful in future research.

2 Materials

2.1 Preparation of Cell Lysates and Immunoprecipitation

1. Cell lysis buffer: 20 mM Tris (pH7.5 adjusted with HCl), 150 mM NaCl, 1 mM Na_2EDTA, 1 mM EGTA, 1 % TritonX-100, 2.5 mM sodium pyrophosphate, 1 mM β-glycerophosphate, 1 mM Na_3VO_4, 1 μg/ml leupeptin, and protease inhibitors.
2. Wash buffer: 50 mM Tris (pH7.4 adjusted with HCl), 150 mM NaCl, 0.5% TritonX-100, and 1 mM Na_3VO_4.
3. Antibodies for research targets and their isotype control antibodies [e.g., an anti-PDGFRα antibody (Santa Cruz Biotechnology, Santa Cruz, CA) in our case].
4. Phosphate-buffered saline (PBS).
5. Sepharose beads conjugated with protein G.

2.2 Extraction of Lipids from Immunoprecipitates

1. Extraction solvent: chloroform mixed with ethanol at a ratio of 2:1 (v/v).
2. Water-based sonicator.

3. Glass bottles.
4. A dry-heating, nitrogen-blowing instrument.

2.3 Thin-Layer Chromatography (TLC)-Immunoblotting

1. Developing solvent: mixture of chloroform, methanol, and 0.22% (w/v) calcium chloride ($CaCl_2$) at a ratio of 55:45:10 (v/v/v).
2. Blotting solvent: mixture of isopropanol, 0.22% (w/v) calcium chloride ($CaCl_2$), and methanol at a ratio of 20:10:3.5 (v/v/v).
3. Blocking buffer: 3% (w/v) bovine serum albumin (BSA) in PBS.
4. Gangliosides standard extracted bovine brains.
5. Primary antibody [e.g., an anti-ganglioside GD3 antibody (clone R24) made in-house in our case].
6. HPR-labeled secondary antibody.
7. Enhanced chemiluminescence (ECL) detection solutions.
8. Phosphate-buffer saline (PBS).
9. High-performance TLC (HPTLC) plate.
10. Filter paper.
11. Pencil.
12. Gas-tight microsyringe [e.g., Hamilton microsyringe provided by the Hamilton Company (Energy Way Reno, NV)].
13. Glass-fiber pad
14. Polyvinylidene difluoride (PVDF) membrane.
15. Glass tank for development.
16. Hair dryer for home use.
17. Thermal Blotter for HPTLC [AC-5970 provided by the ATTO corporation (Tokyo, Japan) in our case].
18. Imager for chemiluminescence [e.g., ChemiDoc Touch MP imaging system provided by the Bio-Rad Laboratories (Hercules, CA)].

3 Methods

3.1 Preparation of Cell Lysates and Immunoprecipitation

1. Wash cells three times with ice-cold PBS.
2. Lyse the cells with the 500 μL of cell lysis buffer. Next, incubate the cells on ice for 5 min. Then scrape the cells for transfer to 1.5 ml sample tubes.
3. The resultant lysates are centrifuged at 17,900 g for 5 min at 4°C (*see* **Note 1**). DO NOT sonicate the lysates as the complexes could be very fragile.

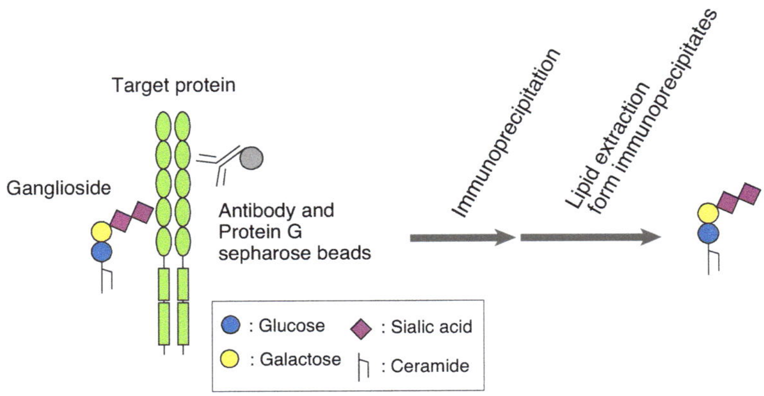

Fig. 1 Workflow of immunoprecipitation and lipid extraction. The complexes containing gangliosides are immunoprecipitated using an antibody against the target protein. The gangliosides in the immunoprecipitate are purified with organic solvents

4. To remove the insoluble materials, transfer the supernatants to new 1.5 ml of sample tubes.
5. Add antibodies and incubate overnight at 4°C with rotation (*see* **Notes 2 and 3**).
6. Add Sepharose beads conjugated to Protein G and incubate for 2 h at 4°C with rotation.
7. Wash the beads three times with wash buffer (*see* **Note 4**). Centrifuge the beads at 700 g for 30 sec at 4°C to pull them down (Fig. 1).

3.2 Extraction of Lipids from Immunoprecipitates

1. One ml of extraction solvent is added to the immunoprecipitates and mixed well (*see* **Note 5**). Sonicate for 5 min with a water-based sonicator.
2. Centrifuge the resultant solvents at 17,900 g for 5 min at room temperature.
3. Transfer the upper layers to glass bottles (*see* **Note 6**).
4. Dry the solvents under a stream of N_2 gas with mild heating (around 45°C) via use of a dry-heating, nitrogen-blowing instrument.
5. Dissolve the obtained lipids in 50 μl of extraction solvent (Fig. 1).

3.3 TLC-Immunoblotting

1. Preheat HPTCL plates at 110°C for more than 30 min.
2. Completely saturate a glass tank with developing solvent and incubate for at least 30 min before performing the next steps (*see* **Note 7**).
3. The sample application points should previously be marked with a pencil on the HPTLC plate.

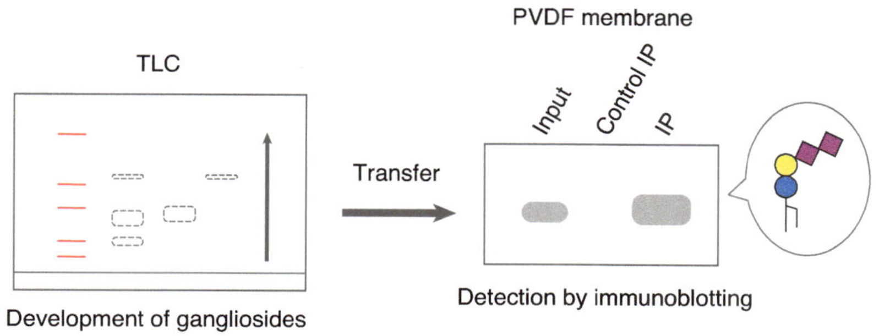

Fig. 2 Workflow of TLC followed by immunoblotting. The purified gangliosides are separated by TLC and transferred onto a PVDF membrane. The gangliosides are detected using antibodies against them

4. Lipid samples prepared as described above should be spotted on the HPTLC plate using a gas-tight microsyringe. As a standard marker, the gangliosides extracted from a bovine brain should be spotted.
5. Dry the HPTLC plate with a hair dryer for at least 10 min.
6. After returning to room temperature, the HPTLC plate is developed by placing it in a glass developing tank.
7. Incubate the HPTLC plate until the developing solvent almost reaches the top of the plate, and then dry the plate at room temperature.
8. Immerse the HPTLC plate in Blotting solvent and incubate for 40 sec (*see* **Note 8**).
9. Place the HPTLC plate on a glass-fiber pad and cover with a PVDF membrane, filter paper, and a glass-fiber pad (*see* **Note 9**).
10. Use a thermal blotter to press the resultant layers while heating at 180°C for 30 s.
11. Incubate the PDVF membrane with blocking buffer while shaking for 1 h at room temperature (*see* **Note 10**).
12. Incubate the membrane with a primary antibody for a ganglioside diluted in blocking buffer.
13. Wash the membrane four times with PBS (*see* **Note 11**).
14. Incubate the membrane with HRP-conjugated secondary antibody diluted with blocking buffer.
15. Wash the membrane four times with PBS.
16. Visualize the membrane during incubation using ELC detection solutions and analyze for chemiluminescence using an imager (Fig. 2).

4 Notes

1. Do not sonicate the lysates as the complexes could be very fragile.
2. Check the stirring up and down under rotation.
3. Immunoprecipitation with antibodies against gangliosides is not recommended because gangliosides are small molecules that could be embedded inside the complexes. The antibodies against gangliosides might not be recognized because the glycan chain could be used to form the complexes.
4. Washing with more than 500 μl of wash buffer is recommended. Transfer to a new tube is also recommended to avoid contamination by materials attached to the cap.
5. Vortexing is recommended for mixing.
6. To avoid the contamination of the middle and lower layers, leave some of the upper layer.
7. Use of filter paper could hasten the saturation. Infiltrate a filter paper of the right size for the glass tank with developing solvent and place it against the wall of the tank.
8. Shaking is unnecessary.
9. Do not dry the HPTLC plate, but remove excess solvent.
10. Do not use skim milk as the blocking buffer. Skim milk contains a large amount of glycan, which is likely to increase background and non-specific reactions.
11. Do not use a detergent containing a wash buffer, because the binding of gangliosides to the PDVF membrane could be very weak.

Acknowledgments

I wish to deeply thank Dr. Koichi Furukawa for his kind support. This experimental method was established when I worked in his laboratory. I would also like to thank Dr. James L. McDonald for English editing. This work was supported by public grants from the Osaka Prefecture (52000093, 52100001).

Conflict of Interest The authors have no conflicts of interest to declare that are relevant to the content of this chapter.

References

1. Schnaar RL, Sandhoff R, Tiemeyer M et al (2022) Glycosphingolipids. In: Varki A, Cummings RD, Esko JD, Stanley P, Hart GW, Aebi M, Mohnen D, Kinoshita T, Packer NH, Prestegard JH, Schnaar RL, Seeberger PH (eds) Essentials of glycobiology. 4th edn. Cold Spring Harbor Laboratory Press, Cold Spring Harbor (NY), http://www.ncbi.nlm.nih.gov/books/NBK579905/ [online] (Accessed September 5, 2024)
2. Markotić A, Lümen R, Marusić A et al (1999) Ganglioside expression in tissues of mice lacking the tumor necrosis factor receptor 1. Carbohydr Res 321:75–87
3. Markotić A, Marusić A, Tomac J et al (2002) Ganglioside expression in tissues of mice lacking β 2-microglobulin. Clin Exp Immunol 128:27–35
4. Svennerholm L, Boström K, Fredman P et al (1989) Human brain gangliosides: developmental changes from early fetal stage to advanced age. Biochim Biophys Acta 1005:109–117
5. Vajn K, Viljetić B, Degmečić IV et al (2013) Differential distribution of major brain gangliosides in the adult mouse central nervous system. PLoS One 8:e75720
6. Traylor TD, Hogan EL (1980) Gangliosides of human cerebral astrocytomas. J Neurochem 34:126–131
7. Ohkawa Y, Zhang P, Momota H et al (2021) Lack of GD3 synthase (St8sia1) attenuates malignant properties of gliomas in genetically engineered mouse model. Cancer Sci 112:3756–3768
8. Kotani N, Gu J, Isaji T et al (2008) Biochemical visualization of cell surface molecular clustering in living cells. Proc Natl Acad Sci U S A 105:7405–7409
9. Ohkawa Y, Momota H, Kato A et al (2015) Ganglioside GD3 enhances invasiveness of gliomas by forming a complex with platelet-derived growth factor receptor α and yes kinase. J Biol Chem 290:16043–16058

Part IV

Behaviors and Biology

Chapter 20

Analysis of Lacto/Neolacto-Series Glycolipids and Gangliosides

Hidemi Arakawa and Akira Togayachi

Abstract

Lacto-*N*-tetraose (LNT) and lacto-*N*-neotetraose (LNnT) are structures expressed in glycoproteins and are key structural components of lacto/neolacto-series glycolipids. These glycans are involved in the expression of carbohydrate antigens, such as blood group antigens, on glycolipids, and are found in leukocytes and other cell types. The β1,3-*N*-acetylglucosaminyltransferase 5 β3GnT5 enzyme, encoded by the *N*, plays a pivotal role in the biosynthesis of lacto/neolacto-series glycolipids by catalyzing the synthesis of lactotriaosylceramide, which serves as a precursor of these glycolipids. β3GnT5 transfers *N*-acetylglucosamine (GlcNAc) from UDP-GlcNAc to lactosylceramide (LacCer). These glycolipids are crucial for various physiological functions, including immune responses, cell–cell interactions, hematopoietic differentiation, and neural development. In this study, we provide an overview of the methods used to analyze lacto- and neolacto-series of glycolipids and gangliosides.

Keywords Lacto/neolacto-series glycolipid, Lactotriaosylceramide (Lc_3Cer) synthase, Enzyme assay, Folch's lipid extraction, Thin layer chromatography (TLC), Mass spectrometry, Flow cytometry

1 Introduction

1.1 Lacto/Neolacto-Series Glycolipid and Gangliosides

Almost all living organisms contain lipids and proteins conjugated to various glycans. Glycolipids comprise both hydrophobic and hydrophilic regions and are classified into glyceroglycolipids (which contain glyceride chains) and glycosphingolipids (which contain ceramides in their hydrophobic domains). Glycolipids are primarily located in the outer leaflet of the plasma membrane where they play essential roles in cell recognition and signal modulation. In mammals, glycolipids are broadly classified into globo-, isoglobo-, ganglio-, lacto/neolacto-, and other glycolipid series (Fig. 1A). Gangliosides are a subgroup of glycosphingolipids containing one or more sialic acid residues. They are predominantly expressed in the brain and play crucial roles in nervous system and cancer biology. For example, ganglioside GM1

Koichi Furukawa (ed.), *Gangliosides*, Methods in Molecular Biology, vol. 3035,
https://doi.org/10.1007/978-1-0716-5288-6_20,

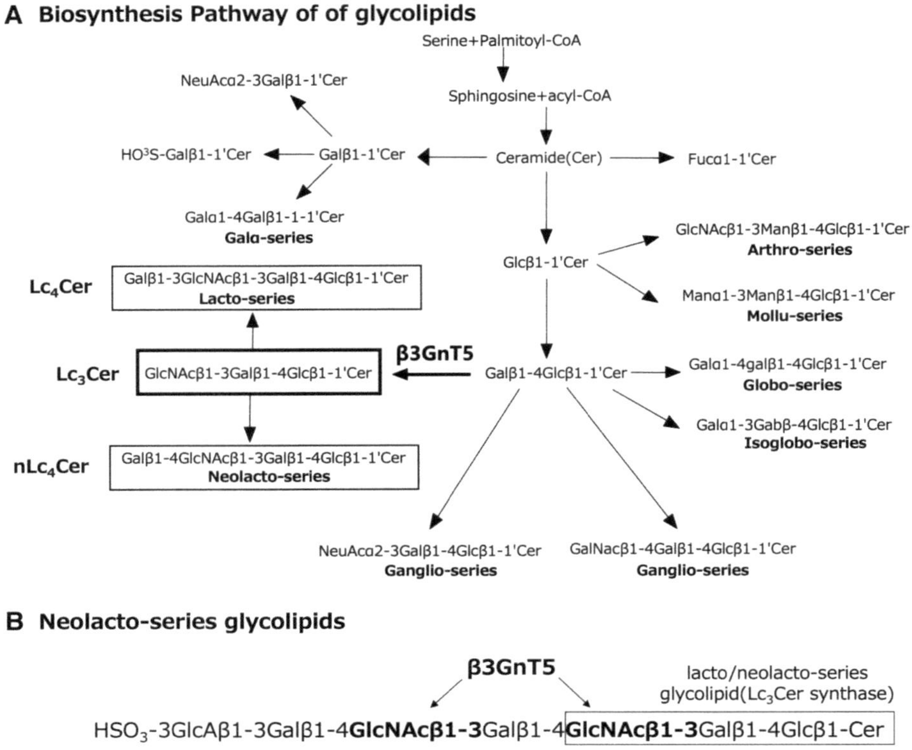

Fig. 1 Biosynthetic pathways of glycolipid series: biosynthesis of lacto/neolacto-series glycolipids. (**A**) Major glycolipid biosynthetic pathways. The β3GnT5 enzyme catalyzes the biosynthesis of lacto- or neolacto-series glycolipids. (**B**) β3GnT5 enzyme transfers GlcNAc to LacCer to synthesize lactotriaosylceramide (Lc_3Cer). Furthermore, it transfers GlcNAc to Paragloboside (neolactotetraosylceramide, nLc_4Cer) to synthesize nLc_5Cer. The glycan structure enclosed in the thick-lined square is Lc_3Cer. (Copyright (2010) National Academy of Sciences [14]).

serves as a receptor for cholera toxin. Gangliosides and related glycosphingolipids (GSLs) are implicated in neural function, inflammation, metabolism, and cancer biology [1, 2]. Studies using gene-knockout mice lacking ganglioside biosynthesis have reported that a lack of complex-type gangliosides causes various abnormalities [3, 4].

Among the various glycolipid classes, lacto- and neolacto-series glycolipids play a significant role in immune regulation, cell-cell interactions, and cancer progression. Lacto-*N*-tetraose (LNT: Galβ 1-3GlcNAcβ1-3Galβ1-4Glc) and lacto-*N*-neotetraose (LNnT: Galβ 1-4GlcNAcβ1-3Galβ1-4Glc) serve as core structural components of lacto/neolacto-series glycolipid and contribute to the expression of various carbohydrate antigens (e.g., ABH/Lewis blood group antigens and HNK-1 antigen). Glycolipids are primarily expressed in leukocytes and other blood cells, and their expression levels are altered in specific cancer tissues [5]. The structural distinction between the lacto- and neolacto-series glycolipids lies in their galactosyl linkages. These glycolipids are involved in immune responses, cell-cell interactions [1], hematopoietic cell differentiation [6], region- and developmental stage-specific expression in the central and peripheral nervous systems [7], and cancer cell invasion and metastasis [8]. Previous studies have identified lacto/neolacto-series glycolipid as crucial biomarkers of non-small cell lung cancer (NSCLC) [9]. Additionally, thin-layer chromatography immunoblotting (TLC-immunoblotting) revealed the diversity of glycolipids in human myeloid cells, demonstrating that neolacto-series glycolipids interact with specific antibodies [10]. Furthermore, β1,3-*N*-acetylglucosaminyltransferase 5 gene (*B3GNT5*) overexpression has been implicated in the prognosis of patients with thymoma [2].

1.2 β3GnT5 (B3GNT5) and Lactotriaosylceramide (Lc_3Cer) Biosynthesis

β1,3-*N*-acetylglucosaminyltransferases (β3GnTs) catalyze the transfer of *N*-acetylglucosamine (GlcNAc) from UDP-GlcNAc to the non-reducing terminal galactose of glycans via β1,3-linkage. To date, eight members of the human β3GnT family have been cloned and characterized [11]. Among them, β1,3-*N*-acetylglucosaminyltransferase 5 (β3GnT5) has been identified as a Lc_3Cer synthase that plays a crucial role in the biosynthesis of lacto- and neolacto-series glycolipids containing gangliosides [12, 13]. The β3GnT5 enzyme facilitates the transfer of GlcNAc from UDP-GlcNAc to lactosylceramide (LacCer) (Fig. 1A). Furthermore, β3GnT5 catalyzes the addition of GlcNAc to paragloboside (PLG; neolactotetraosylceramide, nLc_4Cer), forming neolactopentaosylceramide (nLc_5Cer) (Fig. 1B).

Studies on *B3gnt5* knockout mice have revealed diverse and sometimes contradictory phenotypic outcomes, highlighting the complexity of lacto- and neolacto-series glycolipid functions. In a model established by Togayachi et al. $B3gnt5^{-/-}$ mice were viable, and the absence of lacto/neolacto-series glycolipid resulted in the abnormal formation of lipid rafts (glycolipid-enriched microdomains, GEMs) in B lymphocytes, thereby leading to increased B cell proliferation and sensitivity to stimulation [14]. In contrast, Biellmann et al. reported that $B3gnt5^{-/-}$ mice exhibit embryonic lethality at the pre-implantation stage, likely due to defective cell adhesion, which contributes to embryonic compaction and implantation failure [15]. Using different strategies, Kuan et al. generated two distinct $B3gnt5^{-/-}$ mouse models [16]. Their findings revealed that $B3gnt5^{-/-}$ mice exhibited splenomegaly, lymphadenopathy, and a shortened lifespan. Additional phenotypic abnormalities include reduced B-cell populations, alopecia, obesity, and reproductive dysfunction. Discrepancies in the phenotypes observed in these studies remain unresolved. Many lacto- and neolacto-series glycolipids have been reported to be expressed in mouse embryonic stem cells [17].

2 Materials

2.1 Generation of B3gnt5 Gene Knockout Mice

1. *B3gnt5* gene targeting vector (see Fig. 2 for example).
2. Neomycin resistance gene.
3. Herpes simplex virus thymidine kinase gene.
4. Embryonic stem (ES) cells, clone E14.
5. Mouse embryonic fibroblast (MEF) feeder cell [for example, MEF: REPROCELL Inc., or SNL (STO) feeder cell: Applied Stem Cell Inc.].
6. Dulbecco's modified Eagle medium (DMEM, high glucose; GIBCO).
7. Fetal bovine serum (FBS).
8. Leukemia inhibitory factor (ESGRO; GIBCO).
9. Penicillin (GIBCO).
10. Streptomycin (GIBCO).
11. Non-essential amino acids (GIBCO).
12. L-Glutamine (GIBCO).
13. β-Mercaptoethanol (Sigma-Aldrich).
14. G418 (Geneticin; Invitrogen).

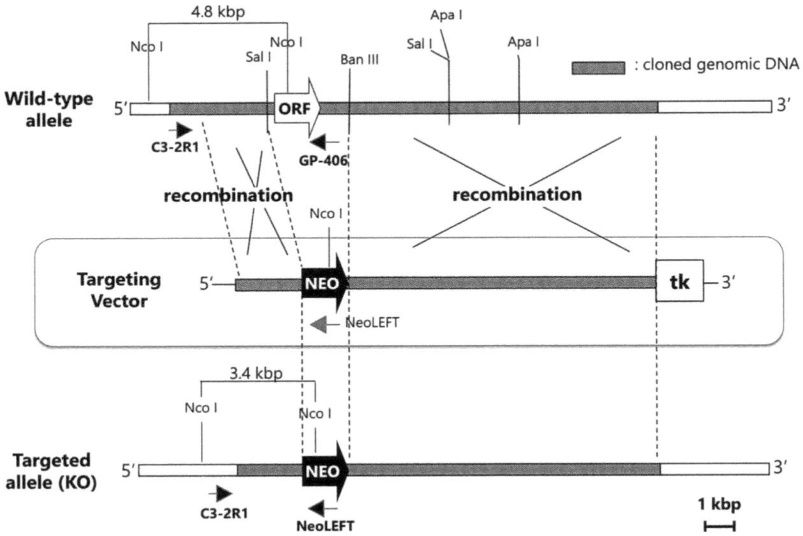

Fig. 2 Schematic representation of *B3gnt5* gene-targeting vector and mouse genome structure [14]. Schematic representation of the mouse *B3GnT5* (*B3gnt5*) targeting strategy. The upper section shows the wild-type genomic structure of mouse *B3gnt5* locus. The entire open reading frame of *B3gnt5* is located within a single exon. The middle section depicts the constructed targeting vector, which includes the replaced neomycin resistance gene. (Copyright (2010) National Academy of Sciences [14]).

15. Ganciclovir (Wako).
16. Nco I restriction enzyme (New England Biolabs).
17. PCR DIG Probe Synthesis Kit (Roche Molecular Biochemicals).
18. C57BL/6N female mice (Charles River Laboratories).
19. LA-Taq DNA polymerase (RR042B, TAKARA BIO Inc.).

2.2 Measurement of Lc_3Cer Biosynthetic Activity (Enzyme Activity)

1. Mouse tissues (target tissues).
2. Cell/Tissue Lysis buffer (with appropriate composition).
3. LacCer (Calbiochem).
4. ^{14}C-labeled UDP-GlcNAc [for example, UDP-[^{14}C]GlcNAc (Amersham Pharmacia Biotech)] is currently not commercially available and must be synthesized.
5. TLC plates (for high-performance thin-layer chromatography; HPTLC), a 20×10 cm plate from Merck (#105641, Glass HPTLC Silica gel 60 plates).
6. TLC glass chamber.
7. TLC developing solvent: Chloroform-Methanol-0.2% $CaCl_2$ (60:35:8, v/v/v).
8. Imaging plate (for Imaging analyzer) or autoradiographic film (Hyperfilm™ ECL, Cytiva).
9. Imaging analyzer (e.g., Fujifilm FLA3000).
10. Positive control for enzyme activity (reaction): Cell extract from HL-60 cells with Lc_3Cer synthase activity.
11. Negative control for enzyme activity (reaction): Cell extract from Jurkat cells lacking Lc_3Cer synthase activity.

2.3 Extraction Method of Glycolipids from Biological Samples, Including Cells and (Gene Knockout) Mouse-Derived Tissues

1. Freeze-dried tissue (brain of a mouse (C57BL/6NJcl) is used as an example; the number of samples will depend on the purpose, but typically it ranges from one to three mice).
2. HPTLC plate (MERCK), a 20×10 cm plate from Merck (#105641, Glass HPTLC Silica gel 60 plates).
3. Chloroform (C).
4. Methanol (M).
5. Water (W; ultrapure water).
6. 0.2% $CaCl_2$ solution.

2.4 Separation and Analysis of Lacto/Neolacto-Series Glycolipid by Thin-Layer Chromatography (TLC)

1. High-performance thin-layer chromatography (HPTLC) provides higher resolution than conventional TLC. Any manufacturer's product can be used; here, a 20×10 cm plate from Merck (#105641, Glass HPTLC Silica gel 60 plates) is used as an example.

2. Glass cutter.
3. TLC Glass chambers for solvent development.
4. Microsyringe (10 mL); capillary tubes can also be used as substitutes for syringes.
5. Hand dryer.
6. Spray bottle.
7. Hot plate with temperature control.
8. Draft chamber for spraying the detection reagents.
9. Solvent systems for glycolipid development:
   For gangliosides (acidic glycolipids), chloroform-methanol-0.2% $CaCl_2$ (55:45:10, v/v/v) or (50:50:11, v/v/v).
   For neutral glycolipids, chloroform-methanol-water (65:25:4, v/v/v) or (60:35:8, v/v/v), depending on the experimental purpose.
10. Detection reagents:

    For gangliosides, resorcinol reagent: mix 2% resorcinol solution (10 mL), concentrated hydrochloric acid (80 mL), 0.1 M $CuSO_4$ (0.25 mL), and distilled water (9.75 mL).
    For neutral glycolipids, orcinol reagent: dissolve 0.2 g orcinol in 11.4 mL of concentrated sulfuric acid (H_2SO_4) and make up to 100 mL with distilled water.

2.5 TLC-Immunoblotting (Immuno-TLC) for the Quantitative Analysis of Lacto/Neolacto-Series Glycolipid

1. Various plastic containers.
2. Wash bottles.
3. Large glass Petri dishes (sufficient to accommodate the HPTLC plates).
4. The chemical detection reagents are the same as those used in the previous TLC Sect. 2.4). The chromogenic substrate is the same as that mentioned in the previous section (e.g., orcinol reagent).
5. Polyisobutyl methacrylate solution (0.1%) and polyisobutyl methacrylate beads (*see* **Note 1**).
6. Clostridium perfringens sialidase (SIGMA).
7. Probes for glycolipid detection. Antibody (monoclonal antibody) against polylactosamine (poly-*N*-acetyllactosamine) glycan structures on lacto/neolacto-series glycolipid: The monoclonal antibody 1B2-1B7 (anti-LacNAc monoclonal antibody, mouse IgM) specifically recognizes polylactosamine structures on neolacto-series glycolipids and can be used in this assay. Culture hybridoma cells (ATCC, TIB-189) producing the 1B2-1B7 antibody in RPMI-1640 medium (GIBCO) supplemented with 15% fetal bovine serum. The 1B2-1B7 hybridoma culture supernatant is used as an example for demonstrating the experiment (*see* **Note 2**).

8. An HRP-labeled secondary antibody (anti-mouse IgM is used in this case, assuming that 1B2-1B7 is employed).
9. VECTASTAIN™EliteABCKit(VECTORLABORATORIES, INC.).
10. Phosphate-buffered saline [PBS(-)] solution.
11. 1% Bovine serum albumin (BSA)-PBS solution.
12. Detection reagent: Matches to the probe label; here, the chemiluminescent reagent Super Signal (Pierce) is used.
13. Chemiluminescent detection equipment: X-ray film (Hyperfilm™ ECL, Cytiva) or chemiluminescence imaging systems such as LAS-1000/LAS-3000 (FUJIFILM) can be used.

2.6 Structural Analysis of Glycolipid Glycan by Mass Spectrometry

Digestion with Ceramide Glycanase and Purification of GSL-Derived Glycans

1. Incubator (37°C).
2. Water bath (boiling apparatus) for reaction termination.
3. Reverse-phase column (Sep-Pak C_{18}, Waters).
4. Freeze dryer.
5. 0.1% Triton X-100.
6. 50 mM Sodium acetate buffer (pH 5.5).
7. Ceramide glycanase (EGCase I, 20 mU) (New England Biolabs or Tokyo Chemical Industry Co., Ltd. [for Recombinant Endoglycoceramidase I]).
8. Water (ultrapure water).

Reduction and Complete Methylation of Glycolipid-Derived Glycans

1. Cation exchange column (Oasis™ MCX, Waters).
2. Speed-Vac (Labconco Corp.).
3. Vacuum dryer.
4. Glass tubes.
5. Reverse-phase column (Oasis™ HLB, Waters).
6. 50 mM NaOH.
7. 500 mM $NaBH_4$.
8. 50% Acetic acid solution.
9. Distilled water.
10. 1% Acetic acid methanol.
11. NaOH/DMSO slurry.
12. Iodomethane (Wako).

13. 50% Acetic acid solution.
14. Acetonitrile.

MALDI MS and Multistage Tandem Mass Spectrometry (MS^n) Analysis of Permethylated Glycans Derived from Glycolipids

1. Mass spectrometry A Reflex IV MALDI-TOF (Bruker Daltonics GmbH) or AXIMA-QIT MALDI quadrupole ion-trap TOF mass spectrometer (Shimadzu Corp., Japan).
2. Mass spectrometer plate.
3. Acetonitrile.
4. 2,5-Dihydroxybenzoic acid (DHB).
5. 30% Ethanol.

2.7 Flow Cytometry Analysis Using Anti-Glycolipid Antibodies and Probes

1. 1% BSA/0.1% NaN_3/PBS(-): Dissolve BSA without stirring, i.e., dissolve naturally, followed by filtering through a 0.22 μm filter, aliquot 500 mL into pre-autoclaved bottles, and store at 4°C until use.
2. PBS(-) buffer (FACS sheath fluid): Filter through a 0.22 μm filter, aliquot 2 L into pre-autoclaved bottles, and store at 4°C until use.
3. 5% Paraformaldehyde/PBS (stock solution): Dissolve 1.5 g of paraformaldehyde in 10 mL of 1 N NaOH, add PBS to make the volume to 80 mL, adjust the pH to 7.2 ± 0.1 with 4 N HCl, and then added PBS to make the final volume to 100 mL.
4. Cell fixation solution (0.5% paraformaldehyde/PBS): 5% paraformaldehyde/PBS tenfold with PBS.
5. Various antibodies and fluorescent probes were used to stain the target cells (e.g., FITC-labeled Lycopersicon esculentum lectin (LEL, tomato lectin), FITC-labeled cholera toxin B subunit (CT-B), PE-labeled anti-mouse BCR antibody, and anti-mouse CD19 antibody).
6. Anti-LacNAc monoclonal antibody 1B2-1B7 for detecting poly-*N*-acetyllactosamine (polyLacNAc, polylactosamine) structures on lacto/neolacto-series glycolipid (*see* **Note 3**).
7. Isotype control antibodies (Use as negative control).
8. Flow Cytometry Instrument and analysis software for flow cytometry data. Flow Cytometry instrument software or FlowJo™ software.

3 Methods

3.1 Generation of B3gnt5 Gene Knockout Mice

1. To elucidate the various biological functions of lacto/neolacto-series glycolipid, generate gene knockout mice for *B3gnt5*, an Lc_3Cer synthase that serves as a key enzyme in

lacto/neolacto-series glycolipid biosynthesis [14], and analyze their phenotypes. Generated *B3gnt5* knockout mice [*B3gnt5*$^{\mathrm{tm1Nari}}$: https://www.informatics.jax.org/allele/MGI:4459891] using a conventional homologous recombination-based gene-targeting method. Here, we present an example using the older generation technique described in [14]. Currently, several more efficient gene-deficiency methods that incorporate genome editing technologies exist, and we recommend their implementation in future studies. To confirm the successful generation of *B3gnt5*-deficient mice, analyze the expression of the *B3gnt5* gene, Lc_3Cer synthase activity, and lacto/neolacto-series glycolipid in the tissues of wild-type (WT) and *B3gnt5*$^{-/-}$ mice.

2. Design a targeting vector to replace the entire open reading frame (ORF) of *B3gnt5* with a neomycin resistance gene (NEO) (Fig. 2). Substitute the exon containing the *B3gnt5* ORF with a neomycin resistance gene. For positive and negative selection, incorporate the herpes simplex virus thymidine kinase (tk) gene at the 3'-end of the targeting vector (*see* **Note 4**).
3. After each step of genetic recombination, perform restriction enzyme analysis and DNA sequencing to confirm the precise gene modifications.
4. Perform homologous recombination using ES derived from C57BL/6N mice. Culture ES cells (clone E14) on a feeder layer of mouse embryonic fibroblasts and transfect with a targeting vector. Introduce the targeting vector into the ES cells via electroporation or lipofection. Specifically, aim to transfect 20 μg of linearized targeting vector DNA into E14 ES cells (*see* **Note 5**) [14].
5. To select ES cells with correctly integrated targeting vector, culture the cells for several days in the presence of geneticin and ganciclovir (*see* **Note 6**). Screen homologous recombination-positive ES cell clones using PCR and Southern blotting. Recently, next-generation sequencing (NGS) has been performed for confirmation.
6. Generate chimeric mice by injecting targeted ES cells into normal C57BL/6 blastocysts and implanting them into the uteri of pseudopregnant female mice. This procedure yields mice harboring cells with inactivated *B3gnt5*. Select chimeric mice selected based on coat color.
7. To achieve germline transmission, cross male chimeric mice carrying *B3gnt5*-inactivated cells with C57BL/6 female mice. Heterozygous offspring carrying *B3gnt5* deletions will be obtained by breeding chimeric mice with wild-type mice.

8. Backcross heterozygous mice with C57BL/6N mice, and then intercross the heterozygous mice to produce homozygous *B3gnt5*-deficient mice.
9. Validate homozygous *B3gnt5*-deficient mice using PCR-based genotyping (described below), confirm homologous recombination using Southern blot analysis, and verify gene inactivation via an assessment of Lc_3Cer synthase activity.
10. Determine the genotypes of individual mice by genotyping via PCR using the following primers: GP-406 (5'-GATTC CCACTTTATTGGCACAG-3'), NeoLEFT (5'-TGCGC TGACAGCCGGAACAC-3'), and C3-2R1 (5'-ACGTTG CCTCAACTTAACTCTG-3') (Fig. 2). Perform the PCR using these primers and LA-Taq DNA polymerase (RR042B; TAKARA BIO Inc.) under the following conditions: the thermal cycling conditions are as follows: an initial denaturation at 94°C for 1 min, followed by 45 cycles of (i) denaturation at 98 °C for 1 s, (ii) annealing at 60°C for 1 min, and (iii) elongation at 68°C for 4 min. Separate the PCR products by 1% agarose gel electrophoresis, and analyze the banding patterns. Amplify a 3.08 kb DNA fragment in the wild-type allele using the C3-2R1 and GP-406 primers, whereas a 2.42 kb fragment in the knockout allele (*B3gnt5*$^{-/-}$) using the C3-2R1 and NeoLEFT primers. Identify mice that display both bands as heterozygous. These primers and conditions will enable clear differentiation between the wild-type and knockout alleles.
11. Examine the phenotypes of the generated *B3gnt5* knockout mice to evaluate the physiological and immunological effects of *B3gnt5* deletion. Using *B3gnt5* gene knockout mice [14], several experimental analyses have been conducted, including protein expression analysis via western blotting, glycan expression analysis via lectin blotting, cell population analysis via flow cytometry, and molecular localization within lipid rafts (glycolipid-enriched microdomains, GEMs) via fractionation. Additionally, analyze cytokine secretion via ELISA, examine lipid raft (GEM) structure via fluorescence microscopy, and assess immune function using B cell proliferation and T cell-independent antibody production assays. These analyses should follow standard methodologies and protocols [14]; therefore, detailed procedural descriptions are omitted.

3.2 Measurement of Lc_3Cer Biosynthetic Activity (Enzyme Activity)

1. Preparation of mouse tissue/cell homogenates: Homogenize target mouse tissues/cells using an appropriate lysis buffer. Select the tissue type and buffer composition based on the experimental goals.
2. Measurement of β3GnT activity on glycolipids. The enzyme reaction mixture should contain 150 mM sodium succinate

buffer (pH 7.2) and 480 μM UDP-GlcNAc (cold) as the donor substrate, 175 nCi of UDP-[^{14}C]GlcNAc (Amersham Pharmacia Biotech) as the donor substrate, 10 mM $MnCl_2$, 0.4% Triton CF-54, 10 nmol of LacCer as the glycolipid acceptor substrate, and 100 μg of protein extract as the enzyme source. Incubate the reaction tube overnight (16 h) at 37°C to facilitate enzymatic activity [13].

3. Reaction Termination: Terminate the enzymatic reaction by heating at 98°C for 5 min to inactivate the enzyme.
4. Separation of the Reaction Products by TLC (TLC method): Spot the reaction mixture onto a TLC plate. Develop the TLC plate in a glass chamber containing a developing solvent.
5. Detection and Quantification of Lc_3Cer: Dry the TLC plate after development. Expose the plate to an imaging plate or an autoradiographic film (X-ray film, Hyperfilm ECL, etc.). Use an imaging analyzer to measure the intensity of the radioactive bands, allowing quantification of the Lc_3Cer signal (Fig. 3) (*see* **Note 7**).

3.3 Extraction Method of Glycolipids from Biological Samples, Including Cells and (Gene Knockout) Mouse-Derived Tissues

1. Thin-layer chromatography (TLC) is a convenient and widely used method for the detection and structural analysis of glycolipids. This method involves the separation of glycolipids on a thin-layer plate followed by chemical detection. In addition, glycolipids can be detected with high sensitivity using specific antibodies or lectins.

   We used neutral and acidic glycolipids (mainly gangliosides) from mouse tissues to obtain structural information. To follow this, develop the extracted glycolipids/gangliosides on a thin-layer plate and detect using chemical methods or TLC immunostaining. Extract the glycolipids from the tissue or cell

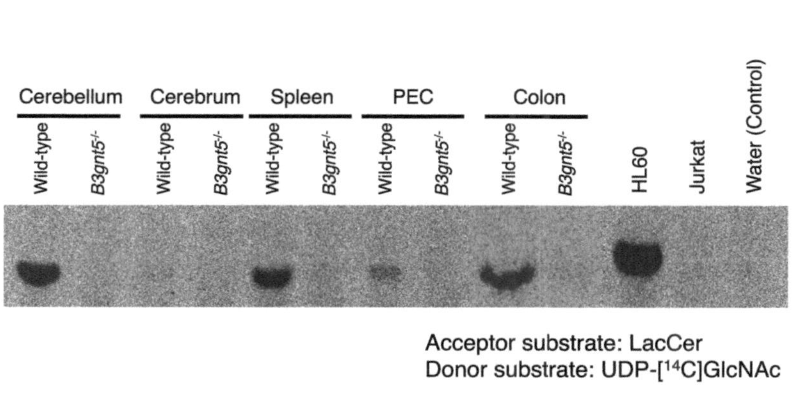

Fig. 3 Detection of lactotriaosylceramide [Lc_3Cer] as a biosynthetic product (enzyme activity) by TLC analysis. Mouse tissue or cell line extracts were used as enzyme sources to measure transferase activity toward the receptor substrate lactosylceramide (LacCer). The biosynthesis of Lc_3Cer, an enzymatic reaction product, was confirmed. *PEC* Peritoneal exudate cell. (Copyright (2010) National Academy of Sciences [14]).

samples of WT and *B3gnt5*$^{-/-}$ knockout mice, and compare their expression profiles via TLC separation. Following TLC separation, analyze the specific glycolipid expression by TLC immunoblotting.

Simple Crude Glycolipid Extraction Method: Freeze-Drying and Lipid Extraction

1. Collect approximately one-fourth of the mouse brain and place it in a glass tube.
2. Add 2 mL of the extraction solvent (chloroform-methanol, volume ratio 1:1) and thoroughly crush the tissue using a small spatula. Mix the suspension thoroughly and allow it to stand at room temperature (25°C) for at least 1 h to extract the lipids.
3. Centrifuge the tubes at 2000 rpm for 10 min.
4. Transfer the supernatant to a new glass tube using a Pasteur pipette. Use the supernatant (crude extracted lipids) for thin layer chromatography (TLC) analysis (*see* **Note 8**).

Extraction and Purification of Neutral and Acidic Glycolipids

1. Use the mouse brain as the sample (depending on the experimental objectives). Grind the tissues finely in liquid nitrogen. Use a homogenizer, if necessary. Collect the ground tissue in a glass tube, and measure its weight in advance.
2. Measure the wet weight, and dissolve the samples in ultrapure water (H_2O) at 0.5 g of sample/10 mL of water. Furthermore, disrupt the tissue using ultrasonic treatment.
3. Use a portion for protein quantification; transfer the remaining portion into glass tubes at 0.1 g wet weight per tube and store at −80°C.
4. Freeze-dry the samples.
5. Add 0.7 mL of methanol to the glass tube and mix using a vortex.
6. Sonicate this mixture for 1 min (*see* **Note 9**).
7. Add 1.4 mL of chloroform and mix using a vortex.
8. Sonicate this mixture for 1 min.
9. Allow this mixture to stand at room temperature (25°C) for 20 min to 1 h or overnight to extract lipids; occasionally, mix using a vortex.
10. Centrifuge at 2500 rpm for 10 min at room temperature.
11. After centrifugation, transfer the supernatant to another tube and dry using an N_2 evaporator (designated as lipid sample [a]). Apply nitrogen gas at approximately 1 cm above the liquid surface (maintain below 40°C) (*see* **Note 10**).
12. Next, add 2 mL of chloroform-methanol-water (CMW 30:60:8, v/v/v) to the precipitate and mix using a vortex.

13. Sonicate this for 1 min ensuring non-development of excessive heat.
14. Allow the mixture to further stand at room temperature (25°C) for 10–30 min to extract lipids (especially long-chain glycolipids).
15. Centrifuge at 2500 rpm for 10 min at room temperature (25°C).
16. After centrifugation, transfer the supernatant to a separate tube and dry using an N_2 evaporator (designated as lipid sample [b]). Dry separately from the supernatant [a] (*see* **Note 11**).
17. Add 0.9 mL of methanol to sample [b], sonicate for 1 min, and transfer to a tube containing sample [a].
18. Add another 0.9 mL of methanol to sample [b], sonicate for 1 min, and transfer to a tube containing sample [a].
19. Suspend and sonicate the mixture for 1 min.
20. Add 0.2 mL of 1 N KOH (final concentration: 0.1 N KOH in methanol) (*see* **Note 12**).
21. Incubate at 40°C for 1–2 h (to degrade and remove phospholipids).
22. Neutralize with 5 M acetic acid.
23. Check the pH using a pH test paper (MERCK Neutralit pH paper: pH 5–10) (*see* **Note 13**).
24. After neutralization, add 2 mL of n-hexane.
25. Centrifuge at 2500 rpm for 5 min at room temperature.
26. Remove the upper layer (*see* **Note 14**).
27. Dry the lower layer (which will contain glycolipids) using an N_2 gas evaporator (*see* **Note 15**).

Folch's Fractionation (Folch's Lipid Extraction Method)

1. Use the Folch's method [18] to separate and extract glycolipids. Folch's lipid extraction method uses differences in solvent polarity to separate substances into those that are more soluble in aqueous layers and those that are more soluble in organic solvents. Using this method, neutral glycolipids, sulfatides, and other glycolipids are recovered from the lower layer (chloroform-methanol). In contrast, acidic glycolipids (i.e., gangliosides) and water-soluble components are recovered from the upper layer (methanol-water). Neutral glycolipids are recovered quantitatively from the lower layer. In addition, water-soluble non-lipid components can be removed (*see* **Note 16**).
2. First, we describe the separation of neutral glycolipids. For this, add 9 mL of chloroform-methanol = 2:1 (v/v) solution and invert the tube slowly for mixing (*see* **Note 17**).

3. Sonicate this solution for 1 min to dissolve lipids.
4. Add 2.25 mL of water and mix slowly. (Mark the position of the solution's surface.)
5. Centrifuge at 2500 rpm for 10 min at room temperature.
6. Transfer the supernatant (acidic fraction) to a separate tube (**Tube 1**). Process the lower layer (neutral glycolipids) further.
7. **<Tube 1>** Mark the surface level and halfway point of the solution in the tube.
8. Evaporate up to the halfway mark using an N_2 evaporator.
9. Restore to the original volume using 0.1 M KCl.
10. Add 112.5 μL of 1 M KCl (*see* **Note 18**).
11. Sonicate this solution for 1 min.
12. This sample corresponds to the **<Tube 1 (Fraction-1)>** sample. Load Fraction-1 onto a Sep-Pak C_{18} column. The procedure for using a C_{18} column is described in a following section.

Continuation of Folch Partitioning: Separation and Desalting of Neutral Glycolipids

1. Add a theoretical upper-phase solution (chloroform: methanol: water = 3:48:47, v/v/v) to the lower layer (neutral glycolipids) up to a marked level.
2. Centrifuge at 2500 rpm for 10 min at room temperature.
3. Transfer the supernatant to a separate tube (set as **Tube 2**).
4. Dry the lower layer using an N_2 gas evaporator. **<Important: This sample is the neutral glycolipid fraction.>**
5. Store this neutral glycolipids fraction at −20°C freezer until use.
6. Perform the processing of **<Tube 2>**. Mark the surface level and halfway point of the solution.
7. Reduce the solution volume to half using an N_2 gas evaporator.
8. Restore the original volume using 0.1 M KCl.
9. Add 225 μL of 1 M KCl.
10. Sonicate the solution for 1 min.
11. Load this mixture onto a Sep-Pak C_{18} column. This sample will correspond to the <**Tube 2 (Fraction-2**)> sample described in the next section.

Continuation of Folch Partitioning: Separation and Desalting of Acidic Glycolipids

1. This section describes the separation of acidic glycolipids. Use a Sep-Pak C_{18} reverse-phase column (Waters) for glycolipid concentration and separation (*see* **Note 19**).
2. Wash the column with 25 mL of chloroform-methanol = 2:1 (v/v) solution.
3. Further, wash it with 6 mL of methanol.

4. Additionally, wash it with 7 mL of 0.1 M KCl solution (dissolved in water) (*see* **Note 20**).
5. Once the Sep-Pak C_{18} column has been conditioned as described above, apply the acidic glycolipid fraction sample prepared in the preceding steps to the column.
6. Load the samples **<Tube 1>** and **<Tube 2>** onto column (*see* **Note 21**).
7. Wash the column with 7 mL of water.
8. Prepare a new collection tube and elute it with 1 mL of methanol.
9. Then, add 4 mL of a chloroform-methanol (2:1, v/v) solution to complete the elution.
10. Dry the sample completely using an N_2 evaporator. **This sample will be the acidic glycolipid fraction.**
11. Store this sample at −20°C (freezer) until the following experiment.

3.4 Separation and Analysis of Lacto/Neolacto-Series Glycolipid by Thin-Layer Chromatography (TLC)

1. Thin-layer chromatography (TLC) is the simplest method for glycolipid analysis. This method provides sufficient separation capability to determine the glycolipids present in the fraction. Target glycolipids can be identified by examining their mobility using thin-layer chromatography (TLC) and comparing them with standard glycolipids. However, TLC alone cannot definitively determine glycan structures, even if the mobility matches that of the standard.
2. Prepare a glass chamber for solvent development. To enhance the solvent vapor saturation within the chamber, lining the inner walls with filter paper is recommended but not mandatory. Pour the developing solvent into the chamber, moisten the filter paper, and tightly seal the chamber. Allow the solution to equilibrate for 20 min–1 h to achieve sufficient saturation.
3. Cut the HPTLC plate to an appropriate size using a glass cutter.
4. Activate (dry) the plate on a hot plate at 120°C for 30 min. Immediately after activation, cover the plate with a clean glass until use (*see* **Note 22**).
5. Mark the sample spotting positions on a plate using a pencil leaving a baseline of 15–20 mm from the bottom edge of the glass plate. Appropriately space and mark the sample loading (spotting) positions with a pencil (*see* **Note 23**) (Fig. 4A).
6. Prepare glycolipid spot (load) samples using a microsyringe. When necessary, a hand dryer with cold air can used to remove the solvent.

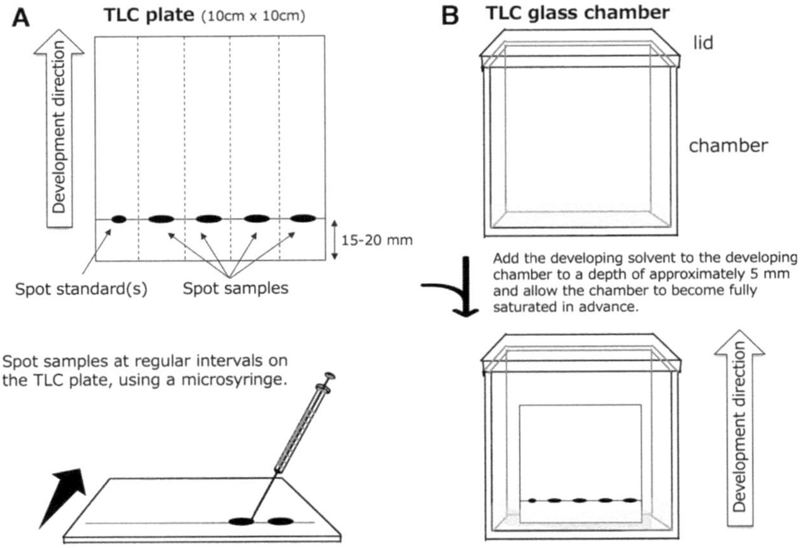

Fig. 4 TLC plate and glycolipid TLC development. (**A**) Sample spotting: A line was drawn with a pencil approximately 15–20 mm from the bottom edge of the HPTLC plate, taking care not to damage the coating. Using a microsyringe, the sample loads (spots) were applied along this line at equal intervals. If multiple samples are applied, the number of lanes is increased to avoid overlap. After loading, the samples were allowed to dry in cold air or left at room temperature. (**B**) Developing chamber: The developing chamber was pre-filled with the developing solvent to a depth of approximately 5 mm, thereby allowing the interior to become saturated with solvent vapor. To enhance saturation, filter paper was placed around the inner walls of the chamber. The HPTLC plate was carefully placed in the correct orientation in the developing chamber to begin the development process

7. Place the HPTLC plate in a development chamber (Fig. 4B). Immediately close the lid to maintain solvent vapor saturation and allow the solvent to develop spots.
8. Once the solvent front reaches the top of the plate, remove the HPTLC plate from the chamber. This will usually take approximately 15–30 min.
9. Dry the plate using a hand dryer.
10. Spray the HPTLC plate with the resorcinol reagent (for ganglioside detection). Immediately cover the HPTLC plate with a clean glass plate and place it upside down on a hot plate at 90°C for a few min to 30 min to develop glycolipid coloration (*see* **Note 24**). For detection using orcinol reagent (for neutral glycolipids), heat the HPTLC with the surface facing upward at 110°C for 15 min.

3.5 TLC-Immunoblotting (Immuno-TLC) for the Quantitative Analysis of Lacto/Neolacto-Series Glycolipid

1. Dry the HPTLC plate using a hand dryer after TLC development, following the previous method.
2. Coat the HPTLC plate.
3. Mix the polyisobutyl methacrylate beads (30 mg) and dissolve them in a glass bottle with 1.2 mL of chloroform (*see* **Note 25**).

4. Mix these polyisobutyl methacrylate beads/chloroform solution with 30 mL of n-hexane while rotating a large glass Petri dish.
5. Immerse the HPTLC plate to ensure even coating, rotate it further, and allow it to sit for 90 s. Remove the plate carefully without touching the surface and allow it to dry (*see* **Note 26**).
6. Transfer the plate to a new Petri dish, and place small pieces of water-moistened absorbent cotton at the corners to prevent drying.
7. Incubate the coated plate in sialidase solution [Clostridium perfringens sialidase (SIGMA); 0.2 U/mL in 40 mM sodium acetate buffer, pH 5.0] at 37 °C for 3 h to remove sialic acids and expose the terminal LacNAc residues,then wash the plate with PBS. This desialylation step is required for detecting sialylated lacto/neolacto-seriesglycolipids, such as sialyllactoparagloboside (SLPG) and sialylparagloboside (SPG), with the 1B2-1B7 mAb. 8. Perform blocking by gently.
8. Perform blocking by gently adding 5–10 mL of 1% BSA/PBS to fully cover the sample surface. Incubate the plate at room temperature for 45 min without drying.
9. Tilt the plate, and remove the excess 1% BSA/PBS using a Pasteur pipette. Then, transfer the plate to a cold room (4°C) or keep it at room temperature (25°C). Add the primary antibody solution using a Pasteur pipette to cover the entire plate surface.
10. Incubate the plate overnight at 4°C or for 1 h at room temperature, depending on the optimal conditions for the probe (first antibody). Use a 1% BSA/PBS solution for antibody dilution and washing.
11. Wash the plate with 25 mL of 1% BSA/PBS.
12. Add the secondary antibody and incubate at room temperature (25°C) for 30 min (*see* **Note 27**). (i) Use HRP-conjugated anti-mouse IgM for detection via ECL (chemiluminescence reagents). (ii) Use biotin-labeled anti-mouse IgM with the VECTASTAIN Elite ABC complex, followed by chromogenic or chemiluminescent reagents. When using ECL, BSA is noted to nonspecifically react, increasing background noise; therefore, thoroughly wash away PBS before reagent application.
13. Wash the plate with 25 mL of 1% BSA/PBS.
14. (When a biotin-labeled antibody is used as the primary antibody) Use the VECTASTAIN Elite ABC Kit according to the manufacturer's instructions, with incubation at room temperature for 30 min (*see* **Note 28**).

15. Wash the plate with 25 mL of 1% BSA/PBS.
16. Apply appropriate chromogenic or chemiluminescent reagents for detection. When using ECL, wash the plate with 1% BSA/PBS, followed by 60 mL PBS, before reagent application
17. Detect the signal from the target glycolipids using X-ray film or an appropriate detection device (Fig. 5).

3.6 Structural Analysis of Glycolipid Glycan by Mass Spectrometry

Digestion of Glycolipids by Ceramide Glycanase and Purification of Glycolipid-Derived Glycans

1. Extract the glycolipids from cells or tissues following the previous protocol.
2. Resuspend the extracted glycolipids in 40 μL of 50 mM sodium acetate buffer (pH 5.5) containing 0.1% Triton X-100.
3. Add ceramide glycanase (20 mU), and incubate the mixture at 37°C for 18 h.
4. Terminate the reaction by boiling (3–5 min).
5. Add a total of 60 μL of water to the reaction mixture, followed by its application to a reverse-phase column (Sep-Pak C18).
6. Collect the unbound fractions containing glycolipid-derived glycans and freeze-dry them.

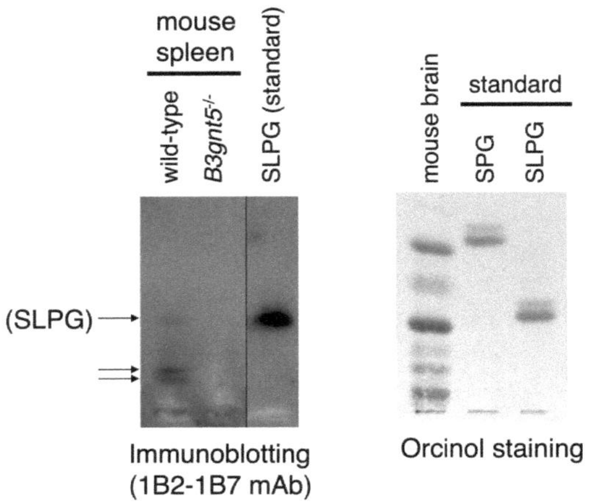

Fig. 5 Detection of $_3$ lacto/neolacto-series GSLs and polylactosamine structures (sialyllactoparagloboside: SLPG) by Immuno-TLC (Adapted from [14]). Long-chain neolacto-series glycolipids (such as SLPG) containing polylactosamine structures observed in the spleen cells of wild-type mice were absent in *B3gnt5*$^{-/-}$ mice. Arrowheads: lacto/neolacto-series glycolipids containing long-chain glycans. *SLPG* sialyllactoparagloboside, *SPG* sialylparagloboside. (Copyright (2010) National Academy of Sciences [14]).

Reduction and Complete Methylation of Glycolipid-Derived Glycans

1. Reduce the free glycans using 50 μL of 500 mM $NaBH_4$ dissolved in 50 mM NaOH and incubate for 12 h at 45°C.
2. Neutralize the reaction with 5 μL of a 50% acetic acid solution.
3. Perform desalting using a cation exchange column (Oasis MCX).
4. Elute alditols with 750 μL of distilled water and dry using a Speed-Vac.
5. Remove residual borate with 50 μL of 1% acetic acid methanol, followed by drying under vacuum.
6. Perform complete methylation of reduced glycolipid-derived glycans using the described method.
7. Add approximately 50 μL of NaOH/DMSO slurry and 50 μL of iodomethane to the sample.
8. Stir the mixture at room temperature for 30 min, and stop the reaction with 150 μL of a 50% acetic acid solution and 800 μL of distilled water.
9. Elute fully methylated alditols with 600 μL of acetonitrile using a reverse-phase column (Oasis HLB) and dry using a Speed-Vac.

Glycolipid mass spectrometry analysis uses mass spectrometers, such as Reflex IV MALDI-TOF and AXIMA-QIT MALDI quadrupole ion-trap TOF, which generate mass and MS^n spectra.

MALDI MS and Multistage Tandem Mass Spectrometry (MS^n) Analysis of Permethylated Glycans Derived from Glycolipids

1. Resuspend the dried permethylated samples in 50 μL of acetonitrile. Perform glycan permethylation following standard methodologies [19–22].
2. Spot a mixture of 0.5 μL of matrix solution (2,5-dihydroxybenzoic acid [DHB] 1 mg/30% ethanol 1 mL) and 0.5 μL of the analyte solution onto a plate.
3. Dry the matrix and analyte mixture, recrystallize with 1 μL of ethanol, and prepare for mass spectrometry analysis.
4. To obtain the mass spectra, perform a sample analysis using MALDI-TOF and quadrupole ion-trap devices. Acquire all MS and MS^n spectra using Na^+ adduct ions.
5. Generate and analyze all MS and MS^n spectra based on Na^+ adduct ions (Fig. 6).

3.7 Flow Cytometry Analysis Using Anti-Glycolipid Antibodies and Probes (Anti-CD antigens, Etc.)

1. Prepare the primary antibody (1B2-1B7 monoclonal antibody) or FITC-labeled cholera toxin B subunit (CT-B) at 100 μg/mL (10× the actual dilution), and add 10 μL of the aliquot into glass tubes.
2. Collect the cells and wash twice with 1% BSA/0.1% NaN_3/PBS(−). Adjust the cell concentration to 1×10^7 cells/mL.

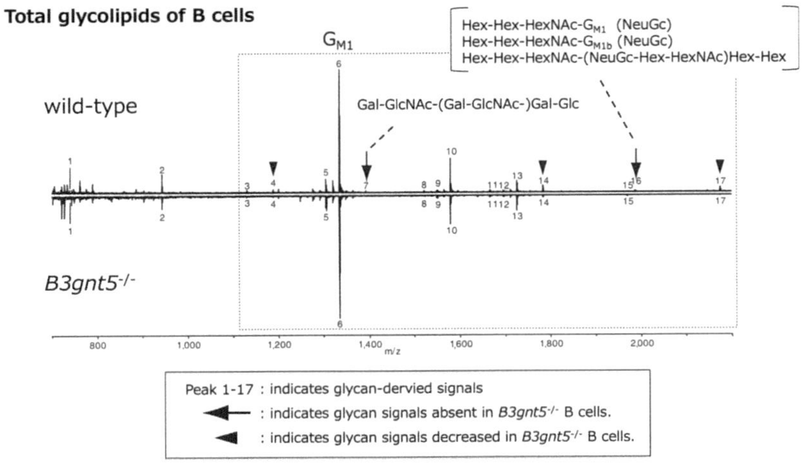

Fig. 6 Example of glycolipid analysis results by mass spectrometry. Mass spectrometry revealed the loss of elongated glycolipids in Lc_3Cer in *B3gnt5*$^{-/-}$ B cells. Signals *7 and 16*, present in wild-type B cells, were absent in *B3gnt5*$^{-/-}$ mice. No. 7 signal was identified as a lacto/neolacto-series glycolipid structure, Galβ 1-4GlcNAcβ1-3(Galβ1-4GlcNAcβ1-6)Galβ1-4Glc. The m/z value of No. 16 (1987.147) suggested three possible candidates, as listed in the figure, which could be ganglio-series or lacto/neolacto-series glycolipids. Since compound 16 was not detected in *B3gnt5*$^{-/-}$ B cells, it is more likely to be a lacto/neolacto-series glycolipid. (Copyright (2010) National Academy of Sciences [14]).

3. Add 100 μL of cells into the glass tubes containing the primary antibody, and shake the tube rack.
4. Incubate the mixture at 4°C in the dark (refrigerator compartment) for 30 min.
5. Add 4 mL of 1% BSA/0.1% NaN_3/PBS to the glass tubes for washing, and gently shake the tube rack to mix the solution.
6. Centrifuged this mixture at 1000 rpm for 5 min.
7. Repeat Steps 5 and 6 once again.
8. Prepare the secondary antibody at 10 μg/mL and aliquot 100 μL into glass tubes. Shake the tube rack for mixing. Use the FITC-labeled cholera toxin B subunit (CT-B: reacts with G_{M1} ganglioside) as the primary probe. In this case, it is not necessary to stain with the secondary antibody before proceeding to the washing procedure (Step 10).
9. Incubate the mixture at 4°C in the dark for 30 min.
10. Add 3 mL of 1% BSA/0.1% NaN_3/PBS to the glass tubes and shake.
11. Centrifuge the mixture at 1000 rpm for 5 min.
12. Repeat Steps 10 and 11 once again.
13. Suspend the cells in 300 μL of 1% BSA/0.1% NaN_3/PBS and filter through a nylon mesh (0.45 μm) (*see* **Note 29**).

14. Pass the cell suspension through the mesh and flow cytometry analysis was performed using a FACS/Cell sorter instrument.
15. Analyze the data using a dedicated software (e.g., FACS instrument software or FlowJo software) (Fig. 7).

4 Notes

1. Polyisobutyl methacrylate can be used at concentrations ranging from 0.1% to 0.4%. A concentration of 0.4% is considered to be relatively high. For 0.4% use, dissolve 2.5% chloroform solution in n-hexane to dilute to 0.4%.
2. The 1B2-1B7 monoclonal antibody is an anti-LacNAc antibody that recognizes poly-*N*-acetyllactosamine (poly-LacNAc: polylactosamine) glycan structures. This antibody reacts with polylactosamine chains present on both glycoproteins and glycolipids. In the case of the extracted glycolipids, all detected signals correspond to the neolacto-series glycolipids. Specifically, the 1B2-1B7 mAb reacts with nLc_4Cer, nLc_6Cer, and other glycolipids containing LacNAc units at their non-reducing ends. Additionally, although it is not a

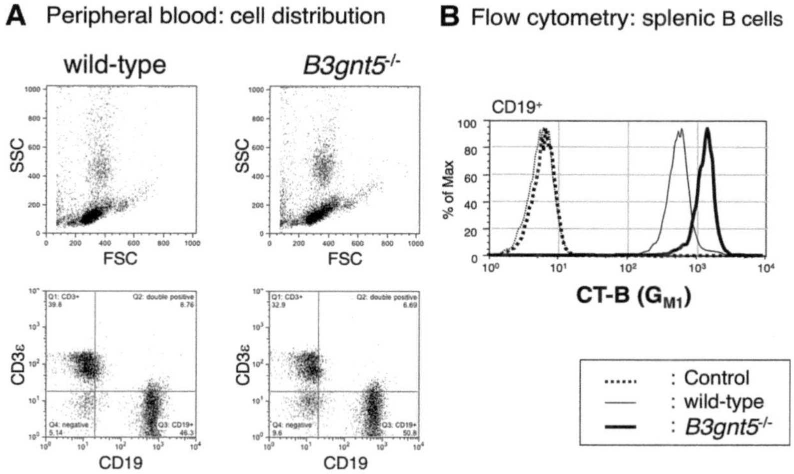

Fig. 7 Example of glycolipid analysis results by flow cytometry: upregulation of cholera toxin B subunit staining as a lipid raft (glycolipid-enriched microdomain, GEM) marker. (**A**) No significant difference was observed in the proportion of B cells (CD19+ cells) or T cells (CD3ε+ cells) in the peripheral blood cells of *B3gnt5*−/− mice compared with that in wild-type mice [14]. (**B**) Flow cytometry (FCM) analysis demonstrated that in B cells from *B3gnt5*−/− mice, the staining intensity of GM1-specific antibody or cholera toxin B subunit (CT-B) was significantly increased compared with that in wild-type mice. In *B3gnt5*−/− mouse splenic B cells (CD19+ cells), the GM1 level was elevated compared with that in wild-type mice [14]. (Copyright (2010) National Academy of Sciences [14]).

lacto/neolacto-series glycolipid, horseradish peroxidase-labeled cholera toxin (CT-B) (Calbiochem) can be used for ganglioside G_{M1} detection in lipid raft analysis.

3. 1B2-1B7 monoclonal antibody is an anti-LacNAc monoclonal antibody that recognizes polylactosamine chains present on both glycoproteins and glycolipids. In the extracted glycolipids, all signals detected by this antibody originate from the neolacto-series glycolipids. However, the contribution of glycoproteins must be considered when using this antibody at the cellular level. Therefore, the use of glycoprotein synthesis inhibitors or alternative antiganglioside antibodies may be necessary.
4. For detailed targeting vector design methodologies, consult specialized books or articles on gene targeting based on the study objectives.
5. Culture ES cells (E14) on a feeder layer of mitomycin C-treated mouse embryonic fibroblasts in DMEM (high glucose) supplemented with FBS, leukemia inhibitory factor, penicillin, streptomycin, non-essential amino acids, L-glutamine, and β-mercaptoethanol. Use standard ES cell culture protocols, including optimal temperatures and CO_2 concentrations. Optimize transfection conditions to improve efficiency.
6. Verifying and adjusting drug concentrations is necessary to ensure that only drug-resistant cells proliferate.
7. β3GnT5 catalyzes the biosynthesis of Lc_3Cer by transferring GlcNAc from UDP-GlcNAc to LacCer. In *B3gnt5* knockout (*B3gnt5*$^{-/-}$) mice, Lc_3Cer biosynthetic activity is undetectable, thereby demonstrating that Lc_3Cer synthase is the sole enzyme responsible for elongating the polylactosamine structure of lacto/neolacto-series glycolipid.
8. As this is a crude purified sample containing a mixture of neutral glycolipids, acidic glycolipids, and other phospholipids, perform a more detailed glycolipid analysis using subsequent extraction and purification methods. For total glycolipid (crude glycolipid) samples extracted from the above tissue quantity, approximately 10 μL is sufficient for chemical detection, while about 1 μL is sufficient for immunohistochemical staining. However, as the required sample volume depends on the target glycolipid concentration, a preliminary experiment should be performed to determine the appropriate sample volume.
9. Maintaining a temperature of 4°C with ice is unnecessary; however, care should be taken to prevent excessive heat generation.

10. If the temperature exceeds 40°C, acidic glycolipids, such as sialic acid-containing glycolipids, are degraded; therefore, temperature control is crucial.
11. This fraction will contain water and may be difficult to dry completely.
12. The sample must be dissolved in methanol instead of water, and sonication is used for dissolution. It should be noted that NaOH is less soluble.
13. Moisten the test paper just before use. The use of a 0.2 μL microsyringe is recommended. A critical point is to ensure slightly acidic conditions (pH 6.5) at this step to prevent the degradation of sialic acid.
14. Under alkaline conditions, fatty acids partition into the aqueous phase rather than the hexane phase; therefore, caution is required.
15. Once dried, the procedure can be temporarily halted.
16. The Folch's lipid extraction method is helpful for the quantitative separation of neutral glycolipids. However, during the separation of acidic glycolipids, especially shorter-chain gangliosides, such as GM4 and GM3, some may flow into the lower layer, leading to poor recovery rates. When obtaining total glycolipid fractions, mixtures of neutral and acidic glycolipids (gangliosides, sulfatides, etc.) are common. Since their migration behavior (mobility) in subsequent thin-layer chromatography is similar, it is preferable to pre-separate the neutral and acidic glycolipids for easier analysis of the neutral glycolipids. One representative method is the preparation of gangliosides using weak anion exchange resins such as DEAE-Sephadex, as described by Ledeen et al. The fraction eluted from the DEAE column using chloroform: methanol: water (30:60:8, v/v/v) contained neutral lipids and other glycolipids. Meanwhile, the fraction eluted with Chloroform-Methanol-0.8 M sodium acetate solution (30:60:8, v/v/v) allows for the quantitative recovery of acidic glycolipids and phospholipids. Ester lipids in the acidic glycolipid fraction can be degraded by alkaline treatment (0.2 N NaOH, 37°C for 2 h), followed by desalting to obtain total acidic glycolipids. Desalting can be performed using dialysis or gel filtration.
17. The solution may become mayonnaise-like if its lipid content is high. If this occurs, evaporate the solvent with nitrogen gas (N_2) and repeat the process with an increased volume.
18. As described later, glycolipids containing sialic acid are not retained on the column unless the salt concentration is at least 0.1 M KCl. Therefore, washing with 0.1 M KCl is necessary at this stage.

19. A previously used but well-maintained column is preferable. However, the column should be thoroughly cleaned after each use. New columns tend to release contaminants; therefore, they must be washed thoroughly before use.
20. Glycolipids containing sialic acid require a minimum salt concentration of 0.1 M KCl for retention on the column. Therefore, washing with 0.1 M KCl is necessary at this stage.
21. Although Fractions 1 and 2 are loaded separately into the column in sequence, mixing the fractions before loading is also possible.
22. If the laboratory environment is not humid and remains dry, this activation step can be omitted.
23. Care should be taken not to scratch the surface layer of the plate.
24. It is recommended to place the glycolipid side facing downward, but the reaction will also occur if it is placed upright.
25. If undissolved, suspend using a Pasteur pipette.
26. If an additional coating is required, repeat the immersion after drying. Alternatively, the polymer concentration can be increased to twice or more.
27. The choice of secondary antibody depends on the target glycolipid antibody and detection method.
28. Prepare the reagent 30 min before use by accurately adding two drops of reagent A to 5 mL of 1% BSA/PBS, followed by two drops of reagent B.
29. Cells may also be suspended in 0.5% paraformaldehyde/PBS for fixation, thereby allowing FACS analysis to be conducted later within a few days.

Acknowledgments

A previous study [14] was supported by the New Energy and Industrial Technology Development Organization (NEDO), Japan. Akira Togayachi is supported by the "Human Glycome Atlas Project" under the "Large-Scale Academic Frontier Promotion Projects" of the Ministry of Education, Culture, Sports, Science and Technology (MEXT). GaLSIC was supported by J-GlycoNet, an official Joint Usage/Research Center of Japan, through MEXT.

Conflict of Interest The authors have no conflicts of interest to declare that are relevant to the content of this chapter.

References

1. Kiguchi K, Henning-Chubb CB, Huberman E (1990) Glycosphingolipid patterns of peripheral blood lymphocytes, monocytes, and granulocytes are cell specific. J Biochem 107:8–14
2. Zhang X, Zeng B, Zhu H et al. (2023) Role of glycosphingolipid biosynthesis coregulators in malignant progression of thymoma. Int J Biol Sci 19:4442–4456
3. Furukawa K, Ohmi Y, Ji S et al. (2017) Glycolipids: essential regulator of neuro-inflammation, metabolism and gliomagenesis. Biochim Biophys Acta Gen Subj 1861:2479–2484
4. Trinchera M, Parini R, Indellicato R et al. (2018) Diseases of ganglioside biosynthesis: an expanding group of congenital disorders of glycosylation. Mol Genet Metab 124:230–237
5. Collins PM, Bum-Erdene K, Yu X et al. (2014) Galectin-3 interactions with glycosphingolipids. J Mol Biol 426:1439–1451
6. Nakamura M, Tsunoda A, Sakoe K et al. (1992) Total metabolic flow of glycosphingolipid biosynthesis is regulated by UDP-GlcNAc:lactosylceramide beta 1-> 3N-acetylglucosaminyltransferase and CMP-NeuAc:lactosylceramide alpha 2->3 sialyltransferase in human hematopoietic cell line HL-60 during differentiation. J Biol Chem 267:23507–23514
7. Jungalwala FB (1994) Expression and biological functions of sulfoglucuronyl glycolipids (SGGLs) in the nervous system—a review. Neurochem Res 19:945–957
8. Martinez-Morales P, Morán Cruz I, Roa-de la Cruz L et al. (2021) Hallmarks of glycogene expression and glycosylation pathways in squamous and adenocarcinoma cervical cancer. PeerJ 9:e12081
9. Meng Q, Hu X, Zhao X et al. (2021) A circular network of coregulated sphingolipids dictates lung cancer growth and progression. EBioMedicine 66:103301
10. Gooi HC, Hounsell EF, Edwards A et al. (1985) Differences in the fine specificities of monoclonal (Class A) antibodies to human myeloid cells. Clin Exp Immunol 60:151–158
11. Togayachi A (2021) Enzyme assay of beta1,3-glycosyltransferase family. In: Nishihara S, Angata K, Aoki-Kinoshita KF, Hirabayashi J (eds) Glycoscience protocols (GlycoPODv2). Research Center for Medical Glycoscience, Soka University, Saitama (JP)
12. Henion TR, Zhou D, Wolfer DP et al. (2001) Cloning of a mouse beta 1,3 N-acetylglucosaminyltransferase GlcNAc(beta 1,3)Gal(beta 1,4)Glc-ceramide synthase gene encoding the key regulator oflacto-series glycolipid biosynthesis. J Biol Chem 276:30261–30269
13. Togayachi A, Akashima T, Ookubo R et al. (2001) Molecular cloning and characterization of UDP-GlcNAc:lactosylceramide beta 1,3-N-acetylglucosaminyltransferase (beta 3Gn-T5), an essential enzyme for the expression of HNK-1 and Lewis X epitopes on glycolipids. J Biol Chem 276:22032–22040
14. Togayachi A, Kozono Y, Ikehara Y et al. (2010) Lack of lacto/neolacto-glycolipids enhances the formation of glycolipid-enriched microdomains, facilitating B cell activation. Proc Natl Acad Sci U S A 107:11900–11905
15. Biellmann F, Hülsmeier AJ, Zhou D et al. (2008) The Lc3-synthase gene B3gnt5 is essential to pre-implantation development of the murine embryo. BMC Dev Biol 8:109
16. Kuan CT, Chang J, Mansson JE et al. (2010) Multiple phenotypic changes in mice after knockout of the B3gnt5 gene, encoding Lc3 synthase—a key enzyme in lacto-neolacto ganglioside synthesis. BMC Dev Biol 10:114
17. Brown DG, Warren VN, Pahlsson P et al. (1993) Carbohydrate antigen expression in murine embryonic stem cells and embryos. I. Lacto and neo-lacto determinants. Histochem J 25:452–463
18. Folch J, Lees M, Sloane Stanley GH (1957) A simple method for the isolation and purification of total lipides from animal tissues. J Biol Chem 226:497–509
19. Ashline D, Singh S, Hanneman A et al. (2005) Congruent strategies for carbohydrate sequencing. 1. Mining structural details by MSn. Anal Chem 77:6250–6262
20. Lapadula AJ, Hatcher PJ, Hanneman AJ et al. (2005) Congruent strategies for carbohydrate sequencing. 3. OSCAR: an algorithm for assigning oligosaccharide topology from MSn data. Anal Chem 77:6271–6279
21. Zhang H, Singh S, Reinhold VN (2005) Congruent strategies for carbohydrate sequencing. 2. FragLib: an MSn spectral library. Anal Chem 77:6263–6270
22. Ito H, Kaji H, Togayachi A et al. (2016) Comparison of analytical methods for profiling N- and O-linked glycans from cultured cell lines: HUPO human disease glycomics/ proteome Initiative multi-institutional study. Glycoconj J 33:405–415

Check for updates

Chapter 21

Ganglioside Functions in Extracellular Vesicles as Revealed by Single-Particle Tracking

Tatsuki Isogai, Miki Kanno, Koichi Furukawa and Kenichi G.N. Suzuki

Abstract

Extracellular vesicles play roles as critical mediators of cell–cell communications. We recently discovered that ganglioside content in extracellular vesicles was approximately 30 times higher than in cells, and GM1 in extracellular vesicles specifically binds to laminin. In this chapter, we describe a method to prepare cells expressing specific gangliosides and introduce in vitro experiments to evaluate the binding ability of extracellular vesicles and liposomes containing gangliosides to the extracellular matrix.

Keywords Gangliosides, Extracellular vesicles, Single-particle tracking, Total internal reflection fluorescence microscopy, Extracellular matrix

1 Introduction

Extracellular vesicles (EVs) have drawn attention in broad research fields as important carriers of cell-cell communications [1–3]. EVs contain various cell components such as proteins, lipids, and miRNAs, which are transported to recipient cells, thereby changing their phenotypes [4, 5]. Previous studies have reported that EVs are bound to recipient cells via specific interactions between integrin heterodimers in EVs and extracellular matrices (ECMs) on recipient cells, and the composition of integrin heterodimers in EVs depends on the cells secreting EVs and influences organ-selective metastasis of cancer in vivo [6].

Gangliosides are one of the representative lipid raft markers and play a wide range of roles in cellular physiological functions [7–10], and the dynamic behaviors of single molecules of gangliosides have been investigated [11–16]. We found that the molecular density of GM1 in extracellular vesicles was approximately 30 times higher than in cells, and GM1 in extracellular vesicles specifically binds to laminin. In this chapter, we introduce protocols to determine the

Koichi Furukawa (ed.), *Gangliosides*, Methods in Molecular Biology, vol. 3035,
https://doi.org/10.1007/978-1-0716-5288-6_21,

ganglioside content in EVs and examine whether gangliosides in EVs have a binding activity to ECMs.

2 Materials

Prepare all solutions using ultrapure water (prepared by purifying deionized water to attain a sensitivity of 18 MΩ at 23°C) and analytical grade reagents.

2.1 Cell Culture

1. 100- and 150-mm tissue culture dish (Iwaki).
2. Mouse melanoma cell line B78 (a subline of B16).
3. Medium: Dulbecco's modified Eagle's medium (DMEM) (Sigma-Aldrich) with 10% fetal bovine serum (FBS), 100 U/ml penicillin, 100 μg/ml streptomycin, and 400 μg/ml G418. Store at 4°C.

2.2 Flow Cytometry

1. Primary antibody against ganglioside: monoclonal antibodies against GM3 (M2590, IgM) (Cosmo Bio), GM2 (10-11, IgM), GD2 (220-51 m IgG1), GD3 (R24, IgG3) (10-11 and R24 antibodies kindly provided by P. O. Livingston and L. J. Old), GD1a (D-266, IgM), or GD1b (370, IgM).
2. Fluorescently labeled secondary antibody: FITC-labeled goat anti-mouse IgM antibodies (Cappel), or goat anti-mouse IgG (H&L chain) antibodies (Cappel).
3. Staining for GM1: biotinylated cholera toxin subunit B (Sigma-Aldrich) and then FITC-labeled avidin (Zymed).
4. Flow cytometry: FACSCalibur (Becton-Dickinson).

2.3 Liposome Preparation

1. Ganglioside: GM1 (bovine brain) (Avanti Plor Lipids), GD1a (bovine brain) (AdipoGen Life Sciences), GD2 (human brain) (Santa Cruz). GM2, GM3, and GD3 are prepared by total chemical synthesis [17, 18].
2. Lipids: dimyristoylphosphatidylcholine (DMPC) (Avanti Polar Lipids), and Bodipy-sphingomyelin (Bodipy-SM) (Invitrogen).
3. VT-2000 evaporator (EYELA).
4. Extruder (LiposoFAST, AVESTIN).
5. 100- and 50-nm pore-sized polycarbonate membrane (AVESTIN).
6. Sialic Acid Assay Kit (MAK314-1KT, Sigma-Aldrich).

2.4 Single-Particle Imaging

1. Triple-well glass base dishes (IWAKI).
2. Hanks' balanced salt solution (HBSS): Hanks' solution (Nissui) with 2 mM PIPES (Dojindo).

3. HaloTag SaraFluor™ 650T ligand (SF650T) (Goryo Chemical).
4. Microscope: IX-83 inverted microscope (Olympus Evident).
5. sCMOS camera: ORCA Flash4.0 V2 (Hamamatsu Photonics).
6. High-speed gated Image intensifier: C9016-02MLG (Hamamatsu Photonics).
7. 647 nm laser (Omicron).
8. Anti-CD81 IgG antibody (JS-81, BD Pharmingen).
9. Extracellular matrices: fibronectin (Roche), laminin (Sigma-Aldrich), collagen type I (Sigma-Aldrich).

3 Methods

3.1 Cell Culture

1. Culture B78 cells in a humidified atmosphere of 5% CO_2 at 37°C.

3.2 Cells Expressing a Specific Ganglioside

1. Transfect B78 cells with pM1T-9 [19] and pM2T1-1 [20] cDNA plasmids for overexpression of GM1, resulting in GM3 and GM2 depletion because these enzymes use GM3 and GM2 as substrates for synthesis of GM1 (Fig. 1).
2. Transfect B78 cells with pM2T1-1 and subsequently with pD3T31 [9] for overexpression of GD2.

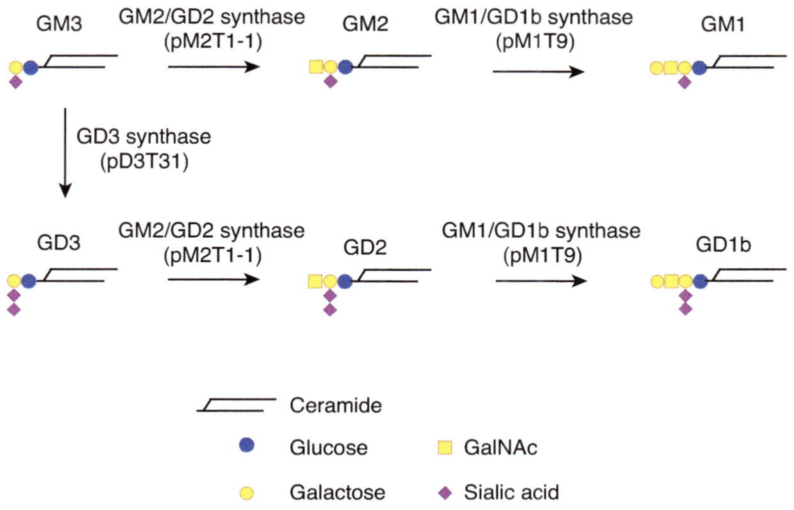

Fig. 1 Schematic representation of the biosynthesis pathway of gangliosides used in our study. B78 cells were transfected with pM2T1-1 (cDNA encoding GM2/GD2 synthase) [20], pM1T9 (cDNA encoding GM1/GD1b synthase) [19], or pD3T31 (cDNA encoding GD3 synthase) [9] to produce cells expressing specific kinds of gangliosides

3. Transfect B78 cells with pD3T31 for overexpression of both GD2 and GD3.

3.3 Flow Cytometry

1. Stain intact and glycosyltransferase-transfected B78 cells by incubation with monoclonal antibodies against GM3, GM2, GD2, GD3, GD1a, or GD1b.
2. Subsequently, stain cells with FITC-labeled goat anti-mouse IgM antibodies or goat anti-mouse IgG (H&L chain) antibodies.
3. Alternatively, stain cells with biotinylated cholera toxin subunit B and then FITC-labeled avidin.
4. Analyze the expression levels of gangliosides on B78 cells with FACSCalibur [9, 20].

3.4 Cells Expressing Halo7tag-Conjugated EVs' Marker Protein

1. Transfect intact and glycosyltransferase-transfected B78 cells with plasmid DNA encoding CD81-Halo7tag [21].
2. Select the cells stably expressing CD81-Halo7 by treatment with 400 μg/ml zeocin (*see* **Note 1**).

3.5 Isolation of EVs

1. Cultivate intact or glycosyltransferase-transfected cells either in two 100-mm dishes or in two 150-mm dishes. Cells are grown to approximately 80% confluence ($0.3–1.6 \times 10^8$ cells/150-mm dish).
2. Replace cell culture medium in the 100- or 150-mm dishes with 10 ml or 30 ml of FBS-free medium, respectively.
3. After 48 h of incubation, harvest the cell culture supernatant.
4. Centrifuge the harvested cell culture supernatant at 300 × *g* for 10 min at 4°C to remove cells, and subsequently centrifuge the supernatant at 2000 × *g* and 4°C for 10 min to remove sediment and eliminate apoptotic bodies [21, 22].
5. Subsequently, centrifuge the supernatant at 10,000 × *g* and 4°C for 30 min (Himac CF16RN, T9A31 angle rotor, 50 ml Falcon tube) to pellet the microvesicles [21, 22].
6. Concentrate the supernatant by ultrafiltration using an Amicon Ultra 15 100 K (Millipore) or a Centricon Plus 70 100 K (Millipore) filter unit [21, 22].
7. Stain CD81-Halo7 on EVs by incubation with 100 nM SF650T-ligand for 1 h at 37°C (*see* **Note 2**).
8. Finally pellet EVs by ultracentrifugation at 200,000 × *g* and 4°C for 4 h (Himac CS100FNX with an S55A2 angle rotor and S308892A microtubes) (*see* **Note 3**) [21, 22].

3.6 Liposome Preparation

1. In an eggplant-shaped flask (*see* **Note 4**), mix 50 μl of 1 mM ganglioside (GM1, GM2, GM3, GD1a, GD2, or GD3) in

methanol, 500 µl of 10 mM DMPC in chloroform, and 0.5 µl of 1 mM Bodipy-SM in methanol in a solution of 99:1:0.01 mol% ratio DMPC: ganglioside: Bodipy-SM [21].

2. Dry the solvent with nitrogen gas and further dry it with an evaporator. Add 5 ml of PBS to the flask, which was incubated at room temperature for 1 h.
3. Completely freeze the solution in liquid nitrogen and then thaw it in hot water (45°C). Repeat the freeze and thaw process five times (*see* **Note 5**).
4. Pass 500 µl of the vesicle suspension through 100-nm pore-sized polycarbonate membrane filters (*see* **Note 6**).
5. Determine the amount of sialic acid in the liposomes using a Sialic Acid Assay Kit [21].

3.7 Single-Particle Imaging

1. Coat the glass windows of triple-well glass-base dishes with 50 µl of 10 µg/ml anti-CD81 antibody in HBSS and incubate for 2 h at 37°C [21].
2. Remove the antibody solution, and coat the glass window with 50 µl of 50 µg/ml casein in HBSS for 1 h at 37°C. This coating process is important for mitigating the nonspecific binding of EVs to the glass surface during subsequent experiments (*see* **Note 7**) [21].
3. Apply either 50 µl of three distinct concentrations of EVs containing CD81-Halo7 conjugated with SF650T in HBSS to glass windows pre-coated with antibodies against CD81 and incubate the sample for 1 h at 37°C (*see* **Note 8**) [21].
4. After removing the EVs and washing twice with HBSS, observe the individual fluorescent particles of the EVs at 30 frames/s by total internal fluorescence microscopy (TIRFM) using an Olympus IX-83 microscope (60×1.49 NA oil objective) equipped with a high-speed gated image intensifier coupled to an sCMOS camera [11, 21–25]. SF650T was excited using a 647 nm laser.
5. Analyze the acquired movies using ImageJ, count the number of EV particles bound to the antibody on glass, and prepare EVs derived from intact and glycosyltransferase-transfected cells at identical concentrations ($0.8–4 \times 10^{10}$ particles/ml) (*see* **Note 9**).
6. Pretreat the glass with 50 µl of 50 µg/ml casein solution in HBSS and subsequently with 50 µl of 100 µg/ml fibronectin, laminin, collagen type I, or casein for 2 h at 37°C (*see* **Note 10**), and then add the EVs, of which concentration was adjusted, and incubate them for 1 h at 37°C [21].
7. Remove the EV suspension, wash with HBSS twice, and visualize individual fluorescent particles of the EVs in HBSS at

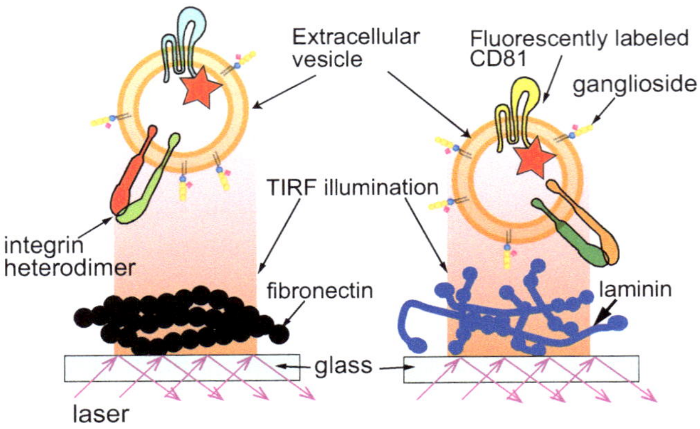

Fig. 2 Single-particle imaging of extracellular vesicles (EVs) containing a different ganglioside to examine their binding activity to the extracellular matrix. Extracellular matrix (fibronectin and laminin) was coated on a glass-based dish, tetraspanin marker protein fluorescently labeled EVs were added, and the bound EVs were observed by total internal reflection fluorescence (TIRF) illumination with single molecule detection sensitivity. EVs were derived from B78 melanoma cells expressing a different kind of ganglioside

37°C with single-molecule detection by TIRFM (Fig. 2) (*see* **Note 11**) [21].

8. Analyze the acquired movies using ImageJ, and count the number of EV particles. The number of EVs bound to ECM-coated glass was estimated by subtracting the number of EVs bound to casein-coated glass in a 1024 pixel × 1024 pixel (81.9 μm × 81.9 μm) image [21].
9. As with EVs, add liposomes of matching concentrations to glass precoated with fibronectin, laminin, and collagen type I, and incubate for 1 h at 37°C [21].
10. After removal of the ganglioside-containing liposome and washing, analyze the number of liposomes containing ganglioside on the glass as well as EVs [21].

4 Notes

1. The transfected cells are treated with zeocin after 1 day because antibiotic resistance is not expressed immediately after transfection. The zeocin treatment is sustained for 1 week to select the cell lines expressing CD81-Halo7 [21].
2. When observing a single particle of EV, it is essential to minimize contamination of the free dye. To achieve this, the absolute amount of dye added to the sample should be reduced as much as possible. In this experiment, a high concentration

of SF650T-ligand (10 μM) was added to 100–200 μl of EV solution concentrated by ultrafiltration [21, 22].

3. The EV pellet obtained by ultracentrifugation is difficult to resuspend using HBSS alone. To facilitate resuspension, the pellet is dissolved in HBSS by gentle rotation at 4°C for 1 h.
4. An eggplant-shaped flask and an extruder should be thoroughly washed and dried at 60°C using a dry heat sterilizer. They should be allowed to reach room temperature just before use.
5. This process generates liposomes by suspending a thin lipid film in PBS. If PBS is not used for suspension, additional freeze-thaw cycles are required.
6. Liposomes of uniform size can be produced by passing the suspension through a filter approximately ten times.
7. Casein powder is suspended in 1 M Tris–HCl (pH 8.0) and heated at 55°C using a block heater to facilitate dissolution. For coating applications, the solution is diluted with HBSS [21].
8. The EV solution needs to be sufficiently diluted because excessively high EV concentrations hinder the observation of individual particles. In this study, 200 μl of EV solution purified from two 150 mm dishes was diluted at ratios of 1:20, 1:60, and 1:180 [21].
9. To reduce noise and enhance image clarity, ten frames from the video (captured at 30 frames/s) are averaged. Bright spots with an intensity lower than that of a single dye molecule are considered potential noise or free dye, while spots with an intensity greater than that of a single molecule are counted as fluorescently labeled EVs [21].
10. Solutions of fibronectin, laminin, and collagen type I are prepared according to the manufacturer's protocols. Laminin is supplied as a solution in 50 mM Tris–HCl (pH 7.4) with 150 mM NaCl, while fibronectin is dissolved in PBS. Since collagen type I is difficult to dissolve in neutral buffers, it is first dissolved in 0.5 M acetic acid. The prepared solutions are then diluted with either HBSS or 0.5 M acetic acid as needed for use.
11. Fibronectin, laminin, and collagen type I were more uniformly coated near the center of the glass surface than at the edges. Therefore, EVs in the central region of the glass were observed and quantified [21].

Acknowledgments

This work was supported by JSPS KAKENHI Grant Numbers JP21H02424 (K.G.N.S.), JP20K21387 (K.G.N.S.), JPJSCCA202000007 (K.G.N.S.), JST CREST (JPMJCR18H2) (K.G.N.S.), JSPS Grant-in-Aid Fellows (23KJ1045) (T.I.), National Cancer Center Research and Development Fund (2023-A-03) (K.G.N.S.), Takeda Science Foundation (K.G.N.S.), The Uehara Memorial Foundation (K.G.N.S.), Mizutani Foundation for Glycoscience (K.G.N.S.), Daiichi Sankyo Foundation of Life Science (K.G.N.S.), Research Foundation for Opto-Science and Technology (K.G.N.S.), The Naito Foundation (K.G.N.S.).

Conflict of Interest The authors have no conflicts of interest to declare that are relevant to the content of this chapter.

References

1. Valadi H, Ekström K, Bossios A et al (2007) Exosome-mediated transfer of mRNAs and microRNAs is a novel mechanism of genetic exchange between cells. Nat Cell Biol 9:654–659
2. Raposo G, Stoorvogel W (2013) Extracellular vesicles: exosomes, microvesicles, and friends. J Cell Biol 200:373–383
3. Van Niel G, D'Angelo G, Raposo G (2018) Shedding light on the cell biology of extracellular vesicles. Nat Rev Mol Cell Biol 19:213–228
4. Peinado H, Alečković M, Lavotshkin S et al (2012) Melanoma exosomes educate bone marrow progenitor cells toward a pro-metastatic phenotype through MET. Nat Med 18:883–891
5. Ono M, Kosaka N, Tominaga N et al (2014) Exosomes from bone marrow mesenchymal stem cells contain a microRNA that promotes dormancy in metastatic breast cancer cells. Sci Signal 7:ra63
6. Hoshino A, Costa-Silva B, Shen TL et al (2015) Tumour exosome integrins determine organotropic metastasis. Nature 527:329–335
7. Fuster MM, Esko JD (2005) The sweet and sour of cancer: glycans as novel therapeutic targets. Nat Rev Cancer 5:526–542
8. Schnaar RL, Gerardy-Schahn R, Hildebrandt H (2014) Sialic acids in the brain: gangliosides and polysialic acid in nervous system development, stability, disease, and regeneration. Physiol Rev 94:461–518
9. Kaneko K, Ohkawa Y, Hashimoto N et al (2016) Neogenin, defined as a GD3-associated molecule by enzyme-mediated activation of radical sources, confers malignant properties via intracytoplasmic domain in melanoma cells. J Biol Chem 291:16630–16643
10. Yesmin F, Bhuiyan RAH, Ohmi Y et al (2021) Ganglioside GD2 enhances the malignant phenotypes of melanoma cells by cooperating with integrins. Int J Mol Sci 23:423
11. Komura N, Suzuki KGN, Ando H et al (2016) Raft-based interactions of gangliosides with a GPI-anchored receptor. Nat Chem Biol 12:402–410
12. Suzuki KGN, Ando H, Komura N et al (2017) Development of new ganglioside probes and unraveling of raft domain structure by single-molecule imaging. Biochim Biophys Acta 1861:2494–2506
13. Suzuki KGN, Ando H, Komura N et al (2018) Revealing the raft domain organization in the plasma membrane by single-molecule imaging of fluorescent ganglioside analogs. Methods Enzymol 598:267–282
14. Konishi M, Komura N, Hirose Y et al (2020) Development of fluorescent ganglioside GD3 and GQ1b analogs for elucidation of raft-associated interactions. J Org Chem 85:15998–16013
15. Takahashi M, Komura N, Yoshida Y et al (2022) Development of lact-series ganglioside fluorescent probe using late-stage sialylation and behavior analysis with single-molecule imaging. RSC Chem Biol 3:868–885
16. Suzuki KGN (2023) Single-molecule imaging of ganglioside probes in living cell plasma membranes. Methods Mol Biol 2613:215–227
17. Koikeda M, Komura N, Tanaka H et al (2019) Synthesis of ganglioside analogs containing fluorescently labeled GalNAc for single-molecule imaging. J Carbohydr Chem 38:509–527
18. Takahashi M, Shirasaki J, Komura N et al (2020) Efficient diversification of GM3 gangliosides via

late-stage sialylation and dynamic glycan structural studies with ^{19}F solid-state NMR. Org Biomol Chem 18:2902–2913

19. Miyazaki H, Miyazaki H, Fukumoto S et al (1997) Expression cloning of rat cDNA encoding UDP-galactose:GD2 beta1,3-galactosyltransferase that determines the expression of GD1b/GM1/GA1. J Biol Chem 272:24794–24799
20. Yamashiro S, Haraguchi M, Furukawa K et al (1995) Substrate specificity of beta 1,4-N-acetylgalactosaminyltransferase in vitro and in cDNA-transfected cells. GM2/GD2 synthase efficiently generates asialo-GM2 in certain cells. J Biol Chem 270:6149–6155
21. Isogai T, Hirosawa KM, Kannno M et al (2025) Extracellular vesicles adhere to cells primarily by interactions of integrins and GM1 with laminin. J Cell Biol 224:e202404064
22. Hirosawa KM, Sato Y, Kasai RS et al (2025) Uptake of small extracellular vesicles by recipient cells is facilitated by paracrine adhesion signaling. Nat Commun 16:2419.
23. Kinoshita M, Suzuki KGN, Matsumori N et al (2017) Raft-based sphingomyelin interactions revealed by new fluorescent sphingomyelin analogs. J Cell Biol 216:1183–1204
24. Morise J, Suzuki KGN, Kitagawa A, Wakazono Y, Takamiya K, Tsunoyama TA, Nemoto YL, Takematsu H, Kusumi A, Oka S (2019) AMPA receptors in the synapse turnover by monomer diffusion. Nat Commun 10(1):5245
25. Kemmoku H, Takahashi K, Mukai K et al (2024) Single-molecule localization microscopy reveals STING clustering at the trans-Golgi network through palmitoylation-dependent accumulation of cholesterol. Nat Commun 15:220.

Chapter 22

Small-Scale Cell Biological Analysis of Lipid Rafts of Blood Platelets and Cerebellar Neurons

Keisuke Komatsuya, Norihito Kikuchi, Tetsuya Hirabayashi and Kohji Kasahara

Abstract

Lipid rafts are dynamic assemblies of glycosphingolipids, sphingomyelin, cholesterol, and proteins that can be stabilized into platforms involved in the regulation of various vital cellular processes. Here, we describe small-scale isolation methods of lipid rafts. Detergent-resistant membrane raft fraction is isolated by a volume of 0.8 mL sucrose density gradient centrifugation from blood platelets. Raft fraction is also isolated from cerebellar neurons by co-immunoprecipitation using anti-glycosphingolipid monoclonal antibodies.

Keywords Lipid rafts, Glycosphingolipids, Sphingomyelin, Signal transduction, Sucrose density gradient, Detergent-resistant membrane, Immunoprecipitation, Platelets, Cerebellar neurons, Mass spectrometry

1 Introduction

Lipid rafts are dynamic assemblies of glycosphingolipids, sphingomyelin, cholesterol, and proteins that can be stabilized into platforms involved in the regulation of various vital cellular processes [1]. The important role of rafts at the cell surface is their function in signal transduction [2]. A number of studies provide considerable evidence that rafts are integral to the regulation of neuronal and immune signaling.

Lipid rafts are also involved in hemostasis and thrombosis [3]. Among blood cells, platelets are critical for maintaining the integrity of the blood coagulation system. Platelet rafts are critical membrane domains in physiological responses such as adhesion, aggregation, and clot retraction [4, 5]. Lipid rafts are isolated as a detergent-resistant membrane (DRM) fraction by sucrose density gradient centrifugation. Platelet DRM is shown to be rich in glycoprotein CD36, and Src family kinases, Src, and Lyn [6, 7]. The localization of the

Koichi Furukawa (ed.), *Gangliosides*, Methods in Molecular Biology, vol. 3035,
https://doi.org/10.1007/978-1-0716-5288-6_22,

adhesion receptor glycoprotein Ib-IX-V complex to lipid rafts is required for platelet adhesion to the vessel wall by binding the von Willebrand factor [8, 9]. In resting platelets, phosphatidylserine (PS) is asymmetrically restricted to the inner leaflet of the plasma membrane. An increase in intracellular Ca^{2+} concentration during platelet activation can lead to the exposure of PS on the outer leaflet. PS forms a procoagulant binding site for tenase and prothrombinase coagulation complexes. Lipid rafts are required for the release of PS-exposing extracellular vesicles from platelets [10]. Thus, lipid rafts are critical membrane domains in platelet activation processes.

We have investigated the signal transduction in lipid rafts of platelets [5, 11]. Previously, we reported that clot retraction is mediated by the coagulation factor XIII-dependent fibrin-integrin α IIbβ3-myosin axis in platelet sphingomyelin-rich membrane rafts [12, 13]. Clot retraction is a process driven by outside-in signaling through the platelet integrin αIIbβ3, resulting in the contraction of the fibrin mesh and the formation of mechanically stable thrombi. To elucidate the function of platelet lipid rafts, we identified DRM-raft-specific proteins from activated platelets. We isolated the DRM raft fraction of platelets treated with thrombin by sucrose gradient centrifugation (Fig. 1). Several specific proteins were present in the DRM fraction of thrombin-stimulated platelets. By mass spectrometry, we identified three proteins of 65, 50, and 47 kDa as fibrins α, β, and γ, respectively. These findings were supported by the results of immunoblot analysis using an anti-fibrinogen/fibrin polyclonal antibody. In resting platelets, fibrinogens Aα (67 kDa), Bβ (52 kDa), and γ (47 kDa) were present in the non-raft fraction. In contrast, fibrins α (65 kDa), β (50 kDa), and γ (47 kDa) were exclusively present in

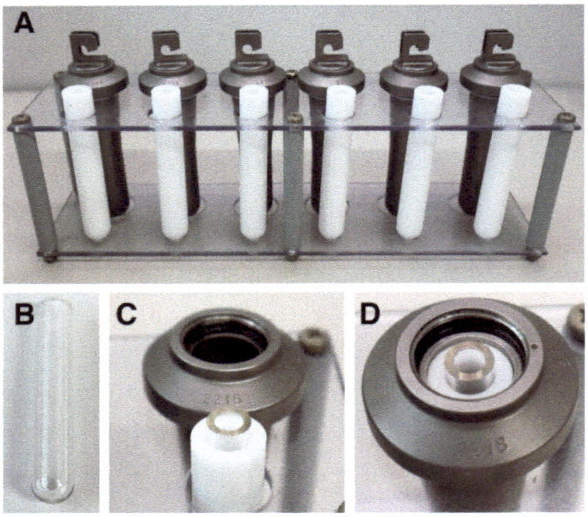

Fig. 1 Buckets, adaptor, and centrifuge tube of swing P40ST rotor. (**A**) Buckets and 2S13 adaptors (white) (**B**) 0.9 PC Thick-walled tube (**C**) Bucket and 2S13 adaptor + 0.9 PC Thick-walled tube (**D**) Bucket + 2S13 adaptor + 0.9 PC Thick-walled tube

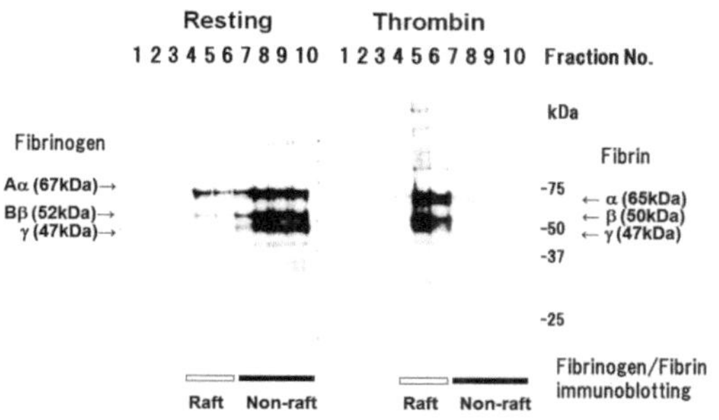

Fig. 2 Fibrin translocation to DRM raft fraction of platelets by thrombin stimulation. Sucrose density gradient analysis in washed human platelets. Resting platelets (left) and platelets stimulated for 3 min with 1 U/mL thrombin (right) were lysed in Triton X-100 and then adjusted to 40% sucrose. A sucrose gradient (5%–30%) in a volume of 6 mL was layered over the lysate (4 mL), and was centrifuged. Ten fractions were collected from top to bottom after centrifugation and subjected to immunoblotting with anti-fibrinogen polyclonal antibody [5]

the DRM fraction of platelets treated with thrombin (Fig. 2) [5, 13]. In Glanzmann's thrombasthenia αIIbβ3-deficient platelets, clot retraction is impaired [13].

The relative amount of raft protein in DRM is dependent on the number of cells extracted. When using low numbers of cells, even caveolin, a raft marker protein, is totally confined to the detergent-soluble fractions [14]. Suitable protein/detergent ratio during the initial homogenization is required for DRM preparation by sucrose density gradient centrifugation. It is difficult to collect a sufficient amount of platelets from a Glanzmann's thrombasthenia patient for DRM preparation. Therefore, we used a small-scale method for sucrose density gradient centrifugation [15]. We found that no fibrin is translocated to the DRM fraction with thrombin stimulation of Glanzmann's thrombasthenia platelets. Interestingly, we found that rapid and transient platelet DRM shifts to a higher density in sucrose density gradients upon thrombin stimulation (Fig. 3), whereas this shift is impaired in Glanzmann's thrombasthenia platelets [15]. Furthermore, we investigated a change in phospholipid composition of the DRM fraction by thrombin stimulation, and found that the level of PS (36:1) is increased and the level of PS (38:4) is decreased (Fig. 4).

We have also investigated the signal transduction in lipid rafts of cerebellar neurons [16, 17]. TAG-1/contactin-2 is a glycosylphosphatidylinositol (GPI)-anchored cell adhesion molecule belonging to the immunoglobulin superfamily [17]. TAG-1 plays important roles in axonal guidance and cellular migration. TAG-1 is transiently expressed in cerebellar granule cells during early stage of cerebellar development. Previously, we demonstrated that TAG-1 transduces intracellular signals to the Src family kinase Lyn and Csk-binding protein Cbp via GD3 rafts of cerebellar granule cells [18, 19]. GD3 is the predominant ganglioside of the early, immature nervous system. We

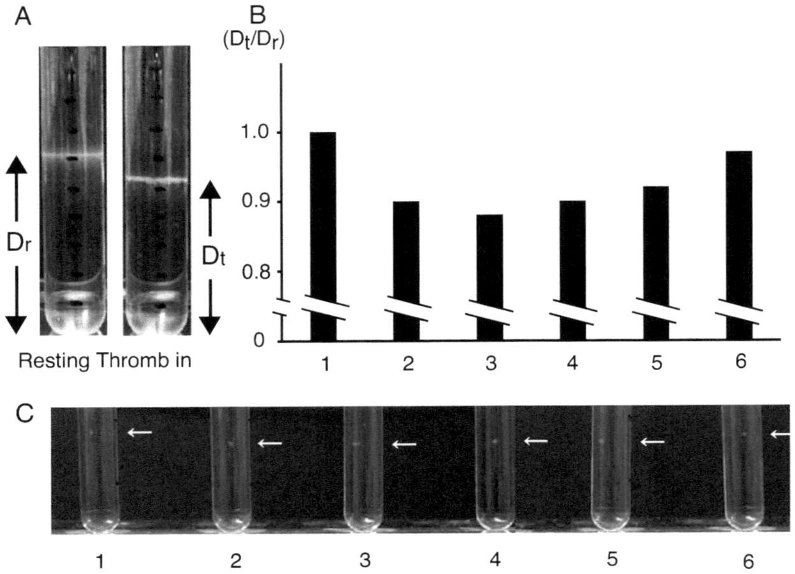

Fig. 3 Transient platelet DRM shifted to a higher density by thrombin stimulation. (**A**) Photo images of DRM in sucrose density gradient analysis of resting platelets (left photo) and platelets stimulated for 3 min with 1 U/mL thrombin (right photo). DRM positions in the sucrose density gradient were shown. Dr, distance from the bottom of the gradient in resting platelets; Dt, distance from the bottom of the gradient in thrombin-stimulated platelets for the indicated times. (**B**) Time-dependent DRM shift in sucrose density gradient analysis shown by the ratio between Dt and Dr (Dt/Dr). Lane 1, 0 min; lane 2, 0.5 min; lane 3, 1 min; lane 4, 5 min; lane 5, 15 min; lane 6, 60 min. (**C**) Photo image of time-dependent DRM shift using small-scale method. Arrows indicate the position of DRM. Lane 1, 0 min; lane 2, 0.5 min; lane 3, 1 min; lane 4, 5 min; lane 5, 15 min; lane 6, 60 min [15]

demonstrated that the anti-ganglioside GD3 monoclonal antibody R24 coprecipitates TAG-1, Lyn, Cbp, heterotrimeric Go α subunit (Fig. 5), and caveolin [18–21], suggesting that immunoprecipitation with anti-glycosphingolipid monoclonal antibody is a simple method for isolating lipid rafts. Consistent with this idea, anti-sulfatide monoclonal antibody O4 immunoprecipitates sulfatide rafts including Fyn, Lyn, and the heterotrimeric G protein α subunit in rat primary immature oligodendrocytes [22]. F3/contactin-1 transduces intracellular signals to the Src family kinase Fyn via sulfatide rafts of oligodendrocytes during myelination. Furthermore, we confirmed that the presence of TAG-1/contactin-2(N = 534), Lyn(N = 745), Cbp/PAG1(N = 1054), Goα(N = 87) in the DRM fraction of primary cerebellar neurons by proteomic analysis (Table 1).

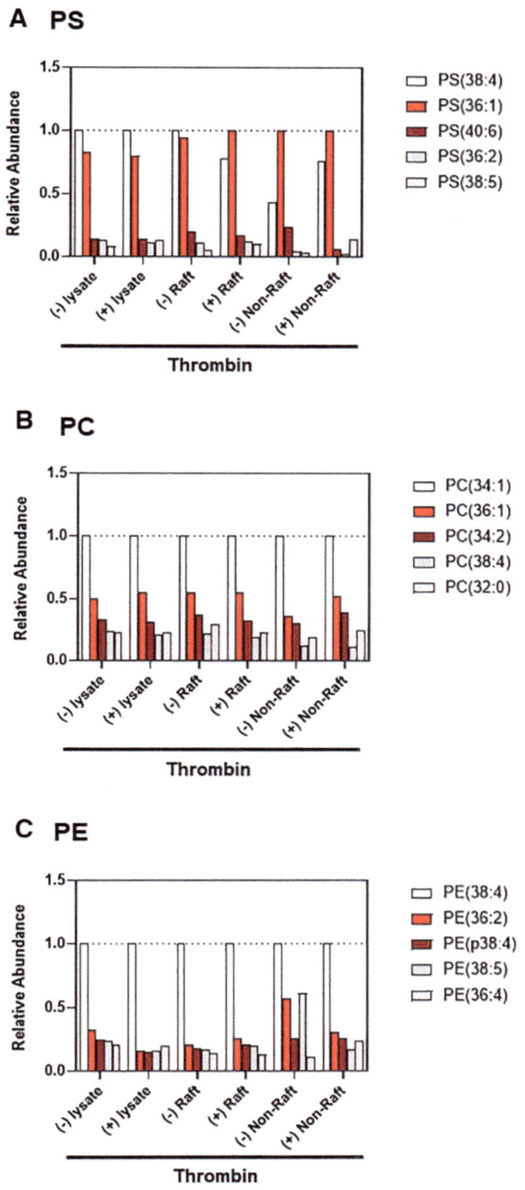

Fig. 4 Relative abundance of phospholipid molecular species in total lysate, DRM raft fraction, and non-raft fraction of human platelets by thrombin stimulation. (**A**) Phosphatidylserines, (**B**) phosphatidylcholines, (**C**) phosphatidylethanolamines. Odd-number lanes; resting platelets. Even number lanes; activated platelets by 0.2 U/mL thrombin stimulation for 10 min [15]

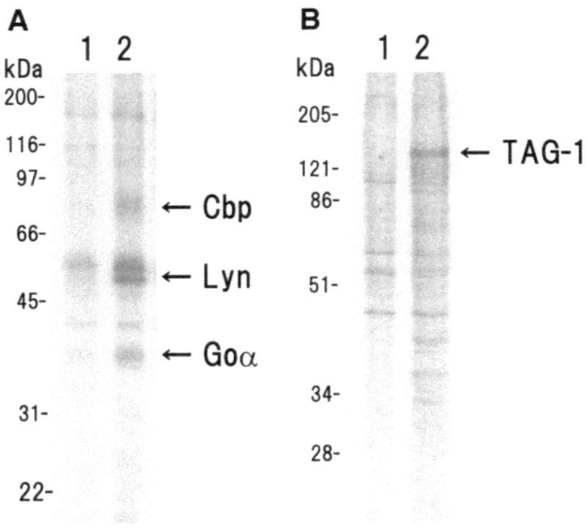

Fig. 5 Anti-ganglioside GD3 antibody R24-precipitated protein kinase activity and protein from primary cerebellar neurons. (**A**) Rat primary cerebellar neurons were solubilized in lysis buffer. Supernatants were immunoprecipitated with anti-ganglioside GD3 antibody R24. Immunoprecipitates were subjected to in vitro kinase assay and SDSPAGE. Phosphorylation was visualized by autoradiography. Lane 1, precipitate with control mouse IgG3; lane 2, precipitate with R24. Arrows indicate Cbp, Lyn, and heterotrimeric Go α subunit. (**B**) Primary cerebellar neurons metabolically labeled with ^{35}S were solubilized in lysis buffer. The supernatants were immunoprecipitated with R24, and the immunoprecipitates were subjected to SDS-PAGE and autoradiography. Lane 1, precipitate with control mouse IgG3; lane 2, precipitate with R24. Arrow indicates TAG-1

2 Materials

2.1 Sucrose Density Gradient Centrifugation

1. TNE/Triton buffer (0.05% Triton X-100, 25 mM Tris-HCl, pH 7.5, 150 mM NaCl, and 1 mM EGTA).
2. 5% sucrose TNE buffer (5% sucrose, 25 mM Tris-HCl, pH 7.5, 150 mM NaCl, and 1 mM EGTA).
3. 30% sucrose TNE buffer (30% sucrose, 25 mM Tris-HCl, pH 7.5, 150 mM NaCl, and 1 mM EGTA).
4. 35% sucrose TNE buffer (35% sucrose, 25 mM Tris-HCl, pH 7.5, 150 mM NaCl, and 1 mM EGTA).
5. P40ST rotor; 13 mL × 6 Tubes (Eppendorf Himac Technologies).
6. 2S13 adaptor; length=93 mm. outer diameter = 15 mm, inner diameter = 9 mm (Eppendorf Himac Technologies, 5720411191) (Fig. 1A).

Table 1

Identification of proteins in detergent-resistant membrane rafts of rat primary cerebellar neurons by mass spectrometry-based proteomic analysis

N	Unused	Total	% Cov (95)	Accession #	Name	Species	Peptides (95%)
1	107.17	107.17	26.3	sp|P16086|SPTA2_RAT	Spectrin alpha chain, brain OS=Rattus norvegicus GN=Sptan1 PE=1 SV=2	RAT	57
2	99.71	99.71	22.2	sp|P29994-8|ITPR1_RAT	Isoform SI-SII of Inositol 1,4,5-trisphosphate receptor type 1 OS=Rattus norvegicus GN=Itpr1	RAT	56
3	82.15	82.15	26.6	sp|P11442|CLH_RAT	Clathrin heavy chain 1 OS=Rattus norvegicus GN=Cltc PE=1 SV=3	RAT	67
4	80.27	80.27	34.5	sp|P06686|AT1A2_RAT	Sodium/potassium-transporting ATPase subunit alpha-2 OS=Rattus norvegicus GN=Atp1a2 PE=1 SV=1	RAT	59
5	68.56	68.56	28.2	sp|Q5XI78|ODO1_RAT	2-oxoglutarate dehydrogenase, mitochondrial OS=Rattus norvegicus GN=Ogdh PE=1 SV=1	RAT	51
6	61	61	36.3	sp|Q9ER34|ACON_RAT	Aconitate hydratase, mitochondrial OS=Rattus norvegicus GN=Aco2 PE=1 SV=2	RAT	47
7	57.96	57.96	15	sp|Q9QWN8|SPTN2_RAT	Spectrin beta chain, brain 2 OS=Rattus norvegicus GN=Sptbn2 PE=1 SV=2	RAT	30
8	49.2	49.2	6.1	sp|P38650|DYHC1_RAT	Cytoplasmic dynein 1 heavy chain 1 OS=Rattus norvegicus GN=Dync1h1 PE=1 SV=1	RAT	24
9	49.16	49.16	43.4	sp|P61765|STXB1_RAT	Syntaxin-binding protein 1 OS=Rattus norvegicus GN=Stxbp1 PE=1 SV=1	RAT	31
10	48.14	48.16	35	sp|P13233|CN37_RAT	2',3'-cyclic-nucleotide 3'-phosphodiesterase OS=Rattus norvegicus GN=Cnp PE=1 SV=2	RAT	45
11	46	46	46.5	sp|P10719|ATPB_RAT	ATP synthase subunit beta, mitochondrial OS=Rattus norvegicus GN=Atp5b PE=1 SV=2	RAT	43

(continued)

Table 1
(continued)

N	Unused	Total	% Cov (95)	Accession #	Name	Species	Peptides (95%)
12	44.21	44.21	35	sp|P63018|HSP7C_RAT	Heat shock cognate 71 kDa protein OS=Rattus norvegicus GN=Hspa8 PE=1 SV=1	RAT	37
13	43.8	43.84	37.1	sp|P15999|ATPA_RAT	ATP synthase subunit alpha, mitochondrial OS=Rattus norvegicus GN=Atp5a1 PE=1 SV=2	RAT	36
14	43.45	76.13	37.3	sp|P06687|AT1A3_RAT	Sodium/potassium-transporting ATPase subunit alpha-3 OS=Rattus norvegicus GN=Atp1a3 PE=1 SV=2	RAT	60
15	42.86	43.21	30.8	sp|Q9QUL6|NSF_RAT	Vesicle-fusing ATPase OS=Rattus norvegicus GN=Nsf PE=1 SV=1	RAT	26
16	42.78	42.81	22.9	sp|P11506-8|AT2B2_RAT	Isoform AICI of Plasma membrane calcium-transporting ATPase 2 OS=Rattus norvegicus GN=Atp2b2	RAT	32
17	42.71	42.75	65.6	sp|P02688|MBP_RAT	Myelin basic protein S OS=Rattus norvegicus GN=Mbp PE=1 SV=3	RAT	48
18	42.35	72.01	29.5	sp|P06685|AT1A1_RAT	Sodium/potassium-transporting ATPase subunit alpha-1 OS=Rattus norvegicus GN=Atp1a1 PE=1 SV=1	RAT	60
19	40.71	40.71	31.1	sp|P63039|CH60_RAT	60 kDa heat shock protein, mitochondrial OS=Rattus norvegicus GN=Hspd1 PE=1 SV=1	RAT	29
20	40.27	40.29	33.1	sp|P48721|GRP75_RAT	Stress-70 protein, mitochondrial OS=Rattus norvegicus GN=Hspa9 PE=1 SV=3	RAT	23
21	39.65	39.68	45.9	sp|P05065|ALDOA_RAT	Fructose-bisphosphate aldolase A OS=Rattus norvegicus GN=Aldoa PE=1 SV=2	RAT	27
22	39.5	39.55	23.5	sp|P53534|PYGB_RAT	Glycogen phosphorylase, brain form (Fragment) OS=Rattus norvegicus GN=Pygb PE=1 SV=3	RAT	22

N	Unused	Total	% Cov (95)	Accession #	Name	Species	Peptides (95%)
23	38.76	38.78	19.4	sp|P05708|HXK1_RAT	Hexokinase-1 OS=Rattus norvegicus GN=Hk1 PE=1 SV=4	RAT	22
24	38.17	38.21	40.3	sp|P10860|DHE3_RAT	Glutamate dehydrogenase 1, mitochondrial OS=Rattus norvegicus GN=Glud1 PE=1 SV=2	RAT	24
25	37.25	37.26	21.6	sp|Q9JI66-3|S4A4_RAT	Isoform rb2NBC of Electrogenic sodium bicarbonate cotransporter 1 OS=Rattus norvegicus GN=Slc4a4	RAT	32
26	37.06	37.07	20.4	sp|Q63198|CNTN1_RAT	Contactin-1 OS=Rattus norvegicus GN=Cntn1 PE=1 SV=2	RAT	18
27	37	37.08	40	sp|P00507|AATM_RAT	Aspartate aminotransferase, mitochondrial OS=Rattus norvegicus GN=Got2 PE=1 SV=2	RAT	33
28	36.25	36.27	10.1	sp|P34926|MAP1A_RAT	Microtubule-associated protein 1A OS=Rattus norvegicus GN=Map1a PE=1 SV=1	RAT	18
29	35.07	35.07	23.1	sp|P82995|HS90A_RAT	Heat shock protein HSP 90-alpha OS=Rattus norvegicus GN=Hsp90aa1 PE=1 SV=3	RAT	20
30	34.62	34.64	19.1	sp|P25286|VPP1_RAT	V-type proton ATPase 116 kDa subunit a isoform 1 OS=Rattus norvegicus GN=Atp6v0a1 PE=2 SV=1	RAT	20
31	34.2	34.2	19.8	sp|P13596|NCAM1_RAT	Neural cell adhesion molecule 1 OS=Rattus norvegicus GN=Ncam1 PE=1 SV=1	RAT	22
32	33.87	33.89	37.2	sp|P26284|ODPA_RAT	Pyruvate dehydrogenase E1 component subunit alpha, somatic form, mitochondrial OS=Rattus norvegicus GN=Pdha1 PE=1 SV=2	RAT	19
33	33.15	33.16	17.1	sp|P18484|AP2A2_RAT	AP-2 complex subunit alpha-2 OS=Rattus norvegicus GN=Ap2a2 PE=1 SV=3	RAT	20

(continued)

Table 1
(continued)

<table>
<tr><th>N</th><th>Unused</th><th>Total</th><th>% Cov (95)</th><th>Accession #</th><th>Name</th><th>Species</th><th>Peptides (95%)</th></tr>
<tr><td>34</td><td>32.02</td><td>32.02</td><td>35.3</td><td>sp|P47942|DPYL2_RAT</td><td>Dihydropyrimidinase-related protein 2 OS=Rattus norvegicus GN=Dpysl2 PE=1 SV=1</td><td>RAT</td><td>20</td></tr>
<tr><td>35</td><td>31.67</td><td>31.68</td><td>17.2</td><td>sp|P52873|PYC_RAT</td><td>Pyruvate carboxylase, mitochondrial OS=Rattus norvegicus GN=Pc PE=1 SV=2</td><td>RAT</td><td>19</td></tr>
<tr><td>36</td><td>31.02</td><td>31.02</td><td>15.5</td><td>sp|Q2TA68-3|OPA1_RAT</td><td>Isoform 7 of Dynamin-like 120 kDa protein, mitochondrial OS=Rattus norvegicus GN=Opa1</td><td>RAT</td><td>17</td></tr>
<tr><td>37</td><td>30.83</td><td>30.85</td><td>17.3</td><td>sp|P21575|DYN1_RAT</td><td>Dynamin-1 OS=Rattus norvegicus GN=Dnm1 PE=1 SV=2</td><td>RAT</td><td>15</td></tr>
<tr><td>38</td><td>30.25</td><td>30.27</td><td>42.7</td><td>sp|P63259|ACTG_RAT</td><td>Actin, cytoplasmic 2 OS=Rattus norvegicus GN=Actg1 PE=1 SV=1</td><td>RAT</td><td>33</td></tr>
<tr><td>39</td><td>30.23</td><td>30.23</td><td>50.6</td><td>sp|P04636|MDHM_RAT</td><td>Malate dehydrogenase, mitochondrial OS=Rattus norvegicus GN=Mdh2 PE=1 SV=2</td><td>RAT</td><td>21</td></tr>
<tr><td>40</td><td>30</td><td>30</td><td>32.4</td><td>sp|P11980|KPYM_RAT</td><td>Pyruvate kinase isozymes M1/M2 OS=Rattus norvegicus GN=Pkm2 PE=1 SV=3</td><td>RAT</td><td>22</td></tr>
<tr><td>41</td><td>29.8</td><td>29.85</td><td>51.7</td><td>sp|P07335|KCRB_RAT</td><td>Creatine kinase B-type OS=Rattus norvegicus GN=Ckb PE=1 SV=2</td><td>RAT</td><td>26</td></tr>
<tr><td>42</td><td>29.7</td><td>29.73</td><td>33.9</td><td>sp|P85108|TBB2A_RAT</td><td>Tubulin beta-2A chain OS=Rattus norvegicus GN=Tubb2a PE=1 SV=1</td><td>RAT</td><td>19</td></tr>
<tr><td>43</td><td>29.29</td><td>29.38</td><td>38.7</td><td>sp|P49432|ODPB_RAT</td><td>Pyruvate dehydrogenase E1 component subunit beta, mitochondrial OS=Rattus norvegicus GN=Pdhb PE=1 SV=2</td><td>RAT</td><td>23</td></tr>
<tr><td>44</td><td>29.27</td><td>29.3</td><td>35.1</td><td>sp|P29101|SYT2_RAT</td><td>Synaptotagmin-2 OS=Rattus norvegicus GN=Syt2 PE=1 SV=1</td><td>RAT</td><td>22</td></tr>
</table>

N	Unused	Total	% Cov (95)	Accession #	Name	Species	Peptides (95%)
45	29.17	29.21	24.5	sp|Q920L2|DHSA_RAT	Succinate dehydrogenase [ubiquinone] flavoprotein subunit, mitochondrial OS=Rattus norvegicus GN=Sdha PE=1 SV=1	RAT	18
46	29.06	29.11	31	sp|P09951|SYN1_RAT	Synapsin-1 OS=Rattus norvegicus GN=Syn1 PE=1 SV=3	RAT	18
47	28.16	28.18	29.2	sp|P47709|RP3A_RAT	Rabphilin-3A OS=Rattus norvegicus GN=Rph3a PE=1 SV=1	RAT	14
48	28.12	28.13	32	sp|Q99NA5|IDH3A_RAT	Isocitrate dehydrogenase [NAD] subunit alpha, mitochondrial OS=Rattus norvegicus GN=Idh3a PE=1 SV=1	RAT	21
49	28.06	28.06	24.9	sp|P51650|SSDH_RAT	Succinate-semialdehyde dehydrogenase, mitochondrial OS=Rattus norvegicus GN=Aldh5a1 PE=1 SV=2	RAT	16
50	27.86	27.86	29.8	sp|P50554|GABT_RAT	4-aminobutyrate aminotransferase, mitochondrial OS=Rattus norvegicus GN=Abat PE=1 SV=3	RAT	19
51	27.77	32.03	24.5	sp|P06761|GRP78_RAT	78 kDa glucose-regulated protein OS=Rattus norvegicus GN=Hspa5 PE=1 SV=1	RAT	20
52	27.49	27.52	40.2	sp|P09606|GLNA_RAT	Glutamine synthetase OS=Rattus norvegicus GN=Glul PE=1 SV=3	RAT	21
53	27.21	27.25	14	sp|Q63633|S12A5_RAT	Solute carrier family 12 member 5 OS=Rattus norvegicus GN=Slc12a5 PE=1 SV=2	RAT	15
54	26.17	26.17	29.4	sp|P56574|IDHP_RAT	Isocitrate dehydrogenase [NADP], mitochondrial OS=Rattus norvegicus GN=Idh2 PE=1 SV=2	RAT	17
55	26.07	26.07	40.6	sp|P61265|STX1B_RAT	Syntaxin-1B OS=Rattus norvegicus GN=Stx1b PE=1 SV=1	RAT	20

(continued)

Table 1 (continued)

N	Unused	Total	% Cov (95)	Accession #	Name	Species	Peptides (95%)
56	26.06	26.06	27.7	sp|Q6P9V9|TBA1B_RAT	Tubulin alpha-1B chain OS=Rattus norvegicus GN=Tuba1b PE=1 SV=1	RAT	22
57	25.97	26.02	18.4	sp|P48037|ANXA6_RAT	Annexin A6 OS=Rattus norvegicus GN=Anxa6 PE=1 SV=2	RAT	14
58	25.86	25.89	17.9	sp|P47860|K6PP_RAT	6-phosphofructokinase type C OS=Rattus norvegicus GN=Pfkp PE=1 SV=2	RAT	16
59	25.83	25.86	40.3	sp|Q09073|ADT2_RAT	ADP/ATP translocase 2 OS=Rattus norvegicus GN=Slc25a5 PE=1 SV=3	RAT	23
60	25.69	40.75	20.3	sp|Q64568-6|AT2B3_RAT	Isoform AIICIII of Plasma membrane calcium-transporting ATPase 3 OS=Rattus norvegicus GN=Atp2b3	RAT	29
61	24.88	24.94	24	sp|P13264|GLSK_RAT	Glutaminase kidney isoform, mitochondrial OS=Rattus norvegicus GN=Gls PE=1 SV=2	RAT	18
62	24.75	24.77	42	sp|P04797|G3P_RAT	Glyceraldehyde-3-phosphate dehydrogenase OS=Rattus norvegicus GN=Gapdh PE=1 SV=3	RAT	26
63	24.49	24.5	41.3	sp|Q68FX0|IDH3B_RAT	Isocitrate dehydrogenase [NAD] subunit beta, mitochondrial OS=Rattus norvegicus GN=Idh3B PE=1 SV=1	RAT	19
64	24.3	24.31	27.4	sp|P35565|CALX_RAT	Calnexin OS=Rattus norvegicus GN=Canx PE=1 SV=1	RAT	13
65	23.7	23.71	23.7	sp|Q64548|RTN1_RAT	Reticulon-1 OS=Rattus norvegicus GN=Rtn1 PE=2 SV=1	RAT	16
66	23.63	23.65	15.5	sp|P24942|EAA1_RAT	Excitatory amino acid transporter 1 OS=Rattus norvegicus GN=Slc1a3 PE=1 SV=2	RAT	32
67	23.63	23.65	65.9	sp|P07936|NEUM_RAT	Neuromodulin OS=Rattus norvegicus GN=Gap43 PE=1 SV=1	RAT	14

N	Unused	Total	% Cov (95)	Accession #	Name	Species	Peptides (95%)
68	23.52	23.55	38.2	sp|Q9Z2L0|VDAC1_RAT	Voltage-dependent anion-selective channel protein 1 OS=Rattus norvegicus GN=Vdac1 PE=1 SV=4	RAT	19
69	23.25	23.27	42.4	sp|P62260|1433E_RAT	14-3-3 protein epsilon OS=Rattus norvegicus GN=Ywhae PE=1 SV=1	RAT	18
70	23.2	23.25	23.3	sp|P62815|VATB2_RAT	V-type proton ATPase subunit B, brain isoform OS=Rattus norvegicus GN=Atp6v1b2 PE=1 SV=1	RAT	15
71	23.13	23.22	4.4	sp|O88778|BSN_RAT	Protein bassoon OS=Rattus norvegicus GN=Bsn PE=1 SV=3	RAT	12
72	23.03	23.06	32.3	sp|P32551|QCR2_RAT	Cytochrome b-c1 complex subunit 2, mitochondrial OS=Rattus norvegicus GN=Uqcrc2 PE=1 SV=2	RAT	19
73	22.96	22.98	31.4	sp|P85834|EFTU_RAT	Elongation factor Tu, mitochondrial OS=Rattus norvegicus GN=Tufm PE=1 SV=1	RAT	12
74	22.48	33.7	24	sp|P09812|PYGM_RAT	Glycogen phosphorylase, muscle form OS=Rattus norvegicus GN=Pygm PE=2 SV=5	RAT	19
75	22.11	22.11	16.3	sp|Q05140|AP180_RAT	Clathrin coat assembly protein AP180 OS=Rattus norvegicus GN=Snap91 PE=1 SV=1	RAT	14
76	21.86	21.88	17.5	sp|Q66HF1|NDUS1_RAT	NADH-ubiquinone oxidoreductase 75 kDa subunit, mitochondrial OS=Rattus norvegicus GN=Ndufs1 PE=1 SV=1	RAT	13
77	21.7	21.7	26.3	sp|P13086|SUCA_RAT	Succinyl-CoA ligase [GDP-forming] subunit alpha, mitochondrial OS=Rattus norvegicus GN=Suclg1 PE=2 SV=2	RAT	17
78	21.6	21.61	11.9	sp|P62944|AP2B1_RAT	AP-2 complex subunit beta OS=Rattus norvegicus GN=Ap2b1 PE=1 SV=1	RAT	10

(continued)

Table 1 (continued)

N	Unused	Total	% Cov (95)	Accession #	Name	Species	Peptides (95%)
79	21.56	21.58	41	sp|P54921|SNAA_RAT	Alpha-soluble NSF attachment protein OS= Rattus norvegicus GN=Napa PE=1 SV=2	RAT	14
80	21.25	21.27	11.3	sp|P11507|AT2A2_ RAT	Sarcoplasmic/endoplasmic reticulum calcium ATPase 2 OS=Rattus norvegicus GN=Atp2a2 PE=1 SV=1	RAT	14
81	20.92	20.94	28.7	sp|P42123|LDHB_RAT	L-lactate dehydrogenase B chain OS=Rattus norvegicus GN=Ldhb PE=1 SV=2	RAT	15
82	20.67	20.71	19.9	sp|P16036|MPCP_RAT	Phosphate carrier protein, mitochondrial OS= Rattus norvegicus GN=Slc25a3 PE=1 SV=1	RAT	14
83	20.64	20.68	30.9	sp|P07340|AT1B1_RAT	Sodium/potassium-transporting ATPase subunit beta-1 OS=Rattus norvegicus GN=Atp1b1 PE=1 SV=1	RAT	16
84	20.22	20.24	27.8	sp|P60203|MYPR_RAT	Myelin proteolipid protein OS=Rattus norvegicus GN=Plp1 PE=1 SV=2	RAT	20
85	20.19	20.2	42.2	sp|P08009|GSTM4_ RAT	Glutathione S-transferase Yb-3 OS=Rattus norvegicus GN=Gstm3 PE=1 SV=2	RAT	10
86	20.15	22.22	14.2	sp|P47858|K6PF_RAT	6-phosphofructokinase, muscle type OS=Rattus norvegicus GN=Pfkm PE=2 SV=3	RAT	12
87	20.04	20.05	33.1	sp|P59215-2|GNAO_ RAT	Isoform Alpha-2 of Guanine nucleotide-binding protein G(o) subunit alpha OS=Rattus norvegicus GN=Gnao1	RAT	14
88	20.01	20.01	10.1	sp|Q05695|L1CAM_ RAT	Neural cell adhesion molecule L1 OS=Rattus norvegicus GN=L1cam PE=2 SV=3	RAT	11
89	19.97	19.97	23.9	sp|P97546|NPTN_RAT	Neuroplastin OS=Rattus norvegicus GN=Nptn PE=1 SV=2	RAT	13

N	Unused	Total	% Cov (95)	Accession #	Name	Species	Peptides (95%)
90	19.6	19.63	26.2	sp|P19511|AT5F1_RAT	ATP synthase subunit b, mitochondrial OS=Rattus norvegicus GN=Atp5f1 PE=1 SV=1	RAT	15
91	19.5	19.55	4.4	sp|P15205|MAP1B_RAT	Microtubule-associated protein 1B OS=Rattus norvegicus GN=Map1b PE=1 SV=2	RAT	9
92	19.3	19.38	23.2	sp|P11598|PDIA3_RAT	Protein disulfide-isomerase A3 OS=Rattus norvegicus GN=Pdia3 PE=1 SV=2	RAT	9
93	19.03	19.05	16.9	sp|P33124|ACSL6_RAT	Long-chain-fatty-acid–CoA ligase 6 OS=Rattus norvegicus GN=Acsl6 PE=1 SV=1	RAT	9
94	18.95	21.86	29.8	sp|P68255|1433T_RAT	14-3-3 protein theta OS=Rattus norvegicus GN=Ywhaq PE=1 SV=1	RAT	13
95	18.66	18.68	35.7	sp|P48500|TPIS_RAT	Triosephosphate isomerase OS=Rattus norvegicus GN=Tpi1 PE=1 SV=2	RAT	16
96	18.42	24.57	30	sp|P09117|ALDOC_RAT	Fructose-bisphosphate aldolase C OS=Rattus norvegicus GN=Aldoc PE=1 SV=3	RAT	18
97	18.36	18.73	28.8	sp|P08050|CXA1_RAT	Gap junction alpha-1 protein OS=Rattus norvegicus GN=Gja1 PE=1 SV=2	RAT	10
98	18.24	18.25	33.1	sp|P97700|M2OM_RAT	Mitochondrial 2-oxoglutarate/malate carrier protein OS=Rattus norvegicus GN=Slc25a11 PE=2 SV=3	RAT	12
99	18.23	18.27	19.3	sp|P84092|AP2M1_RAT	AP-2 complex subunit mu OS=Rattus norvegicus GN=Ap2m1 PE=1 SV=1	RAT	9
100	18.2	18.2	48.1	sp|P05712|RAB2A_RAT	Ras-related protein Rab-2A OS=Rattus norvegicus GN=Rab2a PE=1 SV=1	RAT	14
101	18.13	18.13	25.4	sp|Q5RJQ4|SIRT2_RAT	NAD-dependent deacetylase sirtuin-2 OS=Rattus norvegicus GN=Sirt2 PE=1 SV=1	RAT	10

(continued)

Table 1
(continued)

N	Unused	Total	% Cov (95)	Accession #	Name	Species	Peptides (95%)
102	18.09	18.13	14.6	sp|Q5XIN6|LETM1_RAT	LETM1 and EF-hand domain-containing protein 1, mitochondrial OS=Rattus norvegicus GN=Letm1 PE=1 SV=1	RAT	10
103	18.08	18.12	38.1	sp|P84087|CPLX2_RAT	Complexin-2 OS=Rattus norvegicus GN=Cplx2 PE=1 SV=1	RAT	12
104	18.02	18.03	22.8	sp|Q6P6V0|G6PI_RAT	Glucose-6-phosphate isomerase OS=Rattus norvegicus GN=Gpi PE=1 SV=1	RAT	11
105	18	20	23.7	sp|Q63537|SYN2_RAT	Synapsin-2 OS=Rattus norvegicus GN=Syn2 PE=1 SV=1	RAT	12
106	18	18	59.1	sp|Q5FVI4|CEND_RAT	Cell cycle exit and neuronal differentiation protein 1 OS=Rattus norvegicus GN=Cend1 PE=1 SV=1	RAT	15
107	17.95	18.04	54.7	sp|P31399|ATP5H_RAT	ATP synthase subunit d, mitochondrial OS=Rattus norvegicus GN=Atp5h PE=1 SV=3	RAT	15
108	17.92	18.01	9.2	sp|Q5SGE0|LPPRC_RAT	Leucine-rich PPR motif-containing protein, mitochondrial OS=Rattus norvegicus GN=Lrpprc PE=1 SV=1	RAT	10
109	17.72	17.83	7	sp|Q9WTP0-2|E41L1_RAT	Isoform L of Band 4.1-like protein 1 OS=Rattus norvegicus GN=Epb41l1	RAT	14
110	17.64	17.68	32.7	sp|P25113|PGAM1_RAT	Phosphoglycerate mutase 1 OS=Rattus norvegicus GN=Pgam1 PE=1 SV=4	RAT	16
111	17.05	17.09	17.6	sp|Q8VHF5|CISY_RAT	Citrate synthase, mitochondrial OS=Rattus norvegicus GN=Cs PE=1 SV=1	RAT	12
112	17	17.09	35.6	sp|Q1WIM3|CADM3_RAT	Cell adhesion molecule 3 OS=Rattus norvegicus GN=Cadm3 PE=2 SV=1	RAT	11
113	16.95	29.13	36.8	sp|P21707|SYT1_RAT	Synaptotagmin-1 OS=Rattus norvegicus GN=Syt1 PE=1 SV=3	RAT	25

N	Unused	Total	% Cov (95)	Accession #	Name	Species	Peptides (95%)
114	16.64	16.68	37	sp|Q9Z270|VAPA_RAT	Vesicle-associated membrane protein-associated protein A OS=Rattus norvegicus GN=Vapa PE=1 SV=3	RAT	8
115	16.5	16.63	22	sp|Q561S0|NDUAA_RAT	NADH dehydrogenase [ubiquinone] 1 alpha subcomplex subunit 10, mitochondrial OS=Rattus norvegicus GN=Ndufa10 PE=1 SV=1	RAT	10
116	16.25	16.26	22.3	sp|P50399|GDIB_RAT	Rab GDP dissociation inhibitor beta OS=Rattus norvegicus GN=Gdi2 PE=1 SV=2	RAT	9
117	16.09	16.1	12.6	sp|Q75Q39|TOM70_RAT	Mitochondrial import receptor subunit TOM70 OS=Rattus norvegicus GN=Tomm70a PE=1 SV=1	RAT	9
118	16.01	16.01	50.7	sp|P61107|RAB14_RAT	Ras-related protein Rab-14 OS=Rattus norvegicus GN=Rab14 PE=2 SV=3	RAT	10
119	15.85	15.89	32.5	sp|P81155|VDAC2_RAT	Voltage-dependent anion-selective channel protein 2 OS=Rattus norvegicus GN=Vdac2 PE=1 SV=2	RAT	10
120	15.81	15.85	23.5	sp|Q5BK63|NDUA9_RAT	NADH dehydrogenase [ubiquinone] 1 alpha subcomplex subunit 9, mitochondrial (Fragment) OS=Rattus norvegicus GN=Ndufa9 PE=1 SV=1	RAT	9
121	15.72	15.75	19.2	sp|Q68FY0|QCR1_RAT	Cytochrome b-c1 complex subunit 1, mitochondrial OS=Rattus norvegicus GN=Uqcrc1 PE=1 SV=1	RAT	14
122	15.72	15.75	46.6	sp|P62749|HPCL1_RAT	Hippocalcin-like protein 1 OS=Rattus norvegicus GN=Hpcal1 PE=1 SV=2	RAT	10
123	15.54	15.57	5.1	sp|P23385|GRM1_RAT	Metabotropic glutamate receptor 1 OS=Rattus norvegicus GN=Grm1 PE=1 SV=1	RAT	8

(continued)

Table 1
(continued)

N	Unused	Total	% Cov (95)	Accession #	Name	Species	Peptides (95%)
124	15.49	15.51	35	sp|Q6PCU2|VATE1_RAT	V-type proton ATPase subunit E 1 OS=Rattus norvegicus GN=Atp6v1e1 PE=1 SV=1	RAT	12
125	15.45	15.51	17.5	sp|Q8VBU2|NDRG2_RAT	Protein NDRG2 OS=Rattus norvegicus GN=Ndrg2 PE=1 SV=1	RAT	10
126	15.44	15.5	36.6	sp|Q9R063|PRDX5_RAT	Peroxiredoxin-5, mitochondrial OS=Rattus norvegicus GN=Prdx5 PE=1 SV=1	RAT	12
127	15.14	15.46	15.5	sp|P28480|TCPA_RAT	T-complex protein 1 subunit alpha OS=Rattus norvegicus GN=Tcp1 PE=1 SV=1	RAT	8
128	15.1	15.12	25.2	sp|P18418|CALR_RAT	Calreticulin OS=Rattus norvegicus GN=Calr PE=1 SV=1	RAT	8
129	14.91	14.96	16.8	sp|O88954|MPP3_RAT	MAGUK p55 subfamily member 3 OS=Rattus norvegicus GN=Mpp3 PE=1 SV=3	RAT	8
130	14.79	14.81	20.3	sp|P04764|ENOA_RAT	Alpha-enolase OS=Rattus norvegicus GN=Eno1 PE=1 SV=4	RAT	12
131	14.75	14.78	14.6	sp|P08413|KCC2B_RAT	Calcium/calmodulin-dependent protein kinase type II subunit beta OS=Rattus norvegicus GN=Camk2b PE=1 SV=1	RAT	9
132	14.52	14.57	28.1	sp|O88989|MDHC_RAT	Malate dehydrogenase, cytoplasmic OS=Rattus norvegicus GN=Mdh1 PE=1 SV=3	RAT	7
133	14.5	14.57	13.8	sp|Q5XIM9|TCPB_RAT	T-complex protein 1 subunit beta OS=Rattus norvegicus GN=Cct2 PE=1 SV=3	RAT	8
134	14.47	14.49	8.4	sp|P97685-3|NFASC_RAT	Isoform 155 kDa isoform of Neurofascin OS=Rattus norvegicus GN=Nfasc	RAT	7
135	14.38	22.75	40.4	sp|P63102|1433Z_RAT	14-3-3 protein zeta/delta OS=Rattus norvegicus GN=Ywhaz PE=1 SV=1	RAT	14

<table>
<tr><th>N</th><th>Unused</th><th>Total</th><th>% Cov (95)</th><th>Accession #</th><th>Name</th><th>Species</th><th>Peptides (95%)</th></tr>
<tr><td>136</td><td>14.22</td><td>14.23</td><td>17.5</td><td>sp|O08839|BIN1_RAT</td><td>Myc box-dependent-interacting protein 1 OS=Rattus norvegicus GN=Bin1 PE=1 SV=1</td><td>RAT</td><td>9</td></tr>
<tr><td>137</td><td>14.19</td><td>14.19</td><td>30.2</td><td>sp|P16617|PGK1_RAT</td><td>Phosphoglycerate kinase 1 OS=Rattus norvegicus GN=Pgk1 PE=1 SV=2</td><td>RAT</td><td>12</td></tr>
<tr><td>138</td><td>14.02</td><td>14.02</td><td>18.4</td><td>sp|P21396|AOFA_RAT</td><td>Amine oxidase [flavin-containing] A OS=Rattus norvegicus GN=Maoa PE=1 SV=1</td><td>RAT</td><td>8</td></tr>
<tr><td>139</td><td>14.01</td><td>14.01</td><td>17.7</td><td>sp|Q9JHU0|DPYL5_RAT</td><td>Dihydropyrimidinase-related protein 5 OS=Rattus norvegicus GN=Dpysl5 PE=1 SV=1</td><td>RAT</td><td>8</td></tr>
<tr><td>140</td><td>14.01</td><td>14.01</td><td>27</td><td>sp|P21913|DHSB_RAT</td><td>Succinate dehydrogenase [ubiquinone] iron-sulfur subunit, mitochondrial OS=Rattus norvegicus GN=Sdhb PE=2 SV=2</td><td>RAT</td><td>11</td></tr>
<tr><td>141</td><td>14.01</td><td>14.01</td><td>15.4</td><td>sp|P63329|PP2BA_RAT</td><td>Serine/threonine-protein phosphatase 2B catalytic subunit alpha isoform OS=Rattus norvegicus GN=Ppp3ca PE=1 SV=1</td><td>RAT</td><td>7</td></tr>
<tr><td>142</td><td>14.01</td><td>14.01</td><td>34.7</td><td>sp|Q63716|PRDX1_RAT</td><td>Peroxiredoxin-1 OS=Rattus norvegicus GN=Prdx1 PE=1 SV=1</td><td>RAT</td><td>8</td></tr>
<tr><td>143</td><td>14</td><td>18</td><td>27.3</td><td>sp|P85969|SNAB_RAT</td><td>Beta-soluble NSF attachment protein OS=Rattus norvegicus GN=Napb PE=1 SV=1</td><td>RAT</td><td>11</td></tr>
<tr><td>144</td><td>14</td><td>14</td><td>23.8</td><td>sp|P54311|GBB1_RAT</td><td>Guanine nucleotide-binding protein G(I)/G(S)/G(T) subunit beta-1 OS=Rattus norvegicus GN=Gnb1 PE=1 SV=4</td><td>RAT</td><td>17</td></tr>
<tr><td>145</td><td>14</td><td>14</td><td>26.8</td><td>sp|P26453-2|BASI_RAT</td><td>Isoform 2 of Basigin OS=Rattus norvegicus GN=Bsg</td><td>RAT</td><td>8</td></tr>
<tr><td>146</td><td>14</td><td>14</td><td>21.5</td><td>sp|P17764|THIL_RAT</td><td>Acetyl-CoA acetyltransferase, mitochondrial OS=Rattus norvegicus GN=Acat1 PE=1 SV=1</td><td>RAT</td><td>9</td></tr>
<tr><td>147</td><td>13.87</td><td>13.9</td><td>16.3</td><td>sp|Q6P6R2|DLDH_RAT</td><td>Dihydrolipoyl dehydrogenase, mitochondrial OS=Rattus norvegicus GN=Dld PE=1 SV=1</td><td>RAT</td><td>9</td></tr>
</table>

(continued)

Table 1
(continued)

N	Unused	Total	% Cov (95)	Accession #	Name	Species	Peptides (95%)
148	13.86	14.29	20	sp|Q62951|DPYL4_RAT	Dihydropyrimidinase-related protein 4 (Fragment) OS=Rattus norvegicus GN=Dpysl4 PE=1 SV=1	RAT	10
149	13.75	13.8	7	sp|P55161|NCKP1_RAT	Nck-associated protein 1 OS=Rattus norvegicus GN=Nckap1 PE=2 SV=1	RAT	7
150	13.72	13.78	19.6	sp|Q5FVI6|VATC1_RAT	V-type proton ATPase subunit C 1 OS=Rattus norvegicus GN=Atp6v1c1 PE=2 SV=1	RAT	7
151	13.71	13.76	8.5	sp|Q5QD51-2|AKA12_RAT	Isoform Beta of A-kinase anchor protein 12 OS=Rattus norvegicus GN=Akap12	RAT	9
152	13.55	13.62	13.6	sp|P08461|ODP2_RAT	Dihydrolipoyllysine-residue acetyltransferase component of pyruvate dehydrogenase complex, mitochondrial OS=Rattus norvegicus GN=Dlat PE=1 SV=3	RAT	7
153	13.5	13.56	14	sp|P19332|TAU_RAT	Microtubule-associated protein tau OS=Rattus norvegicus GN=Mapt PE=1 SV=3	RAT	8
154	13.49	13.55	21.2	sp|Q1HCL7|CE033_RAT	UPF0465 protein C5orf33 homolog OS=Rattus norvegicus PE=2 SV=1	RAT	7
155	13.46	13.54	27.3	sp|P97532|THTM_RAT	3-mercaptopyruvate sulfurtransferase OS=Rattus norvegicus GN=Mpst PE=1 SV=3	RAT	7
156	13.35	13.41	18.8	sp|P62630|EF1A1_RAT	Elongation factor 1-alpha 1 OS=Rattus norvegicus GN=Eef1a1 PE=1 SV=1	RAT	9
157	13.29	13.38	12.3	sp|P11915|NLTP_RAT	Non-specific lipid-transfer protein OS=Rattus norvegicus GN=Scp2 PE=1 SV=3	RAT	7
158	13.28	13.33	41.5	sp|P11348|DHPR_RAT	Dihydropteridine reductase OS=Rattus norvegicus GN=Qdpr PE=1 SV=1	RAT	11

N	Unused	Total	% Cov (95)	Accession #	Name	Species	Peptides (95%)
159	13.24	13.38	11.6	sp|P18886|CPT2_RAT	Carnitine O-palmitoyltransferase 2, mitochondrial OS=Rattus norvegicus GN=Cpt2 PE=1 SV=1	RAT	6
160	13.23	13.28	13.9	sp|Q63560|MAP6_RAT	Microtubule-associated protein 6 OS=Rattus norvegicus GN=Map6 PE=1 SV=1	RAT	9
161	13.19	13.27	17.2	sp|P13638|AT1B2_RAT	Sodium/potassium-transporting ATPase subunit beta-2 OS=Rattus norvegicus GN=Atp1b2 PE=2 SV=1	RAT	10
162	13	13.1	13.9	sp|Q924N5|ACBG1_ RAT	Long-chain-fatty-acid–CoA ligase ACSBG1 OS= Rattus norvegicus GN=Acsbg1 PE=1 SV=1	RAT	7
163	12.95	13.14	10.2	sp|Q62915|CSKP_RAT	Peripheral plasma membrane protein CASK OS= Rattus norvegicus GN=Cask PE=1 SV=1	RAT	7
164	12.94	12.98	10.7	sp|P46101|DPP6_RAT	Dipeptidyl aminopeptidase-like protein 6 OS= Rattus norvegicus GN=Dpp6 PE=1 SV=1	RAT	8
165	12.84	12.95	11.7	sp|P50137|TKT_RAT	Transketolase OS=Rattus norvegicus GN=Tkt PE=1 SV=1	RAT	8
166	12.77	31.55	16.1	sp|P11505-6|AT2B1_ RAT	Isoform K of Plasma membrane calcium-transporting ATPase 1 OS=Rattus norvegicus GN=Atp2b1	RAT	19
167	12.77	13.04	19.9	sp|P47728|CALB2_ RAT	Calretinin OS=Rattus norvegicus GN=Calb2 PE=1 SV=1	RAT	8
168	12.76	12.81	15.6	sp|Q01205|ODO2_ RAT	Dihydrolipoyllysine-residue succinyltransferase component of 2-oxoglutarate dehydrogenase complex, mitochondrial OS=Rattus norvegicus GN=Dlst PE=1 SV=2	RAT	9
169	12.7	12.89	25.6	sp|Q9JHY2|SFXN3_ RAT	Sideroflexin-3 OS=Rattus norvegicus GN=Sfxn3 PE=2 SV=1	RAT	7
170	12.63	12.66	42.7	sp|P10111|PPIA_RAT	Peptidyl-prolyl cis-trans isomerase A OS=Rattus norvegicus GN=Ppia PE=1 SV=2	RAT	10

(continued)

Table 1
(continued)

<table>
<tr><th>N</th><th>Unused</th><th>Total</th><th>% Cov (95)</th><th>Accession #</th><th>Name</th><th>Species</th><th>Peptides (95%)</th></tr>
<tr><td>171</td><td>12.62</td><td>12.64</td><td>42.9</td><td>sp|P02091|HBB1_RAT</td><td>Hemoglobin subunit beta-1 OS=Rattus norvegicus GN=Hbb PE=1 SV=3</td><td>RAT</td><td>6</td></tr>
<tr><td>172</td><td>12.61</td><td>13.01</td><td>16.5</td><td>sp|P25809|KCRU_RAT</td><td>Creatine kinase U-type, mitochondrial OS=Rattus norvegicus GN=Ckmt1 PE=1 SV=1</td><td>RAT</td><td>11</td></tr>
<tr><td>173</td><td>12.59</td><td>12.92</td><td>21</td><td>sp|P14604|ECHM_RAT</td><td>Enoyl-CoA hydratase, mitochondrial OS=Rattus norvegicus GN=Echs1 PE=1 SV=1</td><td>RAT</td><td>7</td></tr>
<tr><td>174</td><td>12.58</td><td>12.64</td><td>4</td><td>sp|P04775|SCN2A_RAT</td><td>Sodium channel protein type 2 subunit alpha OS=Rattus norvegicus GN=Scn2a PE=1 SV=1</td><td>RAT</td><td>7</td></tr>
<tr><td>175</td><td>12.58</td><td>12.62</td><td>12.2</td><td>sp|P31596|EAA2_RAT</td><td>Excitatory amino acid transporter 2 OS=Rattus norvegicus GN=Slc1a2 PE=1 SV=2</td><td>RAT</td><td>19</td></tr>
<tr><td>176</td><td>12.4</td><td>12.76</td><td>13.8</td><td>sp|P07323|ENOG_RAT</td><td>Gamma-enolase OS=Rattus norvegicus GN=Eno2 PE=1 SV=2</td><td>RAT</td><td>12</td></tr>
<tr><td>177</td><td>12.26</td><td>12.32</td><td>12.7</td><td>sp|O35814|STIP1_RAT</td><td>Stress-induced-phosphoprotein 1 OS=Rattus norvegicus GN=Stip1 PE=1 SV=1</td><td>RAT</td><td>6</td></tr>
<tr><td>178</td><td>12.21</td><td>12.23</td><td>11.9</td><td>sp|O88600|HSP74_RAT</td><td>Heat shock 70 kDa protein 4 OS=Rattus norvegicus GN=Hspa4 PE=1 SV=1</td><td>RAT</td><td>7</td></tr>
<tr><td>179</td><td>12.2</td><td>12.22</td><td>23.8</td><td>sp|P35435|ATPG_RAT</td><td>ATP synthase subunit gamma, mitochondrial OS=Rattus norvegicus GN=Atp5c1 PE=1 SV=2</td><td>RAT</td><td>7</td></tr>
<tr><td>180</td><td>12.19</td><td>20.36</td><td>16.3</td><td>sp|P34058|HS90B_RAT</td><td>Heat shock protein HSP 90-beta OS=Rattus norvegicus GN=Hsp90ab1 PE=1 SV=4</td><td>RAT</td><td>12</td></tr>
<tr><td>181</td><td>12.04</td><td>14.05</td><td>23.3</td><td>sp|Q9R1Z0|VDAC3_RAT</td><td>Voltage-dependent anion-selective channel protein 3 OS=Rattus norvegicus GN=Vdac3 PE=1 SV=2</td><td>RAT</td><td>8</td></tr>
</table>

N	Unused	Total	% Cov (95)	Accession #	Name	Species	Peptides (95%)
182	12.03	12.04	11.3	sp|P35571|GPDM_RAT	Glycerol-3-phosphate dehydrogenase, mitochondrial OS=Rattus norvegicus GN=Gpd2 PE=1 SV=1	RAT	7
183	12.02	12.02	24.5	sp|Q9Z0V6|PRDX3_RAT	Thioredoxin-dependent peroxide reductase, mitochondrial OS=Rattus norvegicus GN=Prdx3 PE=1 SV=2	RAT	8
184	12.01	12.02	25.1	sp|Q5XIH7|PHB2_RAT	Prohibitin-2 OS=Rattus norvegicus GN=Phb2 PE=1 SV=1	RAT	7
185	12	12	9.3	sp|Q62696|DLG1_RAT	Disks large homolog 1 OS=Rattus norvegicus GN=Dlg1 PE=1 SV=1	RAT	6
186	12	12	12.2	sp|Q499N5|ACSF2_RAT	Acyl-CoA synthetase family member 2, mitochondrial OS=Rattus norvegicus GN=Acsf2 PE=2 SV=1	RAT	6
187	12	12	27.2	sp|P60881|SNP25_RAT	Synaptosomal-associated protein 25 OS=Rattus norvegicus GN=Snap25 PE=1 SV=1	RAT	8
188	11.96	12.32	13.2	sp|P14882|PCCA_RAT	Propionyl-CoA carboxylase alpha chain, mitochondrial (Fragment) OS=Rattus norvegicus GN=Pcca PE=1 SV=2	RAT	8
189	11.59	11.65	14.9	sp|Q5PPJ9|SHLB2_RAT	Endophilin-B2 OS=Rattus norvegicus GN=Sh3glb2 PE=2 SV=2	RAT	6
190	11.59	11.64	12.5	sp|Q704S8|CACP_RAT	Carnitine O-acetyltransferase OS=Rattus norvegicus GN=Crat PE=1 SV=1	RAT	6
191	11.57	19.95	39.9	sp|P04905|GSTM1_RAT	Glutathione S-transferase Mu 1 OS=Rattus norvegicus GN=Gstm1 PE=1 SV=2	RAT	10
192	11.56	11.65	11.4	sp|Q68FS4|AMPL_RAT	Cytosol aminopeptidase OS=Rattus norvegicus GN=Lap3 PE=1 SV=1	RAT	6
193	11.55	11.59	43.5	sp|P21571|ATP5J_RAT	ATP synthase-coupling factor 6, mitochondrial OS=Rattus norvegicus GN=Atp5j PE=1 SV=1	RAT	7

(continued)

Table 1
(continued)

N	Unused	Total	% Cov (95)	Accession #	Name	Species	Peptides (95%)
194	11.54	11.73	7.9	sp|P05197|EF2_RAT	Elongation factor 2 OS=Rattus norvegicus GN=Eef2 PE=1 SV=4	RAT	6
195	11.47	11.53	9.9	sp|Q4V8B0|OXR1_RAT	Oxidation resistance protein 1 OS=Rattus norvegicus GN=Oxr1 PE=1 SV=3	RAT	6
196	11.39	11.52	20.3	sp|Q5XIE6|HIBCH_RAT	3-hydroxyisobutyryl-CoA hydrolase, mitochondrial OS=Rattus norvegicus GN=Hibch PE=1 SV=2	RAT	6
197	11.37	11.58	20.6	sp|Q9JJM9|SEPT5_RAT	Septin-5 OS=Rattus norvegicus GN=Sept5 PE=1 SV=2	RAT	9
198	11.32	11.48	24.5	sp|P07171|CALB1_RAT	Calbindin OS=Rattus norvegicus GN=Calb1 PE=1 SV=2	RAT	6
199	11.27	11.4	56.9	sp|P26772|CH10_RAT	10 kDa heat shock protein, mitochondrial OS=Rattus norvegicus GN=Hspe1 PE=1 SV=3	RAT	6
200	11.27	11.35	6.3	sp|Q5U300|UBA1_RAT	Ubiquitin-like modifier-activating enzyme 1 OS=Rattus norvegicus GN=Uba1 PE=1 SV=1	RAT	5
201	11.25	13.37	9.9	sp|P18596|AT2A3_RAT	Sarcoplasmic/endoplasmic reticulum calcium ATPase 3 OS=Rattus norvegicus GN=Atp2a3 PE=1 SV=1	RAT	7
202	11.24	11.28	36.9	sp|P31044|PEBP1_RAT	Phosphatidylethanolamine-binding protein 1 OS=Rattus norvegicus GN=Pebp1 PE=1 SV=3	RAT	8
203	11.11	11.18	10.2	sp|Q63028|ADDA_RAT	Alpha-adducin OS=Rattus norvegicus GN=Add1 PE=1 SV=2	RAT	6
204	11.05	13.32	12.6	sp|O08838|AMPH_RAT	Amphiphysin OS=Rattus norvegicus GN=Amph PE=1 SV=1	RAT	7

N	Unused	Total	% Cov (95)	Accession #	Name	Species	Peptides (95%)
205	10.78	10.81	21.4	sp|Q64536|PDK2_RAT	[Pyruvate dehydrogenase [lipoamide]] kinase isozyme 2, mitochondrial OS=Rattus norvegicus GN=Pdk2 PE=1 SV=1	RAT	8
206	10.76	10.85	9.3	sp|Q63617|HYOU1_RAT	Hypoxia up-regulated protein 1 OS=Rattus norvegicus GN=Hyou1 PE=1 SV=1	RAT	7
207	10.74	10.79	22.8	sp|P61589|RHOA_RAT	Transforming protein RhoA OS=Rattus norvegicus GN=Rhoa PE=1 SV=1	RAT	7
208	10.73	10.8	14.2	sp|P04762|CATA_RAT	Catalase OS=Rattus norvegicus GN=Cat PE=1 SV=3	RAT	6
209	10.66	16.98	42.5	sp|P63041|CPLX1_RAT	Complexin-1 OS=Rattus norvegicus GN=Cplx1 PE=1 SV=1	RAT	11
210	10.63	10.73	19.4	sp|P09456|KAP0_RAT	cAMP-dependent protein kinase type I-alpha regulatory subunit OS=Rattus norvegicus GN=Prkar1a PE=2 SV=2	RAT	6
211	10.59	10.64	20	sp|Q99PD4|ARC1A_RAT	Actin-related protein 2/3 complex subunit 1A OS=Rattus norvegicus GN=Arpc1a PE=2 SV=1	RAT	7
212	10.49	10.51	23.6	sp|Q6AXX6|CJ058_RAT	Uncharacterized protein C10orf58 homolog OS=Rattus norvegicus PE=1 SV=1	RAT	5
213	10.46	10.54	6.6	sp|Q62717|CAPS1_RAT	Calcium-dependent secretion activator 1 OS=Rattus norvegicus GN=Cadps PE=1 SV=1	RAT	6
214	10.36	10.38	10.1	sp|Q60587|ECHB_RAT	Trifunctional enzyme subunit beta, mitochondrial OS=Rattus norvegicus GN=Hadhb PE=1 SV=1	RAT	5
215	10.33	10.35	9.1	sp|P0C1X8|AAK1_RAT	AP2-associated protein kinase 1 OS=Rattus norvegicus GN=Aak1 PE=1 SV=1	RAT	5
216	10.25	10.3	35.6	sp|P30009|MARCS_RAT	Myristoylated alanine-rich C-kinase substrate OS=Rattus norvegicus GN=Marcks PE=1 SV=2	RAT	8

(continued)

Table 1
(continued)

<table>
<tr><th>N</th><th>Unused</th><th>Total</th><th>% Cov (95)</th><th>Accession #</th><th>Name</th><th>Species</th><th>Peptides (95%)</th></tr>
<tr><td>217</td><td>10.2</td><td>10.26</td><td>6.2</td><td>sp|P15146|MAP2_RAT</td><td>Microtubule-associated protein 2 OS=Rattus norvegicus GN=Map2 PE=1 SV=3</td><td>RAT</td><td>6</td></tr>
<tr><td>218</td><td>10.18</td><td>10.19</td><td>13.7</td><td>sp|P24054|SPRL1_RAT</td><td>SPARC-like protein 1 OS=Rattus norvegicus GN=Sparcl1 PE=2 SV=1</td><td>RAT</td><td>5</td></tr>
<tr><td>219</td><td>10.17</td><td>10.18</td><td>15.3</td><td>sp|P97710|SHPS1_RAT</td><td>Tyrosine-protein phosphatase non-receptor type substrate 1 OS=Rattus norvegicus GN=Sirpa PE=1 SV=1</td><td>RAT</td><td>5</td></tr>
<tr><td>220</td><td>10.16</td><td>10.17</td><td>14.8</td><td>sp|P56603|SCAM1_RAT</td><td>Secretory carrier-associated membrane protein 1 OS=Rattus norvegicus GN=Scamp1 PE=1 SV=1</td><td>RAT</td><td>8</td></tr>
<tr><td>221</td><td>10.12</td><td>10.13</td><td>22.9</td><td>sp|Q63345|MOG_RAT</td><td>Myelin-oligodendrocyte glycoprotein OS=Rattus norvegicus GN=Mog PE=1 SV=1</td><td>RAT</td><td>8</td></tr>
<tr><td>222</td><td>10.11</td><td>10.12</td><td>5.8</td><td>sp|P97536|CAND1_RAT</td><td>Cullin-associated NEDD8-dissociated protein 1 OS=Rattus norvegicus GN=Cand1 PE=1 SV=1</td><td>RAT</td><td>5</td></tr>
<tr><td>223</td><td>10.09</td><td>10.24</td><td>4.9</td><td>sp|Q9JK11|RTN4_RAT</td><td>Reticulon-4 OS=Rattus norvegicus GN=Rtn4 PE=1 SV=1</td><td>RAT</td><td>8</td></tr>
<tr><td>224</td><td>10.09</td><td>10.1</td><td>32.3</td><td>sp|Q6RUV5|RAC1_RAT</td><td>Ras-related C3 botulinum toxin substrate 1 OS=Rattus norvegicus GN=Rac1 PE=1 SV=1</td><td>RAT</td><td>8</td></tr>
<tr><td>225</td><td>10.07</td><td>10.07</td><td>25.9</td><td>sp|P45592|COF1_RAT</td><td>Cofilin-1 OS=Rattus norvegicus GN=Cfl1 PE=1 SV=3</td><td>RAT</td><td>6</td></tr>
<tr><td>226</td><td>10.06</td><td>10.15</td><td>29.6</td><td>sp|Q06647|ATPO_RAT</td><td>ATP synthase subunit O, mitochondrial OS=Rattus norvegicus GN=Atp5o PE=1 SV=1</td><td>RAT</td><td>9</td></tr>
<tr><td>227</td><td>10.06</td><td>10.07</td><td>25.9</td><td>sp|Q5RJL0|ERMIN_RAT</td><td>Ermin OS=Rattus norvegicus PE=2 SV=1</td><td>RAT</td><td>5</td></tr>
</table>

<table>
<tr><th>N</th><th>Unused</th><th>Total</th><th>% Cov (95)</th><th>Accession #</th><th>Name</th><th>Species</th><th>Peptides (95%)</th></tr>
<tr><td>228</td><td>10.04</td><td>10.05</td><td>11.4</td><td>sp|P11884|ALDH2_RAT</td><td>Aldehyde dehydrogenase, mitochondrial OS=Rattus norvegicus GN=Aldh2 PE=1 SV=1</td><td>RAT</td><td>5</td></tr>
<tr><td>229</td><td>10.03</td><td>10.13</td><td>18.1</td><td>sp|Q8CFD0|SFXN5_RAT</td><td>Sideroflexin-5 OS=Rattus norvegicus GN=Sfxn5 PE=2 SV=1</td><td>RAT</td><td>9</td></tr>
<tr><td>230</td><td>10.03</td><td>10.03</td><td>17.6</td><td>sp|P62882|GBB5_RAT</td><td>Guanine nucleotide-binding protein subunit beta-5 OS=Rattus norvegicus GN=Gnb5 PE=2 SV=1</td><td>RAT</td><td>5</td></tr>
<tr><td>231</td><td>10.01</td><td>10.01</td><td>18.3</td><td>sp|P13803|ETFA_RAT</td><td>Electron transfer flavoprotein subunit alpha, mitochondrial OS=Rattus norvegicus GN=Etfa PE=1 SV=4</td><td>RAT</td><td>6</td></tr>
<tr><td>232</td><td>10</td><td>14.79</td><td>31.6</td><td>sp|P61983|1433G_RAT</td><td>14-3-3 protein gamma OS=Rattus norvegicus GN=Ywhag PE=1 SV=2</td><td>RAT</td><td>11</td></tr>
<tr><td>233</td><td>10</td><td>12</td><td>12.5</td><td>sp|P19643|AOFB_RAT</td><td>Amine oxidase [flavin-containing] B OS=Rattus norvegicus GN=Maob PE=1 SV=3</td><td>RAT</td><td>6</td></tr>
<tr><td>234</td><td>10</td><td>10</td><td>32.4</td><td>sp|P62898|CYC_RAT</td><td>Cytochrome c, somatic OS=Rattus norvegicus GN=Cycs PE=1 SV=2</td><td>RAT</td><td>6</td></tr>
<tr><td>235</td><td>10</td><td>10</td><td>13.1</td><td>sp|B2RYW9|FAHD2_RAT</td><td>Fumarylacetoacetate hydrolase domain-containing protein 2 OS=Rattus norvegicus GN=Fahd2 PE=1 SV=1</td><td>RAT</td><td>5</td></tr>
<tr><td>236</td><td>9.92</td><td>12</td><td>8.4</td><td>sp|Q6RJR6|RTN3_RAT</td><td>Reticulon-3 OS=Rattus norvegicus GN=Rtn3 PE=1 SV=1</td><td>RAT</td><td>6</td></tr>
<tr><td>237</td><td>9.88</td><td>9.95</td><td>14.4</td><td>sp|P15650|ACADL_RAT</td><td>Long-chain specific acyl-CoA dehydrogenase, mitochondrial OS=Rattus norvegicus GN=Acadl PE=1 SV=1</td><td>RAT</td><td>5</td></tr>
<tr><td>238</td><td>9.87</td><td>9.91</td><td>17.7</td><td>sp|P07943|ALDR_RAT</td><td>Aldose reductase OS=Rattus norvegicus GN=Akr1b1 PE=1 SV=3</td><td>RAT</td><td>6</td></tr>
<tr><td>239</td><td>9.82</td><td>9.85</td><td>32.4</td><td>sp|P04906|GSTP1_RAT</td><td>Glutathione S-transferase P OS=Rattus norvegicus GN=Gstp1 PE=1 SV=2</td><td>RAT</td><td>8</td></tr>
</table>

(continued)

Table 1
(continued)

N	Unused	Total	% Cov (95)	Accession #	Name	Species	Peptides (95%)
240	9.8	9.86	14.5	sp|Q62813|LSAMP_RAT	Limbic system-associated membrane protein OS=Rattus norvegicus GN=Lsamp PE=1 SV=1	RAT	5
241	9.79	9.88	4	sp|Q05546|TENR_RAT	Tenascin-R OS=Rattus norvegicus GN=Tnr PE=1 SV=1	RAT	6
242	9.78	9.84	36.5	sp|P07895|SODM_RAT	Superoxide dismutase [Mn], mitochondrial OS=Rattus norvegicus GN=Sod2 PE=1 SV=2	RAT	11
243	9.73	9.82	11	sp|P07722-2|MAG_RAT	Isoform S-MAG of Myelin-associated glycoprotein OS=Rattus norvegicus GN=Mag	RAT	5
244	9.7	9.9	4.6	sp|P97846|CNTP1_RAT	Contactin-associated protein 1 OS=Rattus norvegicus GN=Cntnap1 PE=1 SV=1	RAT	5
245	9.69	9.87	25.4	sp|P97576|GRPE1_RAT	GrpE protein homolog 1, mitochondrial OS=Rattus norvegicus GN=Grpel1 PE=1 SV=2	RAT	5
246	9.56	9.72	15.4	sp|B0BNF1|SEPT8_RAT	Septin-8 OS=Rattus norvegicus GN=Sept8 PE=1 SV=1	RAT	5
247	9.55	9.63	12.5	sp|P82471|GNAQ_RAT	Guanine nucleotide-binding protein G(q) subunit alpha OS=Rattus norvegicus GN=Gnaq PE=1 SV=2	RAT	6
248	9.52	9.59	45.4	sp|B0K020|CISD1_RAT	CDGSH iron sulfur domain-containing protein 1 OS=Rattus norvegicus GN=Cisd1 PE=3 SV=1	RAT	9
249	9.41	9.48	27.8	sp|Q9WUS0|KAD4_RAT	Adenylate kinase isoenzyme 4, mitochondrial OS=Rattus norvegicus GN=Ak3l1 PE=2 SV=1	RAT	6
250	9.38	9.45	12.3	sp|Q794F9|4F2_RAT	4F2 cell-surface antigen heavy chain OS=Rattus norvegicus GN=Slc3a2 PE=1 SV=1	RAT	5
251	9.34	9.54	39.5	sp|Q05982|NDKA_RAT	Nucleoside diphosphate kinase A OS=Rattus norvegicus GN=Nme1 PE=1 SV=1	RAT	5

N	Unused	Total	% Cov (95)	Accession #	Name	Species	Peptides (95%)
252	9.34	9.44	33.6	sp|Q63362|NDUA5_RAT	NADH dehydrogenase [ubiquinone] 1 alpha subcomplex subunit 5 OS=Rattus norvegicus GN=Ndufa5 PE=1 SV=3	RAT	5
253	9.26	9.33	5.4	sp|P54290|CA2D1_RAT	Voltage-dependent calcium channel subunit alpha-2/delta-1 OS=Rattus norvegicus GN=Cacna2d1 PE=1 SV=1	RAT	5
254	9.25	9.39	19.2	sp|Q6AYH5|DCTN2_RAT	Dynactin subunit 2 OS=Rattus norvegicus GN=Dctn2 PE=1 SV=1	RAT	5
255	9.19	9.37	12.6	sp|Q641Z6|EHD1_RAT	EH domain-containing protein 1 OS=Rattus norvegicus GN=Ehd1 PE=1 SV=1	RAT	5
256	9.1	9.15	20.6	sp|B2RZ37|REEP5_RAT	Receptor expression-enhancing protein 5 OS=Rattus norvegicus GN=Reep5 PE=1 SV=1	RAT	5
257	9.08	9.19	9.8	sp|Q5I0C3|MCCA_RAT	Methylcrotonoyl-CoA carboxylase subunit alpha, mitochondrial OS=Rattus norvegicus GN=Mccc1 PE=1 SV=1	RAT	5
258	9.06	9.13	15.6	sp|Q58FK9|KAT3_RAT	Kynurenine–oxoglutarate transaminase 3 OS=Rattus norvegicus GN=Ccbl2 PE=2 SV=1	RAT	5
259	9.05	11.24	30.5	sp|P63012|RAB3A_RAT	Ras-related protein Rab-3A OS=Rattus norvegicus GN=Rab3a PE=1 SV=1	RAT	7
260	8.89	9.37	9.2	sp|P49803|RGS7_RAT	Regulator of G-protein signaling 7 OS=Rattus norvegicus GN=Rgs7 PE=2 SV=2	RAT	5
261	8.79	8.84	15.1	sp|Q6AXV4|SAM50_RAT	Sorting and assembly machinery component 50 homolog OS=Rattus norvegicus GN=Samm50 PE=1 SV=1	RAT	5
262	8.74	8.8	10.4	sp|P19627|GNAZ_RAT	Guanine nucleotide-binding protein G(z) subunit alpha OS=Rattus norvegicus GN=Gnaz PE=2 SV=3	RAT	4

(continued)

Table 1
(continued)

N	Unused	Total	% Cov (95)	Accession #	Name	Species	Peptides (95%)
263	8.73	8.78	16.1	sp|Q9Z0W5|PACN1_RAT	Protein kinase C and casein kinase substrate in neurons protein 1 OS=Rattus norvegicus GN=Pacsin1 PE=1 SV=1	RAT	6
264	8.69	8.84	9.5	sp|Q5XHZ0|TRAP1_RAT	Heat shock protein 75 kDa, mitochondrial OS=Rattus norvegicus GN=Trap1 PE=1 SV=1	RAT	6
265	8.59	8.63	2.2	sp|P12785|FAS_RAT	Fatty acid synthase OS=Rattus norvegicus GN=Fasn PE=1 SV=3	RAT	4
266	8.58	8.67	22.8	sp|Q9Z339|GSTO1_RAT	Glutathione S-transferase omega-1 OS=Rattus norvegicus GN=Gsto1 PE=1 SV=2	RAT	5
267	8.53	8.57	6.3	sp|P48768|NAC2_RAT	Sodium/calcium exchanger 2 OS=Rattus norvegicus GN=Slc8a2 PE=2 SV=1	RAT	5
268	8.47	8.7	3.1	sp|Q8CFG6|CA2D2_RAT	Voltage-dependent calcium channel subunit alpha-2/delta-2 OS=Rattus norvegicus GN=Cacna2d2 PE=1 SV=1	RAT	4
269	8.44	8.47	11.6	sp|O55148|GAS7_RAT	Growth arrest-specific protein 7 OS=Rattus norvegicus GN=Gas7 PE=2 SV=2	RAT	4
270	8.39	8.41	11.8	sp|Q08163|CAP1_RAT	Adenylyl cyclase-associated protein 1 OS=Rattus norvegicus GN=Cap1 PE=1 SV=3	RAT	4
271	8.33	8.35	4.6	sp|P19491-2|GRIA2_RAT	Isoform Flip of Glutamate receptor 2 OS=Rattus norvegicus GN=Gria2	RAT	4
272	8.26	8.28	10	sp|P28572|SC6A9_RAT	Sodium- and chloride-dependent glycine transporter 1 OS=Rattus norvegicus GN=Slc6a9 PE=1 SV=2	RAT	6
273	8.24	8.32	7.7	sp|Q62634|VGLU1_RAT	Vesicular glutamate transporter 1 OS=Rattus norvegicus GN=Slc17a7 PE=1 SV=1	RAT	5
274	8.22	8.3	25.5	sp|Q7M0E3|DEST_RAT	Destrin OS=Rattus norvegicus GN=Dstn PE=1 SV=3	RAT	4

N	Unused	Total	% Cov (95)	Accession #	Name	Species	Peptides (95%)
275	8.21	8.23	24.8	sp|P60905|DNJC5_RAT	DnaJ homolog subfamily C member 5 OS=Rattus norvegicus GN=Dnajc5 PE=1 SV=1	RAT	5
276	8.19	14.29	14.2	sp|Q62950|DPYL1_RAT	Dihydropyrimidinase-related protein 1 OS=Rattus norvegicus GN=Crmp1 PE=1 SV=1	RAT	7
277	8.19	8.21	6.9	sp|P10687|PLCB1_RAT	1-phosphatidylinositol-4,5-bisphosphate phosphodiesterase beta-1 OS=Rattus norvegicus GN=Plcb1 PE=1 SV=1	RAT	5
278	8.17	8.18	7.7	sp|P31662|S6A17_RAT	Orphan sodium- and chloride-dependent neurotransmitter transporter NTT4 OS=Rattus norvegicus GN=Slc6a17 PE=2 SV=1	RAT	5
279	8.16	8.21	7.5	sp|O35763|MOES_RAT	Moesin OS=Rattus norvegicus GN=Msn PE=1 SV=3	RAT	5
280	8.12	8.13	19.9	sp|P62762|VISL1_RAT	Visinin-like protein 1 OS=Rattus norvegicus GN=Vsnl1 PE=1 SV=2	RAT	4
281	8.11	8.12	15.2	sp|Q64559|BACH_RAT	Cytosolic acyl coenzyme A thioester hydrolase OS=Rattus norvegicus GN=Acot7 PE=1 SV=4	RAT	4
282	8.07	8.08	8.3	sp|Q9ES21|SAC1_RAT	Phosphatidylinositide phosphatase SAC1 OS=Rattus norvegicus GN=Sacm1l PE=1 SV=1	RAT	4
283	8.06	8.07	7	sp|O35303|DNM1L_RAT	Dynamin-1-like protein OS=Rattus norvegicus GN=Dnm1l PE=1 SV=1	RAT	4
284	8.05	8.07	7.3	sp|Q64428|ECHA_RAT	Trifunctional enzyme subunit alpha, mitochondrial OS=Rattus norvegicus GN=Hadha PE=1 SV=2	RAT	4
285	8.04	8.05	7.7	sp|Q8R500|MFN2_RAT	Mitofusin-2 OS=Rattus norvegicus GN=Mfn2 PE=1 SV=1	RAT	4
286	8.04	8.05	24.8	sp|P56571|ES1_RAT	ES1 protein homolog, mitochondrial OS=Rattus norvegicus PE=1 SV=2	RAT	6

(continued)

Table 1
(continued)

N	Unused	Total	% Cov (95)	Accession #	Name	Species	Peptides (95%)
287	8.03	8.03	11.7	sp|Q6UPE1|ETFD_RAT	Electron transfer flavoprotein-ubiquinone oxidoreductase, mitochondrial OS=Rattus norvegicus GN=Etfdh PE=2 SV=1	RAT	6
288	8.02	10.03	15	sp|P50398|GDIA_RAT	Rab GDP dissociation inhibitor alpha OS=Rattus norvegicus GN=Gdi1 PE=1 SV=1	RAT	5
289	8.02	8.02	10.7	sp|P30839|AL3A2_RAT	Fatty aldehyde dehydrogenase OS=Rattus norvegicus GN=Aldh3a2 PE=1 SV=1	RAT	4
290	8.01	8.02	60.6	sp|P29419|ATP5I_RAT	ATP synthase subunit e, mitochondrial OS=Rattus norvegicus GN=Atp5i PE=1 SV=3	RAT	5
291	8.01	8.01	4.9	sp|Q62910-4|SYNJ1_RAT	Isoform 170 kDa-16AA of Synaptojanin-1 OS=Rattus norvegicus GN=Synj1	RAT	4
292	8.01	8.01	23.9	sp|Q792I0|LIN7C_RAT	Protein lin-7 homolog C OS=Rattus norvegicus GN=Lin7c PE=1 SV=1	RAT	7
293	8.01	8.01	12	sp|P29147|BDH_RAT	D-beta-hydroxybutyrate dehydrogenase, mitochondrial OS=Rattus norvegicus GN=Bdh1 PE=1 SV=2	RAT	4
294	8	24.01	27.1	sp|Q4QRB4|TBB3_RAT	Tubulin beta-3 chain OS=Rattus norvegicus GN=Tubb3 PE=1 SV=1	RAT	17
295	8	14	12.3	sp|P30835|K6PL_RAT	6-phosphofructokinase, liver type OS=Rattus norvegicus GN=Pfkl PE=2 SV=3	RAT	8
296	8	10.03	6.7	sp|Q80ZA5-3|S4A10_RAT	Isoform rb2NCBE of Sodium-driven chloride bicarbonate exchanger OS=Rattus norvegicus GN=Slc4a10	RAT	6
297	8	8	7.9	sp|Q63598|PLST_RAT	Plastin-3 OS=Rattus norvegicus GN=Pls3 PE=2 SV=2	RAT	4

<table>
<tr><th>N</th><th>Unused</th><th>Total</th><th>% Cov (95)</th><th>Accession #</th><th>Name</th><th>Species</th><th>Peptides (95%)</th></tr>
<tr><td>298</td><td>8</td><td>8</td><td>18.3</td><td>sp|Q9JJ19|NHRF1_RAT</td><td>Na(+)/H(+) exchange regulatory cofactor NHE-RF1 OS=Rattus norvegicus GN=Slc9a3r1 PE=1 SV=3</td><td>RAT</td><td>4</td></tr>
<tr><td>299</td><td>8</td><td>8</td><td>8.9</td><td>sp|P14408|FUMH_RAT</td><td>Fumarate hydratase, mitochondrial OS=Rattus norvegicus GN=Fh PE=1 SV=1</td><td>RAT</td><td>4</td></tr>
<tr><td>300</td><td>8</td><td>8</td><td>13.1</td><td>sp|Q1WIM1|CADM4_RAT</td><td>Cell adhesion molecule 4 OS=Rattus norvegicus GN=Cadm4 PE=1 SV=1</td><td>RAT</td><td>4</td></tr>
<tr><td>301</td><td>8</td><td>8</td><td>25.7</td><td>sp|P23928|CRYAB_RAT</td><td>Alpha-crystallin B chain OS=Rattus norvegicus GN=Cryab PE=1 SV=1</td><td>RAT</td><td>4</td></tr>
<tr><td>302</td><td>8</td><td>8</td><td>17</td><td>sp|O35244|PRDX6_RAT</td><td>Peroxiredoxin-6 OS=Rattus norvegicus GN=Prdx6 PE=1 SV=3</td><td>RAT</td><td>5</td></tr>
<tr><td>303</td><td>8</td><td>8</td><td>11</td><td>sp|Q63488|S20A2_RAT</td><td>Sodium-dependent phosphate transporter 2 OS=Rattus norvegicus GN=Slc20a2 PE=1 SV=1</td><td>RAT</td><td>4</td></tr>
<tr><td>304</td><td>8</td><td>8</td><td>15.7</td><td>sp|P45479|PPT1_RAT</td><td>Palmitoyl-protein thioesterase 1 OS=Rattus norvegicus GN=Ppt1 PE=1 SV=1</td><td>RAT</td><td>4</td></tr>
<tr><td>305</td><td>8</td><td>8</td><td>19.4</td><td>sp|P23711|HMOX2_RAT</td><td>Heme oxygenase 2 OS=Rattus norvegicus GN=Hmox2 PE=1 SV=1</td><td>RAT</td><td>4</td></tr>
<tr><td>306</td><td>8</td><td>8</td><td>22.9</td><td>sp|Q6AXT5|RAB21_RAT</td><td>Ras-related protein Rab-21 OS=Rattus norvegicus GN=Rab21 PE=2 SV=1</td><td>RAT</td><td>4</td></tr>
<tr><td>307</td><td>8</td><td>8</td><td>3.8</td><td>sp|P53987|MOT1_RAT</td><td>Monocarboxylate transporter 1 OS=Rattus norvegicus GN=Slc16a1 PE=1 SV=1</td><td>RAT</td><td>4</td></tr>
<tr><td>308</td><td>8</td><td>8</td><td>23.3</td><td>sp|P36860|RALB_RAT</td><td>Ras-related protein Ral-B OS=Rattus norvegicus GN=Ralb PE=2 SV=1</td><td>RAT</td><td>4</td></tr>
<tr><td>309</td><td>8</td><td>8</td><td>14.2</td><td>sp|O70351|HCD2_RAT</td><td>3-hydroxyacyl-CoA dehydrogenase type-2 OS=Rattus norvegicus GN=Hsd17b10 PE=1 SV=3</td><td>RAT</td><td>4</td></tr>
<tr><td>310</td><td>8</td><td>8</td><td>29.2</td><td>sp|Q00981|UCHL1_RAT</td><td>Ubiquitin carboxyl-terminal hydrolase isozyme L1 OS=Rattus norvegicus GN=Uchl1 PE=1 SV=2</td><td>RAT</td><td>5</td></tr>
</table>

(continued)

Table 1
(continued)

N	Unused	Total	% Cov (95)	Accession #	Name	Species	Peptides (95%)
311	8	8	21.3	sp|P09527|RAB7A_RAT	Ras-related protein Rab-7a OS=Rattus norvegicus GN=Rab7a PE=1 SV=2	RAT	4
312	8	8	20.6	sp|Q80W89|NDUAB_RAT	NADH dehydrogenase [ubiquinone] 1 alpha subcomplex subunit 11 OS=Rattus norvegicus GN=Ndufa11 PE=2 SV=1	RAT	5
313	8	8	21.6	sp|P27605|HPRT_RAT	Hypoxanthine-guanine phosphoribosyltransferase OS=Rattus norvegicus GN=Hprt1 PE=1 SV=1	RAT	5
314	7.94	8.05	12.9	sp|Q9Z1E1|FLOT1_RAT	Flotillin-1 OS=Rattus norvegicus GN=Flot1 PE=2 SV=2	RAT	6
315	7.93	8.09	13.9	sp|P20788|UCRI_RAT	Cytochrome b-c1 complex subunit Rieske, mitochondrial OS=Rattus norvegicus GN=Uqcrfs1 PE=1 SV=2	RAT	5
316	7.92	8	13.1	sp|P13234|KCC4_RAT	Calcium/calmodulin-dependent protein kinase type IV OS=Rattus norvegicus GN=Camk4 PE=1 SV=3	RAT	4
317	7.86	8	7.5	sp|Q63564|SV2B_RAT	Synaptic vesicle glycoprotein 2B OS=Rattus norvegicus GN=Sv2b PE=1 SV=1	RAT	8
318	7.85	7.92	6.7	sp|P23978|SC6A1_RAT	Sodium- and chloride-dependent GABA transporter 1 OS=Rattus norvegicus GN=Slc6a1 PE=1 SV=1	RAT	5
319	7.81	8.04	30.2	sp|P12075|COX5B_RAT	Cytochrome c oxidase subunit 5B, mitochondrial OS=Rattus norvegicus GN=Cox5b PE=1 SV=2	RAT	6
320	7.79	7.88	18.8	sp|Q4FZY0|EFHD2_RAT	EF-hand domain-containing protein D2 OS=Rattus norvegicus GN=Efhd2 PE=1 SV=1	RAT	4
321	7.78	7.9	16.2	sp|P52555|ERP29_RAT	Endoplasmic reticulum resident protein 29 OS=Rattus norvegicus GN=Erp29 PE=1 SV=2	RAT	4

N	Unused	Total	% Cov (95)	Accession #	Name	Species	Peptides (95%)
322	7.75	7.83	7.9	sp|Q9EPH8|PABP1_RAT	Polyadenylate-binding protein 1 OS=Rattus norvegicus GN=Pabpc1 PE=2 SV=1	RAT	4
323	7.69	7.8	10.8	sp|Q68FR6|EF1G_RAT	Elongation factor 1-gamma OS=Rattus norvegicus GN=Eef1g PE=1 SV=3	RAT	4
324	7.64	7.72	5.8	sp|Q5M7W5|MAP4_RAT	Microtubule-associated protein 4 OS=Rattus norvegicus GN=Map4 PE=1 SV=1	RAT	4
325	7.6	7.81	13.1	sp|P24329|THTR_RAT	Thiosulfate sulfurtransferase OS=Rattus norvegicus GN=Tst PE=1 SV=3	RAT	6
326	7.53	7.7	13.7	sp|P63086|MK01_RAT	Mitogen-activated protein kinase 1 OS=Rattus norvegicus GN=Mapk1 PE=1 SV=3	RAT	4
327	7.51	7.6	8.7	sp|Q01986|MP2K1_RAT	Dual specificity mitogen-activated protein kinase kinase 1 OS=Rattus norvegicus GN=Map2k1 PE=1 SV=2	RAT	4
328	7.44	7.64	15.5	sp|A1A5P0|BORG5_RAT	Cdc42 effector protein 1 OS=Rattus norvegicus GN=Cdc42ep1 PE=2 SV=1	RAT	4
329	7.44	7.55	8.1	sp|P04785|PDIA1_RAT	Protein disulfide-isomerase OS=Rattus norvegicus GN=P4hb PE=1 SV=2	RAT	4
330	7.41	9.59	5.9	sp|Q63803|GNAS1_RAT	Guanine nucleotide-binding protein G(s) subunit alpha isoforms XLas OS=Rattus norvegicus GN=Gnas PE=1 SV=3	RAT	6
331	7.4	7.51	16.2	sp|P68182|KAPCB_RAT	cAMP-dependent protein kinase catalytic subunit beta OS=Rattus norvegicus GN=Prkacb PE=1 SV=2	RAT	5
332	7.29	7.47	5.6	sp|P56558|OGT1_RAT	UDP-N-acetylglucosamine–peptide N-acetylglucosaminyltransferase 110 kDa subunit OS=Rattus norvegicus GN=Ogt PE=1 SV=1	RAT	4
333	7.22	7.31	3.9	sp|Q9ERH3|WDR7_RAT	WD repeat-containing protein 7 OS=Rattus norvegicus GN=Wdr7 PE=2 SV=1	RAT	4

(continued)

Table 1
(continued)

N	Unused	Total	% Cov (95)	Accession #	Name	Species	Peptides (95%)
334	7.15	7.39	8.8	sp|B2GV06|SCOT1_RAT	Succinyl-CoA:3-ketoacid-coenzyme A transferase 1, mitochondrial OS=Rattus norvegicus GN=Oxct1 PE=1 SV=1	RAT	4
335	7.15	7.37	4.1	sp|Q9QW07|PLCB4_RAT	1-phosphatidylinositol-4,5-bisphosphate phosphodiesterase beta-4 OS=Rattus norvegicus GN=Plcb4 PE=2 SV=2	RAT	4
336	7.12	7.21	15.3	sp|Q9WU34-2|SEPT3_RAT	Isoform G-septin-beta of Neuronal-specific septin-3 OS=Rattus norvegicus GN=Sept3	RAT	4
337	7.05	7.18	6.6	sp|P46462|TERA_RAT	Transitional endoplasmic reticulum ATPase OS=Rattus norvegicus GN=Vcp PE=1 SV=3	RAT	5
338	7	8.45	18.7	sp|P62142|PP1B_RAT	Serine/threonine-protein phosphatase PP1-beta catalytic subunit OS=Rattus norvegicus GN=Ppp1cb PE=1 SV=3	RAT	5
339	6.98	7.1	11.7	sp|Q99068|AMRP_RAT	Alpha-2-macroglobulin receptor-associated protein OS=Rattus norvegicus GN=Lrpap1 PE=1 SV=2	RAT	3
340	6.94	9.63	12.1	sp|B3GNI6|SEP11_RAT	Septin-11 OS=Rattus norvegicus GN=Sept11 PE=1 SV=1	RAT	5
341	6.92	14.01	26.8	sp|P68511|1433F_RAT	14-3-3 protein eta OS=Rattus norvegicus GN=Ywhah PE=1 SV=2	RAT	11
342	6.89	7.08	1.8	sp|Q9JKS6-2|PCLO_RAT	Isoform 2 of Protein piccolo OS=Rattus norvegicus GN=Pclo	RAT	6
343	6.86	6.97	9.5	sp|Q6Q0N1|CNDP2_RAT	Cytosolic non-specific dipeptidase OS=Rattus norvegicus GN=Cndp2 PE=1 SV=1	RAT	3
344	6.75	6.82	5.4	sp|P97686|NRCAM_RAT	Neuronal cell adhesion molecule OS=Rattus norvegicus GN=Nrcam PE=1 SV=2	RAT	4

N	Unused	Total	% Cov (95)	Accession #	Name	Species	Peptides (95%)
345	6.7	6.87	19.8	sp|Q6MG12|CF136_RAT	Uncharacterized protein C6orf136 homolog OS=Rattus norvegicus PE=2 SV=1	RAT	4
346	6.68	6.75	12.8	sp|P04182|OAT_RAT	Ornithine aminotransferase, mitochondrial OS=Rattus norvegicus GN=Oat PE=1 SV=1	RAT	4
347	6.66	6.85	3.7	sp|Q62638|GSLG1_RAT	Golgi apparatus protein 1 OS=Rattus norvegicus GN=Glg1 PE=1 SV=1	RAT	3
348	6.66	6.81	5.8	sp|Q924S5|LONM_RAT	Lon protease homolog, mitochondrial OS=Rattus norvegicus GN=Lonp1 PE=2 SV=1	RAT	4
349	6.62	6.79	14.8	sp|P62828|RAN_RAT	GTP-binding nuclear protein Ran OS=Rattus norvegicus GN=Ran PE=1 SV=3	RAT	4
350	6.62	6.71	4.2	sp|Q9QXQ0|ACTN4_RAT	Alpha-actinin-4 OS=Rattus norvegicus GN=Actn4 PE=1 SV=2	RAT	3
351	6.6	6.77	15.7	sp|Q6AYN4|PHIPL_RAT	Phytanoyl-CoA hydroxylase-interacting protein-like OS=Rattus norvegicus GN=Phyhipl PE=2 SV=2	RAT	4
352	6.58	6.76	12.3	sp|P51635|AK1A1_RAT	Alcohol dehydrogenase [NADP+] OS=Rattus norvegicus GN=Akr1a1 PE=1 SV=2	RAT	4
353	6.58	6.72	18.8	sp|P08082|CLCB_RAT	Clathrin light chain B OS=Rattus norvegicus GN=Cltb PE=1 SV=1	RAT	5
354	6.58	6.63	4.4	sp|Q9QZA2|PDC6I_RAT	Programmed cell death 6-interacting protein OS=Rattus norvegicus GN=Pdcd6ip PE=1 SV=2	RAT	3
355	6.56	6.71	14.7	sp|Q6AY84|SCRN1_RAT	Secernin-1 OS=Rattus norvegicus GN=Scrn1 PE=1 SV=1	RAT	4
356	6.55	23.84	37.3	sp|Q05962|ADT1_RAT	ADP/ATP translocase 1 OS=Rattus norvegicus GN=Slc25a4 PE=1 SV=3	RAT	26
357	6.54	6.6	11.7	sp|Q641Y2|NDUS2_RAT	NADH dehydrogenase [ubiquinone] iron-sulfur protein 2, mitochondrial OS=Rattus norvegicus GN=Ndufs2 PE=1 SV=1	RAT	8

(continued)

Table 1
(continued)

<table>
<tr><th>N</th><th>Unused</th><th>Total</th><th>% Cov (95)</th><th>Accession #</th><th>Name</th><th>Species</th><th>Peptides (95%)</th></tr>
<tr><td>358</td><td>6.54</td><td>6.58</td><td>23.3</td><td>sp|O88767|PARK7_RAT</td><td>Protein DJ-1 OS=Rattus norvegicus GN=Park7 PE=1 SV=1</td><td>RAT</td><td>5</td></tr>
<tr><td>359</td><td>6.53</td><td>6.58</td><td>4.5</td><td>sp|P70627|FOLH1_RAT</td><td>Glutamate carboxypeptidase 2 OS=Rattus norvegicus GN=Folh1 PE=2 SV=1</td><td>RAT</td><td>3</td></tr>
<tr><td>360</td><td>6.42</td><td>6.47</td><td>5.6</td><td>sp|P02770|ALBU_RAT</td><td>Serum albumin OS=Rattus norvegicus GN=Alb PE=1 SV=2</td><td>RAT</td><td>3</td></tr>
<tr><td>361</td><td>6.41</td><td>6.46</td><td>10.7</td><td>sp|P47971|NPTX1_RAT</td><td>Neuronal pentraxin-1 OS=Rattus norvegicus GN=Nptx1 PE=1 SV=1</td><td>RAT</td><td>4</td></tr>
<tr><td>362</td><td>6.4</td><td>6.51</td><td>8.2</td><td>sp|P25122|KCNC1_RAT</td><td>Potassium voltage-gated channel subfamily C member 1 OS=Rattus norvegicus GN=Kcnc1 PE=2 SV=1</td><td>RAT</td><td>3</td></tr>
<tr><td>363</td><td>6.4</td><td>6.45</td><td>17.6</td><td>sp|P20070|NB5R3_RAT</td><td>NADH-cytochrome b5 reductase 3 OS=Rattus norvegicus GN=Cyb5r3 PE=1 SV=2</td><td>RAT</td><td>5</td></tr>
<tr><td>364</td><td>6.39</td><td>6.42</td><td>13.1</td><td>sp|P48303|S1PR1_RAT</td><td>Sphingosine 1-phosphate receptor 1 OS=Rattus norvegicus GN=S1pr1 PE=2 SV=1</td><td>RAT</td><td>5</td></tr>
<tr><td>365</td><td>6.37</td><td>6.41</td><td>16.5</td><td>sp|P12368|KAP2_RAT</td><td>cAMP-dependent protein kinase type II-alpha regulatory subunit OS=Rattus norvegicus GN=Prkar2a PE=1 SV=3</td><td>RAT</td><td>4</td></tr>
<tr><td>366</td><td>6.33</td><td>6.37</td><td>7.5</td><td>sp|Q6MG55|BAT5_RAT</td><td>Protein BAT5 OS=Rattus norvegicus GN=Bat5 PE=2 SV=1</td><td>RAT</td><td>5</td></tr>
<tr><td>367</td><td>6.3</td><td>6.41</td><td>3</td><td>sp|Q63372-6|NRX1A_RAT</td><td>Isoform Alpha-1F of Neurexin-1-alpha OS=Rattus norvegicus GN=Nrxn1</td><td>RAT</td><td>3</td></tr>
<tr><td>368</td><td>6.27</td><td>6.3</td><td>9.2</td><td>sp|Q6P799|SYSC_RAT</td><td>Seryl-tRNA synthetase, cytoplasmic OS=Rattus norvegicus GN=Sars PE=1 SV=3</td><td>RAT</td><td>3</td></tr>
<tr><td>369</td><td>6.25</td><td>8.5</td><td>9.4</td><td>sp|Q9WVC0|SEPT7_RAT</td><td>Septin-7 OS=Rattus norvegicus GN=Sept7 PE=1 SV=1</td><td>RAT</td><td>4</td></tr>
</table>

N	Unused	Total	% Cov (95)	Accession #	Name	Species	Peptides (95%)
370	6.25	6.28	6.1	sp|Q7TPB1|TCPD_RAT	T-complex protein 1 subunit delta OS=Rattus norvegicus GN=Cct4 PE=1 SV=3	RAT	3
371	6.22	6.34	7.9	sp|Q68FP1|GELS_RAT	Gelsolin OS=Rattus norvegicus GN=Gsn PE=1 SV=1	RAT	4
372	6.22	6.25	8	sp|B0BND0|ENPP6_RAT	Ectonucleotide pyrophosphatase/phosphodiesterase family member 6 OS=Rattus norvegicus GN=Enpp6 PE=2 SV=1	RAT	3
373	6.22	6.24	5.3	sp|Q9WU82|CTNB1_RAT	Catenin beta-1 OS=Rattus norvegicus GN=Ctnnb1 PE=1 SV=1	RAT	3
374	6.17	6.19	29.5	sp|Q6PEC4|SKP1_RAT	S-phase kinase-associated protein 1 OS=Rattus norvegicus GN=Skp1 PE=2 SV=3	RAT	3
375	6.16	6.18	4.3	sp|Q01728-4|NAC1_RAT	Isoform SNCX1 of Sodium/calcium exchanger 1 OS=Rattus norvegicus GN=Slc8a1	RAT	3
376	6.12	16.35	35.8	sp|P35213|1433B_RAT	14-3-3 protein beta/alpha OS=Rattus norvegicus GN=Ywhab PE=1 SV=3	RAT	11
377	6.12	6.14	12.9	sp|Q63377|AT1B3_RAT	Sodium/potassium-transporting ATPase subunit beta-3 OS=Rattus norvegicus GN=Atp1b3 PE=2 SV=1	RAT	3
378	6.11	6.24	10.1	sp|Q6P502|TCPG_RAT	T-complex protein 1 subunit gamma OS=Rattus norvegicus GN=Cct3 PE=1 SV=1	RAT	5
379	6.11	6.12	21.2	sp|Q62636|RAP1B_RAT	Ras-related protein Rap-1b OS=Rattus norvegicus GN=Rap1b PE=2 SV=2	RAT	3
380	6.1	6.23	7.4	sp|P10760|SAHH_RAT	Adenosylhomocysteinase OS=Rattus norvegicus GN=Ahcy PE=1 SV=3	RAT	3
381	6.1	6.12	9.7	sp|Q9QZM5|ABI1_RAT	Abl interactor 1 OS=Rattus norvegicus GN=Abi1 PE=2 SV=3	RAT	3
382	6.09	6.23	7.8	sp|P10960|SAP_RAT	Sulfated glycoprotein 1 OS=Rattus norvegicus GN=Psap PE=1 SV=1	RAT	4

(continued)

Table 1 (continued)

<table>
<tr><th>N</th><th>Unused</th><th>Total</th><th>% Cov (95)</th><th>Accession #</th><th>Name</th><th>Species</th><th>Peptides (95%)</th></tr>
<tr><td>383</td><td>6.08</td><td>6.17</td><td>10.9</td><td>sp|O35077|GPDA_RAT</td><td>Glycerol-3-phosphate dehydrogenase [NAD+], cytoplasmic OS=Rattus norvegicus GN=Gpd1 PE=1 SV=4</td><td>RAT</td><td>3</td></tr>
<tr><td>384</td><td>6.07</td><td>6.13</td><td>4.5</td><td>sp|Q07803|EFGM_RAT</td><td>Elongation factor G, mitochondrial OS=Rattus norvegicus GN=Gfm1 PE=1 SV=2</td><td>RAT</td><td>3</td></tr>
<tr><td>385</td><td>6.07</td><td>6.08</td><td>10.8</td><td>sp|P07825|SYPH_RAT</td><td>Synaptophysin OS=Rattus norvegicus GN=Syp PE=1 SV=1</td><td>RAT</td><td>4</td></tr>
<tr><td>386</td><td>6.06</td><td>6.17</td><td>21.5</td><td>sp|P62161|CALM_RAT</td><td>Calmodulin OS=Rattus norvegicus GN=Calm1 PE=1 SV=2</td><td>RAT</td><td>4</td></tr>
<tr><td>387</td><td>6.05</td><td>6.06</td><td>17.9</td><td>sp|O35179|SH3G2_RAT</td><td>Endophilin-A1 OS=Rattus norvegicus GN=Sh3gl2 PE=1 SV=2</td><td>RAT</td><td>5</td></tr>
<tr><td>388</td><td>6.04</td><td>6.04</td><td>9.4</td><td>sp|Q62847-2|ADDG_RAT</td><td>Isoform Short of Gamma-adducin OS=Rattus norvegicus GN=Add3</td><td>RAT</td><td>3</td></tr>
<tr><td>389</td><td>6.02</td><td>6.03</td><td>10.8</td><td>sp|Q4V7C7|ARP3_RAT</td><td>Actin-related protein 3 OS=Rattus norvegicus GN=Actr3 PE=1 SV=1</td><td>RAT</td><td>3</td></tr>
<tr><td>390</td><td>6.02</td><td>6.02</td><td>5.6</td><td>sp|O88339|EPN1_RAT</td><td>Epsin-1 OS=Rattus norvegicus GN=Epn1 PE=1 SV=1</td><td>RAT</td><td>3</td></tr>
<tr><td>391</td><td>6.01</td><td>6.01</td><td>14.4</td><td>sp|P63004|LIS1_RAT</td><td>Platelet-activating factor acetylhydrolase IB subunit alpha OS=Rattus norvegicus GN=Pafah1b1 PE=1 SV=2</td><td>RAT</td><td>4</td></tr>
<tr><td>392</td><td>6.01</td><td>6.01</td><td>14.3</td><td>sp|P41565|IDH3G_RAT</td><td>Isocitrate dehydrogenase [NAD] subunit gamma, mitochondrial OS=Rattus norvegicus GN=Idh3g PE=2 SV=2</td><td>RAT</td><td>5</td></tr>
<tr><td>393</td><td>6</td><td>8.01</td><td>28.5</td><td>sp|P35281|RAB10_RAT</td><td>Ras-related protein Rab-10 OS=Rattus norvegicus GN=Rab10 PE=2 SV=1</td><td>RAT</td><td>4</td></tr>
<tr><td>394</td><td>6</td><td>6.01</td><td>2.6</td><td>sp|Q07310|NRX3A_RAT</td><td>Neurexin-3-alpha OS=Rattus norvegicus GN=Nrxn3 PE=1 SV=1</td><td>RAT</td><td>3</td></tr>
</table>

N	Unused	Total	% Cov (95)	Accession #	Name	Species	Peptides (95%)
395	6	6.01	11.3	sp|Q1WIM2|CADM2_RAT	Cell adhesion molecule 2 OS=Rattus norvegicus GN=Cadm2 PE=1 SV=2	RAT	3
396	6	6	2	sp|Q64605|PTPRS_RAT	Receptor-type tyrosine-protein phosphatase S OS=Rattus norvegicus GN=Ptprs PE=1 SV=2	RAT	3
397	6	6	1.5	sp|Q9ERB4-3|CSPG2_RAT	Isoform Vint of Versican core protein OS=Rattus norvegicus GN=Vcan	RAT	3
398	6	6	11.5	sp|Q9Z2X5|HOME3_RAT	Homer protein homolog 3 OS=Rattus norvegicus GN=Homer3 PE=2 SV=2	RAT	3
399	6	6	8.3	sp|Q9JM53|AIFM1_RAT	Apoptosis-inducing factor 1, mitochondrial OS=Rattus norvegicus GN=Aifm1 PE=1 SV=1	RAT	3
400	6	6	4.5	sp|Q6Q0N0|CSTN1_RAT	Calsyntenin-1 OS=Rattus norvegicus GN=Clstn1 PE=2 SV=1	RAT	3
401	6	6	14.3	sp|P29266|3HIDH_RAT	3-hydroxyisobutyrate dehydrogenase, mitochondrial OS=Rattus norvegicus GN=Hibadh PE=1 SV=3	RAT	3
402	6	6	15	sp|P40241|CD9_RAT	CD9 antigen OS=Rattus norvegicus GN=Cd9 PE=1 SV=2	RAT	3
403	6	6	4.3	sp|P27653|C1TC_RAT	C-1-tetrahydrofolate synthase, cytoplasmic OS=Rattus norvegicus GN=Mthfd1 PE=1 SV=3	RAT	3
404	6	6	18.1	sp|P24368|PPIB_RAT	Peptidyl-prolyl cis-trans isomerase B OS=Rattus norvegicus GN=Ppib PE=2 SV=3	RAT	3
405	6	6	4.4	sp|Q8VII6|CTL1_RAT	Choline transporter-like protein 1 OS=Rattus norvegicus GN=Slc44a1 PE=1 SV=1	RAT	3
406	6	6	9.5	sp|Q6DGG0|PPID_RAT	Peptidyl-prolyl cis-trans isomerase D OS=Rattus norvegicus GN=Ppid PE=1 SV=3	RAT	3
407	6	6	15.2	sp|Q5XI73|GDIR1_RAT	Rho GDP-dissociation inhibitor 1 OS=Rattus norvegicus GN=Arhgdia PE=1 SV=1	RAT	3

(continued)

Table 1 (continued)

N	Unused	Total	% Cov (95)	Accession #	Name	Species	Peptides (95%)
408	6	6	24.6	sp|Q05175|BASP1_RAT	Brain acid soluble protein 1 OS=Rattus norvegicus GN=Basp1 PE=1 SV=2	RAT	3
409	6	6	10.4	sp|P60892|PRPS1_RAT	Ribose-phosphate pyrophosphokinase 1 OS=Rattus norvegicus GN=Prps1 PE=1 SV=2	RAT	3
410	6	6	28.2	sp|P37377-2|SYUA_RAT	Isoform Syn2 of Alpha-synuclein OS=Rattus norvegicus GN=Snca	RAT	4
411	6	6	12.5	sp|P85973|PNPH_RAT	Purine nucleoside phosphorylase OS=Rattus norvegicus GN=Pnp PE=1 SV=1	RAT	3
412	6	6	8.4	sp|O35567|PUR9_RAT	Bifunctional purine biosynthesis protein PURH OS=Rattus norvegicus GN=Atic PE=1 SV=2	RAT	3
413	6	6	32.2	sp|P30904|MIF_RAT	Macrophage migration inhibitory factor OS=Rattus norvegicus GN=Mif PE=1 SV=4	RAT	4
414	6	6	17.1	sp|Q9EPC6|PROF2_RAT	Profilin-2 OS=Rattus norvegicus GN=Pfn2 PE=1 SV=3	RAT	3
415	6	6	8	sp|Q64057|AL7A1_RAT	Alpha-aminoadipic semialdehyde dehydrogenase OS=Rattus norvegicus GN=Aldh7a1 PE=1 SV=2	RAT	4
416	6	6	13.1	sp|Q62703|RCN2_RAT	Reticulocalbin-2 OS=Rattus norvegicus GN=Rcn2 PE=1 SV=2	RAT	3
417	6	6	13.8	sp|P62747|RHOB_RAT	Rho-related GTP-binding protein RhoB OS=Rattus norvegicus GN=Rhob PE=1 SV=1	RAT	3
418	6	6	15.1	sp|P56536|KIF5C_RAT	Kinesin heavy chain isoform 5C (Fragment) OS=Rattus norvegicus GN=Kif5c PE=1 SV=2	RAT	4
419	6	6	26.7	sp|P11240|COX5A_RAT	Cytochrome c oxidase subunit 5A, mitochondrial OS=Rattus norvegicus GN=Cox5a PE=1 SV=1	RAT	7

N	Unused	Total	% Cov (95)	Accession #	Name	Species	Peptides (95%)
420	6	6	12.1	sp|P00787|CATB_RAT	Cathepsin B OS=Rattus norvegicus GN=Ctsb PE=1 SV=2	RAT	4
421	6	6	9.9	sp|O35458|VIAAT_RAT	Vesicular inhibitory amino acid transporter OS=Rattus norvegicus GN=Slc32a1 PE=1 SV=1	RAT	3
422	6	6	9.9	sp|P97544|LPP3_RAT	Lipid phosphate phosphohydrolase 3 OS=Rattus norvegicus GN=Ppap2b PE=1 SV=1	RAT	3
423	6	6	14.9	sp|P01830|THY1_RAT	Thy-1 membrane glycoprotein OS=Rattus norvegicus GN=Thy1 PE=1 SV=1	RAT	4
424	6	6	30.9	sp|Q9WVA1|TIM8A_RAT	Mitochondrial import inner membrane translocase subunit Tim8 A OS=Rattus norvegicus GN=Timm8a PE=1 SV=1	RAT	3
425	5.93	6.02	18.8	sp|P61206|ARF3_RAT	ADP-ribosylation factor 3 OS=Rattus norvegicus GN=Arf3 PE=1 SV=2	RAT	3
426	5.92	6	4.4	sp|P31421|GRM2_RAT	Metabotropic glutamate receptor 2 OS=Rattus norvegicus GN=Grm2 PE=1 SV=1	RAT	3
427	5.85	6.07	14.3	sp|P62994|GRB2_RAT	Growth factor receptor-bound protein 2 OS=Rattus norvegicus GN=Grb2 PE=1 SV=1	RAT	3
428	5.85	6	4.8	sp|P84903|STIM1_RAT	Stromal interaction molecule 1 OS=Rattus norvegicus GN=Stim1 PE=1 SV=1	RAT	3
429	5.83	6	14.8	sp|P62494|RB11A_RAT	Ras-related protein Rab-11A OS=Rattus norvegicus GN=Rab11a PE=1 SV=3	RAT	3
430	5.71	8.41	15.7	sp|P04642|LDHA_RAT	L-lactate dehydrogenase A chain OS=Rattus norvegicus GN=Ldha PE=1 SV=1	RAT	5
431	5.52	5.7	7.9	sp|P30191|GBRA6_RAT	Gamma-aminobutyric acid receptor subunit alpha-6 OS=Rattus norvegicus GN=Gabra6 PE=2 SV=1	RAT	3
432	5.49	5.79	11.3	sp|Q62718|NTRI_RAT	Neurotrimin OS=Rattus norvegicus GN=Ntm PE=1 SV=1	RAT	4

(continued)

Table 1
(continued)

<table>
<tr><th>N</th><th>Unused</th><th>Total</th><th>% Cov (95)</th><th>Accession #</th><th>Name</th><th>Species</th><th>Peptides (95%)</th></tr>
<tr><td>433</td><td>5.49</td><td>5.62</td><td>10</td><td>sp|P15651|ACADS_RAT</td><td>Short-chain specific acyl-CoA dehydrogenase, mitochondrial OS=Rattus norvegicus GN=Acads PE=1 SV=2</td><td>RAT</td><td>3</td></tr>
<tr><td>434</td><td>5.45</td><td>5.74</td><td>9.7</td><td>sp|P14668|ANXA5_RAT</td><td>Annexin A5 OS=Rattus norvegicus GN=Anxa5 PE=1 SV=3</td><td>RAT</td><td>3</td></tr>
<tr><td>435</td><td>5.28</td><td>5.43</td><td>4.8</td><td>sp|Q62888|NLGN2_RAT</td><td>Neuroligin-2 OS=Rattus norvegicus GN=Nlgn2 PE=1 SV=1</td><td>RAT</td><td>3</td></tr>
<tr><td>436</td><td>5.27</td><td>5.52</td><td>9.5</td><td>sp|Q9Z2F5|CTBP1_RAT</td><td>C-terminal-binding protein 1 OS=Rattus norvegicus GN=Ctbp1 PE=1 SV=3</td><td>RAT</td><td>3</td></tr>
<tr><td>437</td><td>5.26</td><td>5.57</td><td>5.6</td><td>sp|Q63531|KS6A1_RAT</td><td>Ribosomal protein S6 kinase alpha-1 OS=Rattus norvegicus GN=Rps6ka1 PE=2 SV=1</td><td>RAT</td><td>3</td></tr>
<tr><td>438</td><td>5.24</td><td>5.36</td><td>10.2</td><td>sp|Q5EB81|NB5R1_RAT</td><td>NADH-cytochrome b5 reductase 1 OS=Rattus norvegicus GN=Cyb5r1 PE=2 SV=1</td><td>RAT</td><td>3</td></tr>
<tr><td>439</td><td>5.21</td><td>5.33</td><td>10.7</td><td>sp|P21533|RL6_RAT</td><td>60S ribosomal protein L6 OS=Rattus norvegicus GN=Rpl6 PE=1 SV=5</td><td>RAT</td><td>3</td></tr>
<tr><td>440</td><td>5.18</td><td>5.38</td><td>9.1</td><td>sp|P97829-2|CD47_RAT</td><td>Isoform 2 of Leukocyte surface antigen CD47 OS=Rattus norvegicus GN=Cd47</td><td>RAT</td><td>3</td></tr>
<tr><td>441</td><td>5.11</td><td>5.21</td><td>10.5</td><td>sp|P47863|AQP4_RAT</td><td>Aquaporin-4 OS=Rattus norvegicus GN=Aqp4 PE=1 SV=1</td><td>RAT</td><td>3</td></tr>
<tr><td>442</td><td>5.11</td><td>5.2</td><td>13.1</td><td>sp|Q6AYR2|NDRG3_RAT</td><td>Protein NDRG3 OS=Rattus norvegicus GN=Ndrg3 PE=2 SV=1</td><td>RAT</td><td>3</td></tr>
<tr><td>443</td><td>5.09</td><td>5.19</td><td>4</td><td>sp|Q5XHY5|SYTC_RAT</td><td>Threonyl-tRNA synthetase, cytoplasmic OS=Rattus norvegicus GN=Tars PE=2 SV=1</td><td>RAT</td><td>3</td></tr>
<tr><td>444</td><td>5.09</td><td>5.19</td><td>16.5</td><td>sp|Q566R0|THEM4_RAT</td><td>Thioesterase superfamily member 4 OS=Rattus norvegicus GN=Them4 PE=2 SV=1</td><td>RAT</td><td>3</td></tr>
</table>

<table>
<tr><th>N</th><th>Unused</th><th>Total</th><th>% Cov (95)</th><th>Accession #</th><th>Name</th><th>Species</th><th>Peptides (95%)</th></tr>
<tr><td>445</td><td>5.06</td><td>5.39</td><td>11.1</td><td>sp|P16446|PIPNA_RAT</td><td>Phosphatidylinositol transfer protein alpha isoform OS=Rattus norvegicus GN=Pitpna PE=1 SV=2</td><td>RAT</td><td>3</td></tr>
<tr><td>446</td><td>5.06</td><td>5.16</td><td>5.3</td><td>sp|P52296|IMB1_RAT</td><td>Importin subunit beta-1 OS=Rattus norvegicus GN=Kpnb1 PE=1 SV=1</td><td>RAT</td><td>3</td></tr>
<tr><td>447</td><td>5.05</td><td>6.77</td><td>12.7</td><td>sp|Q568Z9|PHYIP_RAT</td><td>Phytanoyl-CoA hydroxylase-interacting protein OS=Rattus norvegicus GN=Phyhip PE=2 SV=1</td><td>RAT</td><td>4</td></tr>
<tr><td>448</td><td>4.99</td><td>6.15</td><td>24.7</td><td>sp|Q9Z269|VAPB_RAT</td><td>Vesicle-associated membrane protein-associated protein B OS=Rattus norvegicus GN=Vapb PE=1 SV=3</td><td>RAT</td><td>4</td></tr>
<tr><td>449</td><td>4.99</td><td>5.09</td><td>4.4</td><td>sp|Q9WU70|STXB5_RAT</td><td>Syntaxin-binding protein 5 OS=Rattus norvegicus GN=Stxbp5 PE=1 SV=1</td><td>RAT</td><td>3</td></tr>
<tr><td>450</td><td>4.98</td><td>5.14</td><td>16.1</td><td>sp|Q4QQV3|F162A_RAT</td><td>Protein FAM162A OS=Rattus norvegicus GN=Fam162a PE=2 SV=1</td><td>RAT</td><td>3</td></tr>
<tr><td>451</td><td>4.96</td><td>5.06</td><td>17.3</td><td>sp|Q63610|TPM3_RAT</td><td>Tropomyosin alpha-3 chain OS=Rattus norvegicus GN=Tpm3 PE=1 SV=2</td><td>RAT</td><td>3</td></tr>
<tr><td>452</td><td>4.96</td><td>5.06</td><td>12.1</td><td>sp|P67779|PHB_RAT</td><td>Prohibitin OS=Rattus norvegicus GN=Phb PE=1 SV=1</td><td>RAT</td><td>4</td></tr>
<tr><td>453</td><td>4.91</td><td>5.17</td><td>7.8</td><td>sp|Q91Y81|SEPT2_RAT</td><td>Septin-2 OS=Rattus norvegicus GN=Sept2 PE=1 SV=1</td><td>RAT</td><td>3</td></tr>
<tr><td>454</td><td>4.87</td><td>5.27</td><td>11.9</td><td>sp|Q6JE36|NDRG1_RAT</td><td>Protein NDRG1 OS=Rattus norvegicus GN=Ndrg1 PE=1 SV=1</td><td>RAT</td><td>4</td></tr>
<tr><td>455</td><td>4.86</td><td>4.96</td><td>12.4</td><td>sp|Q6AYR6|HDHD2_RAT</td><td>Haloacid dehalogenase-like hydrolase domain-containing protein 2 OS=Rattus norvegicus GN=Hdhd2 PE=2 SV=1</td><td>RAT</td><td>3</td></tr>
<tr><td>456</td><td>4.83</td><td>4.92</td><td>6.9</td><td>sp|Q9JHZ9|S38A3_RAT</td><td>Sodium-coupled neutral amino acid transporter 3 OS=Rattus norvegicus GN=Slc38a3 PE=2 SV=1</td><td>RAT</td><td>3</td></tr>
</table>

(continued)

Table 1
(continued)

N	Unused	Total	% Cov (95)	Accession #	Name	Species	Peptides (95%)
457	4.81	4.9	5.2	sp|P05508|NU4M_RAT	NADH-ubiquinone oxidoreductase chain 4 OS=Rattus norvegicus GN=Mtnd4 PE=2 SV=2	RAT	5
458	4.78	5.08	36.6	sp|Q7TQ16|QCR8_RAT	Cytochrome b-c1 complex subunit 8 OS=Rattus norvegicus GN=Uqcrq PE=2 SV=1	RAT	4
459	4.78	4.88	19.1	sp|P63159|HMGB1_RAT	High mobility group protein B1 OS=Rattus norvegicus GN=Hmgb1 PE=1 SV=2	RAT	3
460	4.73	4.83	20.1	sp|Q3T1J1|IF5A1_RAT	Eukaryotic translation initiation factor 5A-1 OS=Rattus norvegicus GN=Eif5a PE=1 SV=3	RAT	4
461	4.7	4.79	10.3	sp|P52481|CAP2_RAT	Adenylyl cyclase-associated protein 2 OS=Rattus norvegicus GN=Cap2 PE=2 SV=1	RAT	3
462	4.67	5.73	12.9	sp|P08081|CLCA_RAT	Clathrin light chain A OS=Rattus norvegicus GN=Clta PE=1 SV=1	RAT	4
463	4.64	4.9	3.3	sp|P28023|DCTN1_RAT	Dynactin subunit 1 OS=Rattus norvegicus GN=Dctn1 PE=2 SV=2	RAT	3
464	4.64	4.72	17.4	sp|Q4KM73|KCY_RAT	UMP-CMP kinase OS=Rattus norvegicus GN=Cmpk1 PE=1 SV=2	RAT	3
465	4.63	4.7	4.6	sp|Q02563|SV2A_RAT	Synaptic vesicle glycoprotein 2A OS=Rattus norvegicus GN=Sv2a PE=1 SV=2	RAT	3
466	4.61	4.66	14.2	sp|Q5FVI3|LRC57_RAT	Leucine-rich repeat-containing protein 57 OS=Rattus norvegicus GN=Lrrc57 PE=2 SV=1	RAT	3
467	4.58	4.62	2.2	sp|P54282|CAC1A_RAT	Voltage-dependent P/Q-type calcium channel subunit alpha-1A OS=Rattus norvegicus GN=Cacna1a PE=1 SV=1	RAT	3
468	4.56	4.77	18.1	sp|P24473|GSTK1_RAT	Glutathione S-transferase kappa 1 OS=Rattus norvegicus GN=Gstk1 PE=1 SV=3	RAT	3
469	4.5	4.54	16.4	sp|Q99MZ8|LASP1_RAT	LIM and SH3 domain protein 1 OS=Rattus norvegicus GN=Lasp1 PE=1 SV=1	RAT	2

N	Unused	Total	% Cov (95)	Accession #	Name	Species	Peptides (95%)
470	4.48	4.92	5.6	sp|P41562|IDHC_RAT	Isocitrate dehydrogenase [NADP] cytoplasmic OS=Rattus norvegicus GN=Idh1 PE=1 SV=1	RAT	3
471	4.47	4.56	15	sp|Q8CG45|ARK72_RAT	Aflatoxin B1 aldehyde reductase member 2 OS=Rattus norvegicus GN=Akr7a2 PE=1 SV=2	RAT	4
472	4.47	4.51	9.1	sp|O35952|GLO2_RAT	Hydroxyacylglutathione hydrolase, mitochondrial OS=Rattus norvegicus GN=Hagh PE=1 SV=2	RAT	2
473	4.46	4.5	10.7	sp|Q9Z1N4|BPNT1_RAT	3'(2'),5'-bisphosphate nucleotidase 1 OS=Rattus norvegicus GN=Bpnt1 PE=1 SV=1	RAT	3
474	4.45	4.49	44.8	sp|Q9JJW3|USMG5_RAT	Up-regulated during skeletal muscle growth protein 5 OS=Rattus norvegicus GN=Usmg5 PE=2 SV=1	RAT	2
475	4.42	4.55	4.1	sp|Q66HA8|HS105_RAT	Heat shock protein 105 kDa OS=Rattus norvegicus GN=Hsph1 PE=1 SV=1	RAT	3
476	4.42	4.46	6.3	sp|Q5RKI0|WDR1_RAT	WD repeat-containing protein 1 OS=Rattus norvegicus GN=Wdr1 PE=1 SV=3	RAT	2
477	4.41	4.45	4.8	sp|P32232|CBS_RAT	Cystathionine beta-synthase OS=Rattus norvegicus GN=Cbs PE=1 SV=3	RAT	2
478	4.4	4.83	8.6	sp|Q6AYS7|ACY1A_RAT	Aminoacylase-1A OS=Rattus norvegicus GN=Acy1a PE=1 SV=1	RAT	3
479	4.36	4.51	23.8	sp|P22062|PIMT_RAT	Protein-L-isoaspartate(D-aspartate) O-methyltransferase OS=Rattus norvegicus GN=Pcmt1 PE=1 SV=2	RAT	7
480	4.36	4.4	21.3	sp|P10888|COX41_RAT	Cytochrome c oxidase subunit 4 isoform 1, mitochondrial OS=Rattus norvegicus GN=Cox4i1 PE=1 SV=1	RAT	3
481	4.34	4.39	8.1	sp|P13437|THIM_RAT	3-ketoacyl-CoA thiolase, mitochondrial OS=Rattus norvegicus GN=Acaa2 PE=1 SV=1	RAT	2

(continued)

Table 1
(continued)

N	Unused	Total	% Cov (95)	Accession #	Name	Species	Peptides (95%)
482	4.32	4.37	9.8	sp|P29457|SERPH_RAT	Serpin H1 OS=Rattus norvegicus GN=Serpinh1 PE=1 SV=1	RAT	3
483	4.3	4.34	5.2	sp|P21588|5NTD_RAT	5'-nucleotidase OS=Rattus norvegicus GN=Nt5e PE=1 SV=1	RAT	2
484	4.25	4.28	11	sp|Q5U2X7|TI21L_RAT	TIM21-like protein, mitochondrial OS=Rattus norvegicus PE=2 SV=1	RAT	2
485	4.19	4.39	28	sp|Q03344|ATIF1_RAT	ATPase inhibitor, mitochondrial OS=Rattus norvegicus GN=Atpif1 PE=2 SV=2	RAT	4
486	4.19	4.22	6.4	sp|P0C5X8|TTYH1_RAT	Protein tweety homolog 1 OS=Rattus norvegicus GN=Ttyh1 PE=2 SV=1	RAT	2
487	4.18	4.26	4.5	sp|O35112|CD166_RAT	CD166 antigen OS=Rattus norvegicus GN=Alcam PE=1 SV=1	RAT	2
488	4.17	4.33	10.8	sp|P70645|BLMH_RAT	Bleomycin hydrolase OS=Rattus norvegicus GN=Blmh PE=1 SV=1	RAT	3
489	4.14	10.3	16.3	sp|P04897|GNAI2_RAT	Guanine nucleotide-binding protein G(i) subunit alpha-2 OS=Rattus norvegicus GN=Gnai2 PE=1 SV=3	RAT	6
490	4.13	4.15	8.6	sp|Q8R431|MGLL_RAT	Monoglyceride lipase OS=Rattus norvegicus GN=Mgll PE=1 SV=1	RAT	2
491	4.12	4.14	4.8	sp|P97610|SYT12_RAT	Synaptotagmin-12 OS=Rattus norvegicus GN=Syt12 PE=2 SV=1	RAT	2
492	4.12	4.14	6.9	sp|P05371|CLUS_RAT	Clusterin OS=Rattus norvegicus GN=Clu PE=1 SV=2	RAT	2
493	4.11	4.13	7.1	sp|P62425|RL7A_RAT	60S ribosomal protein L7a OS=Rattus norvegicus GN=Rpl7a PE=1 SV=2	RAT	2

N	Unused	Total	% Cov (95)	Accession #	Name	Species	Peptides (95%)
494	4.09	4.1	11.2	sp|Q63768|CRK_RAT	Adapter molecule crk OS=Rattus norvegicus GN=Crk PE=1 SV=1	RAT	2
495	4.08	4.25	14.7	sp|Q68FT1|COQ9_RAT	Ubiquinone biosynthesis protein COQ9, mitochondrial OS=Rattus norvegicus GN=Coq9 PE=2 SV=2	RAT	4
496	4.08	4.22	5.4	sp|B2GV54|NCEH1_RAT	Neutral cholesterol ester hydrolase 1 OS=Rattus norvegicus GN=Nceh1 PE=2 SV=1	RAT	2
497	4.07	6.12	17.3	sp|Q8CFN2|CDC42_RAT	Cell division control protein 42 homolog OS=Rattus norvegicus GN=Cdc42 PE=1 SV=2	RAT	5
498	4.07	4.08	4.9	sp|Q02253|MMSA_RAT	Methylmalonate-semialdehyde dehydrogenase [acylating], mitochondrial OS=Rattus norvegicus GN=Aldh6a1 PE=1 SV=1	RAT	2
499	4.06	4.17	6.5	sp|O88377|PI42B_RAT	Phosphatidylinositol-5-phosphate 4-kinase type-2 beta OS=Rattus norvegicus GN=Pip4k2b PE=1 SV=1	RAT	2
500	4.06	4.07	5.9	sp|Q63081|PDIA6_RAT	Protein disulfide-isomerase A6 OS=Rattus norvegicus GN=Pdia6 PE=1 SV=2	RAT	2
501	4.05	33.44	20.2	sp|Q9JI66|S4A4_RAT	Electrogenic sodium bicarbonate cotransporter 1 OS=Rattus norvegicus GN=Slc4a4 PE=1 SV=1	RAT	27
502	4.05	4.08	9.8	sp|Q62651|ECH1_RAT	Delta(3,5)-Delta(2,4)-dienoyl-CoA isomerase, mitochondrial OS=Rattus norvegicus GN=Ech1 PE=1 SV=2	RAT	2
503	4.05	4.06	5	sp|Q66HG5|TM9S2_RAT	Transmembrane 9 superfamily member 2 OS=Rattus norvegicus GN=Tm9sf2 PE=2 SV=1	RAT	3
504	4.04	4.06	6	sp|Q63347|PRS7_RAT	26S protease regulatory subunit 7 OS=Rattus norvegicus GN=Psmc2 PE=1 SV=3	RAT	2

(continued)

Table 1
(continued)

N	Unused	Total	% Cov (95)	Accession #	Name	Species	Peptides (95%)
505	4.04	4.05	4.8	sp|P38652|PGM1_RAT	Phosphoglucomutase-1 OS=Rattus norvegicus GN=Pgm1 PE=1 SV=2	RAT	2
506	4.04	4.05	4.5	sp|Q99MZ4|GGT7_RAT	Gamma-glutamyltransferase 7 OS=Rattus norvegicus GN=Ggt7 PE=2 SV=2	RAT	2
507	4.04	4.05	5.1	sp|O35094|TIM44_RAT	Mitochondrial import inner membrane translocase subunit TIM44 OS=Rattus norvegicus GN=Timm44 PE=2 SV=1	RAT	2
508	4.03	4.06	5.1	sp|O08651|SERA_RAT	D-3-phosphoglycerate dehydrogenase OS=Rattus norvegicus GN=Phgdh PE=1 SV=3	RAT	2
509	4.03	4.04	6.8	sp|Q91XU8|CDS2_RAT	Phosphatidate cytidylyltransferase 2 OS=Rattus norvegicus GN=Cds2 PE=2 SV=1	RAT	2
510	4.02	4.13	3.2	sp|Q924S1|PLCD_RAT	1-acyl-sn-glycerol-3-phosphate acyltransferase delta OS=Rattus norvegicus GN=Agpat4 PE=2 SV=1	RAT	2
511	4.02	4.03	12.2	sp|Q68FU3|ETFB_RAT	Electron transfer flavoprotein subunit beta OS=Rattus norvegicus GN=Etfb PE=2 SV=3	RAT	4
512	4.02	4.03	12.2	sp|Q5XIK2|TMX2_RAT	Thioredoxin-related transmembrane protein 2 OS=Rattus norvegicus GN=Tmx2 PE=2 SV=1	RAT	2
513	4.02	4.02	2	sp|Q9WV48|SHAN1_RAT	SH3 and multiple ankyrin repeat domains protein 1 OS=Rattus norvegicus GN=Shank1 PE=1 SV=1	RAT	2
514	4.02	4.02	3.8	sp|Q7TT47|SPG7_RAT	Paraplegin OS=Rattus norvegicus GN=Spg7 PE=2 SV=1	RAT	2
515	4.01	4.02	2.6	sp|Q9QXY2|SRCN1_RAT	SRC kinase signaling inhibitor 1 OS=Rattus norvegicus GN=Srcin1 PE=1 SV=1	RAT	2
516	4.01	4.01	6.8	sp|P12007|IVD_RAT	Isovaleryl-CoA dehydrogenase, mitochondrial OS=Rattus norvegicus GN=Ivd PE=1 SV=2	RAT	3

N	Unused	Total	% Cov (95)	Accession #	Name	Species	Peptides (95%)
517	4.01	4.01	6.3	sp|Q9Z2S9|FLOT2_RAT	Flotillin-2 OS=Rattus norvegicus GN=Flot2 PE=1 SV=1	RAT	2
518	4.01	4.01	11.3	sp|Q5FVQ4|MLEC_RAT	Malectin OS=Rattus norvegicus GN=Mlec PE=2 SV=1	RAT	2
519	4	28	32.6	sp|Q6P9T8|TBB2C_RAT	Tubulin beta-2C chain OS=Rattus norvegicus GN=Tubb2c PE=1 SV=1	RAT	19
520	4	14	12.7	sp|P11730|KCC2G_RAT	Calcium/calmodulin-dependent protein kinase type II subunit gamma OS=Rattus norvegicus GN=Camk2g PE=2 SV=1	RAT	8
521	4	12	23.6	sp|Q64548-2|RTN1_RAT	Isoform S-RexS of Reticulon-1 OS=Rattus norvegicus GN=Rtn1	RAT	12
522	4	6.53	11.9	sp|P62824|RAB3C_RAT	Ras-related protein Rab-3C OS=Rattus norvegicus GN=Rab3c PE=1 SV=1	RAT	3
523	4	6.01	18.5	sp|P63322|RALA_RAT	Ras-related protein Ral-A OS=Rattus norvegicus GN=Rala PE=1 SV=1	RAT	4
524	4	6	7.9	sp|Q62952-2|DPYL3_RAT	Isoform TUC-4a of Dihydropyrimidinase-related protein 3 OS=Rattus norvegicus GN=Dpysl3	RAT	3
525	4	6	7	sp|O35921|EAA4_RAT	Excitatory amino acid transporter 4 OS=Rattus norvegicus GN=Slc1a6 PE=2 SV=1	RAT	3
526	4	6	13.4	sp|P10536|RAB1B_RAT	Ras-related protein Rab-1B OS=Rattus norvegicus GN=Rab1b PE=1 SV=1	RAT	3
527	4	5.37	13.2	sp|Q04970|RASN_RAT	GTPase NRas OS=Rattus norvegicus GN=Nras PE=2 SV=1	RAT	3
528	4	4.02	4.9	sp|P32851|STX1A_RAT	Syntaxin-1A OS=Rattus norvegicus GN=Stx1a PE=1 SV=1	RAT	2
529	4	4.01	6.7	sp|P21531|RL3_RAT	60S ribosomal protein L3 OS=Rattus norvegicus GN=Rpl3 PE=1 SV=3	RAT	2

(continued)

Table 1 (continued)

N	Unused	Total	% Cov (95)	Accession #	Name	Species	Peptides (95%)
530	4	4.01	14.3	sp|P62332|ARF6_RAT	ADP-ribosylation factor 6 OS=Rattus norvegicus GN=Arf6 PE=1 SV=2	RAT	2
531	4	4.01	14	sp|P63029|TCTP_RAT	Translationally-controlled tumor protein OS=Rattus norvegicus GN=Tpt1 PE=1 SV=1	RAT	2
532	4	4	4.3	sp|P97849|S27A1_RAT	Long-chain fatty acid transport protein 1 OS=Rattus norvegicus GN=Slc27a1 PE=2 SV=1	RAT	2
533	4	4	3.8	sp|P97852|DHB4_RAT	Peroxisomal multifunctional enzyme type 2 OS=Rattus norvegicus GN=Hsd17b4 PE=1 SV=3	RAT	2
534	4	4	2.3	sp|P22063|CNTN2_RAT	Contactin-2 OS=Rattus norvegicus GN=Cntn2 PE=1 SV=1	RAT	2
535	4	4	3.9	sp|P0C2X9|AL4A1_RAT	Delta-1-pyrroline-5-carboxylate dehydrogenase, mitochondrial OS=Rattus norvegicus GN=Aldh4a1 PE=1 SV=1	RAT	2
536	4	4	6	sp|Q6AYT7|ABD12_RAT	Monoacylglycerol lipase ABHD12 OS=Rattus norvegicus GN=Abhd12 PE=2 SV=1	RAT	2
537	4	4	4.7	sp|Q5BJU7|WASF1_RAT	Wiskott-Aldrich syndrome protein family member 1 OS=Rattus norvegicus GN=Wasf1 PE=2 SV=1	RAT	2
538	4	4	5.7	sp|Q32Q06|AP1M1_RAT	AP-1 complex subunit mu-1 OS=Rattus norvegicus GN=Ap1m1 PE=1 SV=3	RAT	2
539	4	4	14	sp|Q6AYK6|CYBP_RAT	Calcyclin-binding protein OS=Rattus norvegicus GN=Cacybp PE=1 SV=1	RAT	2
540	4	4	4.4	sp|Q07266|DREB_RAT	Drebrin OS=Rattus norvegicus GN=Dbn1 PE=2 SV=3	RAT	2
541	4	4	9.3	sp|Q9Z254|GIPC1_RAT	PDZ domain-containing protein GIPC1 OS=Rattus norvegicus GN=Gipc1 PE=1 SV=2	RAT	2

N	Unused	Total	% Cov (95)	Accession #	Name	Species	Peptides (95%)
542	4	4	13.9	sp|Q5XIJ4|CR019_RAT	Uncharacterized protein C18orf19 homolog OS=Rattus norvegicus PE=2 SV=1	RAT	2
543	4	4	9.1	sp|Q5I0D1|GLOD4_RAT	Glyoxalase domain-containing protein 4 OS=Rattus norvegicus GN=Glod4 PE=1 SV=1	RAT	2
544	4	4	7	sp|O70593|SGTA_RAT	Small glutamine-rich tetratricopeptide repeat-containing protein alpha OS=Rattus norvegicus GN=Sgta PE=1 SV=1	RAT	2
545	4	4	10.3	sp|Q9Z0W7|CLIC4_RAT	Chloride intracellular channel protein 4 OS=Rattus norvegicus GN=Clic4 PE=1 SV=3	RAT	2
546	4	4	3.1	sp|Q8K4Y5|LGI1_RAT	Leucine-rich glioma-inactivated protein 1 OS=Rattus norvegicus GN=Lgi1 PE=1 SV=1	RAT	2
547	4	4	9	sp|Q68FX9|SIRT5_RAT	NAD-dependent deacetylase sirtuin-5 OS=Rattus norvegicus GN=Sirt5 PE=2 SV=1	RAT	2
548	4	4	14.9	sp|Q5XI72|IF4H_RAT	Eukaryotic translation initiation factor 4H OS=Rattus norvegicus GN=Eif4h PE=1 SV=1	RAT	2
549	4	4	4.3	sp|Q07647|GTR3_RAT	Solute carrier family 2, facilitated glucose transporter member 3 OS=Rattus norvegicus GN=Slc2a3 PE=2 SV=1	RAT	2
550	4	4	6.7	sp|P10499|KCNA1_RAT	Potassium voltage-gated channel subfamily A member 1 OS=Rattus norvegicus GN=Kcna1 PE=1 SV=1	RAT	2
551	4	4	4.5	sp|O54701|NCKX2_RAT	Sodium/potassium/calcium exchanger 2 OS=Rattus norvegicus GN=Slc24a2 PE=2 SV=1	RAT	2
552	4	4	3.2	sp|O35095|NCDN_RAT	Neurochondrin OS=Rattus norvegicus GN=Ncdn PE=1 SV=2	RAT	2
553	4	4	8.3	sp|Q6J4I0|PPR1B_RAT	Protein phosphatase 1 regulatory subunit 1B OS=Rattus norvegicus GN=Ppp1r1b PE=2 SV=1	RAT	3

(continued)

Table 1
(continued)

<table>
<tr><th>N</th><th>Unused</th><th>Total</th><th>% Cov (95)</th><th>Accession #</th><th>Name</th><th>Species</th><th>Peptides (95%)</th></tr>
<tr><td>554</td><td>4</td><td>4</td><td>10.3</td><td>sp|Q63507|RL14_RAT</td><td>60S ribosomal protein L14 OS=Rattus norvegicus GN=Rpl14 PE=1 SV=3</td><td>RAT</td><td>2</td></tr>
<tr><td>555</td><td>4</td><td>4</td><td>30.3</td><td>sp|Q5M9I5|QCR6_RAT</td><td>Cytochrome b-c1 complex subunit 6, mitochondrial OS=Rattus norvegicus GN=Uqcrh PE=2 SV=1</td><td>RAT</td><td>3</td></tr>
<tr><td>556</td><td>4</td><td>4</td><td>11.7</td><td>sp|Q5GFD9|IMPCT_RAT</td><td>Protein IMPACT OS=Rattus norvegicus GN=Impact PE=2 SV=1</td><td>RAT</td><td>2</td></tr>
<tr><td>557</td><td>4</td><td>4</td><td>11.1</td><td>sp|P63100-2|CANB1_RAT</td><td>Isoform 2 of Calcineurin subunit B type 1 OS=Rattus norvegicus GN=Ppp3r1</td><td>RAT</td><td>2</td></tr>
<tr><td>558</td><td>4</td><td>4</td><td>17.1</td><td>sp|P60522|GBRL2_RAT</td><td>Gamma-aminobutyric acid receptor-associated protein-like 2 OS=Rattus norvegicus GN=Gabarapl2 PE=2 SV=1</td><td>RAT</td><td>2</td></tr>
<tr><td>559</td><td>4</td><td>4</td><td>10.1</td><td>sp|P19234|NDUV2_RAT</td><td>NADH dehydrogenase [ubiquinone] flavoprotein 2, mitochondrial OS=Rattus norvegicus GN=Ndufv2 PE=1 SV=2</td><td>RAT</td><td>2</td></tr>
<tr><td>560</td><td>4</td><td>4</td><td>7.6</td><td>sp|P16975|SPRC_RAT</td><td>SPARC OS=Rattus norvegicus GN=Sparc PE=1 SV=4</td><td>RAT</td><td>2</td></tr>
<tr><td>561</td><td>4</td><td>4</td><td>5.7</td><td>sp|P15431|GBRB1_RAT</td><td>Gamma-aminobutyric acid receptor subunit beta-1 OS=Rattus norvegicus GN=Gabrb1 PE=2 SV=1</td><td>RAT</td><td>2</td></tr>
<tr><td>562</td><td>4</td><td>4</td><td>8</td><td>sp|Q9Z311|MECR_RAT</td><td>Trans-2-enoyl-CoA reductase, mitochondrial OS=Rattus norvegicus GN=Mecr PE=2 SV=1</td><td>RAT</td><td>2</td></tr>
<tr><td>563</td><td>4</td><td>4</td><td>6.3</td><td>sp|Q9QZA6|CD151_RAT</td><td>CD151 antigen OS=Rattus norvegicus GN=Cd151 PE=1 SV=2</td><td>RAT</td><td>2</td></tr>
<tr><td>564</td><td>4</td><td>4</td><td>6.8</td><td>sp|Q99P82|CLD11_RAT</td><td>Claudin-11 OS=Rattus norvegicus GN=Cldn11 PE=1 SV=1</td><td>RAT</td><td>4</td></tr>
</table>

N	Unused	Total	% Cov (95)	Accession #	Name	Species	Peptides (95%)
565	4	4	6.5	sp|Q5EB62|S2546_RAT	Solute carrier family 25 member 46 OS=Rattus norvegicus GN=Slc25a46 PE=2 SV=2	RAT	2
566	4	4	13.6	sp|P62963|PROF1_RAT	Profilin-1 OS=Rattus norvegicus GN=Pfn1 PE=1 SV=2	RAT	2
567	4	4	16.6	sp|P47875|CSRP1_RAT	Cysteine and glycine-rich protein 1 OS=Rattus norvegicus GN=Csrp1 PE=2 SV=2	RAT	3
568	4	4	26.8	sp|P01946|HBA_RAT	Hemoglobin subunit alpha-1/2 OS=Rattus norvegicus GN=Hba1 PE=1 SV=3	RAT	2
569	4	4	13.7	sp|P35434|ATPD_RAT	ATP synthase subunit delta, mitochondrial OS=Rattus norvegicus GN=Atp5d PE=1 SV=2	RAT	2
570	4	4	12.3	sp|P30713|GSTT2_RAT	Glutathione S-transferase theta-2 OS=Rattus norvegicus GN=Gstt2 PE=1 SV=3	RAT	2
571	3.96	4	11.5	sp|P05426|RL7_RAT	60S ribosomal protein L7 OS=Rattus norvegicus GN=Rpl7 PE=1 SV=2	RAT	2
572	3.92	4	6.1	sp|Q6AY30|SCPDH_RAT	Probable saccharopine dehydrogenase OS=Rattus norvegicus GN=Sccpdh PE=2 SV=1	RAT	2
573	3.89	4.01	8.4	sp|P63331|PP2AA_RAT	Serine/threonine-protein phosphatase 2A catalytic subunit alpha isoform OS=Rattus norvegicus GN=Ppp2ca PE=1 SV=1	RAT	2
574	3.89	4	9.8	sp|P83941|ELOC_RAT	Transcription elongation factor B polypeptide 1 OS=Rattus norvegicus GN=Tceb1 PE=1 SV=1	RAT	2
575	3.82	4	11	sp|Q5XI32|CAPZB_RAT	F-actin-capping protein subunit beta OS=Rattus norvegicus GN=Capzb PE=1 SV=1	RAT	2
576	3.76	4.13	6.6	sp|Q9QVC8|FKBP4_RAT	Peptidyl-prolyl cis-trans isomerase FKBP4 OS=Rattus norvegicus GN=Fkbp4 PE=1 SV=3	RAT	3
577	3.7	3.89	2.9	sp|Q03348|PTPRA_RAT	Receptor-type tyrosine-protein phosphatase alpha OS=Rattus norvegicus GN=Ptpra PE=2 SV=1	RAT	2

(continued)

Table 1
(continued)

N	Unused	Total	% Cov (95)	Accession #	Name	Species	Peptides (95%)
578	3.68	10.04	17.8	sp|P10824|GNAI1_RAT	Guanine nucleotide-binding protein G(i) subunit alpha-1 OS=Rattus norvegicus GN=Gnai1 PE=1 SV=3	RAT	6
579	3.68	3.85	17.5	sp|Q8R1R5|C99L2_RAT	CD99 antigen-like protein 2 OS=Rattus norvegicus GN=Cd99l2 PE=2 SV=1	RAT	2
580	3.62	3.8	4.5	sp|Q03346|MPPB_RAT	Mitochondrial-processing peptidase subunit beta OS=Rattus norvegicus GN=Pmpcb PE=1 SV=3	RAT	2
581	3.6	3.8	3.5	sp|Q4V8I7|LRC8A_RAT	Leucine-rich repeat-containing protein 8A OS=Rattus norvegicus GN=Lrrc8a PE=2 SV=1	RAT	2
582	3.6	3.8	7.4	sp|P09895|RL5_RAT	60S ribosomal protein L5 OS=Rattus norvegicus GN=Rpl5 PE=1 SV=3	RAT	2
583	3.57	3.93	12.7	sp|P85970|ARPC2_RAT	Actin-related protein 2/3 complex subunit 2 OS=Rattus norvegicus GN=Arpc2 PE=1 SV=1	RAT	3
584	3.57	3.75	12.2	sp|Q9JJ54-3|HNRPD_RAT	Isoform p40 of Heterogeneous nuclear ribonucleoprotein D0 OS=Rattus norvegicus GN=Hnrnpd	RAT	3
585	3.53	3.78	10	sp|O35824|DNJA2_RAT	DnaJ homolog subfamily A member 2 OS=Rattus norvegicus GN=Dnaja2 PE=2 SV=1	RAT	3
586	3.52	3.93	4.9	sp|Q5FVJ0|RUFY3_RAT	Protein RUFY3 OS=Rattus norvegicus GN=Rufy3 PE=1 SV=1	RAT	2
587	3.4	3.57	6.5	sp|Q6P6Q9|CBAA1_RAT	Calcium-binding atopy-related autoantigen 1 OS=Rattus norvegicus GN=Cbara1 PE=2 SV=1	RAT	2
588	3.36	10.18	5.7	sp|P39052-4|DYN2_RAT	Isoform ab of Dynamin-2 OS=Rattus norvegicus GN=Dnm2	RAT	5

N	Unused	Total	% Cov (95)	Accession #	Name	Species	Peptides (95%)
589	3.35	3.64	4.9	sp|P28037|FTHFD_RAT	10-formyltetrahydrofolate dehydrogenase OS=Rattus norvegicus GN=Aldh1l1 PE=1 SV=2	RAT	3
590	3.35	3.52	4.5	sp|Q5BJQ0|ADCK3_RAT	Chaperone activity of bc1 complex-like, mitochondrial OS=Rattus norvegicus GN=Cabc1 PE=2 SV=1	RAT	2
591	3.31	3.59	7.4	sp|P32089|TXTP_RAT	Tricarboxylate transport protein, mitochondrial OS=Rattus norvegicus GN=Slc25a1 PE=1 SV=1	RAT	2
592	3.25	3.44	3.9	sp|Q8R511-5|FNBP1_RAT	Isoform Md of Formin-binding protein 1 OS=Rattus norvegicus GN=Fnbp1	RAT	2
593	3.22	3.4	7	sp|P07633|PCCB_RAT	Propionyl-CoA carboxylase beta chain, mitochondrial OS=Rattus norvegicus GN=Pccb PE=2 SV=1	RAT	3
594	3.2	3.37	4.9	sp|P58295|SC6A5_RAT	Sodium- and chloride-dependent glycine transporter 2 OS=Rattus norvegicus GN=Slc6a5 PE=2 SV=1	RAT	2
595	3.2	3.36	7.2	sp|P04218|OX2G_RAT	OX-2 membrane glycoprotein OS=Rattus norvegicus GN=Cd200 PE=2 SV=1	RAT	2
596	3.19	3.39	11.5	sp|Q3T1K5|CAZA2_RAT	F-actin-capping protein subunit alpha-2 OS=Rattus norvegicus GN=Capza2 PE=1 SV=1	RAT	2
597	3.18	3.34	4.5	sp|Q5I6B8|PI51C_RAT	Phosphatidylinositol-4-phosphate 5-kinase type-1 gamma OS=Rattus norvegicus GN=Pip5k1c PE=1 SV=1	RAT	2
598	3.16	3.4	10.5	sp|Q505J9|ATAD1_RAT	ATPase family AAA domain-containing protein 1 OS=Rattus norvegicus GN=Atad1 PE=2 SV=1	RAT	2
599	3.14	3.32	2.9	sp|P0C5W1|MAP1S_RAT	Microtubule-associated protein 1S OS=Rattus norvegicus GN=Map1s PE=1 SV=1	RAT	2

(continued)

Table 1
(continued)

N	Unused	Total	% Cov (95)	Accession #	Name	Species	Peptides (95%)
600	3.12	3.4	2.4	sp|Q63475|PTPR2_RAT	Receptor-type tyrosine-protein phosphatase N2 OS=Rattus norvegicus GN=Ptprn2 PE=2 SV=1	RAT	2
601	3.12	3.32	10.8	sp|Q6MGB5|DHB8_RAT	Estradiol 17-beta-dehydrogenase 8 OS=Rattus norvegicus GN=Hsd17b8 PE=2 SV=1	RAT	2
602	3.11	3.58	6.7	sp|P62198|PRS8_RAT	26S protease regulatory subunit 8 OS=Rattus norvegicus GN=Psmc5 PE=1 SV=1	RAT	2
603	3.11	3.3	10.9	sp|Q505J6|GHC2_RAT	Mitochondrial glutamate carrier 2 OS=Rattus norvegicus GN=Slc25a18 PE=2 SV=2	RAT	2
604	3.11	3.28	18.2	sp|Q62785|HAP28_RAT	28 kDa heat- and acid-stable phosphoprotein OS=Rattus norvegicus GN=Pdap1 PE=1 SV=1	RAT	2
605	3.11	3.27	6.4	sp|Q4V8B7|HSDL1_RAT	Inactive hydroxysteroid dehydrogenase-like protein 1 OS=Rattus norvegicus GN=Hsdl1 PE=2 SV=1	RAT	2
606	3.1	3.28	14.6	sp|Q5XIG4|OCAD1_RAT	OCIA domain-containing protein 1 OS=Rattus norvegicus GN=Ociad1 PE=2 SV=1	RAT	3
607	3.1	3.27	13.8	sp|P62907|RL10A_RAT	60S ribosomal protein L10a OS=Rattus norvegicus GN=Rpl10a PE=1 SV=2	RAT	2
608	3.09	3.25	4.9	sp|Q9JLJ3|AL9A1_RAT	4-trimethylaminobutyraldehyde dehydrogenase OS=Rattus norvegicus GN=Aldh9a1 PE=1 SV=1	RAT	2
609	3.08	3.24	6.6	sp|P05370|G6PD_RAT	Glucose-6-phosphate 1-dehydrogenase OS=Rattus norvegicus GN=G6pdx PE=1 SV=3	RAT	2
610	3.08	3.24	7.4	sp|P35738|ODBB_RAT	2-oxoisovalerate dehydrogenase subunit beta, mitochondrial OS=Rattus norvegicus GN=Bckdhb PE=1 SV=3	RAT	2

<table>
<tr><th>N</th><th>Unused</th><th>Total</th><th>% Cov (95)</th><th>Accession #</th><th>Name</th><th>Species</th><th>Peptides (95%)</th></tr>
<tr><td>611</td><td>3.07</td><td>3.23</td><td>16.2</td><td>sp|P10818|CX6A1_RAT</td><td>Cytochrome c oxidase subunit 6A1, mitochondrial OS=Rattus norvegicus GN=Cox6a1 PE=1 SV=2</td><td>RAT</td><td>2</td></tr>
<tr><td>612</td><td>3.05</td><td>3.48</td><td>4.1</td><td>sp|Q66HD0|ENPL_RAT</td><td>Endoplasmin OS=Rattus norvegicus GN=Hsp90b1 PE=1 SV=2</td><td>RAT</td><td>2</td></tr>
<tr><td>613</td><td>3.04</td><td>3.22</td><td>5.5</td><td>sp|P20650|PPM1A_RAT</td><td>Protein phosphatase 1A OS=Rattus norvegicus GN=Ppm1a PE=1 SV=1</td><td>RAT</td><td>2</td></tr>
<tr><td>614</td><td>3.03</td><td>3.2</td><td>3.6</td><td>sp|Q05683|DCE2_RAT</td><td>Glutamate decarboxylase 2 OS=Rattus norvegicus GN=Gad2 PE=1 SV=1</td><td>RAT</td><td>2</td></tr>
<tr><td>615</td><td>3.02</td><td>3.19</td><td>2.2</td><td>sp|Q9Z0U4|GABR1_RAT</td><td>Gamma-aminobutyric acid type B receptor subunit 1 OS=Rattus norvegicus GN=Gabbr1 PE=1 SV=1</td><td>RAT</td><td>2</td></tr>
<tr><td>616</td><td>3.02</td><td>3.19</td><td>7.1</td><td>sp|Q6URK4|ROA3_RAT</td><td>Heterogeneous nuclear ribonucleoprotein A3 OS=Rattus norvegicus GN=Hnrnpa3 PE=1 SV=1</td><td>RAT</td><td>2</td></tr>
<tr><td>617</td><td>2.98</td><td>3.14</td><td>9.2</td><td>sp|Q99MG9|KCIP4_RAT</td><td>Kv channel-interacting protein 4 OS=Rattus norvegicus GN=Kcnip4 PE=1 SV=1</td><td>RAT</td><td>2</td></tr>
<tr><td>618</td><td>2.96</td><td>3.48</td><td>4.8</td><td>sp|P68403|KPCB_RAT</td><td>Protein kinase C beta type OS=Rattus norvegicus GN=Prkcb PE=1 SV=3</td><td>RAT</td><td>3</td></tr>
<tr><td>619</td><td>2.95</td><td>3.15</td><td>5.5</td><td>sp|Q920F5|DCMC_RAT</td><td>Malonyl-CoA decarboxylase, mitochondrial OS=Rattus norvegicus GN=Mlycd PE=2 SV=1</td><td>RAT</td><td>2</td></tr>
<tr><td>620</td><td>2.94</td><td>3.12</td><td>2.8</td><td>sp|O88871|GABR2_RAT</td><td>Gamma-aminobutyric acid type B receptor subunit 2 OS=Rattus norvegicus GN=Gabbr2 PE=1 SV=2</td><td>RAT</td><td>2</td></tr>
<tr><td>621</td><td>2.92</td><td>3.3</td><td>3.4</td><td>sp|P12839|NFM_RAT</td><td>Neurofilament medium polypeptide OS=Rattus norvegicus GN=Nefm PE=1 SV=4</td><td>RAT</td><td>2</td></tr>
<tr><td>622</td><td>2.9</td><td>3.09</td><td>7.1</td><td>sp|P02650|APOE_RAT</td><td>Apolipoprotein E OS=Rattus norvegicus GN=Apoe PE=1 SV=2</td><td>RAT</td><td>2</td></tr>
</table>

(continued)

Table 1
(continued)

N	Unused	Total	% Cov (95)	Accession #	Name	Species	Peptides (95%)
623	2.9	3.07	6.4	sp|Q7TP48|APMAP_RAT	Adipocyte plasma membrane-associated protein OS=Rattus norvegicus GN=Apmap PE=2 SV=2	RAT	2
624	2.86	5.45	13	sp|Q63965|SFXN1_RAT	Sideroflexin-1 OS=Rattus norvegicus GN=Sfxn1 PE=2 SV=4	RAT	4
625	2.85	3.01	5.9	sp|P50878|RL4_RAT	60S ribosomal protein L4 OS=Rattus norvegicus GN=Rpl4 PE=1 SV=3	RAT	2
626	2.84	3.05	6.5	sp|P27139|CAH2_RAT	Carbonic anhydrase 2 OS=Rattus norvegicus GN=Ca2 PE=1 SV=2	RAT	2
627	2.83	3	4.4	sp|P20272|CNR1_RAT	Cannabinoid receptor 1 OS=Rattus norvegicus GN=Cnr1 PE=1 SV=1	RAT	2
628	2.82	2.97	2.4	sp|O35142|COPB2_RAT	Coatomer subunit beta' OS=Rattus norvegicus GN=Copb2 PE=1 SV=3	RAT	2
629	2.81	2.97	39.1	sp|P11030|ACBP_RAT	Acyl-CoA-binding protein OS=Rattus norvegicus GN=Dbi PE=1 SV=3	RAT	3
630	2.8	3.2	7.3	sp|P27867|DHSO_RAT	Sorbitol dehydrogenase OS=Rattus norvegicus GN=Sord PE=1 SV=4	RAT	2
631	2.8	2.96	2	sp|P97603|NEO1_RAT	Neogenin (Fragment) OS=Rattus norvegicus GN=Neo1 PE=2 SV=1	RAT	2
632	2.79	3.19	2.5	sp|Q6Q629|DPP10_RAT	Inactive dipeptidyl peptidase 10 OS=Rattus norvegicus GN=Dpp10 PE=1 SV=1	RAT	2
633	2.77	2.93	5	sp|P03889|NU1M_RAT	NADH-ubiquinone oxidoreductase chain 1 OS=Rattus norvegicus GN=Mtnd1 PE=1 SV=2	RAT	2
634	2.76	3	22.1	sp|P38718|BR44_RAT	Brain protein 44 OS=Rattus norvegicus GN=Brp44 PE=2 SV=1	RAT	5
635	2.75	2.91	1.7	sp|O08815|SLK_RAT	STE20-like serine/threonine-protein kinase OS=Rattus norvegicus GN=Slk PE=1 SV=1	RAT	2

N	Unused	Total	% Cov (95)	Accession #	Name	Species	Peptides (95%)
636	2.73	2.89	17	sp|P62870|ELOB_RAT	Transcription elongation factor B polypeptide 2 OS=Rattus norvegicus GN=Tceb2 PE=1 SV=1	RAT	2
637	2.72	2.89	16.4	sp|Q4QR80|RT25_RAT	28S ribosomal protein S25, mitochondrial OS=Rattus norvegicus GN=Mrps25 PE=2 SV=1	RAT	2
638	2.7	2.9	4.4	sp|P41499|PTN11_RAT	Tyrosine-protein phosphatase non-receptor type 11 OS=Rattus norvegicus GN=Ptpn11 PE=1 SV=4	RAT	2
639	2.69	2.86	3.5	sp|P00388|NCPR_RAT	NADPH–cytochrome P450 reductase OS=Rattus norvegicus GN=Por PE=1 SV=3	RAT	2
640	2.68	2.85	14.3	sp|Q5XIF3|NDUS4_RAT	NADH dehydrogenase [ubiquinone] iron-sulfur protein 4, mitochondrial OS=Rattus norvegicus GN=Ndufs4 PE=1 SV=1	RAT	2
641	2.67	2.82	8.6	sp|Q5HZV9|PP1R7_RAT	Protein phosphatase 1 regulatory subunit 7 OS=Rattus norvegicus GN=Ppp1r7 PE=1 SV=1	RAT	2
642	2.65	7.07	5.4	sp|Q63622-7|DLG2_RAT	Isoform 7 of Disks large homolog 2 OS=Rattus norvegicus GN=Dlg2	RAT	4
643	2.6	5.04	6.1	sp|Q68FQ0|TCPE_RAT	T-complex protein 1 subunit epsilon OS=Rattus norvegicus GN=Cct5 PE=1 SV=1	RAT	3
644	2.58	5.33	17.4	sp|P05714|RAB4A_RAT	Ras-related protein Rab-4A OS=Rattus norvegicus GN=Rab4a PE=2 SV=2	RAT	3
645	2.56	2.63	2.7	sp|P16638|ACLY_RAT	ATP-citrate synthase OS=Rattus norvegicus GN=Acly PE=1 SV=1	RAT	2
646	2.53	10.89	32.6	sp|P84076|HPCA_RAT	Neuron-specific calcium-binding protein hippocalcin OS=Rattus norvegicus GN=Hpca PE=1 SV=2	RAT	7
647	2.49	2.89	6.3	sp|Q5QJC9|BAG5_RAT	BAG family molecular chaperone regulator 5 OS=Rattus norvegicus GN=Bag5 PE=1 SV=1	RAT	2

(continued)

Table 1
(continued)

N	Unused	Total	% Cov (95)	Accession #	Name	Species	Peptides (95%)
648	2.45	2.51	5.6	sp|Q6P7B0|SYWC_RAT	Tryptophanyl-tRNA synthetase, cytoplasmic OS=Rattus norvegicus GN=Wars PE=1 SV=2	RAT	2
649	2.44	5.37	21.3	sp|P08644-2|RASK_RAT	Isoform 2B of GTPase KRas OS=Rattus norvegicus GN=Kras	RAT	4
650	2.42	2.48	6.8	sp|B5DEH2|ERLN2_RAT	Erlin-2 OS=Rattus norvegicus GN=Erlin2 PE=1 SV=1	RAT	2
651	2.41	2.52	3.4	sp|Q4V8F9|HSDL2_RAT	Hydroxysteroid dehydrogenase-like protein 2 OS=Rattus norvegicus GN=Hsdl2 PE=2 SV=1	RAT	2
652	2.41	2.47	16.7	sp|Q62658|FKB1A_RAT	Peptidyl-prolyl cis-trans isomerase FKBP1A OS=Rattus norvegicus GN=Fkbp1a PE=2 SV=3	RAT	1
653	2.4	2.76	2.7	sp|P54283|CACB1_RAT	Voltage-dependent L-type calcium channel subunit beta-1 OS=Rattus norvegicus GN=Cacnb1 PE=2 SV=1	RAT	2
654	2.4	2.46	5.9	sp|Q4KLH5|AGFG1_RAT	Arf-GAP domain and FG repeats-containing protein 1 OS=Rattus norvegicus GN=Agfg1 PE=1 SV=1	RAT	2
655	2.38	2.44	2	sp|Q07116|SUOX_RAT	Sulfite oxidase, mitochondrial OS=Rattus norvegicus GN=Suox PE=1 SV=2	RAT	1
656	2.35	2.41	2.5	sp|P62483|KCAB2_RAT	Voltage-gated potassium channel subunit beta-2 OS=Rattus norvegicus GN=Kcnab2 PE=1 SV=1	RAT	1
657	2.35	2.41	7.2	sp|B0BNM9|GLTP_RAT	Glycolipid transfer protein OS=Rattus norvegicus GN=GLTP PE=2 SV=1	RAT	1
658	2.31	2.5	4.5	sp|Q9QX69|LANC1_RAT	LanC-like protein 1 OS=Rattus norvegicus GN=Lancl1 PE=2 SV=2	RAT	2
659	2.31	2.36	0.9	sp|P29476-3|NOS1_RAT	Isoform PNNOS of Nitric oxide synthase, brain OS=Rattus norvegicus GN=Nos1	RAT	1

N	Unused	Total	% Cov (95)	Accession #	Name	Species	Peptides (95%)
660	2.29	2.35	2.4	sp|P56522|ADRO_RAT	NADPH:adrenodoxin oxidoreductase, mitochondrial OS=Rattus norvegicus GN=Fdxr PE=1 SV=1	RAT	1
661	2.26	2.31	5.3	sp|Q62848|ARFG1_RAT	ADP-ribosylation factor GTPase-activating protein 1 OS=Rattus norvegicus GN=Arfgap1 PE=1 SV=1	RAT	1
662	2.25	2.31	2	sp|Q3ZB98-6|BCAS1_RAT	Isoform band83 of Breast carcinoma-amplified sequence 1 homolog OS=Rattus norvegicus GN=Bcas1	RAT	1
663	2.25	2.3	5.8	sp|Q6AY41|CC50A_RAT	Cell cycle control protein 50A OS=Rattus norvegicus GN=Tmem30a PE=2 SV=1	RAT	1
664	2.24	2.3	3.3	sp|O88994|MOSC2_RAT	MOSC domain-containing protein 2, mitochondrial OS=Rattus norvegicus GN=Mosc2 PE=2 SV=1	RAT	1
665	2.24	2.29	2.6	sp|P62813|GBRA1_RAT	Gamma-aminobutyric acid receptor subunit alpha-1 OS=Rattus norvegicus GN=Gabra1 PE=1 SV=1	RAT	1
666	2.23	2.3	1.3	sp|Q63357|MYO1D_RAT	Myosin-Id OS=Rattus norvegicus GN=Myo1d PE=1 SV=3	RAT	1
667	2.23	2.28	3.8	sp|P19139|CSK21_RAT	Casein kinase II subunit alpha OS=Rattus norvegicus GN=Csnk2a1 PE=1 SV=2	RAT	1
668	2.17	2.74	8.7	sp|Q5XIE1|CH055_RAT	UPF0670 protein C8orf55 homolog OS=Rattus norvegicus PE=2 SV=1	RAT	2
669	2.17	2.4	2	sp|Q9ESW0|DDB1_RAT	DNA damage-binding protein 1 OS=Rattus norvegicus GN=Ddb1 PE=2 SV=1	RAT	2
670	2.17	2.21	12.3	sp|Q0ZCA7|CLC2L_RAT	C-type lectin domain family 2 member L OS=Rattus norvegicus GN=Clec2l PE=2 SV=2	RAT	2

(continued)

Table 1
(continued)

N	Unused	Total	% Cov (95)	Accession #	Name	Species	Peptides (95%)
671	2.14	2.17	1.6	sp|P31422|GRM3_RAT	Metabotropic glutamate receptor 3 OS=Rattus norvegicus GN=Grm3 PE=1 SV=1	RAT	1
672	2.13	2.17	5.9	sp|Q5U2R6|K1191_RAT	UPF0498 protein KIAA1191 homolog OS=Rattus norvegicus GN=Aa2-141 PE=2 SV=1	RAT	1
673	2.11	2.16	2.7	sp|P85968|6PGD_RAT	6-phosphogluconate dehydrogenase, decarboxylating OS=Rattus norvegicus GN=Pgd PE=1 SV=1	RAT	1
674	2.1	2.13	2	sp|P85845|FSCN1_RAT	Fascin OS=Rattus norvegicus GN=Fscn1 PE=1 SV=2	RAT	1
675	2.08	2.11	2	sp|O88637|PCY2_RAT	Ethanolamine-phosphate cytidylyltransferase OS=Rattus norvegicus GN=Pcyt2 PE=1 SV=1	RAT	1
676	2.08	2.11	1.6	sp|P11167|GTR1_RAT	Solute carrier family 2, facilitated glucose transporter member 1 OS=Rattus norvegicus GN=Slc2a1 PE=1 SV=1	RAT	1
677	2.08	2.1	9.4	sp|P13668|STMN1_RAT	Stathmin OS=Rattus norvegicus GN=Stmn1 PE=1 SV=2	RAT	2
678	2.07	2.09	5.1	sp|Q5PPN7|CCD51_RAT	Coiled-coil domain-containing protein 51 OS=Rattus norvegicus GN=Ccdc51 PE=2 SV=2	RAT	1
679	2.07	2.09	2.8	sp|O88483|PDP1_RAT	[Pyruvate dehydrogenase [acetyl-transferring]]-phosphatase 1, mitochondrial OS=Rattus norvegicus GN=Pdp1 PE=1 SV=1	RAT	1
680	2.07	2.09	3.5	sp|O35078|OXDA_RAT	D-amino-acid oxidase OS=Rattus norvegicus GN=Dao PE=2 SV=1	RAT	1
681	2.07	2.09	16.6	sp|O35264|PA1B2_RAT	Platelet-activating factor acetylhydrolase IB subunit beta OS=Rattus norvegicus GN=Pafah1b2 PE=1 SV=1	RAT	2

N	Unused	Total	% Cov (95)	Accession #	Name	Species	Peptides (95%)
682	2.06	4.12	6.7	sp|Q9JID2|GNA11_RAT	Guanine nucleotide-binding protein subunit alpha-11 OS=Rattus norvegicus GN=Gna11 PE=2 SV=1	RAT	2
683	2.05	2.07	2.6	sp|Q8VHK2|CSKI1_RAT	Caskin-1 OS=Rattus norvegicus GN=Caskin1 PE=1 SV=1	RAT	2
684	2.05	2.06	3.6	sp|P62193|PRS4_RAT	26S protease regulatory subunit 4 OS=Rattus norvegicus GN=Psmc1 PE=2 SV=1	RAT	1
685	2.05	2.06	1.8	sp|P11661|NU5M_RAT	NADH-ubiquinone oxidoreductase chain 5 OS=Rattus norvegicus GN=Mtnd5 PE=2 SV=2	RAT	1
686	2.04	2.06	4.3	sp|Q6AYT0|QOR_RAT	Quinone oxidoreductase OS=Rattus norvegicus GN=Cryz PE=2 SV=1	RAT	1
687	2.04	2.05	1.1	sp|Q5YLM1|DGLA_RAT	Sn1-specific diacylglycerol lipase alpha OS=Rattus norvegicus GN=Dagla PE=2 SV=1	RAT	1
688	2.04	2.05	2.9	sp|Q6AXS5|PAIRB_RAT	Plasminogen activator inhibitor 1 RNA-binding protein OS=Rattus norvegicus GN=Serbp1 PE=1 SV=2	RAT	1
689	2.03	2.05	1.5	sp|Q02356|AMPD2_RAT	AMP deaminase 2 OS=Rattus norvegicus GN=Ampd2 PE=2 SV=2	RAT	1
690	2.03	2.04	6.9	sp|Q2KMM2|TPPC1_RAT	Trafficking protein particle complex subunit 1 OS=Rattus norvegicus GN=Trappc1 PE=2 SV=1	RAT	1
691	2.03	2.04	8.6	sp|P84817|FIS1_RAT	Mitochondrial fission 1 protein OS=Rattus norvegicus GN=Fis1 PE=1 SV=1	RAT	1
692	2.03	2.04	8.4	sp|B0BN86|TMM11_RAT	Transmembrane protein 11 OS=Rattus norvegicus GN=Tmem11 PE=2 SV=1	RAT	1
693	2.03	2.04	4.2	sp|P67874|CSK2B_RAT	Casein kinase II subunit beta OS=Rattus norvegicus GN=Csnk2b PE=1 SV=1	RAT	1

(continued)

Table 1
(continued)

N	Unused	Total	% Cov (95)	Accession #	Name	Species	Peptides (95%)
694	2.02	8.09	12.9	sp|O35353|GBB4_RAT	Guanine nucleotide-binding protein subunit beta-4 OS=Rattus norvegicus GN=Gnb4 PE=2 SV=4	RAT	8
695	2.02	2.05	3.9	sp|P37285|KLC1_RAT	Kinesin light chain 1 OS=Rattus norvegicus GN=Klc1 PE=2 SV=2	RAT	2
696	2.02	2.03	1.4	sp|Q3KRE0|ATAD3_RAT	ATPase family AAA domain-containing protein 3 OS=Rattus norvegicus GN=Atad3 PE=2 SV=1	RAT	1
697	2.02	2.03	2.5	sp|Q5XI22|THIC_RAT	Acetyl-CoA acetyltransferase, cytosolic OS=Rattus norvegicus GN=Acat2 PE=1 SV=1	RAT	1
698	2.02	2.03	2.9	sp|P85515|ACTZ_RAT	Alpha-centractin OS=Rattus norvegicus GN=Actr1a PE=1 SV=1	RAT	1
699	2.02	2.03	6.7	sp|P0C089|PTPM1_RAT	Protein-tyrosine phosphatase mitochondrial 1 OS=Rattus norvegicus GN=Ptpmt1 PE=2 SV=1	RAT	1
700	2.02	2.03	8.9	sp|Q8K4G6|MACD1_RAT	MACRO domain-containing protein 1 (Fragment) OS=Rattus norvegicus GN=Macrod1 PE=2 SV=2	RAT	1
701	2.02	2.03	4	sp|Q64244|CD38_RAT	ADP-ribosyl cyclase 1 OS=Rattus norvegicus GN=Cd38 PE=2 SV=1	RAT	1
702	2.02	2.02	5.5	sp|P49655|IRK10_RAT	ATP-sensitive inward rectifier potassium channel 10 OS=Rattus norvegicus GN=Kcnj10 PE=1 SV=1	RAT	1
703	2.01	2.07	4.4	sp|P00406|COX2_RAT	Cytochrome c oxidase subunit 2 OS=Rattus norvegicus GN=Mtco2 PE=2 SV=2	RAT	2
704	2.01	2.02	5.1	sp|Q91XR8|GPX42_RAT	Phospholipid hydroperoxide glutathione peroxidase, nuclear OS=Rattus norvegicus GN=Gpx4 PE=2 SV=3	RAT	1

N	Unused	Total	% Cov (95)	Accession #	Name	Species	Peptides (95%)
705	2.01	2.02	3.3	sp|P61150|FGF12_RAT	Fibroblast growth factor 12 OS=Rattus norvegicus GN=Fgf12 PE=1 SV=1	RAT	1
706	2.01	2.02	4.2	sp|Q568Z1|F131B_RAT	Protein FAM131B OS=Rattus norvegicus GN=Fam131b PE=2 SV=1	RAT	1
707	2.01	2.01	1.9	sp|Q9Z1A5|ULA1_RAT	NEDD8-activating enzyme E1 regulatory subunit OS=Rattus norvegicus GN=Nae1 PE=1 SV=1	RAT	1
708	2.01	2.01	3	sp|Q68FS2|CSN4_RAT	COP9 signalosome complex subunit 4 OS=Rattus norvegicus GN=Cops4 PE=2 SV=1	RAT	1
709	2.01	2.01	3.8	sp|Q62698|DC1L2_RAT	Cytoplasmic dynein 1 light intermediate chain 2 OS=Rattus norvegicus GN=Dync1li2 PE=1 SV=1	RAT	1
710	2.01	2.01	6.5	sp|Q6PEB9|CC127_RAT	Coiled-coil domain-containing protein 127 OS=Rattus norvegicus GN=Ccdc127 PE=2 SV=2	RAT	1
711	2.01	2.01	12.4	sp|P39032|RL36_RAT	60S ribosomal protein L36 OS=Rattus norvegicus GN=Rpl36 PE=1 SV=2	RAT	1
712	2.01	2.01	8.6	sp|Q9WVB1|RAB6A_RAT	Ras-related protein Rab-6A (Fragment) OS=Rattus norvegicus GN=Rab6a PE=2 SV=1	RAT	1
713	2.01	2.01	7.6	sp|P50408|VATF_RAT	V-type proton ATPase subunit F OS=Rattus norvegicus GN=Atp6v1f PE=1 SV=1	RAT	1
714	2.01	2.01	7.8	sp|Q9WVM4|ICMT_RAT	Protein-S-isoprenylcysteine O-methyltransferase (Fragment) OS=Rattus norvegicus GN=Icmt PE=1 SV=1	RAT	1
715	2.01	2.01	21.4	sp|P01041|CYTB_RAT	Cystatin-B OS=Rattus norvegicus GN=Cstb PE=1 SV=1	RAT	1
716	2	45.81	40	sp|P61765-2|STXB1_RAT	Isoform HUNC18b of Syntaxin-binding protein 1 OS=Rattus norvegicus GN=Stxbp1	RAT	29
717	2	26	29.3	sp|P69897|TBB5_RAT	Tubulin beta-5 chain OS=Rattus norvegicus GN=Tubb5 PE=1 SV=1	RAT	17

(continued)

Table 1
(continued)

N	Unused	Total	% Cov (95)	Accession #	Name	Species	Peptides (95%)
718	2	17.96	34.9	sp|P97546-1|NPTN_RAT	Isoform np55 of Neuroplastin OS=Rattus norvegicus GN=Nptn	RAT	11
719	2	8.01	8	sp|P20651|PP2BB_RAT	Serine/threonine-protein phosphatase 2B catalytic subunit beta isoform OS=Rattus norvegicus GN=Ppp3cb PE=2 SV=1	RAT	4
720	2	8.01	34.9	sp|P19804|NDKB_RAT	Nucleoside diphosphate kinase B OS=Rattus norvegicus GN=Nme2 PE=1 SV=1	RAT	4
721	2	8	20.3	sp|Q9Z250|LIN7A_RAT	Protein lin-7 homolog A OS=Rattus norvegicus GN=Lin7a PE=1 SV=2	RAT	7
722	2	7.54	14.9	sp|P62138|PP1A_RAT	Serine/threonine-protein phosphatase PP1-alpha catalytic subunit OS=Rattus norvegicus GN=Ppp1ca PE=1 SV=1	RAT	4
723	2	6.01	9.6	sp|P12369|KAP3_RAT	cAMP-dependent protein kinase type II-beta regulatory subunit OS=Rattus norvegicus GN=Prkar2b PE=1 SV=3	RAT	3
724	2	4.4	11.1	sp|P35280|RAB8A_RAT	Ras-related protein Rab-8A OS=Rattus norvegicus GN=Rab8a PE=2 SV=2	RAT	2
725	2	4.01	12.2	sp|P51146|RAB4B_RAT	Ras-related protein Rab-4B OS=Rattus norvegicus GN=Rab4b PE=2 SV=1	RAT	2
726	2	4	5.2	sp|P51647|AL1A1_RAT	Retinal dehydrogenase 1 OS=Rattus norvegicus GN=Aldh1a1 PE=1 SV=3	RAT	2
727	2	2.13	2	sp|Q6P0K8|PLAK_RAT	Junction plakoglobin OS=Rattus norvegicus GN=Jup PE=1 SV=1	RAT	1
728	2	2.01	1.4	sp|Q923S8|PANK4_RAT	Pantothenate kinase 4 OS=Rattus norvegicus GN=Pank4 PE=1 SV=1	RAT	1
729	2	2.01	1.4	sp|Q01992|MIPEP_RAT	Mitochondrial intermediate peptidase OS=Rattus norvegicus GN=Mipep PE=1 SV=1	RAT	1

<table>
<tr><th>N</th><th>Unused</th><th>Total</th><th>% Cov (95)</th><th>Accession #</th><th>Name</th><th>Species</th><th>Peptides (95%)</th></tr>
<tr><td>730</td><td>2</td><td>2.01</td><td>3.4</td><td>sp|Q5RJY4|DRS7B_RAT</td><td>Dehydrogenase/reductase SDR family member 7B OS=Rattus norvegicus GN=Dhrs7b PE=1 SV=1</td><td>RAT</td><td>1</td></tr>
<tr><td>731</td><td>2</td><td>2.01</td><td>2.5</td><td>sp|Q641Y0|OST48_RAT</td><td>Dolichyl-diphosphooligosaccharide–protein glycosyltransferase 48 kDa subunit OS=Rattus norvegicus GN=Ddost PE=2 SV=1</td><td>RAT</td><td>1</td></tr>
<tr><td>732</td><td>2</td><td>2.01</td><td>3</td><td>sp|P63036|DNJA1_RAT</td><td>DnaJ homolog subfamily A member 1 OS=Rattus norvegicus GN=Dnaja1 PE=2 SV=1</td><td>RAT</td><td>1</td></tr>
<tr><td>733</td><td>2</td><td>2</td><td>0.4</td><td>sp|P97924|KALRN_RAT</td><td>Kalirin OS=Rattus norvegicus GN=Kalrn PE=1 SV=3</td><td>RAT</td><td>1</td></tr>
<tr><td>734</td><td>2</td><td>2</td><td>1</td><td>sp|Q7TT49|MRCKB_RAT</td><td>Serine/threonine-protein kinase MRCK beta OS=Rattus norvegicus GN=Cdc42bpb PE=1 SV=1</td><td>RAT</td><td>1</td></tr>
<tr><td>735</td><td>2</td><td>2</td><td>1</td><td>sp|Q8K3M6|ERC2_RAT</td><td>ERC protein 2 OS=Rattus norvegicus GN=Erc2 PE=1 SV=1</td><td>RAT</td><td>1</td></tr>
<tr><td>736</td><td>2</td><td>2</td><td>0.8</td><td>sp|Q7M6Z5|KIF27_RAT</td><td>Kinesin-like protein KIF27 OS=Rattus norvegicus GN=Kif27 PE=2 SV=1</td><td>RAT</td><td>1</td></tr>
<tr><td>737</td><td>2</td><td>2</td><td>0.8</td><td>sp|O08873|MADD_RAT</td><td>MAP kinase-activating death domain protein OS=Rattus norvegicus GN=Madd PE=1 SV=1</td><td>RAT</td><td>1</td></tr>
<tr><td>738</td><td>2</td><td>2</td><td>0.9</td><td>sp|Q99JD4|CLAP2_RAT</td><td>CLIP-associating protein 2 OS=Rattus norvegicus GN=Clasp2 PE=2 SV=1</td><td>RAT</td><td>1</td></tr>
<tr><td>739</td><td>2</td><td>2</td><td>3.2</td><td>sp|Q7TP47|HNRPQ_RAT</td><td>Heterogeneous nuclear ribonucleoprotein Q OS=Rattus norvegicus GN=Syncrip PE=2 SV=1</td><td>RAT</td><td>1</td></tr>
<tr><td>740</td><td>2</td><td>2</td><td>1.8</td><td>sp|P97839|DLGP4_RAT</td><td>Disks large-associated protein 4 OS=Rattus norvegicus GN=Dlgap4 PE=2 SV=1</td><td>RAT</td><td>1</td></tr>
<tr><td>741</td><td>2</td><td>2</td><td>1.5</td><td>sp|Q04462|SYVC_RAT</td><td>Valyl-tRNA synthetase OS=Rattus norvegicus GN=Vars PE=2 SV=2</td><td>RAT</td><td>1</td></tr>
</table>

(continued)

Table 1
(continued)

<table>
<tr><th>N</th><th>Unused</th><th>Total</th><th>% Cov (95)</th><th>Accession #</th><th>Name</th><th>Species</th><th>Peptides (95%)</th></tr>
<tr><td>742</td><td>2</td><td>2</td><td>2.3</td><td>sp|P20595|GCYB1_RAT</td><td>Guanylate cyclase soluble subunit beta-1 OS=Rattus norvegicus GN=Gucy1b3 PE=1 SV=2</td><td>RAT</td><td>1</td></tr>
<tr><td>743</td><td>2</td><td>2</td><td>0.9</td><td>sp|P0C5Y8|ALS2_RAT</td><td>Alsin OS=Rattus norvegicus GN=Als2 PE=2 SV=1</td><td>RAT</td><td>1</td></tr>
<tr><td>744</td><td>2</td><td>2</td><td>1.8</td><td>sp|Q5BJP6|RRF2M_RAT</td><td>Ribosome-releasing factor 2, mitochondrial OS=Rattus norvegicus GN=Gfm2 PE=2 SV=2</td><td>RAT</td><td>1</td></tr>
<tr><td>745</td><td>2</td><td>2</td><td>1.6</td><td>sp|Q07014|LYN_RAT</td><td>Tyrosine-protein kinase Lyn OS=Rattus norvegicus GN=Lyn PE=2 SV=3</td><td>RAT</td><td>1</td></tr>
<tr><td>746</td><td>2</td><td>2</td><td>1.7</td><td>sp|P97878|EXOC5_RAT</td><td>Exocyst complex component 5 OS=Rattus norvegicus GN=Exoc5 PE=1 SV=1</td><td>RAT</td><td>1</td></tr>
<tr><td>747</td><td>2</td><td>2</td><td>3.1</td><td>sp|P51583|PUR6_RAT</td><td>Multifunctional protein ADE2 OS=Rattus norvegicus GN=Paics PE=2 SV=3</td><td>RAT</td><td>1</td></tr>
<tr><td>748</td><td>2</td><td>2</td><td>1.5</td><td>sp|O70277|TRIM3_RAT</td><td>Tripartite motif-containing protein 3 OS=Rattus norvegicus GN=Trim3 PE=1 SV=1</td><td>RAT</td><td>1</td></tr>
<tr><td>749</td><td>2</td><td>2</td><td>1.1</td><td>sp|Q5U2P1|CNNM2_RAT</td><td>Metal transporter CNNM2 OS=Rattus norvegicus GN=Cnnm2 PE=2 SV=1</td><td>RAT</td><td>1</td></tr>
<tr><td>750</td><td>2</td><td>2</td><td>1.1</td><td>sp|Q5RKI8|ABCB8_RAT</td><td>ATP-binding cassette sub-family B member 8, mitochondrial OS=Rattus norvegicus GN=Abcb8 PE=2 SV=1</td><td>RAT</td><td>1</td></tr>
<tr><td>751</td><td>2</td><td>2</td><td>1.4</td><td>sp|Q499S9|RHDF1_RAT</td><td>Rhomboid family member 1 OS=Rattus norvegicus GN=Rhbdf1 PE=2 SV=1</td><td>RAT</td><td>1</td></tr>
<tr><td>752</td><td>2</td><td>2</td><td>7</td><td>sp|P62282|RS11_RAT</td><td>40S ribosomal protein S11 OS=Rattus norvegicus GN=Rps11 PE=1 SV=3</td><td>RAT</td><td>1</td></tr>
<tr><td>753</td><td>2</td><td>2</td><td>2.3</td><td>sp|P19492|GRIA3_RAT</td><td>Glutamate receptor 3 OS=Rattus norvegicus GN=Gria3 PE=1 SV=1</td><td>RAT</td><td>1</td></tr>
<tr><td>754</td><td>2</td><td>2</td><td>1.8</td><td>sp|Q9Z272|GIT1_RAT</td><td>ARF GTPase-activating protein GIT1 OS=Rattus norvegicus GN=Git1 PE=1 SV=1</td><td>RAT</td><td>1</td></tr>
</table>

N	Unused	Total	% Cov (95)	Accession #	Name	Species	Peptides (95%)
755	2	2	2.3	sp|Q8VHF0|MARK3_RAT	MAP/microtubule affinity-regulating kinase 3 OS=Rattus norvegicus GN=Mark3 PE=2 SV=1	RAT	1
756	2	2	3.1	sp|Q6IRE4|TS101_RAT	Tumor susceptibility gene 101 protein OS=Rattus norvegicus GN=Tsg101 PE=1 SV=1	RAT	1
757	2	2	1.2	sp|Q63767|BCAR1_RAT	Breast cancer anti-estrogen resistance protein 1 OS=Rattus norvegicus GN=Bcar1 PE=1 SV=1	RAT	1
758	2	2	3.9	sp|Q63468|KPRA_RAT	Phosphoribosyl pyrophosphate synthase-associated protein 1 OS=Rattus norvegicus GN=Prpsap1 PE=1 SV=1	RAT	1
759	2	2	1.5	sp|Q5U2N3|PITM1_RAT	Membrane-associated phosphatidylinositol transfer protein 1 OS=Rattus norvegicus GN=Pitpnm1 PE=2 SV=1	RAT	1
760	2	2	3.1	sp|Q5M821|PPM1H_RAT	Protein phosphatase 1H OS=Rattus norvegicus GN=Ppm1h PE=2 SV=2	RAT	1
761	2	2	1.5	sp|P49134|ITB1_RAT	Integrin beta-1 OS=Rattus norvegicus GN=Itgb1 PE=2 SV=1	RAT	1
762	2	2	3.5	sp|Q9JKM5|S1PR5_RAT	Sphingosine 1-phosphate receptor 5 OS=Rattus norvegicus GN=S1pr5 PE=2 SV=1	RAT	1
763	2	2	3.7	sp|Q9JHL4|DBNL_RAT	Drebrin-like protein OS=Rattus norvegicus GN=Dbnl PE=1 SV=1	RAT	1
764	2	2	3.1	sp|Q99P39|NFS1_RAT	Cysteine desulfurase, mitochondrial OS=Rattus norvegicus GN=Nfs1 PE=2 SV=1	RAT	1
765	2	2	3.4	sp|Q5RKI1|IF4A2_RAT	Eukaryotic initiation factor 4A-II OS=Rattus norvegicus GN=Eif4a2 PE=1 SV=1	RAT	1
766	2	2	1.6	sp|Q5BJS7|CPNE9_RAT	Copine-9 OS=Rattus norvegicus GN=Cpne9 PE=2 SV=1	RAT	1

(continued)

Table 1
(continued)

<table>
<tr><th>N</th><th>Unused</th><th>Total</th><th>% Cov (95)</th><th>Accession #</th><th>Name</th><th>Species</th><th>Peptides (95%)</th></tr>
<tr><td>767</td><td>2</td><td>2</td><td>0.9</td><td>sp|P97608|OPLA_RAT</td><td>5-oxoprolinase OS=Rattus norvegicus GN=Oplah PE=1 SV=2</td><td>RAT</td><td>1</td></tr>
<tr><td>768</td><td>2</td><td>2</td><td>3</td><td>sp|P61980|HNRPK_RAT</td><td>Heterogeneous nuclear ribonucleoprotein K OS=Rattus norvegicus GN=Hnrnpk PE=1 SV=1</td><td>RAT</td><td>1</td></tr>
<tr><td>769</td><td>2</td><td>2</td><td>2.3</td><td>sp|P61203|CSN2_RAT</td><td>COP9 signalosome complex subunit 2 OS=Rattus norvegicus GN=Cops2 PE=2 SV=1</td><td>RAT</td><td>1</td></tr>
<tr><td>770</td><td>2</td><td>2</td><td>3.8</td><td>sp|P49242|RS3A_RAT</td><td>40S ribosomal protein S3a OS=Rattus norvegicus GN=Rps3a PE=1 SV=2</td><td>RAT</td><td>1</td></tr>
<tr><td>771</td><td>2</td><td>2</td><td>2.1</td><td>sp|P31647|S6A11_RAT</td><td>Sodium- and chloride-dependent GABA transporter 3 OS=Rattus norvegicus GN=Slc6a11 PE=2 SV=1</td><td>RAT</td><td>1</td></tr>
<tr><td>772</td><td>2</td><td>2</td><td>2.4</td><td>sp|P26431|SL9A1_RAT</td><td>Sodium/hydrogen exchanger 1 OS=Rattus norvegicus GN=Slc9a1 PE=1 SV=2</td><td>RAT</td><td>1</td></tr>
<tr><td>773</td><td>2</td><td>2</td><td>1.5</td><td>sp|P10688|PLCD1_RAT</td><td>1-phosphatidylinositol-4,5-bisphosphate phosphodiesterase delta-1 OS=Rattus norvegicus GN=Plcd1 PE=1 SV=1</td><td>RAT</td><td>1</td></tr>
<tr><td>774</td><td>2</td><td>2</td><td>4.1</td><td>sp|Q8VHX0|CCG3_RAT</td><td>Voltage-dependent calcium channel gamma-3 subunit OS=Rattus norvegicus GN=Cacng3 PE=2 SV=1</td><td>RAT</td><td>1</td></tr>
<tr><td>775</td><td>2</td><td>2</td><td>3.7</td><td>sp|Q68FR9-2|EF1D_RAT</td><td>Isoform 2 of Elongation factor 1-delta OS=Rattus norvegicus GN=Eef1d</td><td>RAT</td><td>1</td></tr>
<tr><td>776</td><td>2</td><td>2</td><td>2</td><td>sp|Q5I0D7|PEPD_RAT</td><td>Xaa-Pro dipeptidase OS=Rattus norvegicus GN=Pepd PE=2 SV=1</td><td>RAT</td><td>1</td></tr>
<tr><td>777</td><td>2</td><td>2</td><td>3.1</td><td>sp|P40748|SYT3_RAT</td><td>Synaptotagmin-3 OS=Rattus norvegicus GN=Syt3 PE=1 SV=1</td><td>RAT</td><td>1</td></tr>
</table>

N	Unused	Total	% Cov (95)	Accession #	Name	Species	Peptides (95%)
778	2	2	4.2	sp|P23576|GBRA2_RAT	Gamma-aminobutyric acid receptor subunit alpha-2 OS=Rattus norvegicus GN=Gabra2 PE=2 SV=1	RAT	1
779	2	2	2.1	sp|P20069|MPPA_RAT	Mitochondrial-processing peptidase subunit alpha OS=Rattus norvegicus GN=Pmpca PE=1 SV=1	RAT	1
780	2	2	4.3	sp|P18665|RM03_RAT	39S ribosomal protein L3, mitochondrial OS=Rattus norvegicus GN=Mrpl3 PE=2 SV=1	RAT	1
781	2	2	1.6	sp|P15943|APLP2_RAT	Amyloid-like protein 2 OS=Rattus norvegicus GN=Aplp2 PE=2 SV=2	RAT	1
782	2	2	2.4	sp|P07687|HYEP_RAT	Epoxide hydrolase 1 OS=Rattus norvegicus GN=Ephx1 PE=1 SV=1	RAT	1
783	2	2	6.2	sp|Q9QYW3|MOBL3_RAT	Mps one binder kinase activator-like 3 OS=Rattus norvegicus GN=Mobkl1 PE=1 SV=1	RAT	1
784	2	2	3.7	sp|Q9QUR2-2|DCTN4_RAT	Isoform 2 of Dynactin subunit 4 OS=Rattus norvegicus GN=Dctn4	RAT	1
785	2	2	4.2	sp|Q920G2|NHRF2_RAT	Na(+)/H(+) exchange regulatory cofactor NHE-RF2 OS=Rattus norvegicus GN=Slc9a3r2 PE=1 SV=1	RAT	1
786	2	2	2	sp|Q6PST4|ATLA1_RAT	Atlastin-1 OS=Rattus norvegicus GN=Atl1 PE=1 SV=1	RAT	1
787	2	2	8.5	sp|Q642G4|PEX14_RAT	Peroxisomal membrane protein PEX14 OS=Rattus norvegicus GN=Pex14 PE=1 SV=1	RAT	1
788	2	2	2.1	sp|Q63413|UAP56_RAT	Spliceosome RNA helicase Bat1 OS=Rattus norvegicus GN=Bat1 PE=1 SV=3	RAT	1
789	2	2	2.3	sp|Q62765|NLGN1_RAT	Neuroligin-1 OS=Rattus norvegicus GN=Nlgn1 PE=1 SV=1	RAT	1

(continued)

Table 1
(continued)

<table>
<tr><th>N</th><th>Unused</th><th>Total</th><th>% Cov (95)</th><th>Accession #</th><th>Name</th><th>Species</th><th>Peptides (95%)</th></tr>
<tr><td>790</td><td>2</td><td>2</td><td>3.3</td><td>sp|Q5XIC0|PECI_RAT</td><td>Peroxisomal 3,2-trans-enoyl-CoA isomerase OS=Rattus norvegicus GN=Peci PE=1 SV=1</td><td>RAT</td><td>1</td></tr>
<tr><td>791</td><td>2</td><td>2</td><td>5.6</td><td>sp|Q3T1J9|MOL1B_RAT</td><td>Mps one binder kinase activator-like 1B OS=Rattus norvegicus GN=Mobkl1b PE=1 SV=3</td><td>RAT</td><td>1</td></tr>
<tr><td>792</td><td>2</td><td>2</td><td>1.4</td><td>sp|Q01956|KCNC3_RAT</td><td>Potassium voltage-gated channel subfamily C member 3 OS=Rattus norvegicus GN=Kcnc3 PE=2 SV=1</td><td>RAT</td><td>1</td></tr>
<tr><td>793</td><td>2</td><td>2</td><td>4.2</td><td>sp|P62703|RS4X_RAT</td><td>40S ribosomal protein S4, X isoform OS=Rattus norvegicus GN=Rps4x PE=2 SV=2</td><td>RAT</td><td>1</td></tr>
<tr><td>794</td><td>2</td><td>2</td><td>5</td><td>sp|P37805|TAGL3_RAT</td><td>Transgelin-3 OS=Rattus norvegicus GN=Tagln3 PE=1 SV=2</td><td>RAT</td><td>1</td></tr>
<tr><td>795</td><td>2</td><td>2</td><td>2.9</td><td>sp|P15087|CBPE_RAT</td><td>Carboxypeptidase E OS=Rattus norvegicus GN=Cpe PE=2 SV=1</td><td>RAT</td><td>1</td></tr>
<tr><td>796</td><td>2</td><td>2</td><td>1.5</td><td>sp|O09175|AMPB_RAT</td><td>Aminopeptidase B OS=Rattus norvegicus GN=Rnpep PE=1 SV=2</td><td>RAT</td><td>1</td></tr>
<tr><td>797</td><td>2</td><td>2</td><td>2</td><td>sp|Q9Z1Y3|CADH2_RAT</td><td>Cadherin-2 OS=Rattus norvegicus GN=Cdh2 PE=2 SV=1</td><td>RAT</td><td>1</td></tr>
<tr><td>798</td><td>2</td><td>2</td><td>1.7</td><td>sp|Q9WTW8|S23A2_RAT</td><td>Solute carrier family 23 member 2 OS=Rattus norvegicus GN=Slc23a2 PE=2 SV=2</td><td>RAT</td><td>1</td></tr>
<tr><td>799</td><td>2</td><td>2</td><td>2.1</td><td>sp|Q8VIF7|SBP1_RAT</td><td>Selenium-binding protein 1 OS=Rattus norvegicus GN=Selenbp1 PE=1 SV=1</td><td>RAT</td><td>1</td></tr>
<tr><td>800</td><td>2</td><td>2</td><td>3.5</td><td>sp|Q769K2|NAPEP_RAT</td><td>N-acyl-phosphatidylethanolamine-hydrolyzing phospholipase D OS=Rattus norvegicus GN=Napepld PE=1 SV=1</td><td>RAT</td><td>1</td></tr>
<tr><td>801</td><td>2</td><td>2</td><td>3.2</td><td>sp|Q6AY94|CC001_RAT</td><td>Transmembrane protein C3orf1 homolog OS=Rattus norvegicus PE=2 SV=1</td><td>RAT</td><td>1</td></tr>
</table>

N	Unused	Total	% Cov (95)	Accession #	Name	Species	Peptides (95%)
802	2	2	8.6	sp|Q66HA6|ARL8B_RAT	ADP-ribosylation factor-like protein 8B OS=Rattus norvegicus GN=Arl8b PE=2 SV=1	RAT	2
803	2	2	4.4	sp|Q64640|ADK_RAT	Adenosine kinase OS=Rattus norvegicus GN=Adk PE=1 SV=3	RAT	1
804	2	2	2.9	sp|Q63570|PRS6B_RAT	26S protease regulatory subunit 6B OS=Rattus norvegicus GN=Psmc4 PE=1 SV=1	RAT	1
805	2	2	4.2	sp|Q566C7|NUDT3_RAT	Diphosphoinositol polyphosphate phosphohydrolase 1 OS=Rattus norvegicus GN=Nudt3 PE=1 SV=1	RAT	1
806	2	2	4.6	sp|Q4KM98|MFF_RAT	Mitochondrial fission factor OS=Rattus norvegicus GN=Mff PE=2 SV=1	RAT	1
807	2	2	3.1	sp|Q4FZT0|STML2_RAT	Stomatin-like protein 2 OS=Rattus norvegicus GN=Stoml2 PE=1 SV=1	RAT	1
808	2	2	23.3	sp|P63025|VAMP3_RAT	Vesicle-associated membrane protein 3 OS=Rattus norvegicus GN=Vamp3 PE=2 SV=1	RAT	1
809	2	2	4.4	sp|P46844|BIEA_RAT	Biliverdin reductase A OS=Rattus norvegicus GN=Blvra PE=1 SV=1	RAT	1
810	2	2	4	sp|P29411|KAD3_RAT	GTP:AMP phosphotransferase mitochondrial OS=Rattus norvegicus GN=Ak3 PE=2 SV=2	RAT	1
811	2	2	3.3	sp|P29315|RINI_RAT	Ribonuclease inhibitor OS=Rattus norvegicus GN=Rnh1 PE=1 SV=2	RAT	1
812	2	2	11.5	sp|P23358|RL12_RAT	60S ribosomal protein L12 OS=Rattus norvegicus GN=Rpl12 PE=2 SV=1	RAT	1
813	2	2	8.5	sp|P11250|RL34_RAT	60S ribosomal protein L34 OS=Rattus norvegicus GN=Rpl34 PE=1 SV=3	RAT	1
814	2	2	4.6	sp|B2RYF7|WASH1_RAT	WAS protein family homolog 1 OS=Rattus norvegicus GN=Wash1 PE=2 SV=1	RAT	1

(continued)

Table 1
(continued)

N	Unused	Total	% Cov (95)	Accession #	Name	Species	Peptides (95%)
815	2	2	4.9	sp|Q9Z1B2|GSTM5_RAT	Glutathione S-transferase Mu 5 OS=Rattus norvegicus GN=Gstm5 PE=1 SV=3	RAT	1
816	2	2	2	sp|Q9QZG1|S22AN_RAT	Solute carrier family 22 member 23 OS=Rattus norvegicus GN=Slc22a23 PE=2 SV=2	RAT	1
817	2	2	9.3	sp|Q6PEC1|TBCA_RAT	Tubulin-specific chaperone A OS=Rattus norvegicus GN=Tbca PE=1 SV=1	RAT	1
818	2	2	14.7	sp|Q6P7Q4|LGUL_RAT	Lactoylglutathione lyase OS=Rattus norvegicus GN=Glo1 PE=1 SV=3	RAT	1
819	2	2	1.9	sp|Q63060|GLPK_RAT	Glycerol kinase OS=Rattus norvegicus GN=Gk PE=2 SV=1	RAT	1
820	2	2	8.4	sp|Q5U1X1|ORN_RAT	Oligoribonuclease, mitochondrial OS=Rattus norvegicus GN=Rexo2 PE=2 SV=1	RAT	1
821	2	2	4.1	sp|Q5BJQ2|FA63A_RAT	Protein FAM63A OS=Rattus norvegicus GN=Fam63a PE=2 SV=1	RAT	1
822	2	2	6.4	sp|Q08850|STX4_RAT	Syntaxin-4 OS=Rattus norvegicus GN=Stx4 PE=1 SV=1	RAT	1
823	2	2	16.1	sp|P63055|PCP4_RAT	Purkinje cell protein 4 OS=Rattus norvegicus GN=Pcp4 PE=1 SV=2	RAT	1
824	2	2	13.3	sp|P55053|FABP5_RAT	Fatty acid-binding protein, epidermal OS=Rattus norvegicus GN=Fabp5 PE=1 SV=3	RAT	1
825	2	2	4.2	sp|P21775|THIKA_RAT	3-ketoacyl-CoA thiolase A, peroxisomal OS=Rattus norvegicus GN=Acaa1a PE=2 SV=2	RAT	1
826	2	2	4.6	sp|P18420|PSA1_RAT	Proteasome subunit alpha type-1 OS=Rattus norvegicus GN=Psma1 PE=1 SV=2	RAT	1
827	2	2	2.2	sp|P17164|FUCO_RAT	Tissue alpha-L-fucosidase OS=Rattus norvegicus GN=Fuca1 PE=1 SV=1	RAT	1

<table>
<tr><th>N</th><th>Unused</th><th>Total</th><th>% Cov (95)</th><th>Accession #</th><th>Name</th><th>Species</th><th>Peptides (95%)</th></tr>
<tr><td>828</td><td>2</td><td>2</td><td>2.9</td><td>sp|P15384|KCNA3_RAT</td><td>Potassium voltage-gated channel subfamily A member 3 OS=Rattus norvegicus GN=Kcna3 PE=1 SV=2</td><td>RAT</td><td>1</td></tr>
<tr><td>829</td><td>2</td><td>2</td><td>3.2</td><td>sp|P12711|ADHX_RAT</td><td>Alcohol dehydrogenase class-3 OS=Rattus norvegicus GN=Adh5 PE=1 SV=2</td><td>RAT</td><td>1</td></tr>
<tr><td>830</td><td>2</td><td>2</td><td>4</td><td>sp|O88775|EMB_RAT</td><td>Embigin OS=Rattus norvegicus GN=Emb PE=1 SV=1</td><td>RAT</td><td>1</td></tr>
<tr><td>831</td><td>2</td><td>2</td><td>4.9</td><td>sp|Q9Z142|TMM33_RAT</td><td>Transmembrane protein 33 OS=Rattus norvegicus GN=Tmem33 PE=2 SV=1</td><td>RAT</td><td>1</td></tr>
<tr><td>832</td><td>2</td><td>2</td><td>16.9</td><td>sp|Q9WV97|TIM9_RAT</td><td>Mitochondrial import inner membrane translocase subunit Tim9 OS=Rattus norvegicus GN=Timm9 PE=1 SV=3</td><td>RAT</td><td>1</td></tr>
<tr><td>833</td><td>2</td><td>2</td><td>2.5</td><td>sp|Q9EQV6|TPP1_RAT</td><td>Tripeptidyl-peptidase 1 OS=Rattus norvegicus GN=Tpp1 PE=1 SV=1</td><td>RAT</td><td>1</td></tr>
<tr><td>834</td><td>2</td><td>2</td><td>8.1</td><td>sp|Q78EG7|TP4A1_RAT</td><td>Protein tyrosine phosphatase type IVA 1 OS=Rattus norvegicus GN=Ptp4a1 PE=1 SV=1</td><td>RAT</td><td>1</td></tr>
<tr><td>835</td><td>2</td><td>2</td><td>9.2</td><td>sp|Q642A4|CH082_RAT</td><td>UPF0598 protein C8orf82 homolog OS=Rattus norvegicus PE=2 SV=1</td><td>RAT</td><td>1</td></tr>
<tr><td>836</td><td>2</td><td>2</td><td>3.4</td><td>sp|Q641X9|RM09_RAT</td><td>39S ribosomal protein L9, mitochondrial OS=Rattus norvegicus GN=Mrpl9 PE=2 SV=1</td><td>RAT</td><td>1</td></tr>
<tr><td>837</td><td>2</td><td>2</td><td>7</td><td>sp|Q63228|GMFB_RAT</td><td>Glia maturation factor beta OS=Rattus norvegicus GN=Gmfb PE=1 SV=2</td><td>RAT</td><td>1</td></tr>
<tr><td>838</td><td>2</td><td>2</td><td>9.2</td><td>sp|Q5U318|PEA15_RAT</td><td>Astrocytic phosphoprotein PEA-15 OS=Rattus norvegicus GN=Pea15 PE=1 SV=1</td><td>RAT</td><td>1</td></tr>
<tr><td>839</td><td>2</td><td>2</td><td>5.9</td><td>sp|Q5I0P2|GCSH_RAT</td><td>Glycine cleavage system H protein, mitochondrial OS=Rattus norvegicus GN=Gcsh PE=2 SV=1</td><td>RAT</td><td>1</td></tr>
</table>

(continued)

Table 1
(continued)

N	Unused	Total	% Cov (95)	Accession #	Name	Species	Peptides (95%)
840	2	2	6.2	sp|Q5I0K3|CLYBL_RAT	Citrate lyase subunit beta-like protein, mitochondrial OS=Rattus norvegicus GN=Clybl PE=2 SV=2	RAT	1
841	2	2	2.6	sp|Q4V7F2|CREL1_RAT	Cysteine-rich with EGF-like domain protein 1 OS=Rattus norvegicus GN=Creld1 PE=2 SV=1	RAT	1
842	2	2	16.5	sp|P62890|RL30_RAT	60S ribosomal protein L30 OS=Rattus norvegicus GN=Rpl30 PE=3 SV=2	RAT	1
843	2	2	15.8	sp|P62076|TIM13_RAT	Mitochondrial import inner membrane translocase subunit Tim13 OS=Rattus norvegicus GN=Timm13 PE=2 SV=1	RAT	1
844	2	2	15.6	sp|P62074|TIM10_RAT	Mitochondrial import inner membrane translocase subunit Tim10 OS=Rattus norvegicus GN=Timm10 PE=2 SV=1	RAT	2
845	2	2	5.8	sp|P42533|TF_RAT	Tissue factor OS=Rattus norvegicus GN=F3 PE=2 SV=1	RAT	1
846	2	2	8.2	sp|P41498|PPAC_RAT	Low molecular weight phosphotyrosine protein phosphatase OS=Rattus norvegicus GN=Acp1 PE=1 SV=3	RAT	1
847	2	2	15.7	sp|P35171|CX7A2_RAT	Cytochrome c oxidase subunit 7A2, mitochondrial OS=Rattus norvegicus GN=Cox7a2 PE=1 SV=1	RAT	4
848	2	2	2.4	sp|P25093|FAAA_RAT	Fumarylacetoacetase OS=Rattus norvegicus GN=Fah PE=1 SV=1	RAT	1
849	2	2	7.8	sp|P19814|TGON3_RAT	Trans-Golgi network integral membrane protein TGN38 OS=Rattus norvegicus GN=Ttgn1 PE=1 SV=1	RAT	1

N	Unused	Total	% Cov (95)	Accession #	Name	Species	Peptides (95%)
850	2	2	3.2	sp|P11960|ODBA_RAT	2-oxoisovalerate dehydrogenase subunit alpha, mitochondrial (Fragment) OS=Rattus norvegicus GN=Bckdha PE=1 SV=1	RAT	1
851	2	2	6	sp|P0C546|S2542_RAT	Solute carrier family 25 member 42 OS=Rattus norvegicus GN=Slc25a42 PE=2 SV=1	RAT	1
852	2	2	2.6	sp|Q9Z0J8|NEGR1_RAT	Neuronal growth regulator 1 OS=Rattus norvegicus GN=Negr1 PE=1 SV=1	RAT	1
853	2	2	8.3	sp|Q9WVJ4|SYJ2B_RAT	Synaptojanin-2-binding protein OS=Rattus norvegicus GN=Synj2bp PE=1 SV=2	RAT	1
854	2	2	8.3	sp|Q9QYU2|EFTS_RAT	Elongation factor Ts, mitochondrial OS=Rattus norvegicus GN=Tsfm PE=2 SV=1	RAT	1
855	2	2	4.7	sp|Q923M1|MSRA_RAT	Peptide methionine sulfoxide reductase OS=Rattus norvegicus GN=Msra PE=2 SV=1	RAT	1
856	2	2	8.7	sp|Q91Y78|UCHL3_RAT	Ubiquitin carboxyl-terminal hydrolase isozyme L3 OS=Rattus norvegicus GN=Uchl3 PE=1 SV=1	RAT	1
857	2	2	11.6	sp|Q6XVN8|MLP3A_RAT	Microtubule-associated proteins 1A/1B light chain 3A OS=Rattus norvegicus GN=Map1lc3a PE=1 SV=1	RAT	1
858	2	2	10.2	sp|Q6PEC0|AP4A_RAT	Bis(5'-nucleosyl)-tetraphosphatase [asymmetrical] OS=Rattus norvegicus GN=Nudt2 PE=2 SV=3	RAT	1
859	2	2	6.3	sp|Q6PCT8|DHSD_RAT	Succinate dehydrogenase [ubiquinone] cytochrome b small subunit, mitochondrial OS=Rattus norvegicus GN=Sdhd PE=2 SV=1	RAT	1
860	2	2	5.1	sp|Q62876|SNG1_RAT	Synaptogyrin-1 OS=Rattus norvegicus GN=Syngr1 PE=2 SV=1	RAT	1

(continued)

Table 1
(continued)

N	Unused	Total	% Cov (95)	Accession #	Name	Species	Peptides (95%)
861	2	2	5.5	sp|Q5XIM4|ATP5S_RAT	ATP synthase subunit s, mitochondrial OS=Rattus norvegicus GN=Atp5s PE=2 SV=1	RAT	1
862	2	2	4.2	sp|Q5XII9|FA54B_RAT	Protein FAM54B OS=Rattus norvegicus GN=Fam54b PE=2 SV=1	RAT	1
863	2	2	8.7	sp|Q4KM86|A2LD1_RAT	AIG2-like domain-containing protein 1 OS=Rattus norvegicus GN=A2ld1 PE=2 SV=1	RAT	1
864	2	2	5.6	sp|Q4KM74|SC22B_RAT	Vesicle-trafficking protein SEC22b OS=Rattus norvegicus GN=Sec22b PE=1 SV=3	RAT	1
865	2	2	8.8	sp|Q498E0|TXD12_RAT	Thioredoxin domain-containing protein 12 OS=Rattus norvegicus GN=Txndc12 PE=2 SV=2	RAT	1
866	2	2	3	sp|P84850|D2HDH_RAT	D-2-hydroxyglutarate dehydrogenase, mitochondrial OS=Rattus norvegicus GN=D2hgdh PE=3 SV=1	RAT	1
867	2	2	9.2	sp|P60868|RS20_RAT	40S ribosomal protein S20 OS=Rattus norvegicus GN=Rps20 PE=3 SV=1	RAT	1
868	2	2	13.6	sp|P55051|FABP7_RAT	Fatty acid-binding protein, brain OS=Rattus norvegicus GN=Fabp7 PE=1 SV=2	RAT	1
869	2	2	8.9	sp|P11762|LEG1_RAT	Galectin-1 OS=Rattus norvegicus GN=Lgals1 PE=1 SV=2	RAT	1
870	2	2	16.3	sp|P04631|S100B_RAT	Protein S100-B OS=Rattus norvegicus GN=S100b PE=1 SV=2	RAT	1
871	2	2	4.6	sp|Q9WUH4|FHL1_RAT	Four and a half LIM domains protein 1 OS=Rattus norvegicus GN=Fhl1 PE=2 SV=1	RAT	1
872	2	2	6.5	sp|Q63327|MOBP_RAT	Myelin-associated oligodendrocyte basic protein OS=Rattus norvegicus GN=Mobp PE=1 SV=1	RAT	1
873	2	2	6.8	sp|Q62745|CD81_RAT	CD81 antigen OS=Rattus norvegicus GN=Cd81 PE=1 SV=1	RAT	1

N	Unused	Total	% Cov (95)	Accession #	Name	Species	Peptides (95%)
874	2	2	2.7	sp|Q5XIK1|ACTT1_RAT	Actin-related protein T1 OS=Rattus norvegicus GN=Actrt1 PE=2 SV=1	RAT	1
875	2	2	3.8	sp|Q5RKJ2|CYBR1_RAT	Cytochrome b reductase 1 OS=Rattus norvegicus GN=Cybrd1 PE=2 SV=1	RAT	1
876	2	2	11.3	sp|Q5PR01|PPDPF_RAT	Pancreatic progenitor cell differentiation and proliferation factor OS=Rattus norvegicus GN=Ppdpf PE=2 SV=1	RAT	1
877	2	2	3.5	sp|Q08406|CNTFR_RAT	Ciliary neurotrophic factor receptor subunit alpha OS=Rattus norvegicus GN=Cntfr PE=2 SV=2	RAT	1
878	2	2	13.6	sp|P63301|SELW_RAT	Selenoprotein W OS=Rattus norvegicus GN=Sepw1 PE=1 SV=3	RAT	1
879	2	2	2.2	sp|P58366|ANKH_RAT	Progressive ankylosis protein homolog OS=Rattus norvegicus GN=Ankh PE=2 SV=1	RAT	1
880	2	2	6	sp|P54900|SCN2B_RAT	Sodium channel subunit beta-2 OS=Rattus norvegicus GN=Scn2b PE=1 SV=1	RAT	1
881	2	2	3.7	sp|P38983|RSSA_RAT	40S ribosomal protein SA OS=Rattus norvegicus GN=Rpsa PE=1 SV=3	RAT	1
882	2	2	4	sp|P35704|PRDX2_RAT	Peroxiredoxin-2 OS=Rattus norvegicus GN=Prdx2 PE=1 SV=3	RAT	1
883	2	2	13.9	sp|P02401|RLA2_RAT	60S acidic ribosomal protein P2 OS=Rattus norvegicus GN=Rplp2 PE=1 SV=2	RAT	1
884	2	2	2.6	sp|O54699|S29A2_RAT	Equilibrative nucleoside transporter 2 OS=Rattus norvegicus GN=Slc29a2 PE=2 SV=1	RAT	1
885	2	2	8.4	sp|B5DF51|MMGT1_RAT	Membrane magnesium transporter 1 OS=Rattus norvegicus GN=Mmgt1 PE=2 SV=1	RAT	1
886	2	2	3.4	sp|A6ZIQ8|TM38A_RAT	Trimeric intracellular cation channel type A OS=Rattus norvegicus GN=Tmem38a PE=2 SV=1	RAT	1

(continued)

Table 1
(continued)

N	Unused	Total	% Cov (95)	Accession #	Name	Species	Peptides (95%)
887	2	2	5.4	sp|Q9QXU9|PCSK1_RAT	ProSAAS OS=Rattus norvegicus GN=Pcsk1n PE=1 SV=1	RAT	1
888	2	2	4.7	sp|Q9JKE3|SCAM5_RAT	Secretory carrier-associated membrane protein 5 OS=Rattus norvegicus GN=Scamp5 PE=2 SV=1	RAT	1
889	2	2	19.4	sp|Q6PCU8|NDUV3_RAT	NADH dehydrogenase [ubiquinone] flavoprotein 3, mitochondrial OS=Rattus norvegicus GN=Ndufv3 PE=3 SV=1	RAT	1
890	2	2	6	sp|Q64119|MYL6_RAT	Myosin light polypeptide 6 OS=Rattus norvegicus GN=Myl6 PE=1 SV=3	RAT	1
891	2	2	4.5	sp|Q63524|TMED2_RAT	Transmembrane emp24 domain-containing protein 2 OS=Rattus norvegicus GN=Tmed2 PE=1 SV=1	RAT	1
892	2	2	18.2	sp|P37361|MT3_RAT	Metallothionein-3 OS=Rattus norvegicus GN=Mt3 PE=2 SV=1	RAT	1
893	2	2	9.5	sp|P27274|CD59_RAT	CD59 glycoprotein OS=Rattus norvegicus GN=Cd59 PE=1 SV=2	RAT	1
894	2	2	5.4	sp|P05505|COX3_RAT	Cytochrome c oxidase subunit 3 OS=Rattus norvegicus GN=Mtco3 PE=1 SV=5	RAT	2
895	2	2	2.4	sp|P00159|CYB_RAT	Cytochrome b OS=Rattus norvegicus GN=Mt-Cyb PE=2 SV=2	RAT	1
896	2	2	4.7	sp|A2RUW1|TOLIP_RAT	Toll-interacting protein OS=Rattus norvegicus GN=Tollip PE=2 SV=1	RAT	2
897	1.89	2	5	sp|Q6AY20|MPRD_RAT	Cation-dependent mannose-6-phosphate receptor OS=Rattus norvegicus GN=M6pr PE=2 SV=1	RAT	1
898	1.82	2	6.4	sp|P62752|RL23A_RAT	60S ribosomal protein L23a OS=Rattus norvegicus GN=Rpl23a PE=2 SV=1	RAT	1

<table>
<tr><th>N</th><th>Unused</th><th>Total</th><th>% Cov (95)</th><th>Accession #</th><th>Name</th><th>Species</th><th>Peptides (95%)</th></tr>
<tr><td>899</td><td>1.8</td><td>2</td><td>1.2</td><td>sp|P57097|MERTK_RAT</td><td>Tyrosine-protein kinase Mer OS=Rattus norvegicus GN=Mertk PE=2 SV=1</td><td>RAT</td><td>1</td></tr>
<tr><td>900</td><td>1.77</td><td>2</td><td>4.1</td><td>sp|Q8VIL3|ZWINT_RAT</td><td>ZW10 interactor OS=Rattus norvegicus GN=Zwint PE=1 SV=1</td><td>RAT</td><td>1</td></tr>
<tr><td>901</td><td>1.75</td><td>2</td><td>1.2</td><td>sp|P50475|SYAC_RAT</td><td>Alanyl-tRNA synthetase, cytoplasmic OS=Rattus norvegicus GN=Aars PE=1 SV=3</td><td>RAT</td><td>1</td></tr>
<tr><td>902</td><td>1.7</td><td>2</td><td>1.8</td><td>sp|Q5M7V7|FAKD2_RAT</td><td>FAST kinase domain-containing protein 2 OS=Rattus norvegicus GN=Fastkd2 PE=2 SV=1</td><td>RAT</td><td>1</td></tr>
<tr><td>903</td><td>1.7</td><td>2</td><td>2.9</td><td>sp|P53678|AP3M2_RAT</td><td>AP-3 complex subunit mu-2 OS=Rattus norvegicus GN=Ap3m2 PE=2 SV=1</td><td>RAT</td><td>1</td></tr>
<tr><td>904</td><td>1.64</td><td>4</td><td>4.1</td><td>sp|P31016|DLG4_RAT</td><td>Disks large homolog 4 OS=Rattus norvegicus GN=Dlg4 PE=1 SV=1</td><td>RAT</td><td>2</td></tr>
<tr><td>905</td><td>1.64</td><td>2</td><td>6.7</td><td>sp|P00173|CYB5_RAT</td><td>Cytochrome b5 OS=Rattus norvegicus GN=Cyb5a PE=1 SV=2</td><td>RAT</td><td>1</td></tr>
<tr><td>906</td><td>1.62</td><td>2</td><td>4.3</td><td>sp|P97521|MCAT_RAT</td><td>Mitochondrial carnitine/acylcarnitine carrier protein OS=Rattus norvegicus GN=Slc25a20 PE=1 SV=1</td><td>RAT</td><td>1</td></tr>
<tr><td>907</td><td>1.6</td><td>1.96</td><td>1.3</td><td>sp|Q8K4K5|L2GL1_RAT</td><td>Lethal(2) giant larvae protein homolog 1 OS=Rattus norvegicus GN=Llgl1 PE=2 SV=1</td><td>RAT</td><td>1</td></tr>
<tr><td>908</td><td>1.58</td><td>13.02</td><td>17.3</td><td>sp|P62632|EF1A2_RAT</td><td>Elongation factor 1-alpha 2 OS=Rattus norvegicus GN=Eef1a2 PE=2 SV=1</td><td>RAT</td><td>8</td></tr>
<tr><td>909</td><td>1.55</td><td>1.92</td><td>2.7</td><td>sp|Q5HZE4|MTNA_RAT</td><td>Methylthioribose-1-phosphate isomerase OS=Rattus norvegicus GN=Mri1 PE=2 SV=1</td><td>RAT</td><td>1</td></tr>
<tr><td>910</td><td>1.53</td><td>1.99</td><td>3.9</td><td>sp|Q4FZT8|SPRY4_RAT</td><td>SPRY domain-containing protein 4 OS=Rattus norvegicus GN=Spryd4 PE=2 SV=1</td><td>RAT</td><td>1</td></tr>
<tr><td>911</td><td>1.51</td><td>1.89</td><td>1.4</td><td>sp|O54975|XPP1_RAT</td><td>Xaa-Pro aminopeptidase 1 OS=Rattus norvegicus GN=Xpnpep1 PE=1 SV=1</td><td>RAT</td><td>1</td></tr>
</table>

(continued)

Table 1
(continued)

N	Unused	Total	% Cov (95)	Accession #	Name	Species	Peptides (95%)
912	1.5	2.2	2.4	sp|Q9R080|GPSM1_RAT	G-protein-signaling modulator 1 OS=Rattus norvegicus GN=Gpsm1 PE=1 SV=2	RAT	2
913	1.49	1.99	0.8	sp|Q9JKA7|HCN4_RAT	Potassium/sodium hyperpolarization-activated cyclic nucleotide-gated channel 4 OS=Rattus norvegicus GN=Hcn4 PE=2 SV=1	RAT	1
914	1.49	1.93	2.6	sp|Q9Z2G8|NP1L1_RAT	Nucleosome assembly protein 1-like 1 OS=Rattus norvegicus GN=Nap1l1 PE=2 SV=1	RAT	1
915	1.46	1.82	7	sp|Q9QYU1|PEX19_RAT	Peroxisomal biogenesis factor 19 OS=Rattus norvegicus GN=Pex19 PE=2 SV=1	RAT	1
916	1.44	2	1.6	sp|Q64380|SARDH_RAT	Sarcosine dehydrogenase, mitochondrial OS=Rattus norvegicus GN=Sardh PE=1 SV=2	RAT	1
917	1.44	1.88	6.6	sp|B2RYG6|OTUB1_RAT	Ubiquitin thioesterase OTUB1 OS=Rattus norvegicus GN=Otub1 PE=1 SV=1	RAT	1
918	1.42	1.8	2.6	sp|Q07205|IF5_RAT	Eukaryotic translation initiation factor 5 OS=Rattus norvegicus GN=Eif5 PE=1 SV=1	RAT	1
919	1.41	1.77	5	sp|P04904|GSTA3_RAT	Glutathione S-transferase alpha-3 OS=Rattus norvegicus GN=Gsta3 PE=1 SV=3	RAT	1
920	1.41	1.77	4.4	sp|Q64680|CP2DI_RAT	Cytochrome P450 2D18 OS=Rattus norvegicus GN=Cyp2d18 PE=2 SV=1	RAT	1
921	1.39	1.77	3.4	sp|Q5XIT9|MCCB_RAT	Methylcrotonoyl-CoA carboxylase beta chain, mitochondrial OS=Rattus norvegicus GN=Mccc2 PE=2 SV=1	RAT	1
922	1.38	1.75	4.9	sp|Q3KRD5|TOM34_RAT	Mitochondrial import receptor subunit TOM34 OS=Rattus norvegicus GN=Tomm34 PE=2 SV=1	RAT	1
923	1.35	1.72	3.5	sp|Q3B7U9|FKBP8_RAT	Peptidyl-prolyl cis-trans isomerase FKBP8 OS=Rattus norvegicus GN=Fkbp8 PE=2 SV=1	RAT	1

N	Unused	Total	% Cov (95)	Accession #	Name	Species	Peptides (95%)
924	1.33	2	3.5	sp|P19945|RLA0_RAT	60S acidic ribosomal protein P0 OS=Rattus norvegicus GN=Rplp0 PE=1 SV=2	RAT	1
925	1.33	1.7	2.6	sp|Q4G069|RMD1_RAT	Regulator of microtubule dynamics protein 1 OS=Rattus norvegicus GN=Fam82b PE=2 SV=1	RAT	1
926	1.26	1.63	2.3	sp|P26817|ARBK1_RAT	Beta-adrenergic receptor kinase 1 OS=Rattus norvegicus GN=Adrbk1 PE=2 SV=1	RAT	1
927	1.24	1.6	3.2	sp|Q63159|COQ3_RAT	Hexaprenyldihydroxybenzoate methyltransferase, mitochondrial OS=Rattus norvegicus GN=Coq3 PE=2 SV=2	RAT	1
928	1.24	1.6	2	sp|Q62784|INP4A_RAT	Type I inositol-3,4-bisphosphate 4-phosphatase OS=Rattus norvegicus GN=Inpp4a PE=1 SV=1	RAT	1
929	1.24	1.6	20.6	sp|P43425|GBG7_RAT	Guanine nucleotide-binding protein G(I)/G(S)/G(O) subunit gamma-7 OS=Rattus norvegicus GN=Gng7 PE=2 SV=2	RAT	1
930	1.23	1.61	4.2	sp|Q5RJR8|LRC59_RAT	Leucine-rich repeat-containing protein 59 OS=Rattus norvegicus GN=Lrrc59 PE=1 SV=1	RAT	1
931	1.22	1.59	6.9	sp|Q9JHE8|NINJ2_RAT	Ninjurin-2 OS=Rattus norvegicus GN=Ninj2 PE=2 SV=1	RAT	1
932	1.22	1.59	15.2	sp|Q5BJM7|C42S1_RAT	CDC42 small effector protein 1 OS=Rattus norvegicus GN=Cdc42se1 PE=2 SV=1	RAT	1
933	1.18	1.89	3.5	sp|A0JPM9|EIF3J_RAT	Eukaryotic translation initiation factor 3 subunit J OS=Rattus norvegicus GN=Eif3j PE=2 SV=1	RAT	1
934	1.18	1.73	0.5	sp|Q9JLA3|UGGG1_RAT	UDP-glucose:glycoprotein glucosyltransferase 1 OS=Rattus norvegicus GN=Uggt1 PE=1 SV=2	RAT	1

(continued)

Table 1 (continued)

N	Unused	Total	% Cov (95)	Accession #	Name	Species	Peptides (95%)
935	1.17	1.75	4.1	sp|Q6AYE2|SHLB1_RAT	Endophilin-B1 OS=Rattus norvegicus GN=Sh3glb1 PE=2 SV=1	RAT	1
936	1.16	1.52	1.7	sp|Q9JJ50|HGS_RAT	Hepatocyte growth factor-regulated tyrosine kinase substrate OS=Rattus norvegicus GN=Hgs PE=1 SV=1	RAT	1
937	1.14	1.51	0.9	sp|O54921|EXOC2_RAT	Exocyst complex component 2 OS=Rattus norvegicus GN=Exoc2 PE=1 SV=1	RAT	1
938	1.12	1.66	9.6	sp|P62912|RL32_RAT	60S ribosomal protein L32 OS=Rattus norvegicus GN=Rpl32 PE=1 SV=2	RAT	1
939	1.12	1.48	4.1	sp|Q920P0|DCXR_RAT	L-xylulose reductase OS=Rattus norvegicus GN=Dcxr PE=1 SV=1	RAT	1
940	1.12	1.48	4.4	sp|P05504|ATP6_RAT	ATP synthase subunit a OS=Rattus norvegicus GN=Mt-atp6 PE=1 SV=2	RAT	1
941	1.11	1.47	4.4	sp|P62957|CCG7_RAT	Voltage-dependent calcium channel gamma-7 subunit OS=Rattus norvegicus GN=Cacng7 PE=2 SV=1	RAT	1
942	1.1	1.46	4.6	sp|P70580|PGRC1_RAT	Membrane-associated progesterone receptor component 1 OS=Rattus norvegicus GN=Pgrmc1 PE=1 SV=3	RAT	1
943	1.09	1.44	2.9	sp|Q7TNK0|SERC1_RAT	Serine incorporator 1 OS=Rattus norvegicus GN=Serinc1 PE=1 SV=1	RAT	1
944	1.08	1.43	1.5	sp|Q62825|EXOC3_RAT	Exocyst complex component 3 OS=Rattus norvegicus GN=Exoc3 PE=1 SV=1	RAT	1
945	1.05	1.41	1.5	sp|Q80WF4|T132A_RAT	Transmembrane protein 132A OS=Rattus norvegicus GN=Tmem132a PE=1 SV=1	RAT	1
946	1.04	1.39	3.5	sp|Q4V7E5|RF1ML_RAT	Peptide chain release factor 1-like, mitochondrial OS=Rattus norvegicus GN=Mtrf1l PE=2 SV=1	RAT	1

<table>
<tr><th>N</th><th>Unused</th><th>Total</th><th>% Cov (95)</th><th>Accession #</th><th>Name</th><th>Species</th><th>Peptides (95%)</th></tr>
<tr><td>947</td><td>1.03</td><td>1.6</td><td>16.7</td><td>sp|B0BN94|F136A_RAT</td><td>Protein FAM136A OS=Rattus norvegicus GN=Fam136a PE=2 SV=1</td><td>RAT</td><td>2</td></tr>
<tr><td>948</td><td>1.02</td><td>1.63</td><td>1.3</td><td>sp|Q9QYF3|MYO5A_RAT</td><td>Myosin-Va OS=Rattus norvegicus GN=Myo5a PE=2 SV=1</td><td>RAT</td><td>2</td></tr>
<tr><td>949</td><td>1.01</td><td>1.46</td><td>1.1</td><td>sp|Q9JLT0|MYH10_RAT</td><td>Myosin-10 OS=Rattus norvegicus GN=Myh10 PE=1 SV=1</td><td>RAT</td><td>2</td></tr>
<tr><td>950</td><td>0.95</td><td>1.3</td><td>7.7</td><td>sp|Q75Q41|TOM22_RAT</td><td>Mitochondrial import receptor subunit TOM22 homolog OS=Rattus norvegicus GN=Tomm22 PE=1 SV=1</td><td>RAT</td><td>1</td></tr>
<tr><td>951</td><td>0.92</td><td>1.27</td><td>0.4</td><td>sp|P42346|MTOR_RAT</td><td>Serine/threonine-protein kinase mTOR OS=Rattus norvegicus GN=Mtor PE=1 SV=1</td><td>RAT</td><td>1</td></tr>
<tr><td>952</td><td>0.92</td><td>1.27</td><td>2.3</td><td>sp|Q5XIQ4|MGRN1_RAT</td><td>Probable E3 ubiquitin-protein ligase MGRN1 OS=Rattus norvegicus GN=Mgrn1 PE=2 SV=1</td><td>RAT</td><td>1</td></tr>
<tr><td>953</td><td>0.89</td><td>3.89</td><td>15.1</td><td>sp|Q6NYB7|RAB1A_RAT</td><td>Ras-related protein Rab-1A OS=Rattus norvegicus GN=Rab1A PE=1 SV=3</td><td>RAT</td><td>3</td></tr>
<tr><td>954</td><td>0.88</td><td>1.23</td><td>1.1</td><td>sp|P19493-2|GRIA4_RAT</td><td>Isoform Flip of Glutamate receptor 4 OS=Rattus norvegicus GN=Gria4</td><td>RAT</td><td>1</td></tr>
<tr><td>955</td><td>0.84</td><td>1.26</td><td>5.1</td><td>sp|Q03070|CHIO_RAT</td><td>Beta-chimaerin OS=Rattus norvegicus GN=Chn2 PE=2 SV=1</td><td>RAT</td><td>1</td></tr>
<tr><td>956</td><td>0.83</td><td>1.17</td><td>2</td><td>sp|Q63100|DC1I1_RAT</td><td>Cytoplasmic dynein 1 intermediate chain 1 OS=Rattus norvegicus GN=Dync1i1 PE=1 SV=1</td><td>RAT</td><td>1</td></tr>
<tr><td>957</td><td>0.8</td><td>1.14</td><td>6.9</td><td>sp|Q7M767|UB2V2_RAT</td><td>Ubiquitin-conjugating enzyme E2 variant 2 OS=Rattus norvegicus GN=Ube2v2 PE=1 SV=3</td><td>RAT</td><td>1</td></tr>
<tr><td>958</td><td>0.8</td><td>1.14</td><td>10.5</td><td>sp|P07483|FABPH_RAT</td><td>Fatty acid-binding protein, heart OS=Rattus norvegicus GN=Fabp3 PE=1 SV=2</td><td>RAT</td><td>1</td></tr>
<tr><td>959</td><td>0.79</td><td>1.12</td><td>8.9</td><td>sp|P04166|CYB5B_RAT</td><td>Cytochrome b5 type B OS=Rattus norvegicus GN=Cyb5b PE=1 SV=2</td><td>RAT</td><td>1</td></tr>
</table>

(continued)

Table 1 (continued)

N	Unused	Total	% Cov (95)	Accession #	Name	Species	Peptides (95%)
960	0.78	1.12	1.2	sp|Q9JJ22|ERAP1_RAT	Endoplasmic reticulum aminopeptidase 1 OS=Rattus norvegicus GN=Erap1 PE=2 SV=2	RAT	1
961	0.78	1.12	3.5	sp|O08836|IGBP1_RAT	Immunoglobulin-binding protein 1 OS=Rattus norvegicus GN=Igbp1 PE=2 SV=2	RAT	1
962	0.77	1.11	7	sp|Q99P75|RAB9A_RAT	Ras-related protein Rab-9A OS=Rattus norvegicus GN=Rab9a PE=2 SV=2	RAT	1
963	0.76	1.32	3.3	sp|P63319|KPCG_RAT	Protein kinase C gamma type OS=Rattus norvegicus GN=Prkcg PE=1 SV=1	RAT	2
964	0.76	1.11	0.8	sp|Q9QX74-7|SHAN2_RAT	Isoform E of SH3 and multiple ankyrin repeat domains protein 2 OS=Rattus norvegicus GN=Shank2	RAT	1
965	0.75	1.09	2.3	sp|P14056|ARAF_RAT	Serine/threonine-protein kinase A-Raf OS=Rattus norvegicus GN=Araf PE=2 SV=1	RAT	1
966	0.74	1.08	3.6	sp|Q9EQS0|TALDO_RAT	Transaldolase OS=Rattus norvegicus GN=Taldo1 PE=1 SV=2	RAT	1
967	0.71	1.87	1.5	sp|Q62868|ROCK2_RAT	Rho-associated protein kinase 2 OS=Rattus norvegicus GN=Rock2 PE=1 SV=1	RAT	2
968	0.7	1.11	2	sp|P15178|SYDC_RAT	Aspartyl-tRNA synthetase, cytoplasmic OS=Rattus norvegicus GN=Dars PE=2 SV=1	RAT	1
969	0.7	1.05	2.6	sp|Q5PQX1|TOIP1_RAT	Torsin-1A-interacting protein 1 OS=Rattus norvegicus GN=Tor1aip1 PE=1 SV=1	RAT	1
970	0.69	1.03	3.3	sp|B0BN85|SUGT1_RAT	Suppressor of G2 allele of SKP1 homolog OS=Rattus norvegicus GN=Sugt1 PE=2 SV=1	RAT	1
971	0.69	1.03	9.9	sp|Q5U4F3|F107B_RAT	Protein FAM107B OS=Rattus norvegicus GN=Fam107b PE=2 SV=1	RAT	1

N	Unused	Total	% Cov (95)	Accession #	Name	Species	Peptides (95%)
972	0.67	1	1.6	sp|Q62991|SCFD1_RAT	Sec1 family domain-containing protein 1 OS=Rattus norvegicus GN=Scfd1 PE=1 SV=1	RAT	1
973	0.66	1	3.9	sp|Q5FWY5|AIP_RAT	AH receptor-interacting protein OS=Rattus norvegicus GN=Aip PE=1 SV=1	RAT	1
974	0.66	0.99	4.4	sp|P83868|TEBP_RAT	Prostaglandin E synthase 3 OS=Rattus norvegicus GN=Ptges3 PE=1 SV=2	RAT	1
975	0.64	0.99	2.9	sp|Q9QXU8|DC1L1_RAT	Cytoplasmic dynein 1 light intermediate chain 1 OS=Rattus norvegicus GN=Dync1li1 PE=1 SV=1	RAT	1
976	0.64	0.98	1.3	sp|Q4FZT9|PSMD2_RAT	26S proteasome non-ATPase regulatory subunit 2 OS=Rattus norvegicus GN=Psmd2 PE=2 SV=1	RAT	1
977	0.63	0.97	2.4	sp|B5DF41|SNPH_RAT	Syntaphilin OS=Rattus norvegicus GN=Snph PE=2 SV=1	RAT	1
978	0.62	0.96	0.9	sp|O88588|PACS1_RAT	Phosphofurin acidic cluster sorting protein 1 OS=Rattus norvegicus GN=Pacs1 PE=2 SV=1	RAT	1
979	0.61	0.97	4.3	sp|Q62639|RHEB_RAT	GTP-binding protein Rheb OS=Rattus norvegicus GN=Rheb PE=2 SV=1	RAT	1
980	0.61	0.94	4.1	sp|Q91ZN1|COR1A_RAT	Coronin-1A OS=Rattus norvegicus GN=Coro1a PE=1 SV=3	RAT	1
981	0.6	1.01	2.7	sp|Q99MS0|S14L2_RAT	SEC14-like protein 2 OS=Rattus norvegicus GN=Sec14l2 PE=1 SV=1	RAT	1
982	0.59	0.92	1.7	sp|Q64303|PAK2_RAT	Serine/threonine-protein kinase PAK 2 OS=Rattus norvegicus GN=Pak2 PE=1 SV=1	RAT	1
983	0.58	0.9	0.8	sp|P19490-2|GRIA1_RAT	Isoform Flip of Glutamate receptor 1 OS=Rattus norvegicus GN=Gria1	RAT	1

(continued)

Table 1
(continued)

N	Unused	Total	% Cov (95)	Accession #	Name	Species	Peptides (95%)
984	0.57	0.89	1.4	sp|O88382|MAGI2_RAT	Membrane-associated guanylate kinase, WW and PDZ domain-containing protein 2 OS=Rattus norvegicus GN=Magi2 PE=1 SV=1	RAT	1
985	0.56	3.3	5.7	sp|P63138|GBRB2_RAT	Gamma-aminobutyric acid receptor subunit beta-2 OS=Rattus norvegicus GN=Gabrb2 PE=1 SV=1	RAT	2
986	0.55	0.89	1.5	sp|O54715|VAS1_RAT	V-type proton ATPase subunit S1 OS=Rattus norvegicus GN=Atp6ap1 PE=2 SV=1	RAT	1
987	0.55	0.88	4.4	sp|Q6QBQ4|PLS3_RAT	Phospholipid scramblase 3 OS=Rattus norvegicus GN=Plscr3 PE=2 SV=1	RAT	1
988	0.55	0.87	1.1	sp|Q99JE6|PLCB3_RAT	1-phosphatidylinositol-4,5-bisphosphate phosphodiesterase beta-3 OS=Rattus norvegicus GN=Plcb3 PE=1 SV=2	RAT	1
989	0.53	0.85	1.2	sp|Q5XIR4|CCD33_RAT	Coiled-coil domain-containing protein 33 OS=Rattus norvegicus GN=Ccdc33 PE=2 SV=1	RAT	1
990	0.51	0.82	2.4	sp|Q9EPQ0|NCKX3_RAT	Sodium/potassium/calcium exchanger 3 (Fragment) OS=Rattus norvegicus GN=Slc24a3 PE=2 SV=1	RAT	1
991	0.5	0.85	1.5	sp|Q641Y8|DDX1_RAT	ATP-dependent RNA helicase DDX1 OS=Rattus norvegicus GN=Ddx1 PE=2 SV=1	RAT	1
992	0.5	0.82	2.3	sp|P21451|EDNRB_RAT	Endothelin B receptor OS=Rattus norvegicus GN=Ednrb PE=2 SV=2	RAT	1
993	0.46	0.76	13	sp|P05506|NU3M_RAT	NADH-ubiquinone oxidoreductase chain 3 OS=Rattus norvegicus GN=Mtnd3 PE=3 SV=2	RAT	1
994	0.45	0.75	17.7	sp|Q4G012|CG055_RAT	UPF0562 protein C7orf55 homolog OS=Rattus norvegicus PE=2 SV=1	RAT	1

N	Unused	Total	% Cov (95)	Accession #	Name	Species	Peptides (95%)
995	0.44	0.74	3.7	sp|Q91ZW1|TFAM_RAT	Transcription factor A, mitochondrial OS=Rattus norvegicus GN=Tfam PE=2 SV=1	RAT	1
996	0.44	0.73	2.2	sp|P36876|2ABA_RAT	Serine/threonine-protein phosphatase 2A 55 kDa regulatory subunit B alpha isoform OS=Rattus norvegicus GN=Ppp2r2a PE=2 SV=1	RAT	1
997	0.42	0.79	0.7	sp|Q64560|TPP2_RAT	Tripeptidyl-peptidase 2 OS=Rattus norvegicus GN=Tpp2 PE=2 SV=3	RAT	1
998	0.42	0.53	1.7	sp|Q9R1N3-3|S4A7_RAT	Isoform NBCn1C of Sodium bicarbonate cotransporter 3 OS=Rattus norvegicus GN=Slc4a7	RAT	2
999	0.42	0.52	5.2	sp|O35165|GOSR2_RAT	Golgi SNAP receptor complex member 2 OS=Rattus norvegicus GN=Gosr2 PE=1 SV=2	RAT	1
1000	0.41	0.52	2.5	sp|P27615|SCRB2_RAT	Lysosome membrane protein 2 OS=Rattus norvegicus GN=Scarb2 PE=1 SV=2	RAT	1
1001	0.41	0.52	3.5	sp|Q9R1T5|ACY2_RAT	Aspartoacylase OS=Rattus norvegicus GN=Aspa PE=1 SV=2	RAT	1
1002	0.4	0.51	3.1	sp|P03994|HPLN1_RAT	Hyaluronan and proteoglycan link protein 1 OS=Rattus norvegicus GN=Hapln1 PE=1 SV=2	RAT	2
1003	0.38	2.87	16.1	sp|Q63754|SYUB_RAT	Beta-synuclein OS=Rattus norvegicus GN=Sncb PE=1 SV=1	RAT	2
1004	0.37	0.49	3.3	sp|Q03555|GEPH_RAT	Gephyrin OS=Rattus norvegicus GN=Gphn PE=1 SV=3	RAT	2
1005	0.37	0.47	4.3	sp|Q7TSE9|HAX1_RAT	HCLS1-associated protein X-1 OS=Rattus norvegicus GN=Hax1 PE=1 SV=1	RAT	1
1006	0.37	0.47	8.4	sp|P07632|SODC_RAT	Superoxide dismutase [Cu-Zn] OS=Rattus norvegicus GN=Sod1 PE=1 SV=2	RAT	1

(continued)

Table 1 (continued)

N	Unused	Total	% Cov (95)	Accession #	Name	Species	Peptides (95%)
1007	0.37	0.47	12.9	sp|P14841|CYTC_RAT	Cystatin-C OS=Rattus norvegicus GN=Cst3 PE=1 SV=2	RAT	1
1008	0.36	0.53	2.4	sp|Q04631|FNTA_RAT	Protein farnesyltransferase/geranylgeranyltransferase type-1 subunit alpha OS=Rattus norvegicus GN=Fnta PE=1 SV=1	RAT	1
1009	0.35	0.46	3.8	sp|Q8R478|WBP2_RAT	WW domain-binding protein 2 OS=Rattus norvegicus GN=Wbp2 PE=1 SV=1	RAT	1
1010	0.34	0.44	2.1	sp|P08592|A4_RAT	Amyloid beta A4 protein OS=Rattus norvegicus GN=App PE=1 SV=2	RAT	1
1011	0.33	0.43	3.4	sp|P81795|IF2G_RAT	Eukaryotic translation initiation factor 2 subunit 3 OS=Rattus norvegicus GN=Eif2s3 PE=1 SV=2	RAT	1
1012	0.32	0.42	1.1	sp|Q63881|KCND2_RAT	Potassium voltage-gated channel subfamily D member 2 OS=Rattus norvegicus GN=Kcnd2 PE=1 SV=1	RAT	1
1013	0.31	0.41	1.7	sp|Q9QYJ2|RASA3_RAT	Ras GTPase-activating protein 3 OS=Rattus norvegicus GN=Rasa3 PE=2 SV=2	RAT	1
1014	0.29	17.23	25.5	sp|P68136|ACTS_RAT	Actin, alpha skeletal muscle OS=Rattus norvegicus GN=Acta1 PE=2 SV=1	RAT	20
1015	0.29	0.75	2.7	sp|Q63054|ICA69_RAT	Islet cell autoantigen 1 OS=Rattus norvegicus GN=Ica1 PE=2 SV=2	RAT	1
1016	0.28	4.09	14.1	sp|P62836|RAP1A_RAT	Ras-related protein Rap-1A OS=Rattus norvegicus GN=Rap1a PE=1 SV=1	RAT	2
1017	0.28	0.38	1.3	sp|O08560|DGKZ_RAT	Diacylglycerol kinase zeta OS=Rattus norvegicus GN=Dgkz PE=2 SV=1	RAT	1
1018	0.28	0.38	11.3	sp|B0BN72|F195B_RAT	Protein FAM195B OS=Rattus norvegicus GN=Fam195b PE=3 SV=1	RAT	1

N	Unused	Total	% Cov (95)	Accession #	Name	Species	Peptides (95%)
1019	0.27	0.37	4.2	sp|P32215|PACR_RAT	Pituitary adenylate cyclase-activating polypeptide type IA receptor OS=Rattus norvegicus GN=Adcyap1r1 PE=2 SV=2	RAT	1
1020	0.27	0.36	2.7	sp|Q63569|PRS6A_RAT	26S protease regulatory subunit 6A OS=Rattus norvegicus GN=Psmc3 PE=2 SV=1	RAT	1
1021	0.26	2.14	2.9	sp|P81377|KAP1_RAT	cAMP-dependent protein kinase type I-beta regulatory subunit OS=Rattus norvegicus GN=Prkar1b PE=2 SV=2	RAT	1
1022	0.26	0.36	5.5	sp|Q63692|CDC37_RAT	Hsp90 co-chaperone Cdc37 OS=Rattus norvegicus GN=Cdc37 PE=1 SV=2	RAT	1
1023	0.26	0.36	3.4	sp|P51868|CASQ2_RAT	Calsequestrin-2 OS=Rattus norvegicus GN=Casq2 PE=1 SV=2	RAT	1
1024	0.24	0.34	2.2	sp|P13221|AATC_RAT	Aspartate aminotransferase, cytoplasmic OS=Rattus norvegicus GN=Got1 PE=1 SV=3	RAT	1
1025	0.24	0.34	5.4	sp|O08776|NDUF3_RAT	NADH dehydrogenase [ubiquinone] 1 alpha subcomplex assembly factor 3 OS=Rattus norvegicus GN=Ndufaf3 PE=2 SV=1	RAT	1
1026	0.24	0.33	1.3	sp|Q5M9G9|TBRG4_RAT	Protein TBRG4 OS=Rattus norvegicus GN=Tbrg4 PE=2 SV=1	RAT	1
1027	0.22	0.33	0.3	sp|P30427-2|PLEC1_RAT	Isoform 2 of Plectin-1 OS=Rattus norvegicus GN=Plec1	RAT	1
1028	0.22	0.31	3.3	sp|Q68FT3|PYRD2_RAT	Pyridine nucleotide-disulfide oxidoreductase domain-containing protein 2 OS=Rattus norvegicus GN=Pyroxd2 PE=2 SV=1	RAT	1
1029	0.22	0.31	3.3	sp|Q63083|NUCB1_RAT	Nucleobindin-1 OS=Rattus norvegicus GN=Nucb1 PE=1 SV=1	RAT	1

(continued)

Table 1
(continued)

N	Unused	Total	% Cov (95)	Accession #	Name	Species	Peptides (95%)
1030	0.22	0.31	27.6	sp|P63249|IPKA_RAT	cAMP-dependent protein kinase inhibitor alpha OS=Rattus norvegicus GN=Pkia PE=1 SV=2	RAT	1
1031	0.21	0.3	5.1	sp|Q6P7R8|DHB12_RAT	Estradiol 17-beta-dehydrogenase 12 OS=Rattus norvegicus GN=Hsd17b12 PE=2 SV=1	RAT	1
1032	0.21	0.29	14.9	sp|P60841|ENSA_RAT	Alpha-endosulfine OS=Rattus norvegicus GN=Ensa PE=1 SV=1	RAT	1
1033	0.2	0.29	10.3	sp|P39069|KAD1_RAT	Adenylate kinase isoenzyme 1 OS=Rattus norvegicus GN=Ak1 PE=1 SV=3	RAT	2
1034	0.18	0.37	2	sp|Q5I0D6|NOSTN_RAT	Nostrin OS=Rattus norvegicus GN=Nostrin PE=2 SV=1	RAT	1
1035	0.18	0.33	1.7	sp|Q5PPI9|TTLL1_RAT	Probable tubulin polyglutamylase TTLL1 OS=Rattus norvegicus GN=Ttll1 PE=2 SV=1	RAT	1
1036	0.18	0.26	1.6	sp|Q63065|PDK1_RAT	[Pyruvate dehydrogenase [lipoamide]] kinase isozyme 1, mitochondrial OS=Rattus norvegicus GN=Pdk1 PE=1 SV=1	RAT	1
1037	0.18	0.26	0	sp|Q5XIM5|CDV3_RAT	Protein CDV3 homolog OS=Rattus norvegicus GN=Cdv3 PE=2 SV=1	RAT	1
1038	0.18	0.26	6.6	sp|Q4G009|MCTS1_RAT	Malignant T cell-amplified sequence 1 OS=Rattus norvegicus GN=Mcts1 PE=2 SV=1	RAT	1
1039	0.17	0.26	4.2	sp|P29410|KAD2_RAT	Adenylate kinase 2, mitochondrial OS=Rattus norvegicus GN=Ak2 PE=2 SV=2	RAT	1
1040	0.17	0.25	1.9	sp|Q8K3P6|SCMC2_RAT	Calcium-binding mitochondrial carrier protein SCaMC-2 OS=Rattus norvegicus GN=Slc25a25 PE=1 SV=1	RAT	1
1041	0.16	0.23	0	sp|Q5NDF0|AGO61_RAT	Uncharacterized glycosyltransferase AGO61 OS=Rattus norvegicus GN=Ago61 PE=2 SV=1	RAT	0

<table>
<tr><th>N</th><th>Unused</th><th>Total</th><th>% Cov (95)</th><th>Accession #</th><th>Name</th><th>Species</th><th>Peptides (95%)</th></tr>
<tr><td>1042</td><td>0.15</td><td>2.43</td><td>4.6</td><td>sp|Q66HF8|AL1B1_RAT</td><td>Aldehyde dehydrogenase X, mitochondrial OS=Rattus norvegicus GN=Aldh1b1 PE=1 SV=1</td><td>RAT</td><td>2</td></tr>
<tr><td>1043</td><td>0.15</td><td>0.25</td><td>12.9</td><td>sp|P52504|NDUS6_RAT</td><td>NADH dehydrogenase [ubiquinone] iron-sulfur protein 6, mitochondrial OS=Rattus norvegicus GN=Ndufs6 PE=3 SV=1</td><td>RAT</td><td>3</td></tr>
<tr><td>1044</td><td>0.15</td><td>0.24</td><td>0</td><td>sp|P32198|CPT1A_RAT</td><td>Carnitine O-palmitoyltransferase 1, liver isoform OS=Rattus norvegicus GN=Cpt1a PE=1 SV=2</td><td>RAT</td><td>0</td></tr>
<tr><td>1045</td><td>0.15</td><td>0.23</td><td>0</td><td>sp|P45953|ACADV_RAT</td><td>Very long-chain specific acyl-CoA dehydrogenase, mitochondrial OS=Rattus norvegicus GN=Acadvl PE=1 SV=1</td><td>RAT</td><td>0</td></tr>
<tr><td>1046</td><td>0.15</td><td>0.22</td><td>3.2</td><td>sp|A1L134|AUP1_RAT</td><td>Ancient ubiquitous protein 1 OS=Rattus norvegicus GN=Aup1 PE=2 SV=1</td><td>RAT</td><td>1</td></tr>
<tr><td>1047</td><td>0.14</td><td>4.73</td><td>4.8</td><td>sp|P11275|KCC2A_RAT</td><td>Calcium/calmodulin-dependent protein kinase type II subunit alpha OS=Rattus norvegicus GN=Camk2a PE=1 SV=1</td><td>RAT</td><td>2</td></tr>
<tr><td>1048</td><td>0.14</td><td>0.22</td><td>0</td><td>sp|Q8VHK0|ACOT8_RAT</td><td>Acyl-coenzyme A thioesterase 8 OS=Rattus norvegicus GN=Acot8 PE=2 SV=1</td><td>RAT</td><td>0</td></tr>
<tr><td>1049</td><td>0.14</td><td>0.22</td><td>0</td><td>sp|O08589|PLM_RAT</td><td>Phospholemman OS=Rattus norvegicus GN=Fxyd1 PE=1 SV=2</td><td>RAT</td><td>0</td></tr>
<tr><td>1050</td><td>0.14</td><td>0.21</td><td>0</td><td>sp|P61354|RL27_RAT</td><td>60S ribosomal protein L27 OS=Rattus norvegicus GN=Rpl27 PE=2 SV=2</td><td>RAT</td><td>0</td></tr>
<tr><td>1051</td><td>0.13</td><td>0.22</td><td>0</td><td>sp|Q05764|ADDB_RAT</td><td>Beta-adducin OS=Rattus norvegicus GN=Add2 PE=2 SV=4</td><td>RAT</td><td>0</td></tr>
<tr><td>1052</td><td>0.13</td><td>0.21</td><td>0</td><td>sp|Q5XIG0|NUDT9_RAT</td><td>ADP-ribose pyrophosphatase, mitochondrial OS=Rattus norvegicus GN=Nudt9 PE=2 SV=1</td><td>RAT</td><td>0</td></tr>
<tr><td>1053</td><td>0.13</td><td>0.21</td><td>0</td><td>sp|P11608|ATP8_RAT</td><td>ATP synthase protein 8 OS=Rattus norvegicus GN=Mt-atp8 PE=1 SV=1</td><td>RAT</td><td>0</td></tr>
</table>

(continued)

Table 1
(continued)

N	Unused	Total	% Cov (95)	Accession #	Name	Species	Peptides (95%)
1054	0.13	0.2	3.8	sp|Q9JM80|PAG1_RAT	Phosphoprotein associated with glycosphingolipid-enriched microdomains 1 OS=Rattus norvegicus GN=Pag1 PE=1 SV=1	RAT	1
1055	0.12	0.2	0	sp|B2GV87-2|PTPRE_RAT	Isoform cyt-PTPe of Receptor-type tyrosine-protein phosphatase epsilon OS=Rattus norvegicus GN=Ptpre	RAT	0
1056	0.12	0.2	6.6	sp|Q9EQX9|UBE2N_RAT	Ubiquitin-conjugating enzyme E2 N OS=Rattus norvegicus GN=Ube2n PE=1 SV=1	RAT	1
1057	0.12	0.19	0	sp|Q4G064|COQ5_RAT	Ubiquinone biosynthesis methyltransferase COQ5, mitochondrial OS=Rattus norvegicus GN=Coq5 PE=2 SV=1	RAT	0
1058	0.11	0.23	7.5	sp|P83883|RL36A_RAT	60S ribosomal protein L36a OS=Rattus norvegicus GN=Rpl36a PE=1 SV=2	RAT	1
1059	0.11	0.2	0	sp|Q9QWM1|NR5A2_RAT	Nuclear receptor subfamily 5 group A member 2 OS=Rattus norvegicus GN=Nr5a2 PE=2 SV=1	RAT	0
1060	0.11	0.18	0	sp|P02683|NSG1_RAT	Neuron-specific protein family member 1 OS=Rattus norvegicus GN=Nsg1 PE=2 SV=2	RAT	0
1061	0.11	0.17	0	sp|Q4V7D2|ROGDI_RAT	Protein rogdi homolog OS=Rattus norvegicus GN=Rogdi PE=2 SV=1	RAT	0
1062	0.1	0.17	0	sp|P55068|PGCB_RAT	Brevican core protein OS=Rattus norvegicus GN=Bcan PE=1 SV=2	RAT	0
1063	0.1	0.17	0	sp|B2RYT9|TACO1_RAT	Translational activator of cytochrome c oxidase 1 OS=Rattus norvegicus GN=Taco1 PE=2 SV=1	RAT	0
1064	0.1	0.17	0	sp|Q62889|NLGN3_RAT	Neuroligin-3 OS=Rattus norvegicus GN=Nlgn3 PE=1 SV=1	RAT	0

N	Unused	Total	% Cov (95)	Accession #	Name	Species	Peptides (95%)
1065	0.09	0.16	0	sp|P05765|RS21_RAT	40S ribosomal protein S21 OS=Rattus norvegicus GN=Rps21 PE=1 SV=1	RAT	0
1066	0.09	0.15	0	sp|P24155|THOP1_RAT	Thimet oligopeptidase OS=Rattus norvegicus GN=Thop1 PE=1 SV=4	RAT	0
1067	0.09	0.15	0	sp|P97612|FAAH1_RAT	Fatty-acid amide hydrolase 1 OS=Rattus norvegicus GN=Faah PE=1 SV=1	RAT	0
1068	0.09	0.14	0	sp|P0C1S9|DGLB_RAT	Sn1-specific diacylglycerol lipase beta OS=Rattus norvegicus GN=Daglb PE=1 SV=1	RAT	0
1069	0.08	0.34	0	sp|Q9JI12|VGLU2_RAT	Vesicular glutamate transporter 2 OS=Rattus norvegicus GN=Slc17a6 PE=1 SV=1	RAT	0
1070	0.08	0.14	0	sp|Q6KC51|ABLM2_RAT	Actin-binding LIM protein 2 OS=Rattus norvegicus GN=Ablim2 PE=2 SV=1	RAT	0
1071	0.08	0.14	0	sp|Q7TNZ6|STRAA_RAT	STE20-related kinase adapter protein alpha OS=Rattus norvegicus GN=Strada PE=1 SV=1	RAT	0
1072	0.08	0.13	0	sp|Q4KM49|SYYC_RAT	Tyrosyl-tRNA synthetase, cytoplasmic OS=Rattus norvegicus GN=Yars PE=2 SV=3	RAT	0
1073	0.08	0.13	0	sp|Q5BK81|PTGR2_RAT	Prostaglandin reductase 2 OS=Rattus norvegicus GN=Ptgr2 PE=2 SV=2	RAT	0
1074	0.07	38.45	65.1	sp|P02688-2|MBP_RAT	Isoform 18.5 kDa of Myelin basic protein S OS=Rattus norvegicus GN=Mbp	RAT	45
1075	0.07	0.15	0	sp|O88277|FAT2_RAT	Protocadherin Fat 2 OS=Rattus norvegicus GN=Fat2 PE=1 SV=1	RAT	0
1076	0.07	0.13	0	sp|Q5M7U6|ARP2_RAT	Actin-related protein 2 OS=Rattus norvegicus GN=Actr2 PE=1 SV=1	RAT	0
1077	0.07	0.12	0	sp|Q62796|RBP1_RAT	RalA-binding protein 1 OS=Rattus norvegicus GN=Ralbp1 PE=1 SV=3	RAT	0

(continued)

Table 1 (continued)

N	Unused	Total	% Cov (95)	Accession #	Name	Species	Peptides (95%)
1078	0.07	0.12	0	sp|P53042|PPP5_RAT	Serine/threonine-protein phosphatase 5 OS=Rattus norvegicus GN=Ppp5c PE=2 SV=1	RAT	0
1079	0.07	0.12	0	sp|Q9Z1Z3|EPN2_RAT	Epsin-2 OS=Rattus norvegicus GN=Epn2 PE=1 SV=1	RAT	0
1080	0.07	0.12	0	sp|A7VJC2|ROA2_RAT	Heterogeneous nuclear ribonucleoproteins A2/B1 OS=Rattus norvegicus GN=Hnrnpa2b1 PE=1 SV=1	RAT	0
1081	0.07	0.12	0	sp|Q0VGK4|GDPD1_RAT	Glycerophosphodiester phosphodiesterase domain-containing protein 1 OS=Rattus norvegicus GN=Gdpd1 PE=2 SV=1	RAT	0
1082	0.06	2.53	8.7	sp|P70550|RAB8B_RAT	Ras-related protein Rab-8B OS=Rattus norvegicus GN=Rab8b PE=1 SV=1	RAT	2
1083	0.06	0.12	0	sp|Q71RP1|HPSE_RAT	Heparanase OS=Rattus norvegicus GN=Hpse PE=2 SV=1	RAT	0
1084	0.06	0.12	0	sp|Q5PPN4|CAH8_RAT	Carbonic anhydrase-related protein OS=Rattus norvegicus GN=Ca8 PE=1 SV=3	RAT	0
1085	0.06	0.11	0	sp|O54735|PDE5A_RAT	cGMP-specific 3',5'-cyclic phosphodiesterase OS=Rattus norvegicus GN=Pde5a PE=2 SV=1	RAT	0
1086	0.06	0.11	0	sp|O88370|PI42C_RAT	Phosphatidylinositol-5-phosphate 4-kinase type-2 gamma OS=Rattus norvegicus GN=Pip4k2c PE=1 SV=1	RAT	0
1087	0.06	0.11	0	sp|P62271|RS18_RAT	40S ribosomal protein S18 OS=Rattus norvegicus GN=Rps18 PE=1 SV=3	RAT	0
1088	0.06	0.11	0	sp|Q5RKH7|CB018_RAT	Transmembrane protein C2orf18 homolog OS=Rattus norvegicus PE=2 SV=1	RAT	0

<table>
<tr><th>N</th><th>Unused</th><th>Total</th><th>% Cov (95)</th><th>Accession #</th><th>Name</th><th>Species</th><th>Peptides (95%)</th></tr>
<tr><td>1089</td><td>0.06</td><td>0.1</td><td>0</td><td>sp|O35760|IDI1_RAT</td><td>Isopentenyl-diphosphate Delta-isomerase 1 OS=Rattus norvegicus GN=Idi1 PE=2 SV=2</td><td>RAT</td><td>0</td></tr>
<tr><td>1090</td><td>0.05</td><td>6.51</td><td>11.3</td><td>sp|P08753|GNAI3_RAT</td><td>Guanine nucleotide-binding protein G(k) subunit alpha OS=Rattus norvegicus GN=Gnai3 PE=1 SV=3</td><td>RAT</td><td>4</td></tr>
<tr><td>1091</td><td>0.05</td><td>2.21</td><td>5.5</td><td>sp|Q5U316|RAB35_RAT</td><td>Ras-related protein Rab-35 OS=Rattus norvegicus GN=Rab35 PE=2 SV=1</td><td>RAT</td><td>1</td></tr>
<tr><td>1092</td><td>0.05</td><td>0.1</td><td>0</td><td>sp|Q6GMN2|BAIP2_RAT</td><td>Brain-specific angiogenesis inhibitor 1-associated protein 2 OS=Rattus norvegicus GN=Baiap2 PE=1 SV=1</td><td>RAT</td><td>0</td></tr>
<tr><td>1093</td><td>0.05</td><td>0.1</td><td>0</td><td>sp|Q76K24|ANR46_RAT</td><td>Ankyrin repeat domain-containing protein 46 OS=Rattus norvegicus GN=Ankrd46 PE=2 SV=1</td><td>RAT</td><td>0</td></tr>
<tr><td>1094</td><td>0.05</td><td>0.09</td><td>0</td><td>sp|O88902|PTN23_RAT</td><td>Tyrosine-protein phosphatase non-receptor type 23 (Fragment) OS=Rattus norvegicus GN=Ptpn23 PE=1 SV=2</td><td>RAT</td><td>0</td></tr>
<tr><td>1095</td><td>0.05</td><td>0.09</td><td>0</td><td>sp|P68101|IF2A_RAT</td><td>Eukaryotic translation initiation factor 2 subunit 1 OS=Rattus norvegicus GN=Eif2s1 PE=1 SV=2</td><td>RAT</td><td>0</td></tr>
<tr><td>1096</td><td>0.05</td><td>0.09</td><td>0</td><td>sp|P23749|MRGRF_RAT</td><td>Mas-related G-protein coupled receptor member F OS=Rattus norvegicus GN=Mrgprf PE=2 SV=1</td><td>RAT</td><td>0</td></tr>
</table>

7. 0.9 PC Thick-walled tube; length=50 mm. outer diameter = 8 mm, inner diameter=6 mm (Eppendorf Himac Technologies, 5720411036) (Fig. 1B).

2.2 Immunoprecipitation by Anti-Glycosphingolipid Antibody

1. Lysis buffer (1% Triton X-100, 50 mM Tris-HCl, pH 7.4, 150 mM NaCl, 1 mM Na_3VO_4, 1 mM EGTA, 1 mM phenylmethylsulfonyl fluoride, 5 μg/mL leupeptin, 5 μg/mL pepstatin A).
2. Kinase buffer (30 mM HEPES, pH 7.5, 10 mM $MgCl_2$, 2 mM $MnCl_2$).
3. [γ-^{32}P]ATP (111 TBq/mmol, NEN Life Science Products).
4. Laemmli sample buffer.
5. Protein G-Sepharose.

3 Methods

3.1 Platelets Preparation

1. Blood is collected into 3.8% sodium citrate at a ratio of 9:1.
2. The blood is centrifuged (140 × *g*) to prepare platelet-rich plasma (PRP).
3. To prepare washed platelets, we incubate PRP with 4 mM citric acid and wash with Tyrode's buffer (137 mM NaCl, 2.7 mM KCl, 3.75 mM NaH_2PO_4, 5 mM N-2hydroxyethylpiperazine-N9-2-ethanesulfonic acid (HEPES), 0.35% bovine serum albumin, and 5 mM glucose; pH 6.8) containing 1 mM PGE1 and 1 U/mL heparin. Finally, the platelets are resuspended in Tyrode's buffer (137 mM NaCl, 2.7 mM KCl, 3.75 mM NaH2PO4, 5 mM HEPES, 0.35% bovine serum albumin, and 5 mM glucose; pH7.4) containing 1 mM CaCl2 and 1 mM $MgCl_2$.

3.2 Sucrose Density Gradient Analysis

1. Platelets (6×10^9) are homogenized using a Teflon glass homogenizer (1000 rpm, 10 strokes) in 2 mL of TNE/Triton buffer (0.05% Triton X-100, 25 mM Tris-HCl, pH 7.5, 150 mM NaCl, and 1 mM EGTA).
2. The sucrose content is then adjusted to 40% by adding 2 mL of 80% sucrose/TNE.
3. A sucrose gradient (5%–30%) in 6 mL of TNE without Triton X-100 is layered over the lysate.
4. The contents are centrifuged for 17 h at 39,000 rpm at 4°C in a swing P40ST rotor (Eppendorf Himac Technologies).
5. The ten fractions (1 mL each) are collected from the top to bottom of the gradient (*see* **Note 1**).

3.3 Small-Scale Sucrose Density Gradient Analysis

1. Platelets (6×10^8) are homogenized using a Teflon glass homogenizer in 100 μL of TNE/Triton buffer (0.05% Triton X-100, 25 mM Tris-HCl, pH 7.5, 150 mM NaCl, and 1 mM EGTA).
2. The sucrose content of the homogenate is then adjusted to 40% by adding 100 μL of 80% sucrose/TNE, placed at the bottom of an ultracentrifuge tube (0.9 PC Thick-walled tube, Eppendorf Himac Technologies 5720411036), and overlaid with 35% and then 5% sucrose in 300 μL of TNE without Triton X-100. The discontinuous gradient is centrifuged for 17 h at 35,000 rpm (153,000 × *g*) at 4°C in a P40ST rotor with 2S13 adaptor (Eppendorf Himac Technologies, 5720411191) (*see* **Note 2**).

3.4 The Targeted Lipidomics of Platelet DRM Raft and Non-Raft Fractions by Mass Spectrometry

1. Lipid Extraction: The platelet samples are homogenized with chloroform/methanol (1:2, v/v). Phospholipids (PLs) are extracted using the Bligh and Dyer method [23]. The total lipid extract is dried under a gentle stream of nitrogen, dissolved in chloroform/methanol (1:1, v/v), and stored at −20°C.
2. The targeted lipidomics is performed via direct infusion at a flow rate of 15 μL/min. Mobile phase consisted of methanol/acetonitrile/H_2O (19/19/2) containing 0.1% acetic scanning and neutral loss scanning of polar head groups [24] using a TSQ Vantage triple quadrupole mass spectrometer (ThermoElectron Corp., San Jose, CA, USA), coupled with a LC-10ADVPμ HPLC system (Shimadzu, Kyoto, Japan). Phosphatidylserines (PSs) and Phosphatidylethanolamines (PEs) are measured through neutral loss scanning of polar head groups (PS, 185.1 amu; PE 141.1 amu) in the positive ion mode. Phosphatidylcholines (PCs) are measured with precursor ion scanning of polar head groups-related fragment ion (m/z 184) in the positive ion mode. Relative quantification of PLs is calculated via intensity ratio (compared to the molecule with the highest intensity). quadrupole mass spectrometer (ThermoElectron Corp., San Jose, CA, USA), coupled with a LC-10ADVPμHPLCsystem (Shimadzu, Kyoto, Japan). Phosphatidylserines (PSs) and Phosphatidylethanolamines (PEs) are measured through neutral loss scanning of polar head groups (PS, 185.1 amu; PE 141.1 amu) in the positive ion mode. Phosphatidylcholines (PCs) are measured with precursor ion scanning of polar head groups-related fragment ion (m/z 184) in the positive ion mode. Relative quantification of PLs is calculated via intensity ratio (compared to the molecule with the highest intensity)

3.5 Primary Culture of Cerebellar Neurons

1. Five cerebella are dissected from Wistar rats of either sex sacrificed by cervical dislocation at 7 days of age, and are placed in 4°C Hanks Balanced Salt Solution, while meninges are removed.
2. Cerebellar cells are prepared using a dissociation solution (DS Pharma Biomedical).
3. The dissociated cells are plated onto 35-mm poly-L-lysine-coated plastic dishes (Becton Dickinson Labware) at a density of 2.5×10^6 cells in 2 mL of Neurobasal medium (Gibco BRL) containing 25 mM KCl, 2 mM glutamine, and B27 supplement (Gibco BRL). Most of the cells are cerebellar granule cells.

3.6 Immuno-precipitation with Anti-Glycosphingolipid Monoclonal Antibody, and In Vitro Kinase Assay

1. Primary centrifuged cultured cells are solubilized in lysis buffer (1% Triton X-100, 50 mM Tris-HCl, pH 7.4, 150 mM NaCl, 1 mM Na3VO4, 1 mM EGTA, 1 mM phenylmethylsulfonyl fluoride, 5 μg/mL leupeptin, 5 μg/mL pepstatin A) at 4°C for 20 min.
2. The postnuclear supernatants are collected after centrifugation at 14,000 rpm for 3 min. Aliquots (0.5 mL, 750 μg of protein) of the supernatants are precleared with protein G-Sepharose (7.5 μL) and then incubated with anti-GD3 antibody R24 (2.5 μg) for 1 h and precipitated with protein G-Sepharose (7.5 μL).
3. Following immunoprecipitation, the beads are washed three times with lysis buffer, washed once with kinase buffer (30 mM HEPES, pH 7.5, 10 mM $MgCl_2$, 2 mM $MnCl_2$), and resuspended in 20 μL of kinase buffer. The reaction is started by the addition of 185 kBq of [γ-^{32}P]ATP (111 TBq/mmol, NEN Life Science Products) and incubated for 10 min at room temperature.
4. Phosphorylation is stopped by the addition of Laemmli sample buffer, and samples are subjected to sodium dodecyl sulfate-polyacrylamide gel electrophoresis (SDS-PAGE), followed by autoradiography.
5. In a re-immunoprecipitation experiment, after the kinase reaction, the samples are boiled for 5 min in lysis buffer with 1% SDS, diluted 10-fold with lysis buffer, and then re-immunoprecipitated with protein G-Sepharose conjugated antibodies against the Src family kinase Lyn (Lyn8), Csk-binding protein Cbp, and heterotrimeric Go α subunit (GC/2) (*see* **Note 3**).

3.7 Metabolic Labeling and Immunoprecipitation with Anti-Glycosphingolipid Monoclonal Antibody

1. Primary cerebellar cultures are labeled with the EXPRE^{35}S^{35}S Protein Labeling Mix (PerkinElmer Life Sciences) for 3 h in a methionine-/cysteine-depleted medium, at a final concentration of 3.7 MBq/mL.
2. The cells are solubilized in lysis buffer (1% Triton X-100, 50 mM Tris-HCl (pH 7.4), 150 mM NaCl, 1 mM Na_3VO_4, 1 mM EGTA, 1 mM phenylmethylsulfonyl fluoride) at 4°C for 20 min.
3. The postnuclear supernatants are collected after centrifugation at 14,000 rpm for 3 min.
4. Aliquots (0.5 mL, 750 μg of protein) of the supernatants are precleared with protein G-Sepharose (7.5 μL) and then incubated with the anti-ganglioside GD3 antibody, R24 (2.5 μg), for 1 h and precipitated with protein G-Sepharose (7.5 μL).
5. Following immunoprecipitation, the beads are washed three times with lysis buffer, and after the addition of the Laemmli sample buffer, the samples are subjected to SDS-PAGE (10% acrylamide) followed by autoradiography.
6. In the sequential immunoprecipitation experiment, the immunoprecipitates are boiled for 5 min in lysis buffer with 1% SDS, diluted 10-fold with lysis buffer, and then re-immunoprecipitated with anti-TAG-1/contactin-2 monoclonal antibody 3.1C12 (*see* **Note 4**).

3.8 Proteomics of DRM Raft Fraction of Rat Cerebellar Neurons

1. The DRM raft of rat cerebellar neurons is lysed octyl-glucoside lysis buffer (2%(w/v) octyl-glucoside, 50 mM Tris–HCl, 150 mM NaCl, 1 mM EGTA) at room temperature for 20 min.
2. The solubilized DRM is centrifuged at 15,000 × *g* at 4°C for 20 min. The supernatants are digested by trypsin.
3. The digests are reduced by Tris(2-carboxyethyl)phosphine, alkylated by methyl methanethiosulfonate, acidified, desalted, and concentrated by using Zip tip SCX and C18 (Millipore).
4. MS/MS spectra data are acquired using Triple TOF 5600 (SCIEX) equipped with a Nano LC Ultra 1D Plus system. Raw data are analyzed against Uniprot database restricted to Rattus norvegicus. Proteins are identified using Protein Pilot™ software 5.0.1 (SCIEX). Detected proteins threshold is applied with <0.05.

4 Notes

1. Immunoblotting demonstrated that CD36, a raft marker of platelets, and Lyn are present in fractions 4, 5, 6, and transferrin receptor, a non-raft marker, and most of cellular proteins are present in fractions 7, 8, 9, 10 [13].
2. Immunoblotting demonstrated that Lyn is present in DRM raft fraction.
3. In vitro kinase reaction resulted in phosphorylation of several proteins of 40, 53, 56, and 80 kDa, as judged by SDS-PAGE (Fig. 5A). In the sequential immunoprecipitation experiment, 40, 53/56, and 80 kDa proteins are re-immunoprecipitated by anti-heterotrimeric Go α subunit [21], Src family kinase Lyn [20], and Csk-binding protein Cbp antibodies [19], respectively. These observations suggest that there is a specific association of ganglioside GD3 with heterotrimeric Go α subunit, Src family kinase Lyn, and Csk-binding protein Cbp in the rat cerebellar neuron membrane.

   Although anti-ganglioside GD3 monoclonal antibody R24 is IgG3, anti-glycosphingolipids monoclonal antibodies are often IgM [25]. Anti-sulfatide monoclonal antibody O4 is IgM. Immunoprecipitation with an anti-sulfatide monoclonal antibody O4 and an in vitro kinase assay were performed as described previously [22]. In brief, 120,000 primary immature oligodendrocytes were solubilized in lysis buffer (0.5 % Triton X-100, 50 mM Tris–HCl, pH 7.4, 150 mM NaCl, 1 mM Na_3VO_4,1 mM EGTA, 1 mM phenylmethylsulfonylfluoride, 5 μg/mL leupeptin, and 5 μg/mL pepstatin A) at 4°C for 20 min. Aliquots (0.5 mL) of the supernatants were incubated with O4 (2.5 μg/mL; R&D Systems MAB1326) or normal mouse IgM for 1 h, and then anti-mouse IgM goat antibody (2.5 μg/mL; Bethyl Laboratories A90-101A) for 1 h, and precipitated with protein G-Sepharose. In vitro kinase reaction resulted in the phosphorylation of several proteins of 40, 53, 56, 59, and 80 kDa, as determined by SDS-PAGE. In the sequential immunoprecipitation experiment, 40, 53/56, and 59 kDa proteins were re-immunoprecipitated with the α subunit of the heterotrimeric G protein internal (40–54) antibodies (Calbiochem), antibodies to Lyn (Lyn8) and Fyn (Fyn301), respectively.

   Other investigators also reported co-immunoprecipitation of lipid raft fraction using monoclonal antibodies against lactosylceramide [26], ganglioside GM3 [27], ganglioside GD3 [28], and globotriaosyl ceramide Gb3 [29].
4. Anti-ganglioside GD3 monoclonal antibody R24 coimmunoprecipitated 135 kDa protein, as revealed by SDS-PAGE

(Fig. 5B). In the sequential immunoprecipitation experiment, 135 kDa protein is re-immunoprecipitated by anti-TAG-1 monoclonal antibody 3.1C12. This observation suggests that there is a specific association of ganglioside GD3 with TAG-1 in the rat cerebellar neuron membrane.

Acknowledgments

This work was supported by JSPS KAKENHI Grant Number JP22K19301 for Challenging Research (Exploratory) (to K. Kasahara) and JP23K14149 for Early-Career Scientists (to K. Komatsuya).

Conflict of Interest The authors have no conflicts of interest to declare that are relevant to the content of this chapter.

References

1. Simons K, Gerl MJ (2010) Revitalizing membrane rafts: new tools and insights. Nat Rev Mol Cell Biol 11:688–699
2. Kasahara K, Sanai Y (2000) Functional roles of glycosphingolipids in signal transduction via lipid rafts. Glycoconj J 17:153–162
3. López JA, del Conde I, Shrimpton CN (2005) Receptors, rafts, and microvesicles in thrombosis and inflammation. J Thromb Haemost 3:1737–1744
4. Bodin S, Tronchère H, Payrastre B (2003) Lipid rafts are critical membrane domains in blood platelet activation processes. Biochim Biophys Acta 1610:247–257
5. Komatsuya K, Kaneko K, Kasahara K (2020) Function of platelet glycosphingolipid microdomains/lipid rafts. Int J Mol Sci 21:5539
6. Dorahy DJ, Lincz LF, Meldrum CJ et al. (1996) Biochemical isolation of a membrane microdomain from resting platelets highly enriched in the plasma membrane glycoprotein CD36. Biochem J 319:67–72
7. Dorahy DJ, Burns GF (1998) Active Lyn protein tyrosine kinase is selectively enriched within membrane microdomains of resting platelets. Biochem J 333:373–379
8. Shrimpton CN, Borthakur G, Larrucea S et al. (2002) Localization of the adhesion receptor glycoprotein Ib-IX-V complex to lipid rafts is required for platelet adhesion and activation. J Exp Med 196:1057–1066
9. Jin W, Inoue O, Tamura N et al. (2007) A role for glycosphingolipid-enriched microdomains in platelet glycoprotein Ib-mediated platelet activation. J Thromb Haemost 5:1034–1040
10. Wei H, Malcor JM, Harper MT (2018) Lipid rafts are essential for release of phosphatidylserine-exposing extracellular vesicles from platelets. Sci Rep 8: 9987
11. Ohtsuka H, Iguchi T, Hayashi M et al. (2017) SDF-1α/CXCR4 signaling in lipid rafts induces platelet aggregation via PI3 kinase-dependent akt phosphorylation. PLoS One 12:e0169609
12. Kasahara K, Souri M, Kaneda M et al. (2010) Impaired clot retraction in factor XIII A subunit-deficient mice. Blood 115:1277–1279
13. Kasahara K, Kaneda M, Miki T et al. (2013) Clot retraction is mediated by factor XIII-dependent fibrin-αIIbβ3-myosin axis in platelet sphingomyelin-rich membrane rafts. Blood 122:3340–3348
14. Millán J, Puertollano R, Fan L et al. (1997) Caveolin and MAL, two protein components of internal detergent-insoluble membranes, are in distinct lipid microenvironments in MDCK cells. Biochem Biophys Res Commun 233:707–712
15. Komatsuya K, Ishikawa M, Kikuchi N et al. (2023) Integrin-dependent transient density increase in detergent-resistant membrane rafts in platelets activated by thrombin. Biomedicines 12:69
16. Komatsuya K, Iguchi T, Fukuyama M et al. (2022) Phosphacan acts as a repulsive cue in murine and rat cerebellar granule cells in a TAG-1/GD3 rafts-dependent manner. J Neurochem 163:375–390
17. Komatsuya K, Kikuchi N, Hirabayashi T et al. (2023) The regulatory roles of cerebellar

glycosphingolipid microdomains/lipid rafts. Int J Mol Sci 24:5566
18. Kasahara K, Watanabe K, Takeuchi K et al. (2000) Involvement of gangliosides in glycosylphosphatidylinositol-anchored neuronal cell adhesion molecule TAG-1 signaling in lipid rafts. J Biol Chem 275:34701–34709
19. Sekino-Suzuki N, Yuyama K, Miki T et al. (2013) Involvement of gangliosides in the process of Cbp/PAG phosphorylation by Lyn in developing cerebellar growth cones. J Neurochem 124:514–522
20. Kasahara K, Watanabe Y, Yamamoto T et al. (1997) Association of Src family tyrosine kinase Lyn with ganglioside GD3 in rat brain. Possible regulation of Lyn by glycosphingolipid in caveolae-like domains. J Biol Chem 272:29947–29953
21. Yuyama K, Sekino-Suzuki N, Sanai Y et al. (2007) Translocation of activated heterotrimeric G protein Galpha(o) to gangliosideenriched detergent-resistant membrane rafts in developing cerebellum. J Biol Chem 282:26392–26400
22. Miki T, Kaneda M, Iida K et al. (2013) An anti-sulfatide antibody O4 immunoprecipitates sulfatide rafts including Fyn, Lyn and the G protein α subunit in rat primary immature oligodendrocytes. Glycoconj J 30:819–823
23. Bligh EG, Dyer WJ (1959) A rapid method of total lipid extraction and purification. Can J Biochem Physiol 37:911–917
24. Taguchi R, Houjou T, Nakanishi H et al. (2005) Focused lipidomics by tandem mass spectrometry. J Chromatogr B Analyt Technol Biomed Life Sci 823:26–36
25. Kotani M, Tai T (1997) An immunohistochemical technique with a series of monoclonal antibodies to gangliosides: their differential distribution in the rat cerebellum. Brain Res Brain Res Protoc 1:152–156
26. Iwabuchi K, Nagaoka I (2002) Lactosylceramide-enriched glycosphingolipid signaling domain mediates superoxide generation from human neutrophils. Blood 100:1454–1464
27. Iwabuchi K, Yamamura S, Prinetti A et al. (1998) GM3-enriched microdomain involved in cell adhesion and signal transduction through carbohydrate-carbohydrate interaction in mouse melanoma B16 cells. J Biol Chem 273:9130–9138
28. Prinetti A, Prioni S, Chigorno V et al. (2001) Immunoseparation of sphingolipid-enriched membrane domains enriched in Src family protein tyrosine kinases and in the neuronal adhesion molecule TAG-1 by anti-GD3 ganglioside monoclonal antibody. J Neurochem 78:1162–1167
29. Katagiri YU, Mori T, Nakajima H et al. (1999) Activation of Src family kinase yes induced by Shiga toxin binding to globotriaosyl ceramide (Gb3/CD77) in low density, detergent-insoluble microdomains. J Biol Chem 274:35278–35282

Check for updates

Chapter 23

Gangliosides and Membrane Microdomains

Motohide Murate, Noriko Yokoyama, Nario Tomishige, Nicolas Humbert, Ludovic Richert, Yves Mély, Kazuhisa Iwabuchi and Toshihide Kobayashi

Abstract

Monosialoganglioside GM3 is the simplest ganglioside involved in various cellular signaling via plasma membrane. Elucidating the precise distribution of GM3 helps to understand its function. Although several anti-GM3 monoclonal antibodies are established, the results from the immunostaining of the antibody on the cell surface need to be carefully evaluated: the antibody binds to GM3 in sparse but not in confluent cells, and the binding is temperature dependent. Our model membrane study evidenced that GM3 makes clusters in the less fluid membrane environment, where monoclonal anti-GM3 antibodies showed stronger binding. These results suggest that anti-GM3 antibody senses GM3 clustering and the number and/or size of GM3 clusters differ in different membrane environments. This chapter describes the detailed protocols to examine the binding of anti-GM3 antibodies to different lipid mixtures by solid phase reaction, as well as to high- and low-density melanoma cells. This chapter also describes the methods to study the clustering of fluorescent GM3 analogs in different lipid compositions at various temperatures.

Keywords GM3, Lipid domains, Monoclonal anti-lipid antibody, Crypticity, Fluorescent lipid analogs

1 Introduction

Ganglioside belonging to glycosphingolipids consists of ceramide and at least one sialic acid, N-acetylneuraminic acid (NANA) in the sugar chain. More than 1500 molecules are listed in LIPID MAPS (http://www.lipidmaps.org) as gangliosides with different sugar chains with different numbers and positions of sialic acid and different ceramides with various fatty acids. Analysis of the localization of each ganglioside in cells or tissues is critical to understand the significance of extensive variation of the lipids in our body. For this purpose, specific antibodies against various gangliosides have been developed and characterized. These antibodies enabled to clarify the actual distribution in a cell- or tissue-specific manner.

The simplest ganglioside, monosialoganglioside, GM3 (NANA-galactose-glucose-ceramide) is synthesized in the Golgi

Koichi Furukawa (ed.), *Gangliosides*, Methods in Molecular Biology, vol. 3035,
https://doi.org/10.1007/978-1-0716-5288-6_23,

apparatus [1] and serves as a precursor for the biosynthesis of the other complex gangliosides in various tissues and cells [2]. GM3 itself is transported from the trans-Golgi network (TGN) to the outer leaflet of the plasma membrane to play important roles in cellular signaling by modulating the activities of integrin [3], insulin receptor [4], and epidermal growth factor receptor [5, 6]. Monoclonal and polyclonal antibodies revealed heterogeneous distributions of GM3 on the plasma membrane of various cells, such as human T cells [7], Madin–Darby canine kidney cells [8, 9], and mouse fibroblasts [10]. However, careful attention has to be paid to interpret the results from immunostaining by antibodies against membrane gangliosides due to the crypticity of the antigen. A "cryptic antigen" is an antigen that is not detected by the antibody. The capability to bind to the antigen in anti-ganglioside antibodies is thought to be affected by the steric hindrance caused by adjacent gangliosides [11, 12] and by the conformational changes of the sugar chain induced by cholesterol [13].

Previously, it was reported that the reactivity of plasma membrane GM3 with the specific antibody changed markedly depending on the confluency of the B16 mouse melanoma cells and hamster embryonic fibroblasts [14]. At low density, the punctuated spots appeared on the cell surface, but in the confluent culture, the antibody failed to label cells. These cells contain GM3 as a single ganglioside. Thus, the cross-reactivity with other gangliosides by the antibody in sparse cells and the steric hindrance of the antibody binding by other gangliosides are unlikely in high-density culture. It has been shown that the amount of GM3 in cells in confluent culture is slightly lower than in low-density culture. However, the amount of GM3 was sufficient to be recognized by the antibody in both cell cultures. Similar results were obtained in KG-1C human glioma cells [15] and MEB4 mouse melanoma cells [16]. The amount of cell surface GM3 in KG-1C cells was not affected by the cell confluency since similar amount of GM3 was chemically labeled in sparse and confluent cells [15]. Model membrane study showed that the antibody used in the aforementioned experiments gave stronger binding when GM3 was included in a less fluid membrane environment [16]. Moreover, fluorescence lifetime imaging showed that the fluidity of the cell surface is higher in confluent melanoma cells than in sparse-density cells [16]. Studies using fluorescent GM3 analogs suggested that GM3 was clustered in less fluid membranes [16]. These results imply that the anti-GM3 antibody used in these studies detects the different clustering states of GM3 which depend on the number of molecules and/or the intermolecular distances regulated by membrane fluidity.

This chapter describes methods for measuring the binding of anti-GM3 monoclonal antibodies to synthetic lipids by solid-phase reaction using enzyme-linked immunosorbent assay (ELISA) and surface plasmon resonance (SPR), as well as to cells grown at different confluencies. This chapter also shows the methods to study the clustering of fluorescent GM3 analogs in different liposomes.

2 Materials

2.1 Binding of Anti-GM3 Antibody to GM3-Containing Lipid Mixture (ELISA)

1. Immulon 1B 96-well flat-bottom microtiter plate (Thermo Fisher Scientific, Waltham, MA).
2. Microplate reader, filters at 490 and 630 nm.
3. Lipids: GM3 (Neu5Acα2,3Galβ1,4Glcβ1,1′-ceramide) from bovine buttermilk can be purchased from Matreya LLC (State College, PA). The following lipids can be purchased from Avanti Polar Lipids (Alabaster, AL): 1,2-dimyristoyl-*sn*-glycero-3-phosphocholine (DMPC), 1,2-dioleoyl-*sn*-glycero-3-phosphocholine (DOPC), 1,2-dipalmitoyl-*sn*-glycero-3-phosphocholine (DPPC), and 1-palmitoyl-2-oleoyl-*sn*-glycero-3-phosphocholine (POPC). Store at −20 °C as a chloroform/methanol (2/1) solution for GM3 and a chloroform solution for the phospholipids at the concentration of 1–10 mM, shading from light to avoid oxidation (*see* **Notes 1 and 2**).
4. Ethanol.
5. Primary antibodies: anti-GM3 monoclonal IgM clone GMR6 (TCI Europe, Zwijndrecht, Belgium).
6. Secondary antibodies: horseradish peroxidase (HRP)-labeled anti-mouse IgM (Sigma, St. Louis, MI).
7. Tris-buffered saline (TBS): 20 mM Tris–HCl, pH 7.4, 150 mM NaCl.
8. Blocking buffer: 3% (w/v) bovine serum albumin (BSA) fraction V, 0.1% (w/v) NaN_3 in TBS. Store at −20°C.
9. Dilution buffer: 1% (w/v) BSA in TBS. Store at −20°C.
10. Phosphate–citrate buffer: 25 mM citric acid, 50 mM $Na_2HPO_4 \cdot 12H_2O$, pH 5.0.
11. Coloring solution: Dissolve one tablet of o-phenylenediamine dihydrochloride (P6787, Sigma) in 5 mL of phosphate–citrate buffer and add 2.5 μL of 30% hydrogen peroxidase solution prior to use. Prepare 100 μL for one well (*see* **Note 2**).
12. 4 N H_2SO_4 solution: Dilute ninefold concentrated H_2SO_4 with pure water. Put water in a beaker in an ice box on a stirrer, then add conc. H_2SO_4 slowly to water with stirring. Wait for the solution to come to room temperature. Store at room temperature (*see* **Note 2**).

2.2 Binding of Anti-GM3 Antibody to GM3-Containing Lipid Mixture (SPR)

1. Biacore 3000 analytical system (Cytiva, North Logan, UT).
2. Biosensor Chip (L1, Cytiva, North Logan, UT) (*see* **Note 3**).
3. Nitrogen gas.
4. Vortex mixer.

5. Sonorex Super ultrasonic bath (Bandelin Electronic GmbH & Co., Berlin, Germany).
6. Water bath.
7. GM3 from bovine buttermilk. Store at −20°C as 1 mg/mL solution of in chloroform/methanol (2/1).
8. DOPC and DPPC. Store at −20°C as 10 mg/mL solutions in chloroform/methanol (2/1).
9. Antibody: anti-GM3 monoclonal IgM clone GMR6.
10. 40 mM octylglucoside: dissolve in ultrapure water and filter through a 0.22 μm polyvinylidene difluoride membrane (PVDF, Millipore Co., Burlington, MA). Store at 4°C.
11. Equilibrating buffer: HBS-N buffer (10 mM Hepes, pH 7.4, 150 mM NaCl) from Cytiva (North Logan, UT), degassed.
12. 100 mg/mL fatty acid-free BSA (Sigma-Aldrich, St. Louis, MO) in HBS-N buffer. Filtered through a 0.22 μm PVDF membrane. Store at 4°C.
13. 50 mM NaOH.
14. Chloroform.
15. Methanol.

2.3 Formation of Clusters of Fluorescent GM3 Analogs in Model Membranes

1. Pyrex disposable round-bottom glass tubes, rimless (99445-13, Corning, Corning, NY).
2. Nitrogen gas.
3. Vortex mixer.
4. Desiccator and vacuum pump.
5. Slide-A-Lyzer Dialysis Cassette (66380, Thermo Fisher Scientific).
6. Quartz Suprasil (QS) microcells (115-F-10-40; light path: 10 mm; Hellma, Müllheim, Germany).
7. Spectrofluorimeter (Fluoromax Plus, HORIBA Jobin Yvon, Palaiseau, France) equipped with a Peltier-based heated/cooled cuvette holder. The optical specifications of the instrument are characterized by a source (150 W ozone-free xenon arc lamp), two monochromators (Czerny–Turner design with plane gratings for optimized focus at all wavelengths and minimum stray light, 1200 grooves/mm blazed at 330 and 500 nm for excitation and emission grating, respectively), continuously adjustable bandpass, and an emission detector photomultiplier (R928P, photon-counting mode with a range of measurement from 230 to 850 nm).
8. Lipids: DOPC, DPPC, and C11 TopFluor-GM3 from Avanti Polar Lipids. ATTO594-GM3: prepare as described [17] (Fig. 1).

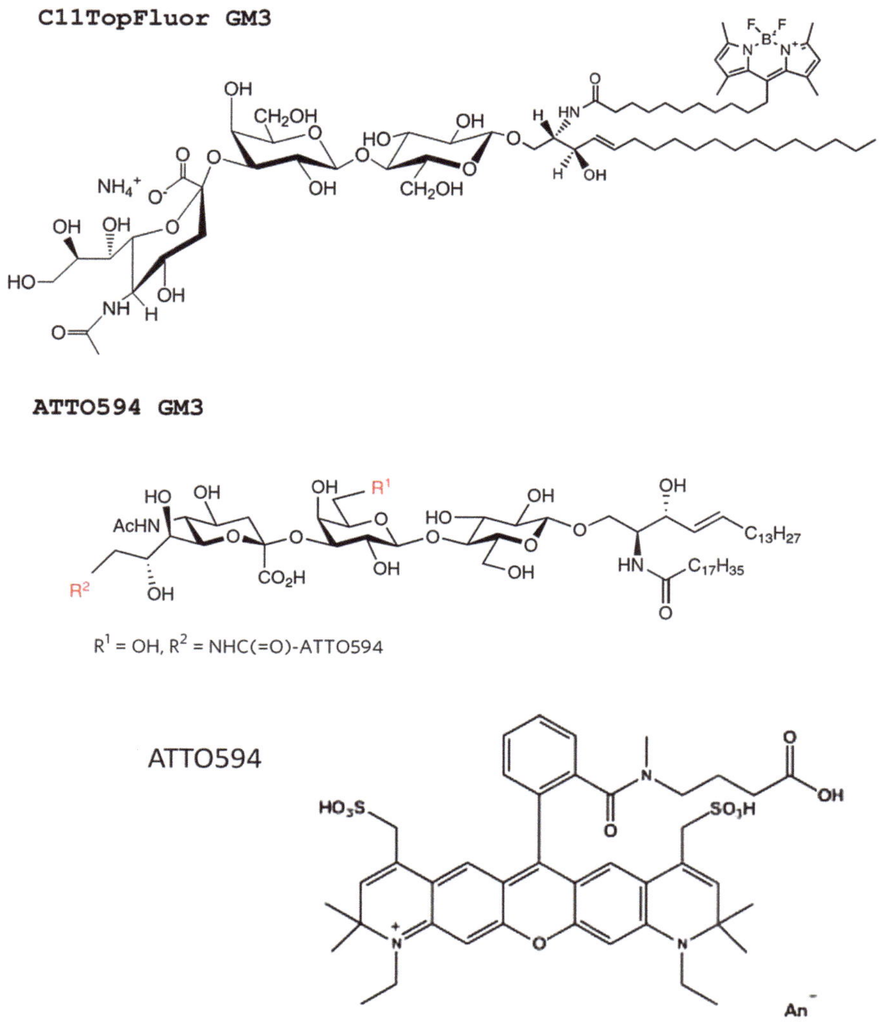

Fig. 1 Structures of fluorescent GM3 analogs

9. Chloroform.
10. Ethanol.
11. Phosphate-buffered saline (PBS): 137 mM NaCl, 1.5 mM KH_2PO_4, 2.7 mM KCl, 8.1 mM Na_2HPO_4, pH 7.4.
12. Triton X-100.

2.4 Cell Density-Dependent Labeling of Cellular GM3 by Anti-GM3 Antibody

1. Round micro coverslips (diameter 15 mm, thickness 0.13–0.17 mm), sterilized.
2. Microscope slides.
3. 12-well cell culture plate.

4. Plastic-bottom dishes (μ-Dish 35 mm diameter (diameter of observation area is 21 mm); high height (12 mm); Ibidi, Gräfelfing, Germany).
5. Fine tweezers.
6. LSM700 confocal microscope equipped with a C-Apochromat 63×W Korr (1.2 NA) objective (Carl Zeiss, Oberkochen, Germany).
7. Phosphate-buffered saline (PBS).
8. Paraformaldehyde (PFA) solution: dilute 20% (w/v) PFA solution (Electron Microscopy Sciences, Hatfield, PA) with PBS (*see* **Note 2**).
9. Blocking buffer: 3% (w/v) BSA fraction V, 0.1% (w/v) NaN_3 in PBS. Store at −20°C.
10. Antibodies: anti-GM3 monoclonal IgM clone GMR6 and Alexa488-labeled goat anti-mouse IgM (Molecular Probes, Eugene, OR).
11. Antibody reaction buffer: 1% (w/v) BSA fraction V, 0.1% (w/v) NaN_3 in PBS. Store at −20°C.
12. Alexa546-labeled phalloidin (Molecular Probes) (*see* **Note 2**).
13. Phalloidin-labeling buffer: 1% (w/v) BSA fraction V, 0.2% Triton X-100 in PBS. Store at −20°C.
14. Mounting media (ProLong Diamond Antifade Mountant, Invitrogen, Carlsbad, CA).
15. MEB4 from RIKEN BioResorce Research Center Cell Bank. Maintain cells at 37°C with 5% CO_2 in Dulbecco's modified Eagle's medium (DMEM) supplemented with 10% (v/v) fetal bovine serum (FBS), 100 U/mL penicillin, and 100 μg/mL streptomycin.

3 Methods

Prepare all aqueous solutions using ultrapure water (Milli-Q system, Merck Millipore, Molsheim, France). Prepare and store all reagents at room temperature unless indicated. All manipulations are performed at room temperature except where noted.

3.1 Binding of Anti-GM3 Antibody to GM3-Containing Lipid Mixture (ELISA) (Fig. 2)

Day 1

1. Prepare 10 μM lipid mixtures by diluting stock lipids with ethanol. Add 1–10 μL of lipid stock solution (10-1 mM in chloroform) to 1 mL of ethanol and mix well (*see* **Notes 1 and 2**).

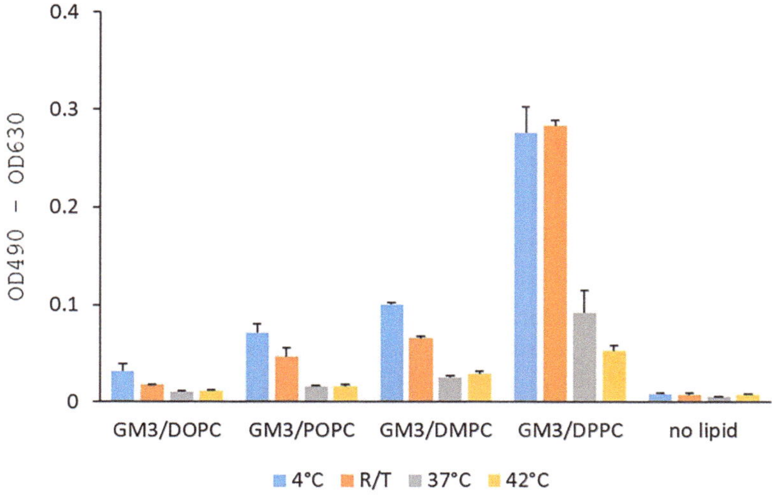

Fig. 2 Binding of anti-GM3 antibody (GMR6) to GM3/partner lipids measured by ELISA. Binding of 0.37 μg/mL of anti-GM3 antibody to GM3/DOPC (gel-to-liquid crystalline phase transition temperature, $T_m = -17°C$)(1/9), GM3/POPC($T_m = -2°C$)(1/9), GM3/DMPC($T_m = 24°C$) (1/9), or GM3/DPPC ($T_m = 41°C$) (1/9) (0.5 nmol GM3/well) at indicated temperature measured by ELISA. Efficient binding occurred below phase transition temperature. Data are means ±SD of three experiments

2. Add 50 μL of the lipid solution into each well in an Immulon 1B 96-well plate (*see* **Note 4**). Perform experiments in triplicate.
3. Add 50 μL of ethanol into three additional wells as a negative control.
4. Cover the plate with aluminum foil to prevent dust contamination.
5. Leave the plate overnight to allow the solvent to evaporate.

Day 2

6. Add 200 μL of blocking buffer per well and incubate for 2 h (*see* **Note 5**).
7. Discard the blocking buffer (*see* **Note 6**).
8. Wash wells by adding TBS and discarding it three times (*see* **Note 7**).
9. Remove excess TBS thoroughly by tapping the plate to a paper towel several times.

10. Add 50 µL of anti-GM3 antibody in dilution buffer per well and incubate for 2 h. Incubation temperature is 4°C, room temperature, 37°C or 42°C.
11. Wash wells five times with TBS (see step 7 - 9).
12. Add 50 µL of an HRP-conjugated secondary antibody in dilution buffer per well and incubate for 2 h at room temperature.
13. Wash wells five times with TBS (see step 7 - 9).
14. Add 100 µL of coloring solution per well and incubate for 10 min under aluminum foil (*see* **Notes 8 and 9**).
15. Stop the chromogenic reaction by adding 50 µL of H_2SO_4 solution per well (*see* **Notes 8 and 10**).
16. Measure the absorption at 490 nm as a result and 630 nm as a subtraction reference in each well using a microplate reader.
17. Subtract the value of ethanol control from that of the lipids.

3.2 Binding of Anti-GM3 Antibody to GM3-Containing Lipid Mixture (SPR) (Fig. 3)

Preparation of Liposomes

1. Mix 50 nmol of GM3 with 250 nmol of either DOPC or DPPC in a chloroform/methanol (2/1) solution (*see* **Note 2**).
2. Dry under a nitrogen stream.
3. Dissolve in 200 µL of HBS-N buffer by mixing vigorously using a vortex mixer at 41°C.
4. Prepare GM3-containing liposomes by sonication at 1300 W using Sonoprex Super ultrasonic bath for 5 min three times at 41°C and stored at 41°C.
5. Filter liposome solutions through a 0.22 µm PVDF membrane and dilute ten times with HBS-N buffer.

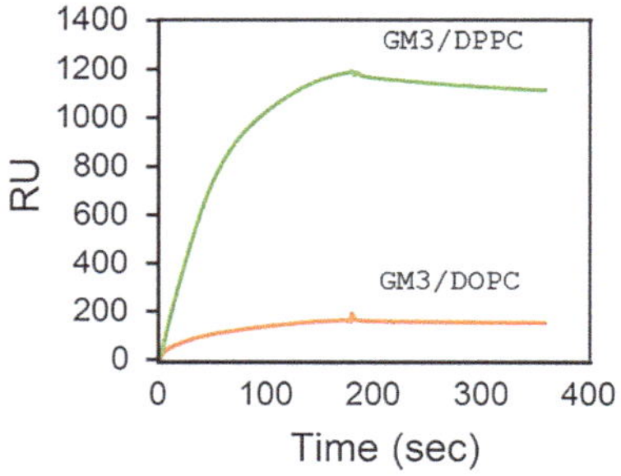

Fig. 3 Binding of anti-GM3 antibody (10 ng/mL) to GM3/DOPC (1:5) (orange) or GM3/DPPC (1:5) (green) liposomes measured by SPR. The antibody binds GM3/DPPC but not GM3/DOPC

Surface Plasmon Resonance (SPR)

SPR is performed as described [18, 19] with slight modifications.

6. Insert biosensor L1 chip into the instrument.
7. Pre-equilibrate the chip with HBS-N buffer until the baselines become flat.
8. Wash the chip with 40 mM octylglucoside.
9. Immobilize the liposomes on the chip under a flow rate of 10 μL/min at 25°C.
10. Wash the chip surface with 50 mM NaOH for 1 min at 5 μL/min.
11. Block with 100 μg/mL BSA in HBS-N buffer for 1 min at 5 μL/min.
12. Equilibrate the chip with HBS-N buffer until the baseline stabilizes.
13. Apply anti-GM3 antibody in HBS-N buffer (10 ng/mL) to the chip for 3 min at a flow rate of 10 μL/min at 25°C.
14. Change the channel and run HBS-N buffer for 3 min at a flow rate of 10 μL/min.
15. Analyze data with the BIAevaluation 4.1.1 software program (Biacore, Cytiva).

3.3 Formation of Clusters of Fluorescent GM3 Analogs in Model Membranes (Fig. 4)

Preparation of Liposomes

Day 1

1. Take 100 μL of chloroform and 50 μL of ethanol in a glass tube (*see* **Note 2**).
2. Add C11 TopFluor-GM3 (2.5 nmol), ATTO594-GM3 (2.5 nmol), and DOPC or POPC (495 nmol) and vortex the tube (*see* **Note 11**).
3. Evaporate the solvent by nitrogen gas flow (*see* **Note 12**).
4. Set the glass tubes in a desiccator, and aspirate with a vacuum pump for 2 h to remove organic solvent.
5. Add 37.5 μL of ethanol and vortex.
6. Add the liposome solution slowly by drops into 500 μL PBS while vortexing (*see* **Note 13**).
7. Dialyze the liposome solution in 500 mL of PBS using a Slide-A-Lyzer Dialysis Cassette overnight.

Day 2

8. Dilute the liposome solution ten times with PBS.

Self-Quenching of C11 TopFluor-GM3 and ATTO594-GM3 in Liposomes

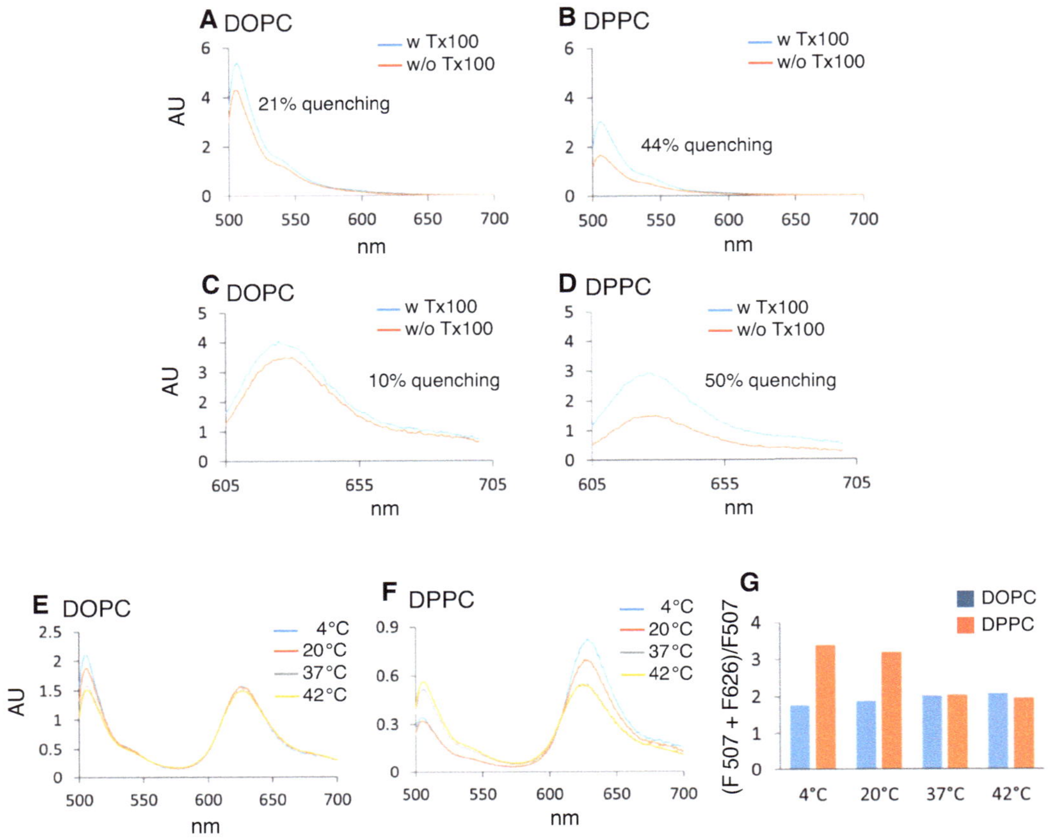

Fig. 4 Fluorescent GM3 analogs form clusters in DPPC. (**A, B**) Emission spectra of DOPC/C11 TopFluor-GM3 (99.5/0.5 mol/mol) (**A**) or DPPC/C11 TopFluor-GM3 (99.5/0.5 mol/mol) (**B**) liposomes in the presence and absence of Triton X-100 at 20°C. Excitation wavelength at 495 nm. (**C, D**) Emission spectra of DOPC/C11 TopFluor-GM3/ATTO594-GM3 (99/0.5/0.5 mol/mol/mol)(**C**) or DPPC/C11 TopFluor-GM3/ATTO594-GM3 (99/0.5/0.5 mol/mol/mol) (**D**) liposomes in the presence and absence of Triton X-100 at 20°C. Excitation wavelength at 603 nm. (**E–G**) Emission spectra of DOPC/C11 TopFluor-GM3/ATTO594-GM3 (99/0.5/0.5 mol/mol/mol)(**E**) or DPPC/C11 TopFluor-GM3/ATTO594-GM3 (99/0.5/0.5 mol/mol/mol) (**F**) liposomes at different temperatures. Excitation wavelength at 495 nm. (**G**) FRET efficiency of (**E**) and (**F**) was calculated by (Emission of donor (F_{507}) + Emission of acceptor (F_{626}))/Emission of donor (F_{507}). Data are means ±SD of three experiments. Both C11 TopFluor-GM3 (**A, B**) and ATTO594-GM3 (**C, D**) showed higher fluorescence quenching in DPPC than in DOPC, suggesting that fluorescent GM3 analogs are clustered in DPPC. Figure (**G**) shows higher FRET in DPPC than in DOPC at 4 and 20°C, suggesting that fluorescent GM3 is clustered at low temperature

9. Record emission spectra of DOPC or DPPC/C11 TopFluor-GM3/ ATTO594-GM3 liposomes for an excitation wavelength fixed at 495 or 603 nm at 20°C (*see* **Note 14**).
10. Add Triton X-100 (final concentration 1%) to the liposome mixtures and mix well.
11. Record the emission spectra as step 9.
12. Evaluate the percentage of self-quenching in liposomes by calculating the ratio at the maximum intensity of the emission spectra in the presence and absence of Triton X-100.

Temperature Dependence of the Förster Resonance Energy Transfer (FRET) Efficiency of C11 TopFluor-GM3/ ATTO594-GM3 in Liposomes

13. Record the fluorescence emission spectra from 500 to 700 nm (slit: 1 nm, increment: 1 nm) for an excitation wavelength fixed at 495 nm (slit: 2 nm) with an integration time of 0.2 s every 1 nm at 4, 20, 37, and 42°C (*see* **Note 14**).
14. FRET efficiency was calculated as: (Emission of donor (F_{507}) + Emission of acceptor (F_{626}))/(Emission of donor (F_{507})).

3.4 Cell Density-dependent Labeling of Cellular GM3 by Anti-GM3 Antibody (Fig. 5)Please replace Fig. 5 to New Fig. 5 in attachments.

Day 1

1. Seed cells at a concentration of 1 × 10^4 cells/mL (low density) or 3 × 10^5 cells/mL (high density) in 1 mL of culture medium on a coverslip in a 12-well culture plate and incubate overnight.

Day 2

2. Wash cells twice with 1 mL of culture medium and once with PBS.
3. Fix cells with 1 mL of PFA solution for 20 min (*see* **Note 2**).
4. Change the solution to PBS. Incubate for 5 min. Then, remove PBS. Repeat this step for four times.
5. Change PBS to 500 μL of blocking buffer and incubate the cells for 1 h.
6. Change the blocking buffer to anti-GM3 antibody diluted in a blocking buffer and incubate the cells for 1 h at 4°C, room temperature, 37°C or 42°C (*see* **Note 15**).
7. Wash four times with PBS as step 4.
8. Change PBS to Alexa488-labeled secondary antibody in a blocking buffer and incubate the cells for 2 h under aluminum foil (*see* **Note 16**). Protect samples from light hereafter.
9. Wash four times with PBS as step 4.
10. Fix the cells with 1 mL of PFA solution for 5 min(*see* **Note 2**).

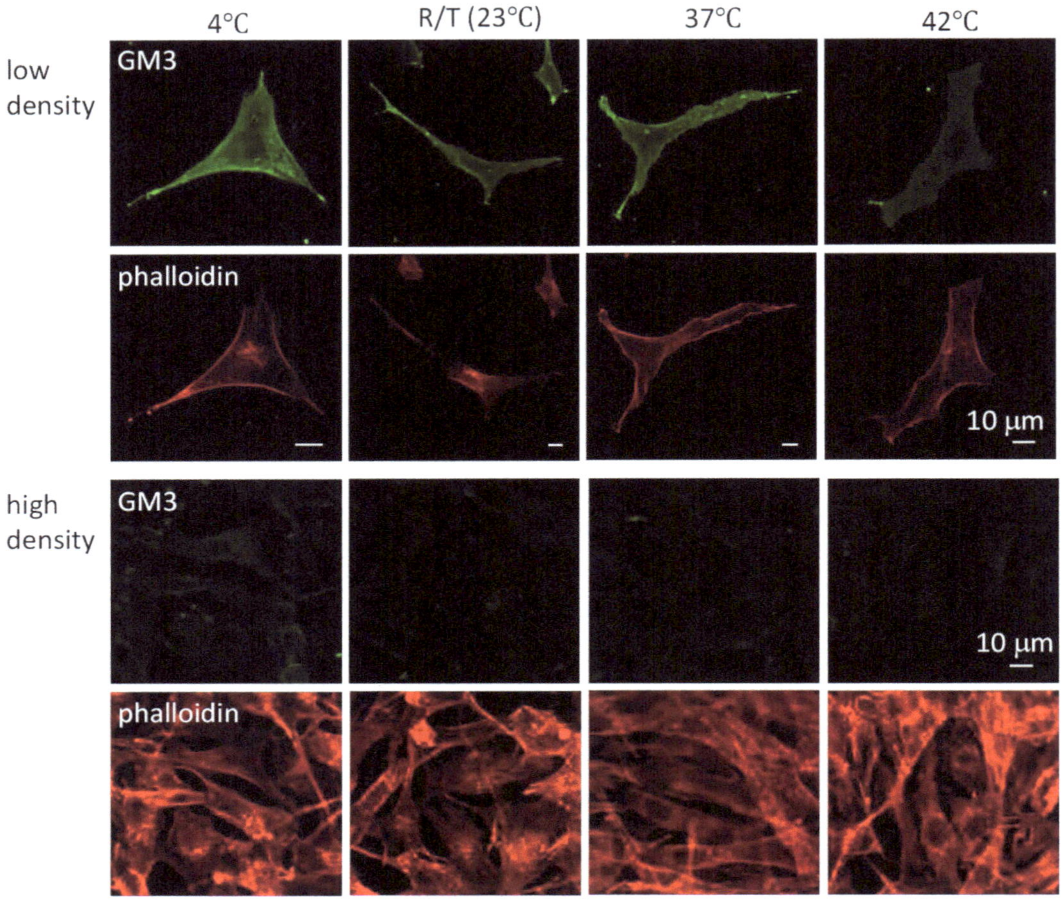

Fig. 5 Cell density-dependent labeling of cellular GM3 by anti-GM3 antibody. Low-density ($1.4 \pm 0.4 \times 10^4$ cells/15 mmϕ coverslip ($n = 3$)) and high-density ($2.8 \pm 0.6 \times 10^5$ cells/15 mmϕ coverslip ($n = 4$)) MEB4 cells were fixed and labeled with anti-GM3 antibody. Cells were doubly labeled with phalloidin to identify cells. Only low-density cells were labeled at 4°C, room temperature and 37°C. Both low- and high-density cells were not labeled at 42°C

11. Wash four times with PBS as step 4 (*see* **Note 16**).
12. Change PBS to Alexa546-labeled phalloidin in a phalloidin-labeling buffer and incubate the cells for 1 h (*see* **Notes 2 and 15**).
13. Wash four times with PBS as step 4.
14. Pick up the coverslip with tweezers and place it quickly once down and up in pure water.
15. Mount the sample on a microscope slide and keep it protected from light at room temperature overnight (*see* **Note 17**).

Day 3

16. Observe and acquire images with a confocal laser scanning microscope (*see* **Note 18**).

4 Notes

1. Keep the stock solutions of lipids in glass tubes with Teflon-sealed caps at −20°C because the lipids are in organic solvent containing chloroform. Open the cap after waiting until the temperature of the lipid solution reaches room temperature to escape from aqueous vapor contamination. To take lipid solution, handling with microsyringes, not micropipettes, is recommended to avoid plastic contamination to organic solvent.
2. Several poisonous and deleterious chemicals are used in these experiments. Take appropriate countermeasures, e.g., wearing gloves, lab coat, and/or eye protection, using a fume hood, following the rules for disposal, etc.
3. Dextran matrixes are used to capture liposomes and mimic lipid bilayers. The L1 chip contains hydrophobic alkanethiol chains with an exposed polar head group. The lipid bilayer allows the study of membrane-inserted proteins and transmembrane proteins. The L1 chip is reusable.
4. Do not use the most peripheral two rows (1 and 12) and two lines (A and H) of 96 wells because the reaction rates of the samples in these slots differ from the others.
5. After adding the blocking buffer into all wells, gently tap the side or bottom of the plate several times to spread the solution over the well.
6. Invert the plate on a sink and discard the solution by shaking the plate vigorously.
7. Multichannel pipette is helpful to treat many wells.
8. The solution is added as quickly as possible to reduce the variation of reaction time.
9. The color of the liquid turns yellow as the peroxidase reaction progresses.
10. The color of the liquid turns orange by the addition of H_2SO_4. After adding the H_2SO_4 into all wells, tap gently the side or bottom of the plate several times to mix well.
11. As a negative control for FRET measurement, 5 nmol of C11 TopFluor-GM3 or ATTO594-GM3 with 495 nmol of DOPC or POPC liposomes are prepared.

12. A glass tube is continuously rotated manually until the solvent is evaporated by nitrogen gas flow to make uniform lipid film. We also use rotary evaporator to prepare lipid film.
13. For DPPC liposomes, PBS is pre-warmed to 40–41 °C.
14. Record emission spectra of PBS to correct for buffer emission and lamp fluctuations as well as for wavelength-dependent response of the optics and detectors in the emission pathway.
15. To save antibodies, drop 40 μL of antibody solution on parafilm attached to a sheet of wet filter paper in a glass Petri dish and slowly place the coverslip with the cells face down. Make sure to avoid trapping air bubbles.
16. Samples can be kept at 4°C overnight to stop the experiment temporarily.
17. Warm the mounting media to room temperature prior to use. Pour one drop onto a microscope slide and carefully place the coverslip over the liquid with the cells facing down to avoid trapping air bubbles. Do not push the coverslip to remove air bubbles. Keep the samples at room temperature for 18–24 h to harden the mounting media. For longer storage, place the slides at 4 °C.
18. The samples can be kept at 4°C for a while. However, it is recommended to examine by a microscope as soon as possible due to the gradual quenching of the fluorescence.

Acknowledgments

This work was supported by grants from Mizutani Foundation for Glycoscience (https://mizutanifdn.or.jp), ITMO Cancer of Aviesan within the framework of the 2021–2030 Cancer Control Strategy, on funds administered by Inserm (https://itcancer.inserm.fr), Vaincre les maladies lysosomales (19/LBPH/S44; http://www.vml-asso.org/), Agence Nationale pour la Recherche (ANR-19-CE16-0012-02 and ANR-22-CE44-0019-01; https://anr.fr), Seed Money, Assemblée du groupement européen de coopération territoriale (GECT) Eucor (https://www.eucor-uni.org/en/seed-money/), RIKEN Integrated Lipidology Program, and RIKEN Glycolipidologue Initiative. We also acknowledge funds from the Institut National de Santé et de Recherche Médicale (http://www.inserm.fr/), the Centre National de la Recherche Scientifique (http://www.cnrs.fr/), and the Université de Strasbourg (http://www.unistra.fr). The fluorescence spectra measurement was performed on the PSSP facility (UMR 7021) supported by the University of Strasbourg (https://lbp.unistra.fr/services-communs/plateforme-de-spectroscopie-et-de-synthese-peptidique).

Conflict of Interest The authors have no conflicts of interest to declare that are relevant to the content of this chapter.

References

1. Giraudo CG, Maccioni HJ (2003) Ganglioside glycosyltransferases organize in distinct multienzyme complexes in CHO-K1 cells. J Biol Chem 78:40262–40271
2. Inamori KI, Inokuchi JI (2022) Ganglioside GM3 synthase deficiency in mouse models and human patients. Int J Mol Sci 23:5368
3. Zheng M, Fang H, Tsuruoka T et al (1993) Regulatory role of GM3 ganglioside in alpha 5 beta 1 integrin receptor for fibronectin-mediated adhesion of FUA169 cells. J Biol Chem 268:2217–2222
4. Kabayama K, Sato T, Saito K et al (2007) Dissociation of the insulin receptor and caveolin-1 complex by ganglioside GM3 issn the state of insulin resistance. Proc Natl Acad Sci USA 104:13678–13683
5. Coskun Ü, Grzybek M, Drechsel D, Simons K (2011) Regulation of human EGF receptor by lipids. Proc Natl Acad Sci USA 108:9044–9048
6. Nakano M, Hanashima S, Hara T et al (2021) FRET detects lateral interaction between transmembrane domain of EGF receptor and ganglioside GM3 in lipid bilayers. Biochim Biophys Acta Biomembr 1863:183623
7. Gómez-Móuton C, Abad JL, Mira E et al (2001) Segregation of leading-edge and uropod components into specific lipid rafts during T cell polarization. Proc Natl Acad Sci USA 98:9642–9647.
8. Janich P, Corbeil D (2007) GM1 and GM3 gangliosides highlight distinct lipid microdomains within the apical domain of epithelial cells. FEBS Lett 581:1783–1787
9. Chigorno V, Palestini P, Sciannamblo M et al (2000) Evidence that ganglioside enriched domains are distinct from caveolae in MDCK II and human fibroblast cells in culture. Eur J Biochem 267:4187–4197
10. Fujita A, Cheng J, Hirakawa M et al (2007) Gangliosides GM1 and GM3 in the living cell membrane form clusters susceptible to cholesterol depletion and chilling. Mol Biol Cell 18:2112–2122
11. Lloyd KO, Gordon CM, Thampoe IJ, DiBenedetto C (1992) Cell surface accessibility of individual gangliosides in malignant melanoma cells to antibodies is influenced by the total ganglioside composition of the cells. Cancer Res 52:4948–4953
12. Greenshields KN, Halstead SK, Zitman FM et al (2009) The neuropathic potential of anti-GM1 autoantibodies is regulated by the local glycolipid environment in mice. J Clin Invest 119:595–610
13. Lingwood D, Binnington B, Róg T et al (2011) Cholesterol modulates glycolipid conformation and receptor activity. Nat Chem Biol 7:260–262
14. Sakiyama H, Takahashi T, Hirabayashi Y, Taniguchi M (1987) Change in the topographical distribution of GM3 during cell spreading and growth: immunostaining with monoclonal antibody against GM3. Cell Struct Funct 12:93–105
15. Tatewaki K, Yamaki T, Maeda Y et al (1997) Cell density regulates crypticity of GM3 ganglioside on human glioma cells. Exp Cell Res 233:145–154
16. Murate M, Yokoyama N, Tomishige N et al (2023) Cell density-dependent membrane distribution of ganglioside GM3 in melanoma cells. Cell Mol Life Sci 80:167
17. Komura N, Suzuki KG, Ando H et al (2016) Raft-based interactions of gangliosides with a GPI-anchored receptor. Nat Chem Biol 12:402–410
18. Shima S, Kawamura N, Ishikawa T et al (2014) Anti-neutral glycolipid antibodies in encephalomyeloradiculoneuropathy. Neurology 82:114–118
19. Iwabuchi K, Masuda H, Kaga N et al (2015) Properties and functions of lactosylceramide from mouse neutrophils. Glycobiology 25:655–668

Check for updates

Chapter 24

Metabolic Analysis: Tracing Gangliosides and Precursors in Foods

Tetsuya Okuda

Abstract

We describe herein a practical method for analyzing gangliosides in food and in the blood and tissues of mice that have ingested the food. In this method, the oligosaccharide moiety of gangliosides is enzymatically released, labeled with a fluorescent substance, and analyzed using an in-line fluorescence detector in high-performance liquid chromatography. By using a high-performance liquid chromatography system equipped with an amide column that can separate oligosaccharides with different structures, the amount and ratio of ganglioside molecular species in samples can be analyzed. By comparing the data obtained using this method, we can investigate the influence of a diet on the ganglioside content of different tissues in the body.

Keywords Ganglioside extraction, Food, Serum, Liver, Endoglycoceramidase, HPLC analysis, Fluorescent labeling

1 Introduction

Gangliosides are a series of acidic glycosphingolipids that contain sialic acid in the oligosaccharide moiety [1–3]. They are found in the plasma membrane of mammalian cells, and they regulate membrane functions, such as signal transduction [4–9] and molecular transport [10]. Previous studies have also shown that aberrant expression of gangliosides induces cellular pathologies and leads to the development of human diseases, such as cancer [9, 11], diabetes [6], and neurological diseases [3, 12]. Thus, appropriate amounts of gangliosides in tissues are thought to play an important role in maintaining body homeostasis.

The beneficial effects of dietary gangliosides on human health are not well understood due to the lack of easy and practical methods for analyzing gangliosides [2]. Gangliosides are digested in the intestinal tract, and most ingested food-derived gangliosides are not directly absorbed into the bloodstream and tissues [13].

Koichi Furukawa (ed.), *Gangliosides*, Methods in Molecular Biology, vol. 3035,
https://doi.org/10.1007/978-1-0716-5288-6_24,

However, the degradation products can be absorbed, and they affect the expression of gangliosides in tissues [2]. Our previous studies also showed that the ingestion of certain foods alters the expression levels of ganglioside metabolism-related genes, resulting in changes in the ganglioside content of tissues [14].

In this chapter, we describe a practical method based on high-performance liquid chromatography (HPLC) for analyzing the gangliosides in food, and in the blood and tissues of animals that have ingested the food [14]. In this method, the oligosaccharide moiety of gangliosides is enzymatically released with endoglycosylceramidase I (EGCase I) [15] and labeled with a fluorescent substance, anthranilic acid (2-AA) [16]. The 2-AA–labeled oligosaccharides with different structures are separated using an HPLC system equipped with an amide column (Amide-80), and the amount and ratio of ganglioside molecular species in samples are analyzed based on the fluorescence intensity detected by an in-line fluorescence detector ($Ex_\lambda = 360$ nm, and $Em_\lambda = 425$ nm) in HPLC. As this method enables the analysis of gangliosides prepared from a small amount of sample, it can be applied to functional food research using small laboratory animals, such as mice. By comparing the data obtained using this method, the influence of a diet on the ganglioside content of different tissues in the body can be investigated.

2 Materials

Prepare all solutions using ultrapure water and analytical-grade reagents. Prepare and store all reagents at room temperature.

2.1 Ganglioside Extraction

1. Chloroform/methanol at 2:1 (v/v).
2. Chloroform/methanol/water at 1:2:0.8 (v/v/v).
3. Chloroform/methanol/water at 3:48:47 (v/v/v).
4. Methanol.
5. Screw-cap glass test tube (7–10 ml volume) and Pasteur pipette.
6. Swing-type centrifuge.
7. Vortex or sonicator.
8. Vacuum dryer or nitrogen gas blowdown concentrator.

2.2 Degradation of Phospholipids and Triglycerides in Samples by Saponification

1. Chloroform/methanol at 2:1 (v/v).
2. 6 N NaOH.
3. 2 N acetic acid.
4. Thermostatic bath.
5. Angle- or swing-type centrifuge.

2.3 Reverse-Phase Column Purification of Extracted Gangliosides

1. MonoSpin® C18 column (GL Sciences, Tokyo, Japan).
2. Methanol.
3. Chloroform/methanol at 2:1 (v/v).
4. Angle- or swing-type centrifuge.
5. Vacuum dryer or nitrogen gas blowdown concentrator.

2.4 Normal-Phase Column Purification of Extracted Gangliosides and Neutral Glycosphingolipids

1. Silica gel, particle size 60 μm, average pore diameter 80 Å (Iatrobeads, 6RS-8060, PHC, Tokyo, Japan) (*see* **Note 1**).
2. 3-ml empty reservoir (InertSep No. 5010-60121, GL Sciences).
3. Methanol.
4. Chloroform/methanol at 98:2 (v/v).
5. Chloroform/methanol at 1:1 (v/v).
6. Chloroform.
7. Acetone/methanol at 9:1 (v/v).
8. Vacuum dryer or nitrogen gas blowdown concentrator.

2.5 Oligosaccharide Fluorescent Labeling

1. EGCase I from *Rhodococcus triatomae* (New England Biolabs, Ipswich, MA).
2. EGCase I reaction buffer (New England Biolabs).
3. 2-AA (30 mg/ml) dissolved in a solution of 4% sodium acetate trihydrate (w/v) and boric acid (2% w/v) in methanol.
4. 45 mg/ml sodium cyanoborohydride dissolved in the 2-AA solution.
5. MonoSpin® Amide column (GL Sciences).
6. Acetonitrile/water at 9:1 (v/v).
7. Thermostatic bath.
8. Angle- or swing-type centrifuge.

2.6 Thin-Layer Chromatography (TLC) Analysis

1. Chloroform/methanol at 2:1 (v/v).
2. Developing solvent of chloroform/methanol/water at 60:35:8 (v/v/v).
3. HPTLC Silica gel 60 glass plates, size 10 cm × 10 cm (Merck, Darmstadt, Germany).
4. TLC developing tank and filter paper.
5. Orcinol sulfate reagent (*see* **Note 2**).
6. Resorcinol reagent (*see* **Note 3**).
7. Glass spray bottle.
8. Temperature-controllable hot plate.
9. Scanner and image analysis software.

2.7 HPLC Analysis of Fluorescently Labeled Oligosaccharides

1. Solvent A: 20% 100 mM ammonium acetate (pH 3.85) and 80% acetonitrile.
2. Solvent B: 20% 100 mM ammonium acetate (pH 3.85), 60% water, and 20% acetonitrile.
3. TSKgel Amide-80 column, particle size 5 μm, 4.6 mm in diameter × 25 cm in length (Tosoh, Tokyo, Japan).
4. HPLC gradient elution system composed of a TSKgel Amide80 column and fluorescence detector.

3 Methods

Carry out all procedures at room temperature unless otherwise specified. Ganglioside extraction, oligosaccharide fluorescent labeling, and HPLC analysis are performed as reported previously [14]. As gangliosides can be detected with high sensitivity in this method, analysis can be performed using less than one-tenth the amount of sample required for TLC analysis using common orcinol sulfate [17] and resorcinol reagent [18] chromogenic methods. This HPLC analysis method also has the advantage of being able to clearly separate gangliosides containing *N*-glycolylneuraminic acid (Neu5Gc) from those containing *N*-acetylneuraminic acid (Neu5Ac) [14, 19].

3.1 Ganglioside Extraction

1. For solid food and tissue samples, pre-homogenize the sample with 100 μl of water.
2. Mix 50–100 mg of the food or tissue sample, or 50 μl of serum sample with 1.5 ml of chloroform/methanol at 2:1, and vortex (or sonicate) for 10 min.
3. Add 0.5 ml of methanol, vortex for 30 s, and centrifuge at 1000 × *g* for 4 min (*see* **Note 4**). Collect the supernatant with a Pasteur pipette. Leave the sample residue behind.
4. Mix the sample residue with 0.95 ml of chloroform/methanol/water at 1:2:0.8 and vortex (or sonicate) for 10 min.
5. Centrifuge at 1000 × *g* for 4 min and collect the supernatant. Leave the sample residue behind.
6. Combine the collected supernatants, add 1.75 ml of chloroform and 0.825 ml of water, and vortex for 30 s.
7. Centrifuge at 1000 × *g* for 4 min. After centrifugation, the supernatant will separate into two phases: the upper phase contains the polar solvent, and the lower phase contains the non-polar solvent. Mark the top of the solvent and collect the upper phase.
8. Add chloroform/methanol/water at 3:48:47 to the mark and vortex for 30 s.

9. Centrifuge at 1000 × *g* for 4 min and collect the upper phase.
10. Combine the collected upper phases and evaporate the solvent using a vacuum dryer or nitrogen gas blowdown concentrator to obtain a dried sample (*see* **Note 5**). The dried sample is optionally subjected to a reverse-phase column purification step (Sect. 3.3) if necessary (*see* **Note 6**).
11. The solvent in the lower phase is also evaporated using a vacuum dryer or nitrogen gas blowdown concentrator to obtain a dried sample (*see* **Note 5**). The dried sample is optionally subjected to a normal-phase column purification step (Sect. 3.4) if necessary (*see* **Note 7**).

3.2 Degradation of Phospholipids and Triglycerides in Samples by Saponification (see Note 8)

1. Dissolve the dried sample in 500 μl of chloroform/methanol at 2:1 by vortexing.
2. Add 16.7 μl of 6 N NaOH and vortex for 10 s.
3. Incubate at 37°C for 1 h with vortexing for 10 s every 20 min.
4. Add 25 μl of 2 N acetic acid and vortex for 10 s.
5. Evaporate the solvent using a vacuum dryer or nitrogen gas blowdown concentrator.

3.3 Reverse-Phase Column Purification of Extracted Gangliosides (see Note 9)

1. Add 100 μl of methanol to a MonoSpin® C18 column and centrifuge at 5000 × *g* for 2 min to activate the column.
2. Add 100 μl of water to the column and centrifuge at 5000 × *g* for 2 min to equilibrate the column.
3. Dissolve the dried sample in 800 μl of water (*see* **Notes 10 and 11**).
4. Add 800 μl of sample to the C18 column and centrifuge at 5000 × *g* for 2 min.
5. Add 300 μl of water to the column and centrifuge at 5000 × *g* for 2 min to wash the column (*see* **Note 12**).
6. Add 100 μl of methanol to the column and centrifuge at 5000 × *g* for 2 min. Collect the flow-through.
7. Add 200 μl of chloroform/methanol at 2:1 to the column and centrifuge at 5000 × *g* for 2 min. Collect the flow-through.
8. Combine the collected flow-throughs and evaporate the solvent using a vacuum dryer or nitrogen gas blowdown concentrator.

3.4 Normal-Phase Column Purification of Extracted Gangliosides and Neutral Glycosphingolipids

1. Pack the empty reservoir with 250 μl of silica gel (*see* **Note 13**), and allow the solvent to flow-through the column under the influence of gravity (*see* **Note 14**).
2. Add 2.5 ml of chloroform/methanol at 98:2 to equilibrate the column (*see* **Note 15**).
3. Dissolve the sample in 40 μl of chloroform/methanol at 1:1, then add 960 μl of chloroform, and vortex thoroughly.

4. Apply the sample to the column and allow it to flow to the top of the column bed.
5. Add 2 ml of chloroform/methanol at 98:2 to wash out the simple lipids, such as triglycerides.
6. Add 2 ml of acetone/methanol at 9:1 to elute the glycosphingolipids (*see* **Note 16**).
7. Evaporate the solvent using a vacuum dryer or nitrogen gas blowdown concentrator.

3.5 Fluorescent Labeling of Oligosaccharides

1. Dissolve the dried sample in 19 μl of reaction buffer appropriately diluted with water.
2. Add 1 μl (6 mU) of EGCase I and incubate at 37°C for 16 h in a thermostatic bath (*see* **Note 17**).
3. Prepare the 45 mg/ml sodium cyanoborohydride solution dissolved in 2-AA solution.
4. Add 80 μl of the sodium cyanoborohydride solution to the EGCase I-digested sample.
5. Mix these solutions and incubate at 80°C for 1 h.
6. Add 200 μl of acetonitrile/water at 9:1 to a MonoSpin® Amide column and centrifuge at 10,000 × *g* for 1 min to activate the column.
7. Add 700 μl of acetonitrile/water at 9:1 to the sample, mix, and apply to the column.
8. Centrifuge at 10,000 × *g* for 1 min. Discard the flow-through.
9. Add 200 μl of acetonitrile/water at 9:1 to the column. Centrifuge at 10,000 × *g* for 1 min to wash the column.
10. Add 200 μl of water to the column. Centrifuge at 10,000 × *g* for 1 min to elute the 2-AA–labeled oligosaccharides (*see* **Note 18**).

3.6 TLC Analysis (see Note 19)

1. Dissolve the sample in chloroform/methanol at 2:1.
2. Apply the sample onto a TLC plate (*see* **Note 20**) and air-dry to evaporate the sample solvent.
3. Add developing solvent to the developing tank and leave at room temperature for more than 10 min to allow the solvent to saturate the tank (*see* **Note 21**).
4. Place the TLC plate with the sample into the developing tank, and allow to develop until the solvent reaches about 1 cm from the top of the TLC plate (*see* **Note 22**).
5. Remove the TLC plate from the development tank and air-dry to evaporate the developing solvent.
6. Apply orcinol sulfate reagent evenly to the TLC plate using a spray bottle (*see* **Note 23**).

7. Heat the TLC plate sprayed with orcinol sulfate reagent to 110°C on a hotplate with the sample side facing up. Glycosphingolipids turn reddish-purple after about 10 min of heating (*see* **Note 24**).
8. Scan the TLC plate with a scanner to obtain image data, and analyze the intensity of the detected bands using image analysis software (*see* **Note 25**).

3.7 HPLC Analysis of Fluorescently Labeled Oligosaccharides

1. Perform HPLC at 30°C with the following gradient conditions (*see* **Note 26**).
   a. 0 min with 86% solvent A/14% solvent B (0.8 ml/min).
   b. 6 min with 86% solvent A/14% solvent B (0.8 ml/min).
   c. 35 min with 54.7% solvent A/45.3% solvent B (0.8 ml/min).
   d. 37 min with 5% solvent A/95% solvent B (1 ml/min).
   e. 39 min with 5% solvent A/95% solvent B (1 ml/min).
   f. 41 min with 86% solvent A/14% solvent B (1 ml/min).
   g. 42 min with 86% solvent A/14% solvent B (1.2 ml/min).
   h. 54 min with 86% solvent A/14% solvent B (1.2 ml/min).
2. Inject 10 μl of the 2-AA–labeled oligosaccharide samples (*see* **Note 27**).
3. Measure the intensity of the fluorescence signal of 2-AA ($Ex_\lambda = 360$ nm and $Em_\lambda = 425$ nm). Record the data from 0 to 45 min after sample injection.
4. Identify the elution peak for the target ganglioside from the retention time of the 2-AA–labeled oligosaccharide prepared from the ganglioside standard (Fig. 1).
5. Calculate the content of the target ganglioside from the area of the fluorescence peak (*see* **Note 28**).

4 Notes

1. Activate by stirring 1 g of Iatorobeads in 10 ml of methanol for 10 min. Wash the gel at least three times by spinning down, discarding the supernatant (methanol), adding fresh methanol up to the original position,and stirring.
2. Dissolve 0.2 g of orcinol in 88.6 ml of water in a glass bottle. Place the bottle on ice, slowly add 11.4 ml of sulfuric acid, and stir. Please note that adding sulfuric acid too quickly may cause bumping.

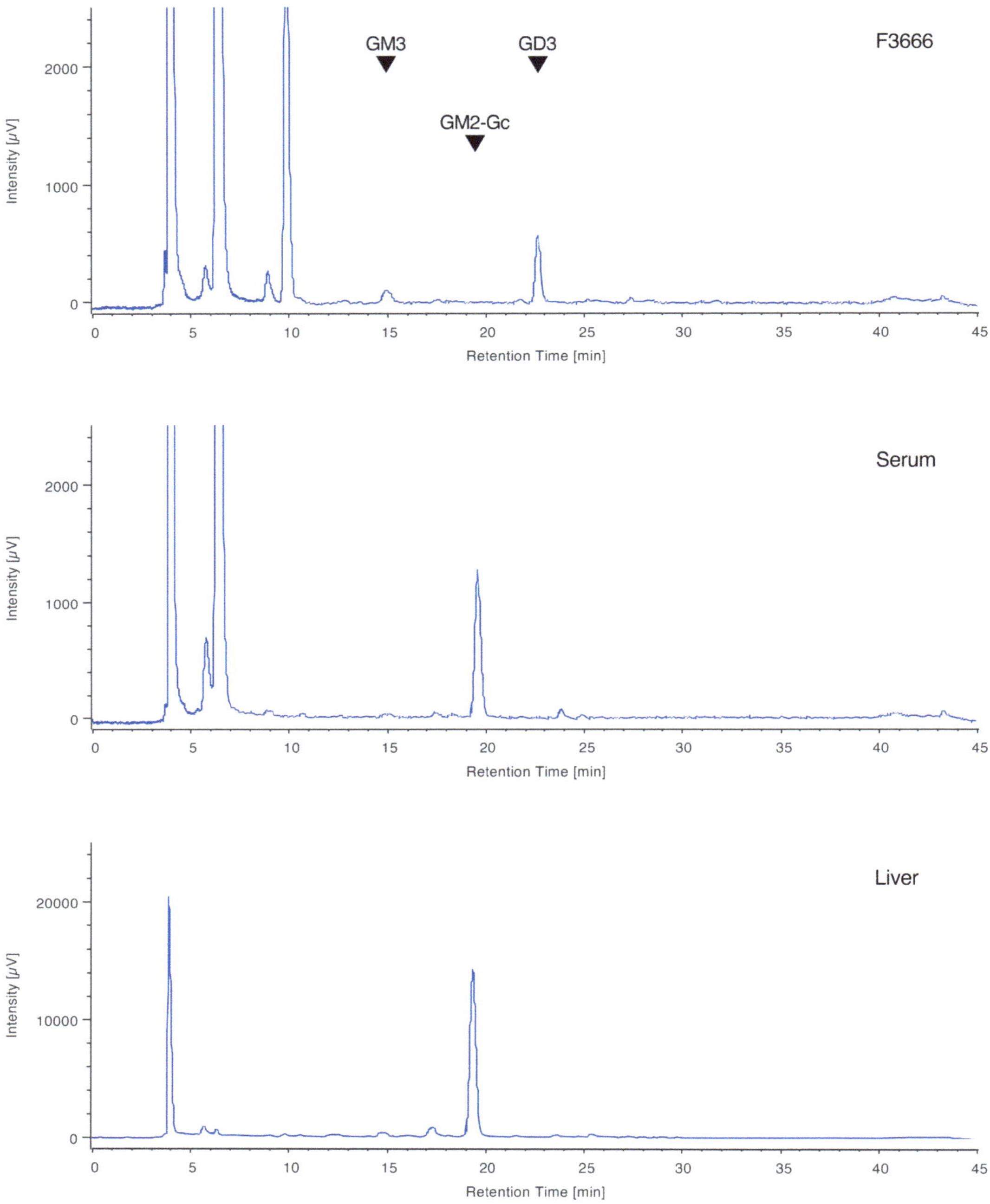

Fig. 1 HPLC elution patterns of gangliosides extracted from an animal ketogenic diet, F3666 (Bio-serv, Frenchtown, NJ), and the serum and liver of a mouse fed F3666. We used a 5-week-old female mouse that was homozygous for the obese spontaneous mutation (B6.Cg-*Lep*^{ob}/J, Charles River Laboratories Japan, Yokohama, Japan). The mouse was fed F3666, which comprised lard, butter fat, dextrose, casein, fiber, corn oil, mineral mix, and a vitamin mix, for a period of 7 weeks, then serum and liver samples were collected. The 2-AA oligosaccharides prepared from the gangliosides in amounts corresponding to those extracted from 1.5 mg of F3666 (top panel), 1.25 μl of serum (middle panel), and 1.5 mg of the liver (bottom panel) were analyzed. Arrowheads indicate the elution times for 2-AA oligosaccharides prepared from GM3, GM2-Gc, and GD3 standards. Abbreviations: GM3 Neu5Acα2,3Galβ1,4Glcβ 1,1ceramide, GM2-Gc GalNAcβ1,4 (Neu5Gcα2,3)Galβ1,4Glcβ1,1ceramide, GD3 NeuAcα2,8NeuAcα2,3Galβ1,4Glcβ 1,1ceramide. (Reproduced from [14] under a Creative Commons license)

3. Dissolve 0.2 g of resorcinol in 20 ml of water. Add 80 ml of hydrochloric acid and stir, then add 0.25 ml of 0.1 M copper sulfate aqueous solution and stir.
4. Increasing the proportion of methanol will help to spin down the tissue residue.
5. The glass test tube containing the sample can be heated to 50°C to facilitate evaporation of the solvent.
6. Gangliosides are contained in the upper phase.
7. Neutral glycosphingolipids are contained in the lower phase. Highly hydrophobic gangliosides, such as GM3 with very long-chain fatty acids, also tend to be distributed in the lower phase.
8. This step is optional. Saponification degrades lipids with ester bonds in their molecules, such as phospholipids and triglycerides that interfere with the TLC analysis of glycosphingolipids. Perform this step before reverse-phase or normal-phase column purification steps, if necessary.
9. This step removes hydrophilic contaminants, such as salts and oligosaccharides, from the sample. This step can be omitted if you want to also analyze the oligosaccharides present in the sample, or if the sample does not contain oligosaccharides that would interfere with the 2-AA–labeled oligosaccharide analysis by HPLC.
10. The dried samples may also contain water-insoluble lipids. If so, vortex thoroughly (or sonicate), collect the supernatant, and proceed to the next step.
11. If a large amount of dried lipid has been obtained in the previous step, divide the sample into aliquots and purify them using multiple columns.
12. If all of the water does not pass through the column during this time, increase the centrifugation time.
13. Mark the 250-μl position in the empty reservoir, then inject the silica gel, let it settle by gravity, and pack the silica gel so that the top surface of the gel reaches the mark.
14. Do not drain the solvent below the top of the column bed to avoid gel drying.
15. Inject in several doses.
16. Other lipids, such as phospholipids, remain adsorbed on the column.
17. TLC analysis of a replicate sample prepared at the same time (Sect. 3.6) can confirm that the oligosaccharide moieties of gangliosides are completely released by the enzymatic reaction. When checking the sample by TLC, the sample should be pre-purified using a MonoSpin® C18 column (Sect. 3.3).

18. If the sample is not used immediately, store it tightly sealed at 4°C to prevent solvent evaporation. For long-term storage, store at −20°C.
19. This step is optional.
20. Using a pencil, draw lines approximately 7 mm in length located 1.5 cm from the bottom edge of the TLC plate that run parallel to the bottom edge of the plate, and apply the sample evenly over this line.
21. Insert a filter paper slightly smaller than the wall of the developing tank and soak it in the developing solvent. The solvent soaked into the filter paper evaporates, saturating the tank with solvent. This process is essential to the development of an accurate sample on the TLC plate. Stick the filter paper to the wall so that it does not touch the TLC plate.
22. Pre-developing hexane to the top edge of the TLC plate with the sample mobilizes only simple lipids and reduces their interference in the ganglioside analysis by TLC.
23. The orcinol sulfate reagent detects all types of glycolipids. In contrast, the resorcinol reagent detects only gangliosides (glycolipids containing sialic acids).
24. When using resorcinol reagent, heat the TLC plate with the sample side facing down at 95°C for about 15–30 min until gangliosides can be clearly detected. The resorcinol reagent reacts with sialic acids and turns them blue-purple.
25. When storing the TLC plate for a long period of time, cover the sample application surface with a glass plate of the same size as the TLC plate and seal the edges.
26. The gradient conditions were reported by Neville et al. [16].
27. The amount of 2-AA–labeled oligosaccharides in the prepared sample depends on the amount of gangliosides contained in the original sample. Thus, it may need to be diluted with water for an appropriate amount to be injected for analysis.
28. The content of gangliosides in a sample is calculated as the ratio of the peak area of each ganglioside to the area of 2-AA–labeled oligosaccharides prepared from ganglioside standards analyzed separately.

Acknowledgments

This work was supported by the Japan Society for the Promotion of Science (JSPS KAKENHI Grant Numbers 22H03546 and 23K24803).

Conflict of Interest The authors have no conflicts of interest to declare that are relevant to the content of this chapter.

References

1. Kolter T, Proia RL, Sandhoff K (2002) Combinatorial ganglioside biosynthesis. J Biol Chem 277:25859–25862
2. Okuda T (2020) Dietary control of ganglioside expression in mammalian tissues. Int J Mol Sci 21:177
3. Furukawa K, Ohmi Y, Yesmin F et al. (2020) Novel molecular mechanisms of gangliosides in the nervous system elucidated by genetic engineering. Int J Mol Sci 21:1906
4. Hakomori S, Yamamura S, Handa AK (1998) Signal transduction through glyco(sphingo) lipids. Introduction and recent studies on glyco(sphingo)lipid-enriched microdomains. Ann N Y Acad Sci 845:1–10
5. Ledeen RW, Wu G (2002) Ganglioside function in calcium homeostasis and signaling. Neurochem Res 27:637–647
6. Inokuchi J (2014) GM3 and diabetes. Glycoconj J 31:193–197
7. Nordström V, Willershäuser M, Herzer S et al. (2013) Neuronal expression of glucosylceramide synthase in central nervous system regulates body weight and energy homeostasis. PLoS Biol 11:e1001506
8. Herzer S, Meldner S, Gröne HJ et al. (2015) Fasting-induced lipolysis and hypothalamic insulin signaling are regulated by neuronal glucosylceramide synthase. Diabetes 64:3363–3376
9. Furukawa K, Ohmi Y, Ohkawa Y et al. (2019) New era of research on cancer-associated glycosphingolipids. Cancer Sci 110:1544–1551
10. Furukawa K, Takamiya K, Ohmi Y et al. (2023) Disordered testosterone transport in mice lacking the ganglioside GM2/GD2 synthase gene. FEBS Open Bio 13:1615–1624
11. Kannagi R, Cai BH, Huang HC et al. (2018) Gangliosides and tumors. Methods Mol Biol 1804:143–171
12. Schengrund CL (2015) Gangliosides: glycosphingolipids essential for normal neural development and function. Trends Biochem Sci 40:397–406
13. Sugawara T (2022) Sphingolipids as functional food components: benefits in skin improvement and disease prevention. J Agric Food Chem 70:9597–9609
14. Okuda T (2019) A low-carbohydrate ketogenic diet promotes ganglioside synthesis via the transcriptional regulation of ganglioside metabolism-related genes. Sci Rep 9: 7627
15. Albrecht S, Vainauskas S, Stöckmann H et al. (2016) Comprehensive profiling of glycosphingolipid glycans using a novel broad specificity endoglycoceramidase in a high-throughput workflow. Anal Chem 88:4795–4802
16. Neville DC, Coquard V, Priestman DA et al. (2004) Analysis of fluorescently labeled glycosphingolipid-derived oligosaccharides following ceramide glycanase digestion and anthranilic acid labeling. Anal Biochem 331:275–282
17. Okuda T, Morita N (2016) A very low carbohydrate ketogenic diet increases hepatic glycosphingolipids related to regulation of insulin signalling. J Funct Foods 21:70–74
18. Okuda, Nakayama K (2008) Identification and characterization of the human Gb3/CD77 synthase gene promoter. Glycobiology 18:1028–1035
19. Okuda T (2019) Data set for characterization of the glycosylation status of hepatic glycoproteins in mice fed a low-carbohydrate ketogenic diet. Data Brief 27:104604

Check for updates

Chapter 25

Flow Cytometry Analysis of Gangliosides Expressed on Extracellular Vesicles

Yuhsuke Ohmi, Hiroka Koyanagi, Mohammad Abul Hasnat and Keiko Furukawa

Abstract

In flow cytometry analysis, molecules on cells can be detected by labeling with antibodies or other reagents. However, extracellular vesicles (EVs) released by cells are of a small particle size, making it difficult to detect molecules on EVs using conventional flow cytometry. Here, we describe a method for isolating EVs secreted by cells and report a flow cytometry analysis for gangliosides on EVs. By using Tim4-conjugated beads, which bind specifically to EVs, molecules expressed on EVs can be easily detected even with flow cytometry.

Keywords Extracellular vesicles, Ganglioside, GD2, GD3, Tim4, Flow cytometry

1 Introduction

Extracellular vesicles (EVs) are lipid bilayer-bound vesicles with a diameter of 40–160 nm, formed within the endosome, and containing various cellular components. It has been reported that EVs secreted from a variety of cells, including cancer cells, play an important role in intercellular communication [1]. Similar to the plasma membrane, EV membranes express a variety of molecules, including acidic glycosphingolipids such as gangliosides [2]. In this chapter, we describe the most common method for isolating EVs from cell culture supernatant, namely, ultracentrifugation [3]. Unlike purification methods that require interactions with vesicle molecules, such as antibody-based methods, ultracentrifugation allows EVs to be collected in an intact state, without altering their composition. By varying the centrifugation speed in steps, ultracentrifugation enables the sequential separation and removal of impurities, ultimately leading to the collection of highly concentrated EVs. Furthermore, in this method, fetal calf serum (FCS) is removed from the culture medium prior to EV collection and replaced with insulin–transferrin–selenium

Koichi Furukawa (ed.), *Gangliosides*, Methods in Molecular Biology, vol. 3035,
https://doi.org/10.1007/978-1-0716-5288-6_25,

(ITS) to minimize protein contamination in the supernatant [2]. The EVs were quantified using the bicinchoninic acid (BCA) assay for protein concentration [4]. However, it should be noted that the measurement of EV yield by protein quantification includes contaminating substances, and this quantification provides an indirect estimation of EV quantity under standardized culture conditions, as described in this study [5]. Next, we describe the method for detecting gangliosides on both isolated EVs and the parent cells using flow cytometry with anti-ganglioside antibodies. Due to the small size of EVs, magnetic beads conjugated with Tim4 were used to collect EVs, followed by the detection of gangliosides on these EVs-binding beads [6]. Using this method, we successfully and consistently detected the gangliosides GD2 and GD3 on EVs collected from GD2/GD3-expressing glioma cell lines. Thus, this method is demonstrated to be suitable for the detection of gangliosides on EVs isolated from cell culture supernatants [7].

2 Materials

All cell cultures should be conducted within a clean bench under sterile conditions. Cell cultures should be maintained at 5% CO_2 and 37°C unless otherwise specified. All reagents should be prepared and stored at room temperature unless specifically instructed otherwise. When disposing of waste, it is imperative to adhere to all waste disposal regulations.

2.1 Collection of EVs

1. Cells: We used the GT cell line, which stably expresses the GD3 synthase gene in the U251 glioma cell line and highly expressed GD2 and GD3, as well as the CV cell line, which was prepared as a vector control and shows no expression of GD2 or GD3 [8].
2. 15 cm dish: Petri Dish, 145 mm × 20 mm (Greiner Bio-One).
3. 10% FCS-medium: Dulbecco's Modified Eagle Medium (DMEM) supplemented with 10% fetal calf serum (FCS) was used for cell cultures. The medium was pre-warmed to 37°C prior to use.
4. ITS-medium: DMEM containing 1% ITS premix (Corning). ITS premix is a mixed reagent consisting of insulin, transferrin, and selenious acid. The medium was pre-warmed to 37°C prior to use (*see* **Note 1**).
5. Ultracentrifugation: A centrifuge capable of reaching 100,000 × g was used for the ultracentrifugation process. In this study, the Optima XPN-100 Ultracentrifuge (Beckman Coulter) was used. The rotor employed was the Beckman SW32Ti rotor (Beckman Coulter). For ultracentrifugation, Beckman polypropylene ultracentrifuge tubes (Beckman Coulter) were used.

6. Filtration: 0.22 μm Sartolab® RF-150 (Sartorius) (*see* **Note 2**).
7. 1.5 mL microtube: Reaction tube 1.5 mL, No. 616261 (Greiner Bio-One) (*see* **Note 3**).

2.2 Quantification of EVs

1. Protein quantification reagents: Pierce™ BCA Protein Assay Kits (Thermo Scientific) include reagent A and reagent B. These are mixed in a 50:1 ratio of A to B.
2. Standard control: Dilute the Albumin Standard (2 mg/mL) (Thermo Scientific) in PBS to obtain five concentrations: 250, 125, 62.5, 31.25, and 15.625 μg/mL. Standard controls can be stored frozen at −30°C for long-term preservation.
3. Plate reader: Use a spectrophotometer capable of measuring absorbance at 540–590 nm (*see* **Note 4**). Here, the Multiskan FC (Thermo Scientific) was utilized.

2.3 Detection of Gangliosides Expressed on EVs

1. Primary antibodies (*see* **Note 5**): Anti-GD3 monoclonal antibody, R24, was provided by L.J. Old at the Memorial Sloan Kettering Cancer Center (New York) [9]. R24 can also be purchased from a supplier. Anti-GD2 mAb, 220-51, was generated in our laboratory [10]. In addition to anti-GD2 mAb, 14G2a is also available for purchase.
2. Secondary antibody: Fluorescein-conjugated Affinipure Goat Anti-Mouse IgG(H+L) (Proteintech) (*see* **Note 5**).
3. Exosome Capture Beads: Included in the PS Capture Exosome Flow Cytometry Kit (Fujifilm) (*see* **Note 6**).
4. Washing Buffer (10×): Included in the PS Capture Exosome Flow Cytometry Kit (Fujifilm) (*see* **Note 7**).
5. Exosome Binding Enhancer (100×): Included in the PS Capture Exosome Flow Cytometry Kit (Fujifilm) (*see* **Note 7**).
6. Magnetic stand: Magnet stand (Fujifilm) (*see* **Note 8**).
7. Mesh: We use mesh size with an opening of 57 μm.
8. Flowcytometry: Magnetic beads were gated based on side scatter (SSC) and forward scatter (FSC), and FITC detection was performed using flow cytometry (*see* **Note 9**). Here, the BD Accuri™ C6 (BD) was used.

3 Methods

3.1 Collection of EVs

1. Adherent cells derived from the glioma cell line GT or CV were seeded onto a 15 cm dish to reach 60%–70% confluence with 15 mL of 10% FCS-medium, and the cells were cultured at 37°C in a 5% CO_2 environment. The cells were evenly

distributed and deposited on the dish before placing the dish in the incubator (*see* **Note 10**).

2. After 24 h of culture, the cells were washed three times with 7–10 mL of PBS, then replaced with 15 mL of ITS-medium, and cultured at 37°C in a 5% CO_2 environment (*see* **Note 11**).
3. When the cells reach approximately 100% confluence (after about 72 h of culture), the supernatant is completely collected into a 50 mL conical tube (*see* **Note 12**). The collected supernatant is centrifuged at 500 × *g*, 4°C, for 10 min (*see* **Note 13**). The supernatant is then transferred to a new 50 mL conical tube and centrifuged at 20,000 × *g*, 4°C, for 20 min. Transfer the supernatant to a new 50 mL conical tube and centrifuge at 20,000 × *g*, 4°C, for 20 min (*see* **Note 14**).
4. After centrifugation, the supernatant is passed through a 0.22 μ m filter (Sartorius) (*see* **Note 15**), then transferred to an ultracentrifuge tube and ultracentrifuged at 175,000 × *g*, 4°C, for 84 min (*see* **Note 15**). After ultracentrifugation, decant the supernatant and resuspend the EV pellets by vortex mixer at maximum power for 10 s. Add PBS to reach a final volume of 37.5 mL (*see* **Note 16**) and ultracentrifuge at 175,000 × *g*, 4°C, for 84 min. After ultracentrifugation, decant the supernatant, resuspend the EV pellets by vortex mixer at maximum power for 30 s, then add 200 μL of PBS and resuspend the EV pellets by vortex mixer again at maximum power for 30 s (*see* **Note 17**). Finally, transfer the entire liquid to 1.5 mL microtubes and store at 4°C (*see* **Note 18**).

3.2 Quantification of EVs

1. Reagent A and reagent B included in the Pierce™ BCA Protein Assay Kits (Thermo Scientific™) are mixed at a ratio of 50:1 and resuspended. The mixture (A+B) can be stored at room temperature.
2. Prepare standard controls at concentrations of 250, 125, 62.5, 31.25, and 15.625 μg/mL, and add 10 μL of each to a 96-well plate (*see* **Note 19**). Also, add 10 μL of the EV samples collected in Sect. 3.1.4 to the 96-well plate.
3. Add 200 μL of the mixture (A+B) prepared in Sect. 3.2.1 to each well, shake the plate for 30 s, and cover the plate. Incubate it at 37°C for 30 min. After 30 min, bring the samples back to room temperature.
4. Measure the absorbance at 540–590 nm using a plate reader. Create a standard curve and use it to quantify the protein content of the EVs.

3.3 Detection of Gangliosides Expressed on EVs

1. An overview of the experimental procedure is summarized in Fig. 1.
2. Prepare WB + Enhancer by diluting Washing Buffer (10×) tenfold in ultrapure water and Exosome Binding Enhancer (100×) 100-fold (*see* **Note 20**).
3. Then, add 300 μL of WB + Enhancer to a new 1.5 mL microtubes.
4. Suspend the vial containing Exosome Capture Beads for more than 10 s by vortex mixer, then add 30 μL of Exosome Capture Beads to the 1.5 mL microtube containing 300 μL of WB + Enhancer. Samples are suspended by vortex mixer and then spin down.
5. Place the microtube containing the Exosome Capture Beads in a magnetic stand and let it sit for 1 min. The Exosome Capture Beads will accumulate on the wall, allowing the liquid to be removed by pipetting and discarded (*see* **Note 21**).
6. Add 300 μL of WB + Enhancer, vortex mix, and spin down. Repeat Sect. 3.3, step 4 twice (Wash) (*see* **Note 22**).
7. Add 100 μL of EVs adjusted to 2.5 μg/mL (*see* **Note 23**), vortex mix, spin down, and rotate on a rotator for 1 h at room temperature to bind the EVs to the Exosome Capture Beads (*see* **Note 24**).
8. After 1 h, samples are suspended by vortex mixer and then spin down, and place the microtube in a magnetic stand, and let it sit for 1 min. The Exosome Capture Beads will accumulate on

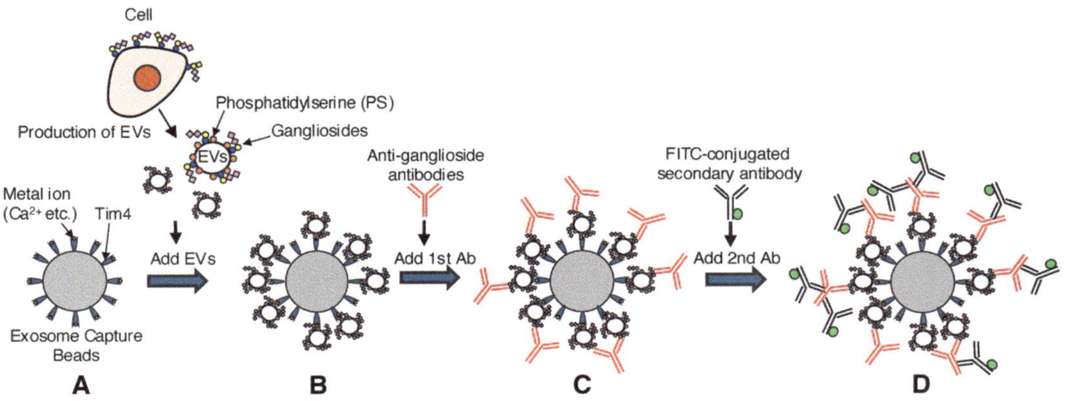

Fig. 1 Overview of the detection of gangliosides expressed on EVs. (**a**) Pre-treatment of Exosome Capture Beads with WB + Enhancer (Washing Buffer + Exosome Binding Enhancer; *see* **Note 7**). (**b**) Binding of EVs to Exosome Capture Beads via the interaction between PS on EVs and Tim4 on beads. (**c**) Binding of anti-ganglioside antibody to gangliosides in EVs. (**d**) Binding of FITC-conjugated secondary antibody and flow cytometry analysis

the wall, allowing the liquid to be removed by pipetting and discarded.

9. Add 300 μL of WB + Enhancer, vortex, and spin down. Repeat Sect. 3.3, step 8 twice (Wash).
10. Add 200 μL of blocking solution (*see* **Note 25**), vortex mix, spin down, and rotate on a rotator for 30 min at room temperature.
11. During the blocking reaction, prepare the primary antibodies, anti-GD3 antibody (R24) and anti-GD2 antibody (220-51), at a concentration of 10 μg/mL in the blocking solution, and store on ice until use (*see* **Note 26**).
12. After 30 min, place the samples in a magnetic stand and let it sit for 1 min. Then, remove and discard the liquid by pipetting, and add 100 μL of the primary antibody. For the negative control, use the blocking solution instead. Incubate the sample on ice for 45 min, tapping every 15 min to mix the sample.
13. During the primary antibody reaction, prepare the FITC-labeled secondary antibody by diluting it 1:200 in the blocking solution and store on ice until use.
14. After the primary antibody reaction, samples are suspended by vortex mixer and then spin down, and place the microtube in a magnetic stand, and let it sit for 1 min. The Exosome Capture Beads will accumulate on the wall, allowing the liquid to be removed by pipetting and discarded.
15. Add 300 μL of WB + Enhancer, vortex, and spin down. Repeat Sect. 3.3, step 14 twice (Wash).
16. Add 100 μL of the secondary antibody and incubate for 30 min on ice in the dark. Tap the sample every 15 min to mix.
17. After secondary antibody reaction, samples are suspended by vortex mixer, then spin down, and place the microtube in a magnetic stand, and let it sit for 1 min. The Exosome Capture Beads will accumulate on the wall, allowing the liquid to be removed by pipetting and discarded.
18. Add 300 μL of WB + Enhancer, vortex, and spin down. Repeat Sect. 3.3, step 17 twice (Wash).
19. Add 300 μL of WB + Enhancer, pass through a mesh to remove aggregates.
20. Flow cytometry is used to detect gangliosides on EVs bound to beads, and the results are shown in Fig. 2. In this way, both GD2 and GD3 are expressed in EVs derived from GT cells (blue line), while neither GD2 nor GD3 is detected in EVs derived from CV cells (red line).

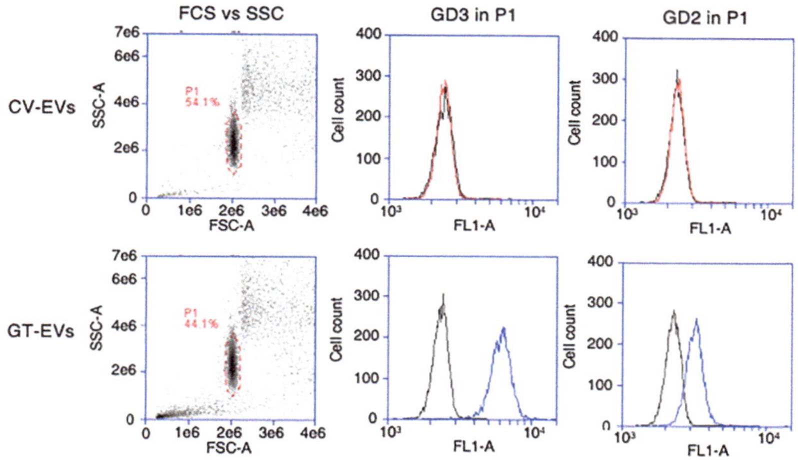

Fig. 2 Detection of gangliosides expressed on EVs. EVs bound to beads (upper, CV-EVs; lower, GT-EVs) are identified with the *x*-axis as FSC-A and the *y*-axis as SSC-A, defining the EV-bound bead population (P1) (left). The number of events within the P1 gate is set to 5000, and a histogram of FITC fluorescence intensity is detected for events within the P1 gate. Histograms display the *x*-axis as FITC intensity (FL1-A) and the *y*-axis as the number of events, with the negative control (black line) as the baseline, and GD3 and GD2 shown (middle and right, respectively)

4 Notes

1. When using the ultracentrifugation method, it is crucial to be mindful of potential contamination from impurities. When cultured with FCS, proteins derived from FCS may be recovered in substantial amounts. Therefore, to minimize the influence of proteins during the recovery of EVs, ITS is employed as the medium.
2. 0.22 μm Sartolab® RF-150 is used for the removal of larger impurities at the microbial level.
3. Reaction tubes are used to minimize the adsorption of EVs to the tube surface.
4. The EVs recover from a single dish are approximately 2–5 μ g. Therefore, a method capable of measuring protein concentrations at this level should be employed. In this manuscript, a BCA protein assay will be conducted, requiring absorbance measurement at 562 nm. However, due to issues with the filters used in the laboratory, it is acceptable to evaluate absorbance in the vicinity of 540–590 nm, even if direct measurement at this wavelength is not possible.
5. Anti-GD3 monoclonal antibody (R24) and anti-GD2 monoclonal antibody (220-51 or 14G2a) are used at a concentration of 2–10 μg/mL. Since both R24 and 220-51 are

mouse-derived monoclonal antibodies, FITC-conjugated goat anti-mouse IgG (H+L) is used as a secondary antibody at 5 μg/mL. Antibodies against other gangliosides can also be considered. Antibodies for detecting other gangliosides include GD1b antibody (GGR12), GT1b antibody (GMR5), GM3 antibody (M2590), GM2 antibody (MKl-16), and GD1a antibody (GMR17). These antibodies are also available for purchase and can be used in flow cytometry. Additionally, for detecting GM1, cholera toxin B subunit (CTB) can be used. CTB binds strongly to GM1. However, since CTB may also bind weakly to asialo-GM1, caution should be taken.

6. The Exosome Capture Beads are reagents included in the PS Capture Exosome Flow Cytometry Kit (Fujifilm) and are magnetic beads that adsorb Tim4, which specifically binds to the anionic phospholipid, phosphatidylserine (PS). Since EVs have a high content of PS on their membrane surface, they are captured by Tim4-magnetic beads. The magnetic beads can be magnetically collected using a magnetic stand. Exosome Capture Beads should be stored at −20°C after thawing.

7. The Washing Buffer (10×) and Exosome Binding Enhancer (100×) are reagents included in the PS Capture Exosome Flow Cytometry Kit (Fujifilm) and are used by diluting with ultrapure water and mixing (Washing Buffer + Exosome Binding Enhancer: WB + Enhancer). After thawing, the Washing Buffer (10×) and Exosome Binding Enhancer (100×) should be stored at 2–10°C after thawing.

8. A magnetic stand designed for the collection of magnetic beads that can accommodate 1.5 mL microtubes should be used.

9. Flow cytometry is a widely used technique in cell biology that enables the counting, sorting, and characterization of cells within heterogeneous mixtures. In this method, cells and magnetic beads suspended in a flowing liquid are aligned and illuminated by laser light, resulting in interactions with the light that are detected as scatter and fluorescence intensity by an electrical detection system. By employing fluorescently labeled specific antibodies, it is possible to isolate biological samples effectively. Samples can be plotted based on forward scatter (FSC), which detects particle size, and side scatter (SSC), which assesses the complexity of internal structures. This allows for the gating of fractions of magnetic beads and cells. Gating refers to the selection of specific populations within the sample for further analysis. The gated population can subsequently be analyzed for fluorescence signals, such as fluorescein isothiocyanate (FITC). In this experiment, the number of magnetic beads within the gate is set

to 5000 events, and the fluorescence intensity of these FITC signals (excitation wavelength 495 nm, emission wavelength 519 nm) is measured.

10. The number of dishes used depends on the amount of EVs needed. For flow cytometry, 0.25 μg of EVs (in terms of protein content) is used per reaction. Since it is possible to collect 2–5 μg of EVs from one dish, adjust the number of cells and dishes according to the cell type and experimental scale. The ultracentrifuge tubes used for separating the culture supernatant have a capacity of 38 mL. To fill the tubes with the culture supernatant, use three dishes, ensuring sufficient volume for the EV collection procedure.
11. In this experiment, protein quantification is used to calculate the yield of EVs. If DMEM containing FCS is used, proteins derived from FCS may be co-purified during centrifugation, which would affect the calculation of EV yield. Therefore, FCS was replaced with ITS to eliminate the influence of contamination from FCS-derived proteins.
12. If you wish to collect EVs from the remaining cells again, it is possible to add 15 mL of ITS-DMEM to the cells at 100% confluence, culture for another 48 h, and then collect the supernatant again.
13. This centrifugation step removes large debris, such as dead cells.
14. This centrifugation step removes intermediate-sized components, such as intracellular organelles and large vesicles from dead cells. Since EVs are small, they remain in the supernatant at this level of centrifugation.
15. The 0.22 μm filter is used to remove bacteria and other microorganisms.
16. At this stage, if the volume of liquid in the ultracentrifuge tube is too low and the air space is large, the tube may deform due to the centrifugal force. Therefore, the liquid should be filled up to within 3 mm of the rim of the tube, ensuring a full volume (37.5 mL).
17. When decanting the supernatant, a small amount of liquid will remain, which is used to resuspend the pellet by vortex mixer. At this stage, partial resuspension is achieved, and then PBS is added to fully resuspend the entire EV pellet.
18. Since EVs may be damaged by freeze-thaw cycles, they are stored at 4°C. Additionally, we use them for experiments within a few days after collection.
19. Preparing standard controls in duplicates or triplicates ensures a stable standard curve. Since this reaction is conducted at

37°C, it is recommended to add distilled water to the wells at the edge of the 96-well plate to stabilize the results.

20. In this experiment, a total of eight samples are used to detect GD3 and GD2 in EVs derived from GT and CV cell lines, along with their respective negative controls. Therefore, WB + Enhancer is prepared by adding 4 mL of Washing Buffer (10×) and 400 μL of Exosome Binding Enhancer (100×) to 35.6 mL of ultrapure water. Adjust the volume of WB + Enhancer according to the experimental scale.
21. Exosome Capture Beads are very small and require some time to accumulate magnetically. Additionally, since the beads may be difficult to see, carefully remove the waste liquid without touching the magnetic surface to avoid discarding the beads.
22. Equilibrate the Exosome Capture Beads by washing them with WB + Enhancer.
23. Add WB + Enhancer to the EV sample to reach a concentration of 2.5 μg/mL. In this experiment, to detect GD2 and GD3 on EVs derived from GT and CV cell lines, prepare EV samples for GD2 and GD3, as well as the respective negative controls for each.
24. Be cautious with high rotation speeds, as they may result in insufficient binding of extracellular vesicles to the magnetic beads. Alternatively, instead of rotating the sample on a rotator, it can be left stationary and suspend by vortex mixer for about 5 s every 20 min, for a total of three times.
25. Using the blocking solution inhibits nonspecific antibody binding to EVs + beads. The best blocking effect is observed with 5% skim milk for the anti-GD2 antibody (220-51) and 5% BSA for the anti-GD3 antibody (R24). The blocking solution was prepared by mixing skim milk or BSA into WB + Enhancer. The choice of blocking solution should be considered based on the antibody used.
26. It is also possible to detect other gangliosides using the other anti-ganglioside antibodies (*see* **Note 4**).

Conflict of Interest

The authors have no conflicts of interest to declare that are relevant to the content of this chapter.

References

1. Hoshino A, Costa-Silva B, Shen TL et al (2015) Tumour exosome integrins determine organotropic metastasis. Nature 527:329–335
2. Yesmin F, Furukawa K, Kambe M et al (2023) Extracellular vesicles released from ganglioside GD2-expressing melanoma cells enhance the malignant properties of GD2-negative melanomas. Sci Rep 13: 4987
3. Momen-Heravi F, Balaj L, Alian S et al (2013) Current methods for the isolation of extracellular vesicles. Biol Chem 394:1253–1262
4. Smith PK, Krohn RI, Hermanson GT et al (1985) Measurement of protein using bicinchoninic acid. Anal Biochem 150:76–85
5. Théry C, Witwer KW, Aikawa E et al (2018) Minimal information for studies of extracellular vesicles 2018 (MISEV2018): a position statement of the International Society for Extracellular Vesicles and update of the MISEV guidelines. J Extracell Vesicles 7:1535750
6. Nakai W, Yoshida T, Diez D et al (2016) A novel affinity-based method for the isolation of highly purified extracellular vesicles. Sci Rep 6: 33935
7. Hasnat MA, Ohmi Y, Yesmin F et al (2024) Crucial roles of exosomes secreted from ganglioside GD3/GD2-positive glioma cells in enhancement of the malignant phenotypes and signals of GD3/GD2-negative glioma cells. Nagoya J Med Sci 86:435–451
8. Iwasawa T, Zhang P, Ohkawa Y et al (2018) Enhancement of malignant properties of human glioma cells by ganglioside GD3/GD2. Int J Oncol 52:1255–1266
9. Real FX, Houghton AN, Albino AP et al (1985) Surface antigens of melanomas and melanocytes defined by mouse monoclonal antibodies: specificity analysis and comparison of antigen expression in cultured cells and tissues. Cancer Res 45:4401–4411
10. Zhao J, Furukawa K, Fukumoto S et al (1999) Attenuation of interleukin 2 signal in the spleen cells of complex ganglioside-lacking mice. J Biol Chem 274:13744–13747

Part V

Diseases and Tissue Development

Check for updates

Chapter 26

Methods for Evaluation of the Differentiation to Osteoblasts and Osteoclasts: Roles of Gangliosides in Bone Metabolisms

Hisataka Kondo, Gyosuke Sakashita, Mayu Nagao and Kazunori Hamamura

Abstract

Recent research has demonstrated that gangliosides play a role in the regulation of bone metabolism. Bone formation and resorption are occurring continuously, and resulting that new bones are formed. This is called bone remodeling and bone volume is maintained by a balance between bone formation by mature osteoblasts and bone resorption by osteoclasts. This chapter will introduce the three experimental techniques: alizarin red staining, alkaline phosphatase (ALP) assay, and tartrate-resistant acid phosphatase (TRAP) staining. These methods are useful for studying the relationship between gangliosides and the differentiation of osteoblasts and osteoclasts, which are involved in bone metabolism.

Keywords Osteoblasts, Osteoclasts, Alkaline phosphatase assay, Alizarin red staining, TRAP staining

1 Introduction

Osteoblasts are cells that differentiate from mesenchymal stem cells and subsequently become mature osteoblasts. On the other hand, osteoclasts are differentiated from bone marrow-derived macrophages. It has recently become evident that gangliosides play a pivotal role in both of these processes [1].

To investigate the role of gangliosides in osteoblast differentiation, osteoblasts manipulated glycosyltransferase genes are induced to differentiate into mature osteoblasts in vitro. It is also necessary to check the degree of the differentiation. In such cases, alizarin red staining and alkaline phosphatase (ALP) assay are commonly used. Alizarin red is a red dye that binds to metal groups and can stain calcium deposits in osteogenic and calcified cells [2]. In contrast, ALP assay quantifies the enzyme activity of alkaline phosphatase, an enzyme marker of osteoblasts whose expression increases during osteoblast differentiation, using p-nitrophenyl phosphate (pNPP), a non-proteinaceous substrate [3]. In both methods, the process begins

Koichi Furukawa (ed.), *Gangliosides*, Methods in Molecular Biology, vol. 3035,
https://doi.org/10.1007/978-1-0716-5288-6_26,

with the differentiation of osteoblasts into mature osteoblasts using ascorbic acid and β-glycerophosphate. Following the induction of the differentiation, alizarin red staining method requires the cells to be washed, fixed, and subsequently stained. In the case of an alkaline phosphatase assay, the subsequent steps are cell washing, lysis, substrate addition, enzyme reaction, and the addition of a reaction stop solution.

To investigate the role of gangliosides in differentiation of macrophages to osteoclasts, macrophages manipulated glycosyltransferase genes are induced to differentiate into osteoclasts, it is also necessary to check the progress of the differentiation. In such cases, tartrate-resistant acid phosphatase (TRAP) staining is commonly used as the appropriate methodology. This method allows for the degree of osteoclasts differentiation to be determined by measuring the expression of TRAP, which is representative marker of osteoclasts. This method consists of inducing differentiation of macrophages to osteoclasts using recombinant RANKL, fixing the cells, and staining the cells using TRAP solution.

2 Materials

2.1 Osteoblast Cell Culture

Osteoblasts are responsible for bone formation in bone tissue. MC3T3-E1 cell line has been commonly used for the functional analysis of osteoblasts [4]. This cell line was established from the calvaria of newborn mice. When cultured for an extended period in the presence of ascorbic acid and β-glycerophosphate, the cells differentiate into osteoblasts with high alkaline phosphatase (ALP) activity and mineralization ability [5]. The action of ALP increases the local concentration of phosphate required for mineralization, resulting in the formation of bone-like nodules in vitro.

1. MC3T3-E1 cells (ATCC Number: TIB-67).
2. Alpha Minimum Essential Medium (α-MEM), 10% fetal bovine serum (FBS), 1% antibiotics (10,000 U/mL penicillin, 10,000 μg/mL streptomycin).
3. 0.25% trypsin–ethylenediaminetetraacetic acid (EDTA).
4. Phosphate-buffered saline (PBS), without Ca^{2+} and Mg^{2+}.
5. 5 mg/mL ascorbic acid.
6. 500 mM β-glycerophosphate.
7. 6- to 24-well cell culture plate.
8. Osteoblast differentiation medium:
    - 1 mL of 5 mg/mL ascorbic acid (final concentration 50 μg/mL).

- 1 mL of 500 mM β-glycerophosphate (final concentration 5 mM).
- 98 mL of α-MEM, 10% FBS, 1% antibiotics, total 100 mL (*see* **Note 1**).

2.2 Alizarin Red Staining

Osteoblasts are responsible for bone formation and induce bone matrix proteins and mineralization. For this assay, a 6- to 24-well cell culture plate is used.

1. PBS, without Ca^{2+} and Mg^{2+}.
2. 60% isopropanol.
3. 1% alizarin red solution.
4. 6- to 24-well cell culture plate.

2.3 ALP Assay

ALP is a glycoprotein present in the cell membrane of osteoblasts and is an enzyme that decomposes phosphate esters into inorganic phosphorus and alcohol under alkaline conditions (pH 9–11). The ALP present in bone tissue is called bone-specific alkaline phosphatase (BAP) and is produced from osteoblasts during bone formation.

ALP activity was determined kinetically by monitoring the conversion of pNPP to p-nitrophenol [3]. ALP assay is performed in a 96-well plate, with regular tissue culture plates. All samples, including blanks, and bovine serum albumin (BSA) for the protein standard, are tested in duplicate.

1. Cell lysis buffer solution:
   - 5 mL of 10 mM Tris–HCl (pH 7.5).
   - 5 mL of 0.5 mM $MgCl_2$.
   - 5 mL of 0.1% Triton X-100, total 15 mL (can be stored in the refrigerator for 1–2 weeks).
2. 0.9% NaCl (saline solution).
3. ALP assay mix:
   - 2 mL of 20 mM $MgCl_2$.
   - 2 mL of 20 mM p-nitrophenyl phosphate (pNPP).
   - 2 mL of 1 M 2-amino-2-methyl-1-propanol (AMP) (pH 10.5).
   - 12 mL of distilled H_2O, total 18 mL (can be stored in the refrigerator for 1–2 weeks).
4. Coomassie Blue G reagent (Bio-Rad 500-0006) 5× diluted.
5. 1 mL of p-nitrophenol (1 nmol/μL): Make a 1:10 dilution of 10 mM stock in cell lysis buffer solution (*see* **Note 2**).
6. 10 mL of 1 mg/mL BSA solution.

7. Preparation of alkaline phosphatase standard curves (see Table 1).
8. Preparation of protein standard curves (see Table 2).

2.4 Osteoclast Cell Culture

Osteoclasts are responsible for bone resorption in bone tissue, originating from monocyte/macrophage precursor cells. RAW264.7 cell line, macrophages derived from the ascites of BALB/c mice, has been widely used to study function and differentiation of osteoclast [6]. When cultured in the presence of receptor activator of nuclear

Table 1
Preparation of alkaline phosphatase standard curves. To prepare 100 μL of standard solution, mix 1 nmol/μL of p-nitrophenol with cell lysis buffer solution as shown in the table, and transfer 20 μL of the standard solution to each of the assay well (in duplicate or triplicate)

nmol/well (p-nitrophenol)	μL (p-nitrophenol: 1 nmol/L)	μL (cell lysis buffer solution)
2	10	90
4	20	80
6	30	70
8	40	60
10	50	50
12	60	40
16	80	20
20	100	0

Table 2
Preparation of protein standard curves. To prepare 100 μL of standard solution, mix 1 mg/mL BSA solution with distilled H_2O as shown in the table, and transfer 20 μL of the standard solution to each of the assay well (in duplicate or triplicate)

μg/well (BSA)	μL (BSA solution: 1 mg/mL)	μL (dH_2O)
2	10	90
4	20	80
6	30	70
8	40	60
10	50	50
12	60	40
16	80	20
20	100	0

factor-kappa B ligand (RANKL), the cells fuse and differentiate into osteoclasts capable of bone resorption [7].

1. RAW264.7 (ATCC Number: TIB-71).
2. α-MEM, 10% FBS, 1% antibiotics (10,000 U/mL penicillin, 10,000 μg/mL streptomycin).
3. 0.05% trypsin–EDTA.
4. Recombinant murine sRANK Ligand (final concentration 10–100 ng/mL).

2.5 Tartrate-Resistant Acid Phosphatase (TRAP) Staining for Osteoclast Culture

TRAP is an enzyme that exists exclusively in osteoclasts and is released with increased bone resorption. TRAP staining leverages this characteristic to stain osteoclasts. While the marker enzyme for osteoblasts is alkaline phosphatase (ALP), the marker enzyme for osteoclasts is TRAP. Osteoclast cells are detected as TRAP-positive multinucleated cells [8].

1. TRAP buffer (pH 5.0):
   - Sodium acetate trihydrate 6.8 g (final concentration 100 mM).
   - Sodium tartrate dihydrate 5.8 g (final concentration 50 mM).
   - Adjust the pH to 5.0 using acetic acid 1–3 mL.
   - Distilled H_2O mess up to 500 mL.
2. Naphthol AS-MX phosphate.
3. N,N-dimethylformamide (DMF).
4. Fast Red Violet LB salt.
5. 0.1 M $MnCl_2 \cdot 4H_2O$.

TRAP stain solution (*see* **Note 3**):

1. Dissolve 5 mg of Naphthol AS-MX phosphate in 250 μL of DMF.
2. Add 50 mL TRAP buffer.
3. Add 30 mg Fast Red Violet LB salt
4. Add 200 μL of 0.1 M $MnCl_2$.
5. Stirring in dark for 30 min.

Total 50 mL

3 Methods

3.1 Osteoblast Differentiation

MC3T3-E1 cells produce collagen and a mineralized matrix when cultured for 7–28 days after being changed to osteoblast differentiation medium.

1. Select wells of appropriate size for experiments after cell culturing. For example, for alizarin red staining, use 6- or 12-well cell culture plate. For ALP assay, use 96-well cell culture plate.
2. MC3T3-E1 cells are grown in α-MEM (10% FBS, 1% antibiotics) in a 5% CO_2 incubator at 37°C.
3. When cells reach to confluent, replace to osteoblast differentiation medium.
4. Culture in a 5% CO_2 incubator at 37°C for 7–28 days.
5. Replace the culture medium every other day or 2 days.

3.2 Alizarin Red Staining for Nodule Formation Assay of Osteoblast Culture

1. Remove the medium from the cell culture dish and gently wash three times with PBS.
2. Fixed in 60% isopropanol for 1 min at room temperature.
3. Rehydrate with distilled H_2O for 3 min at room temperature.
4. Add 1% alizarin red stain solution and stain for 5 min at room temperature.
5. Repeatedly wash with distilled H_2O until no stain is present in the wash.

3.3 Alkaline Phosphatase Assay for Osteoblast Culture

1. Aspirate medium from 96-well culture plate.
2. Rinse each well two times with 0.9% NaCl.
3. Add 100 μL of cell lysis buffer solution in each well (*see* **Note 4**).
4. Incubate at room temperature for 5 min.
5. Scrape with a scraper (to detach and break the cells).
6. Pipetting and collect the samples into microtubes (The samples can be stored at −20°C).
7. Freeze and thaw the microtubes two times (to completely break cell membrane).
8. Mix the cell lysates by pipetting.
9. Transfer 20 μL of the cell lysates to a 96-well assay plate.
10. Set up "blank" wells containing the same amount of lysis buffer.
11. Set up standard wells containing 2–20 nM p-nitrophenol per well.
12. Add 80 μL of ALP assay mix to each well and incubate the plates at 37°C for 5–60 min.
13. As the reaction proceeds, the wells will become bright yellow. Stop the reaction by adding 100 μL of 0.5 M NaOH to each well.
14. Measure the absorbance at 405 nm.
15. Determine concentration of p-nitrophenol from the standard curve (*see* **Note 5**).

16. Determine concentration of protein using the Coomassie Blue G reagent. Use the same volume of lysate as you used for the ALP assay. For example, 20 μL of the cell lysate, 140 μL of the distilled H_2O, and 40 μL of the Coomassie Blue G reagent (total 200 μL). Determine concentration of protein from the standard curve.
17. Calculate ALP specific activity.

On an Excel spreadsheet, divide the concentrations of p-nitrophenol by the protein concentrations and the number of minutes that the samples were incubated. The data will be expressed as μM p-nitrophenol/μg protein/min.

3.4 Osteoclast Differentiation

RAW264.7 cells produce tartrate-resistant acid phosphatase (TRAP) when cultured for 3–5 days after the addition of RANKL.

1. RAW264.7 cells are grown in α-MEM (10% FBS, 1% antibiotics) at 37°C with 5% CO_2.
2. To differentiate RAW264.7 cells into osteoclasts, seed them at 2.5–5.0 × 10^3 cells/cm^2 on cell culture plates containing α-MEM with 10–100 ng/mL RANKL, 10% FBS, and 1% antibiotics.
3. Culture in a 5% CO_2 incubator at 37°C for 3–5 days.
4. Replace the culture medium every other day to 2 days.

3.5 TRAP Staining for Osteoclast Culture

1. Remove the medium from the cell culture dish.
2. Fixed in 10% formalin neutral buffer solution for 5 min at room temperature.
3. Remove formalin neutral buffer solution and wash with distilled H_2O.
4. Add TRAP staining solution and stain for 15–60 min at 37°C.
5. Wash with distilled H_2O.
6. If necessary, nuclei can be stained by 1% methyl green for counterstain.

4 Notes

1. Medium with the added reagents can be stored at 4°C for approximately 1 month.
2. A 10 mM stock solution is prepared by dissolving p-nitrophenol in distilled H_2O.
3. Use within 12 h.

4. The amount of cell lysis buffer solution per well should be 100 μL for 96-well plates, 300 μL for 48-well plates, and 600 μL for 24-well plates.
5. The ALP activity must be normalized by protein as an indication of the cell number.

Conflict of Interest

The authors have no conflicts of interest to declare that are relevant to the content of this chapter.

References

1. Hamamura K, Nagao M, Furukawa K (2024) Regulation of glycosylation in bone metabolism. Int J Mol Sci 25:3568
2. Puchtler H, Meloan SN, Terry MS (1969) On the history and mechanism of alizarin and alizarin red S stains for calcium. J Histochem Cytochem 17:110–124
3. Sabokbar A, Millett PJ, Myer B et al (1994) A rapid, quantitative assay for measuring alkaline phosphatase activity in osteoblastic cells *in vitro*. Bone Miner 27:57–67
4. Sudo H, Kodama HA, Amagai Y et al (1983) *In vitro* differentiation and calcification in a new clonal osteogenic cell line derived from newborn mouse calvaria. J Cell Biol 96:191–198
5. Bellows CG, Aubin JE, Heersche JN et al (1986) Mineralized bone nodules formed *in vitro* from enzymatically released rat calvaria cell populations. Calcif Tissue Int 38:143–154
6. Ralph P, Nakoinz I (1977) Antibody-dependent killing of erythrocyte and tumor targets by macrophage-related cell lines: enhancement by PPD and LPS. J Immunol 119:950–954
7. Hsu H, Lacey DL, Dunstan CR et al (1999) Tumor necrosis factor receptor family member RANK mediates osteoclast differentiation and activation induced by osteoprotegerin ligand. Proc Natl Acad Sci USA 96:3540–3545
8. Lv Y, Wang G, Xu W et al (2015) Tartrate-resistant acid phosphatase 5b is a marker of osteoclast number and volume in RAW 264.7 cells treated with receptor-activated nuclear κB ligand. Exp Ther Med 9:143–146

Chapter 27

Ganglioside Expression in Melanocytes: Modulation by Intrinsic and Extrinsic Factors

Keiko Furukawa and Orie Tajima

Abstract

Melanocytes are a unique cell lineage and have possible functions in skin tissue. Gangliosides GD3 and/or GD2 have been rigorously studied in malignant melanomas that derive from melanocytes. Recently, ganglioside expression on normal melanocytes have been paid attention with a focus on their functions and responses to various stimulations. These results suggest potential meaning of gangliosides in the activation status of cells or "omen" for malignant transformation of normal melanocytes. Here, ways for melanocyte culture, treatment by ultraviolet exposure and by various cytokines such as TNFα and IL-6 to analyze responses of melanocytes are explained. The ways for measurement of changes in glycosyltransferase genes and caused signaling responses as well as ganglioside expression are also explained.

Keywords Melanocyte, Ganglioside, Glycosyltransferase, UV irradiation, Inflammatory cytokine

1 Introduction

Gangliosides are sialic acid-containing glycosphingolipids mainly expressed in nervous systems of vertebrates [1]. However, they are expressed in almost all tissues and cells in our bodies [2] and are considered to play important roles in the regulation of their expressing cells [3]. Ganglioside expression profiles alter according to development, differentiation, activation, degeneration, and malignant transformation [4]. Some of them have been paid attention due to cancer-associated expression patterns and have been used as tumor markers and/or therapeutic targets [5]. In human malignant melanomas derived from skin melanocytes, gangliosides GD3 and/or GD2 have been rigorously studied [6], and those in normal melanocytes have been also analyzed with a focus on their functions and responses to various stimulations. These results suggest potential meaning of gangliosides in the activation status of cells or "omen" for malignant transformation of normal melanocytes [7]. Here, methods of melanocyte culture and immunological and biochemical detection of gangliosides are summarized.

Koichi Furukawa (ed.), *Gangliosides*, Methods in Molecular Biology, vol. 3035,
https://doi.org/10.1007/978-1-0716-5288-6_27,

2 Materials

2.1 Melanocyte Sources, Related Cell Lines, and Culture Medium

1. Commercial sources of human melanocytes and culture medium:

   HEMn-LP (lightly pigmented human melanocyte line) and HEMn-MP (moderately pigmented normal human melanocyte line) (Kurabo, Osaka, Japan) [8, 9] (*see* **Note 1**).

   Culture medium: Medium 254 supplemented with Human Growth Supplement™ (HMGS) (Thermo Fisher Scientific, Yokohama, Japan). When 70% confluency was reached, cells were cultured in Ham's F10 medium with 7% FBS, 1% penicillin–streptomycin, 1 mM N6,2′-O-dibutyryladenosine 3′,5′-cyclic monophosphate sodium salt (dcAMP), 0.1 mM 3-isobutyl-1-methylxanthine (IBMX), 1 μM Na_3VO_4, and 50 ng/mL phorbol 12-myristate 13-acetate (PMA) [9] (*see* **Note 2**).
2. Human melanoma cell lines, kindly provided by LJ Old (Memorial Sloan-Kettering Cancer Center, NY, USA), are cultured in DMEM with 7.5% FBS.
3. A human keratinocyte line, HaCaT cell (from Dr. Sugiura, Nagoya University Department of Dermatology) is cultured in DMEM with 10% FBS (*see* **Note 3**).
4. Culture flasks, plates, and trypsin–EDTA

   1. Plastic flasks (75/35 cm^2; CELLSTAR™, Greiner Bio-One GmbH, Frickenhausen, Germany) for culture of all these cells. Cell culture plastic dishes, 10 (3.5 or 6 cm diameter; CELLSTAR™, Greiner Bio-One).
   2. Trypsin–EDTA.
   3. Microtest plates (60-well, Terasaki, Greiner Bio-One) for plating cells.

2.2 Materials for Melanocyte Stimulation

TNFα, IL-6, and IL-1β (Sigma-Aldrich, St. Louis, MO, USA).

Phorbol 12-myristate 13-acetate (PMA; Life Technologies, Carlsbad, CA, USA).

UVB irradiation (DNA-FIX™, ATTO, Tokyo, Japan).

IBMX: an inhibitor of cAMP phosphodiesterase and dcAMP (Sigma-Aldrich).

2.3 mAbs Reactive with Gangliosides and Other Cellular Molecules

Main gangliosides to be examined and current mAbs being used [10, 11] are as follows:

GM3: M2590, GD3: R24, GM2: 10-11, GM1: 381, GD1b: 370, GD1a: D266, GT1b: 549 (*see* **Note 4**; see Fig. 1 for the individual ganglioside structures and synthetic pathway).

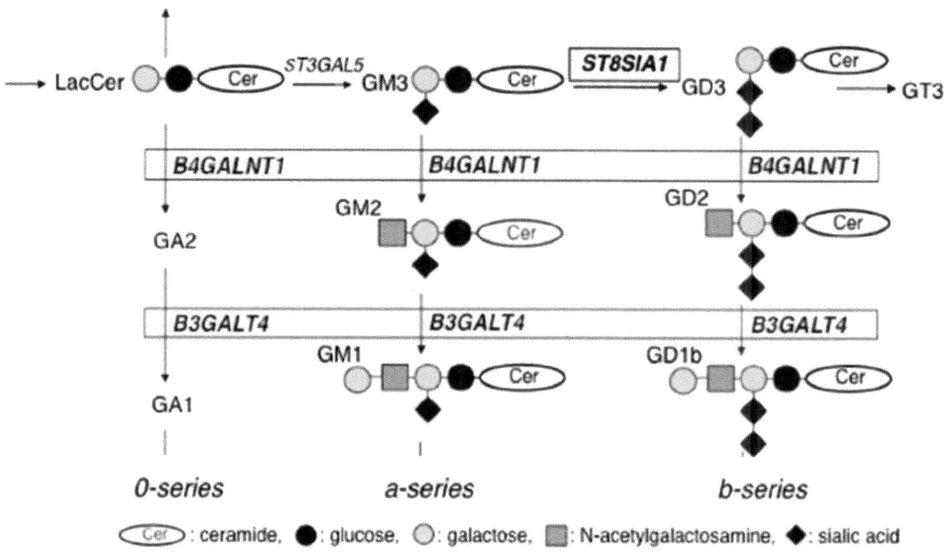

Fig. 1 Synthetic pathway of gangliosides and genes of related glycosyltransferases in melanocytes and melanoma cell lines. Glycosyltransferase genes involved in the synthesis of gangliosides are shown in italics. (Adapted from [7])

2.4 Analysis of Ganglioside Expression and Glycosyltransferase Genes

Immunocytochemistry: fluorescence microscopy, flow cytometry.

RT-PCR: RNeasy Plus Mini™ Kit (QIAGEN, Hilden, Germany) for total RNA preparation; iScript RT (Bio-Rad Laboratories, Hercules, CA) for RT; iTaq™ Universal SYBR Green Supermix™ (Bio-Rad) for PCR; PCR machine (Applied Biosystems) for DNA amplification.

Primers for PCR of *UGCG*, *B4GALT5*, *B4GALT6*, *ST8SIA1*, *B4GALNT1*, *ST3GAL5*, *B3GALT4*, *βACT*, and *ST3GAL2* cDNAs are summarized in Table 1.

Immunochemistry: 4% paraformaldehyde in phosphate-buffered saline; Alexa 594-conjugated donkey anti-mouse IgG (Thermo Fisher Scientific, Waltham, MA, USA).

Flow cytometry: Accuri™ C6 flow cytometry (BD Biosciences, San Diego, CA, USA) or similar apparatus is acceptable, mAbs as described in Sect. 2.3, and FITC-labeled anti-mouse IgG (H+L) as secondary antibody.

2.5 Materials for Signals in Melanocytes

PKA inhibitor: InSolution™ H-89 dihydrochloride.

IKK inhibitor: Wedelolactone™ (Merck Millipore, Burlington, MA, USA).

α-MSH ([Nle4, D-Phe7]-alpha-melanocyte-stimulating hormone trifluoroacetate salt) (Sigma-Aldrich).

2.6 Measurement of Cytokines

Culture supernatants from UVB-irradiated cells were collected after 24 h. Levels of cytokines (IL-8, IL-1β, IL-6, IL-10 TNFα, and IL-12p70) are determined using BD™ Cytometric Bead Array (CBA) Human Inflammation Kit (BD Biosciences) according to the manufacturer's instructions.

Table 1
Primer sequences for qRT-PCR of glycosyltransferase genes

Primer sequences for genes analyzed in this study			
Enzymes etc.	**Genes**	**Forward**	**Reverse**
Glc-Cer syn	UGCG	gctgccttacgtagcagaca	tcttggatgtgaagttccaaaata
Lac-Cer syn	B4GALT5	ttttgcaaccaaattggataag	ctcactccgccaaagaactc
Lac-Cer syn	B4GALT6	tctgattccaatgctccaga	atcgcacggttaaaaggttg
GM3 syn	ST3GAL5	ctgcctttgacatccttcagt	cgattgtggggacgttctta
GD3 syn	ST8SIA1	ggaaatggtgggattctgaag	tgacaaaggagggagattgc
GM2/GD2 syn	B4GALNT1	ccaactcaacaggcaactacaa	atgtccctcggtggagaac
GM1/GD1b syn	B3GALT4	tgctgcagttgttctctcaag	aagtttattgaggagcttgacacc
GD1a syn	ST3GAL2	gtccagaggtggtggatgat	cagcacctcattggtgttgt
Gal-Cer syn	UGT8	ttgtttatgtaggaggaatcctaacc	accatttacccatctttggaga
β-Actin	ACTB	ccaaccgcgagaagatga	ccagaggcgtacagggatag

Adapted from [12]

2.7 Analysis of Signals and Fates of Melanocytes by Western Immunoblotting

1. Cell lysis buffer (20 mM Tris-HCl, 150 mM NaCl, 1 mM Na_2EDTA, 1 mM EGTA, 1% Triton X-100, 2.5 mM sodium pyrophosphate, 1 mM β-glycerophosphate, 1 mM Na_3VO_4, 1 mM leupeptin) added with 1 mM PMSF or RIPA lysis buffer (50 mM Tris-HCl, 1% NP-40, 0.1% SDS, 0.5% sodium deoxycholate, 150 mM NaCl, 1 mM EDTA, 1 mM Na_3VO_4, and 1 mM NaF) added with 1 mM PMSF and Protease Inhibitor Cocktail Set I (Merck Millipore).
2. SDS-PAGE apparatus as currently used.
3. Blotting apparatus and current methods for Western blotting.
4. Antibodies for glycosyltransferases and representative signals.

Antibodies for Western immunoblotting:

Representative antibodies are as follows: anti-IKKα, anti-IKKβ antibody, or anti-phospho-IKKα/β (Ser176/180) antibody (Cell Signaling Technology). CREB (48H2) and phospho-CREB (Ser133, 87G3) (Cell Signaling Technology). For detection of glycosyltransferases, i.e., GD3 synthase (B11), GM2/GD2 synthase (C-5) (Santa Cruz Biotechnology, Inc., Dallas, TX, USA), GM1/GD1b synthase (polyclonal antibodies, Proteintech Group Inc., Rosemont, IL, USA).

3 Methods

Expression profiles of gangliosides and glycosyltransferase genes in melanocytes.

3.1 Cell Culture

1. Start culture of melanocytes at 3–6 × 10^5 cells in 3.5-cm culture dishes (HEMn-LP or HEMn-MP) in Medium 254 supplemented with HMGS. For HaCaT cells, 1–2 × 10^6 in 6-cm dishes using DMEM with 10% FBS.
2. After culture for 2–4 days, replace medium with PBS containing 7.5% FBS and incubate for 30 min (*see* **Note 5**).
   Then, expose cells to a UVB irradiation (312 nm; 7, 14, or 20 mJ/cm^2), and change medium to the original ones. After 2–3 days, harvest the culture supernatants and/or cells for measurement of secreted cytokines and/or gene expression in cells, respectively [12].
3. Measurement of cytokines (HaCaT), gangliosides, and genes:

Culture melanocytes in Ham's F-10 supplemented with 7.5% FBS, 1% penicillin–streptomycin, dcAMP, IBMX, 1 μM Na_3VO_4, and PMA (50 ng/mL) (F10-A medium) for 4–7 days, treat cells by various cytokines such as TNF (1.0 or 10.0 ng/mL), IL-6 (15 or 150 ng/mL), IL-1β (0.4 ng/mL), or other cytokines for 2–3 days, then collect cells for extraction of mRNA or for expression analysis of gangliosides and glycosyltransferase genes.

3.2 Immunofluorescence Assay and Immunocytochemical Detection

Prepare melanocytes 6 × 10^5 cells in 3.5-cm dishes, then incubate and treat by cytokines as described above.

1. Fix melanocytes in 4% paraformaldehyde for 30 min at room temperature.
2. Block the fixed cells with 2% BSA overnight at 4°C.
3. Stain fixed cells using various mAbs and Alexa 594-conjugated donkey anti-mouse IgG (Thermo Fisher Scientific) (see Fig. 2).
4. Analyze stained patterns by confocal fluorescence microscopy.

3.3 Flow Cytometry

1. Detach treated cells by trypsin–EDTA as usual, and prepare cell suspension at about 1 × 10^6 cells/mL or less in PBS.
2. Analyze expression profiles of gangliosides by flow cytometry.

3.4 RT-PCR for Ganglioside Synthase Genes

1. Extract RNA and apply for reverse transcription according to the company's instructions.
2. Perform PCR for the individual genes using primers summarized in Table 1.
3. Perform RT-PCR according to the machine's instructions (see Fig. 3).

3.5 Analysis of Signals and Fates of Melanocytes by Western Immunoblotting

1. Prepare melanocytes treated by various stimulations and signal inhibitors.
2. Lyse cells with cell lysis buffer added with 1 mM PMSF, or RIPA lysis buffer.

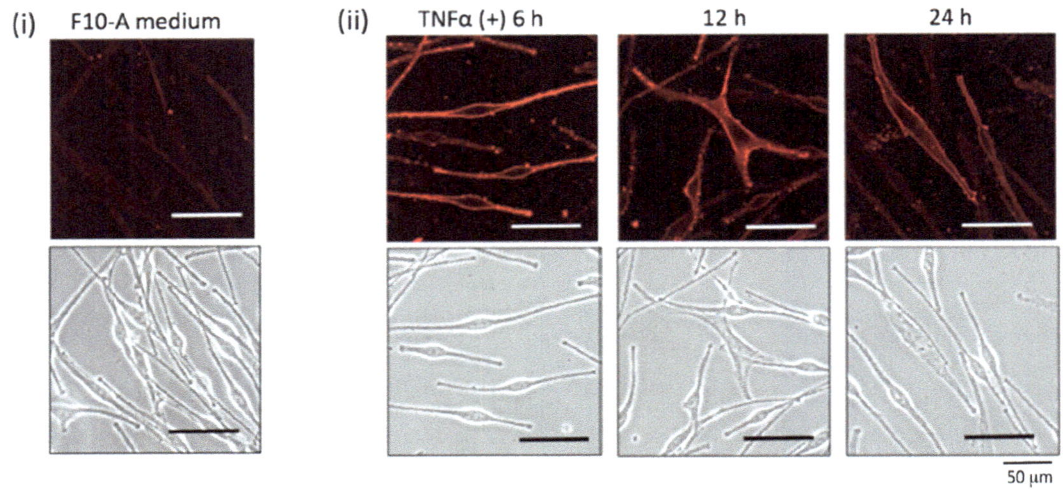

Fig. 2 Induction of GD3 ganglioside on melanocytes by various stimulations. Induction of GD3 by TNFα analyzed by immunofluorescence method is shown. After 6–24 h of TNFα stimulation, cells were fixed in 4% paraformaldehyde and stained by mAb R24 (anti-GD3) and subsequently by Alexa 594-conjugated anti-mouse IgG. (Adapted from [7]

3. Perform Western immunoblotting using SDS-PAGE apparatus, blotting apparatus, and current methods for Western blotting.
4. Use antibodies for glycosyltransferases and representative signals as shown in Methods.
5. Wash the blots, and analyze the results by ImageQuant™ LAS 4000 (GE Healthcare Bio-Sciences AB, Uppsala, Sweden) or Amersham Imager 680™ (GE Healthcare UK Ltd., Buckinghamshire, UK).

All these protocols are for human melanocytes. For mouse melanocytes (*see* **Note 6**).

4 Notes

1. Originally, human melanocytes were isolated from resected foreskin and facial skin containing keratinocytes, melanocytes, and fibroblasts, etc. Melanocytes are isolated by repeating trypsinization and transferring to new flasks based on the morphology and dopamine as a marker [9].
2. Effects of PMA and cholera toxin on melanocyte growth were described in details in [9, 10, 13]. Namely, PMA alone does not enhance GD3 expression, but effectively increases GD3 expression when added with TNFα by enhancing GM3 synthesis [10]. PMA is toxic to human keratinocytes [9], being useful for epidermal single cells. Melanocytes grow best in the presence of

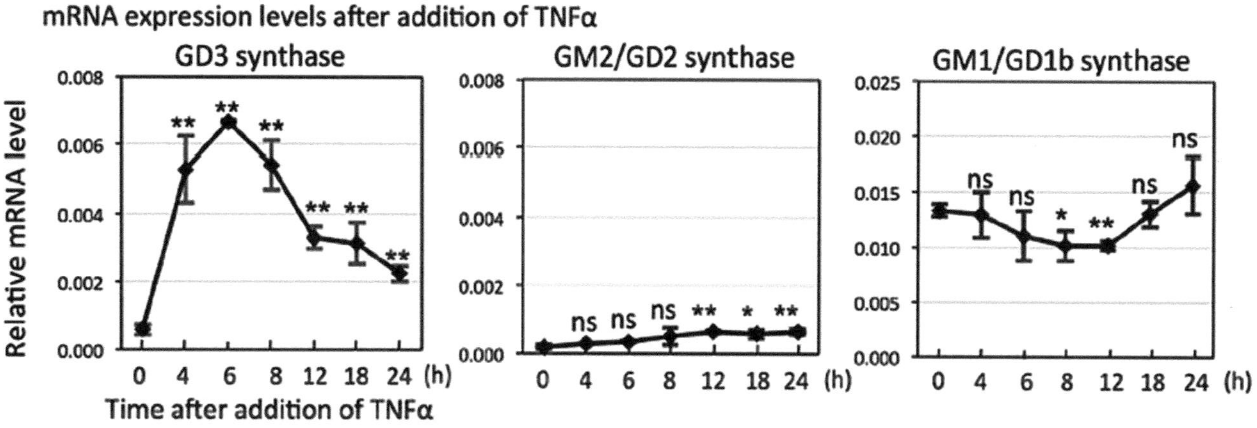

Fig. 3 Changes in the expression levels of mRNAs of glycosyltransferase genes in melanocytes treated by various cytokines. mRNA levels were analyzed by real-time RT-PCR using RNAs from melanocytes treated with TNFα for 0–24 h. GD3 synthase, GM2/GD2 synthase, and GM1/GD1b synthase genes were analyzed. (Adapted from [7])

PMA and cholera toxin, since cholera toxin inhibits the growth of fibroblasts [9]. Thus, PMA and cholera toxin are advantageous to eliminate contamination of keratinocytes and fibroblasts. Furthermore, cholera toxin is a potent inducer of cAMP, possibly exerting suppressive effects on GD3 expression despite its action for growth of melanocytes and keratinocytes [9].

3. Cytokine concentration can be measured by current commercial kit. Consequently, direct exposure of melanocytes to UV-B does not cause secretion of inflammatory cytokines. Expression levels of ganglioside synthase genes in melanocytes are not affected by UV-B. Actually, keratinocytes irradiated by UV-B secrete TNFα, IL-6, IL-1β, etc., as reported in [12]. Thus, effects of direct treatment of melanocytes by TNFα and IL-6 are examined in [12] (see Fig. 4).
4. mAbs toward individual gangliosides can be obtained from commercial services with fairly high reliability.

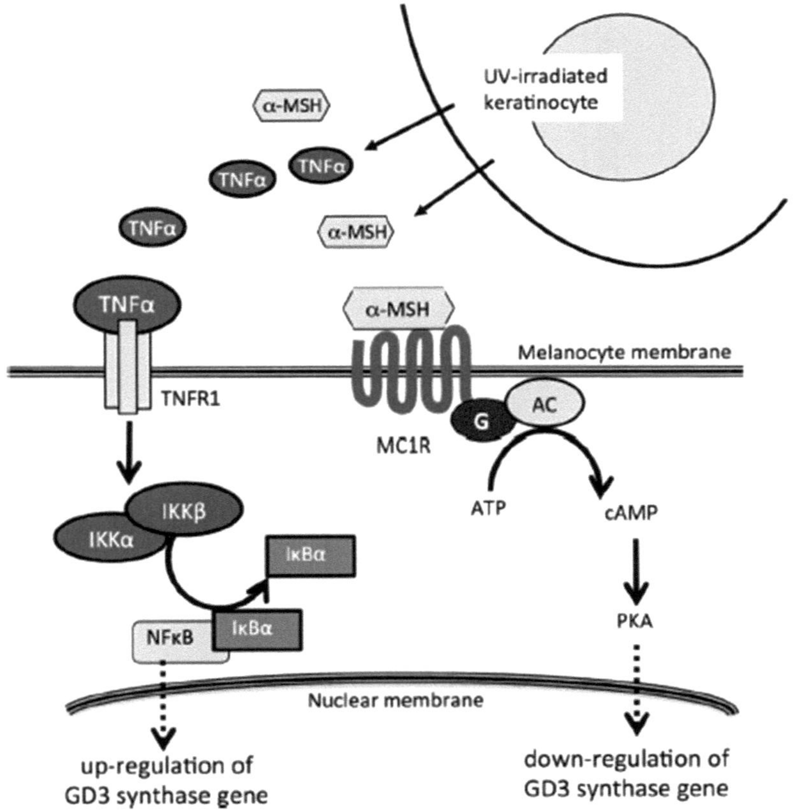

Fig. 4 Summary of ganglioside expression in melanocytes. By UVB irradiation of keratinocytes, inflammatory cytokines are released. When TNFα and IL-6 are added to melanocytes, GD3 synthase and GD3 were induced [12]. On the other hand, cAMP-PKA signals suppressed GD3 and GD3 synthase gene expression [7]. (Adapted from [7])

5. Differential roles of melanocytes and keratinocytes in the skin tissue were deeply described in [13]. Namely, interactions between keratinocytes and melanocytes are advantageous to regulate the growth and differentiation of melanocytes [13]. Direct contact and cytokine-mediated communication between these cells seem to be important.
6. As for culture and handling of mouse melanocytes, refer to the unique report from Sao Paulo [14].

Acknowledgments

We would like to thank Drs. M. Miyata and R. Takeuchi for excellent experimental work, Ms. M. Kambe for nice experimental help, and Ms. M. Kojima for nice secretarial help. We would also thank Grant-in-Aid support from MEXT, Japan (KAKENHI Grant Numbers 23590371, 24390078, 24791182, 26460404, 15K15080, and 15H04696).

Conflict of Interest The authors have no conflicts of interest to declare that are relevant to the content of this chapter.

References

1. Yu RK, Macala LJ, Taki T et al (1988) Developmental changes in ganglioside composition and synthesis in embryonic rat brain. J Neurochem 50:1825–1829
2. Wiegandt H (1985) Gangliosides. In: Wiegandt H (Ed.)Eds. Glycolipids. Elsevier, Amsterdam, 199–260
3. Hakomori S, Yamamura S, Handa AK (1998) Signal transduction through glyco(sphingo) lipids. Introduction and recent studies on glyco(sphingo)lipid-enriched microdomains. Ann N Y Acad Sci 845:1–10
4. Hamamura K, Furukawa K, Hayashi T et al (2005) Ganglioside GD3 promotes cell growth and invasion through p130Cas and paxillin in malignant melanoma cells. Proc Natl Acad Sci USA 102:11041–11046
5. Lloyd KO, Furukawa K (1998) Biosynthesis and functions of gangliosides: recent advances. Glycoconj J 15:627–636
6. Furukawa K, Ohmi Y, Tajima O et al (2018) Gangliosides in inflammation and neurodegeneration. Prog Mol Biol Transl Sci 156:265–287
7. Takeuchi R, Kambe M, Miyata M et al (2019) TNFα-signal and cAMP-mediated signals oppositely regulate melanoma-associated ganglioside GD3 synthase gene in human melanocytes. Sci Rep 9: 14740
8. Furukawa K, Arita Y, Satomi N et al (1990) Tumor necrosis factor enhances GD3 ganglioside expression in cultured human melanocytes. Arch Biochem Biophys 281:70–75
9. Eisinger M, Marko O (1982) Selective proliferation of normal human melanocytes *in vitro* in the presence of phorbol ester and cholera toxin. Proc Natl Acad Sci USA 79:2018–2022
10. Furukawa K, Lloyd KO (1990) Gangliosides in melanoma. In: Ferrone S (Ed.)Eds. Human melanoma: from basic research to clinical application. Springer, Heidelberg, 15–30
11. Zhao J, Furukawa K, Fukumoto S et al (1999) Attenuation of interleukin 2 signal in the spleen cells of complex ganglioside-lacking mice. J Biol Chem 274:13744–13747
12. Miyata M, Ichihara M, Tajima O et al (2014) UVB-irradiated keratinocytes induce melanoma-associated ganglioside GD3 synthase gene in melanocytes via secretion of tumor necrosis factor α and interleukin 6. Biochem Biophys Res Commun 445:504–510
13. Zhang C, Zhou L, Huang J et al (2018) A preliminary study of growth characteristics of melanocytes co-cultured with keratinocytes *in vitro*. J Cell Biochem 119:6173–6180
14. Otake AH, de Freitas Saito R, Duarte APM et al (2019) GD3 ganglioside-enriched extracellular vesicles stimulate melanocyte migration. Biochim Biophys Acta Mol Cell Biol Lipids 1864:422–432

Check for updates

Chapter 28

Gangliosides and Autoimmune Neuropathies

Motoi Kuwahara and Susumu Kusunoki

Abstract

Antibodies against gangliosides are frequently detected in autoimmune polyneuropathies such as Guillain-Barré syndrome, Miller Fisher syndrome, multifocal motor neuropathy, and IgM paraproteinemic polyneuropathy. These antibodies, which can be examined by thin-layer chromatography immunostaining, enzyme-linked immunosorbent assay, and glycoarray, are useful diagnostic markers and possible disease pathogenetic factors. Here, we describe the methods for antibody detection in detail.

Keywords Ganglioside, Autoantibody, Guillain-Barré syndrome, Miller Fisher syndrome, Multifocal motor neuropathy, IgM paraproteinemic polyneuropathy

1 Introduction

Gangliosides are acidic glycolipids containing sialic acid in their carbohydrate sequence and are named GM1, GD1b, GT1a, GQ1b, etc, according to the number and position of their sialic acid (s).

Each ganglioside is uniquely localized in its respective central and peripheral nervous system. Antibodies against gangliosides are often detected in autoimmune polyneuropathies such as Guillain-Barré syndrome (GBS), Miller Fisher syndrome (MFS), multifocal motor neuropathy (MMN), and IgM paraproteinemic polyneuropathy (IgM-PN) [1–6]. The clinical symptoms of patients with anti-ganglioside antibodies are closely associated with the localization of the respective target ganglioside antigens within the peripheral nervous system [7] (Table 1). In particular, antibodies against GQ1b, which is localized in the paranodal region of the oculomotor, trochlear, and abducens nerves as well as the large neurons of the dorsal root ganglia in humans, are specifically associated with ophthalmoplegia and ataxia in MFS and GBS [2, 7]. These anti-ganglioside antibodies are valuable diagnostic markers in such autoimmune neuropathies, and some participate in the disease's pathogenetic mechanisms. Moreover, some antibodies are associated with the severity and prognosis of the disease, particularly in GBS [8–10]. In addition, specific antibodies

Koichi Furukawa (ed.), *Gangliosides*, Methods in Molecular Biology, vol. 3035,
https://doi.org/10.1007/978-1-0716-5288-6_28,

Table 1
Anti-ganglioside antibodies and autoimmune neuropathies

Glycolipids	Localization in the nervous system	Ig class	Autoimmune neuropathies
GM1	Nodes of Ranvier	IgM	MMN
		IgG	GBS (AMAN)
GM1b	N.D.	IgG	GBS (AMAN)
GD1a	N.D.	IgG	GBS (AMAN)
GD1b	Paranodal myelin and large neurons in the DRG	IgM	IgM-PN (CANOMAD)
		IgG	GBS (AIDP, ataxic form)
GT1a	Cranial nerves IX and X	IgG	GBS (PCB variant, AOP)
GQ1b	Paranodal myelin of cranial nerves III, IV, and VI, and some large neurons in the DRG	IgG	GBS with ophthalmoplegia MFS BBE
GalNAc-GD1a	Periaxonal membrane	IgG	GBS (AMAN)
LM1	Myelin in the PNS	IgG	GBS (AIDP)
Gal-C	Myelin in the CNS and PNS	IgG	GBS (AIDP)
SGPG	Myelin in the PNS	IgM	IgM-PN (MAG neuropathy)

Antibodies to glycolipids (LM1, Gal-C, and SGPG) other than gangliosides are also included in this table *N.D.* not determined, *DRG* dorsal root ganglia, *PNS* peripheral nervous system, *CNS* central nervous system, *Ig* immunoglobulin, *MMN* multifocal motor neuropathy, *GBS* Guillain-Barré syndrome, *AMAN* acute motor axonal neuropathy, *AIDP* acute inflammatory demyelinating polyneuropathy, *IgM-PN* IgM paraproteinemic polyneuropathy, *CANOMAD* chronic ataxic neuropathy, ophthalmoplegia, IgM paraprotein, cold agglutinins, and disialosyl antibodies, *MFS* Miller Fisher syndrome, *BBE* Bickerstaff brainstem encephalitis, *PCB variant* pharyngeal-cervical-brachial variant, *AOP* acute oropharyngeal palsy, *MAG* myelin-associated glycoprotein

that bind to ganglioside complexes, which are formed by two different gangliosides, are sometimes identified in patients with GBS, MFS, chronic inflammatory demyelinating polyradiculoneuropathy, MMN, and IgM-PN [11–15]. Many anti-ganglioside antibodies have been identified in autoimmune neuropathies, and thin-layer chromatography (TLC) immunostaining and enzyme-linked immunosorbent assay (ELISA) are usually employed for the detection of anti-ganglioside (complex) antibodies. Because there are numerous combinations of ganglioside complexes, measuring antibodies against all of these combinations using TLC immunostaining and ELISA would be time-consuming and require a great deal of effort. Conversely, a combinatorial glycoarray [16] can examine numerous anti-ganglioside complex antibodies simultaneously, requiring a small volume of serum sample and ganglioside antigens, and can be used for analysis of many samples [17, 18]. Here, we describe the experimental methods of TLC immunostaining, ELISA, and glycoarray.

2 Materials

2.1 Serum Samples

Serum samples obtained from patients with autoimmune neuropathy are stored at −80°C.

2.2 Ganglioside Antigens

1. Gangliosides (GM1, GM2, GM3, GD1a, GD1b, GD3, GT1a, GT1b, GQ1b, and GalNAc-GD1a) are obtained commercially (Sigma-Aldrich, USA, and Hytest, Finland). Other glycolipids, such as Gal-C, LM1, and SGPG, are sometimes used as target antigens in the same way as gangliosides.
2. Each ganglioside is dissolved in a solution of chloroform and methanol (1:1) at a concentration of 1 mg/mL, which is stored at −20°C. This solution (Solution A) is used for TLC.
3. Solution A is diluted with ethanol to a concentration of 4 ng/μL, which is stored at −20°C. This solution (Solution B) is used for ELISA. An equivalent volume of Solution B of two different gangliosides is combined (Solution C), which is used for the examination of antibodies against ganglioside complexes with an ELISA.
4. Solution A of each ganglioside is diluted with methanol to a concentration of 100 μg/mL (Solution D). An equivalent volume of Solution D of two different gangliosides is combined (Solution E). Solutions D and E are stored at −20°C and used for the glycoarray.

2.3 Reagents for TLC Immunostaining

1. A total of 1.6 g of polyisobutylmethacrylate (Sigma-Aldrich, USA) is dissolved in 64 mL of chloroform, to which *n*-hexane is added to bring the volume to 400 mL.
2. The TLC-developing solution is made by combining 50 mL of chloroform, 45 mL of methanol, and 10 mL of 0.2% $CaCl_2$.
3. A total of 25 mg of 3,3′-diaminobenzidine tetrahydrochloride (DOJINDO, Japan) is dissolved in 50 mL of phosphate-buffered saline (PBS) as the coloring solution, and 15 μL of H_2O_2 is added to the solution immediately before the reaction.

2.4 Reagents for ELISA

1. A total of 13 mL of 0.2 M $Na_2HPO_4 \cdot 12\ H_2O$, 12 mL of 0.1 M citric acid, 25 mL of ultra-pure water, 0.02 g of *o*-phenylendiamine dihydrochloride (Sigma-Aldrich, USA), and 10 μL of H_2O_2 is required as the reaction solution, for two plates. H_2O_2 is added to the reaction solution immediately before the reaction.

3 Methods

3.1 TLC Immunostaining

1. One μL of Solution A, containing 1 μg of each ganglioside, is loaded on a plastic-backed TLC plate (MACHEREY-NAGEL, Germany) (*see* **Note 1**).
2. The plate is developed with the developing solvent consisting of chloroform, methanol, and 0.2% $CaCl_2$ (50:45:10) for 25 min at room temperature.
3. After air-drying the plate, it is coated with a solution of 0.4% polyisobutylmethacrylate in *n*-hexane for 1 min and air-dried for a further 30 min.
4. Thereafter, 400 μL of 1% bovine serum albumin (BSA) in PBS per lane is applied as a blocking solution to the plate and incubated for 30 min at room temperature. The plate is then washed with PBS and dried (*see* **Note 2**).
5. Each lane of the plate is incubated for 90 min with the respective serum sample diluted 1:40 with 400 μL of 1% BSA in PBS (*see* **Note 3**).
6. The plate is washed with PBS and incubated for 90 min with anti-human IgM or IgG horseradish peroxidase (HRP)-conjugated antibody (Jackson ImmunoResearch, USA), diluted 1:100 or 1:200 with 1% BSA in PBS, respectively.
7. The plate is then washed with PBS, and immunoreactivity is visualized with 3,3′-diaminobenzidine tetrahydrochloride solution as described above (Sect. 2.3).
8. After the color change is confirmed, the reactions are stopped with distilled water (Fig. 1).

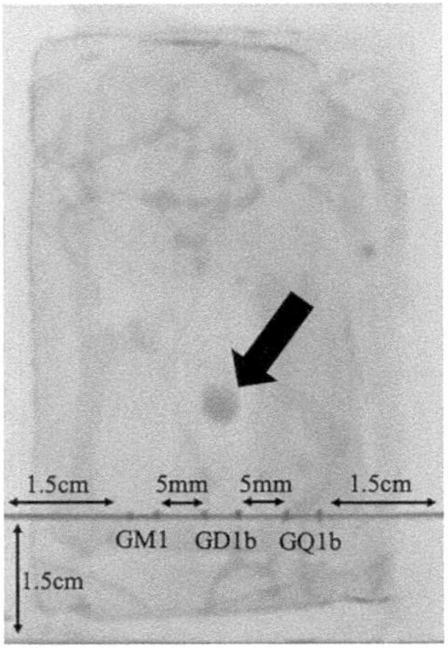

Fig. 1 Antibody detection by TLC immunostaining. Serum sample from a patient with Guillain-Barré syndrome strongly reacts with the antigen on the GD1b lane

3.2 ELISA

1. Each well of a 96-well polystyrene ELISA plate (Corning, USA) is incubated overnight with 50 μL of solution B, containing 200 ng of a ganglioside (*see* **Note 4**), at room temperature.
2. The wells are incubated with 200 μL per well of 1% BSA in PBS, as a blocking solution, for 30 min at room temperature.
3. After removing the blocking solution, 50 μL of the serum sample, diluted 1:40 with 1% BSA in PBS, is applied to each well and incubated for 90 min at room temperature (*see* **Note 5**).
4. After removing the diluted serum, each well is washed three times with 300 μL of 0.1% BSA in PBS.
5. Thereafter, 50 μL of HRP-conjugated anti-human IgM or IgG antibody, which is diluted 1:10,000 with 1% BSA in PBS, is added to each well as a secondary antibody (Jackson ImmunoResearch, USA) and then incubated for 90 min at room temperature.
6. After removing the diluted secondary antibody, each well is washed three times with 300 μL of 0.1% BSA in PBS.
7. Thereafter, 200 μL of the reaction solution as described above (Sect. 2.4) is applied to each well and incubated for 2 min at room temperature, and 50 μL of 8 N H_2SO_4 is added to each well to stop the reaction.
8. Optical densities (OD, 490 nm) are measured using a microplate reader. Each measured OD value is corrected by subtracting the OD of the control well (*see* **Note 6**) (Fig. 2).

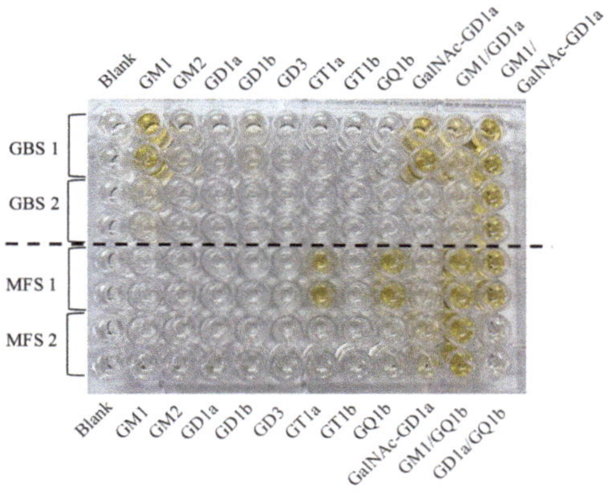

Fig. 2 Antibody detection by ELISA. While the serum sample from Guillain-Barré syndrome (GBS) 1 reacts to GM1 and GalNAc-GD1a and the complexes containing them, the serum sample from GBS 2 specifically reacts with the GM1/GalNAc-GD1a complex. This is known as an "anti-ganglioside complex antibody." Similarly, while the serum sample from Miller Fisher syndrome (MFS) 1 reacts to GT1a, GQ1b, and GQ1b-containing complexes, the serum sample from MFS 2 reacts only with the GM1/GQ1b complex but not with the individual GM1 and GQ1b

3.3 Glycoarray

1. A sheet of polyvinylidene difluoride (PVDF) membrane (Merk Millipore, Germany) is cut to the size of a glass slide (76×26 mm, Matsunami Glass Ind, Japan) and then attached with spray glue (3 M, USA).
2. Glass slides with the PVDF membrane are placed in an ATS4 automatic TLC sampler (Camag, Switzerland), and 12 slides can be set together in the TLC sampler.
3. A total of 55 antigens, including 10 individual gangliosides (Solution D) and 45 combinations consisting of two different gangliosides (Solution E), are automatically spotted in 11×11 lanes on a glass slide by a TLC sampler according to the program of winCATS software (Camag, Switzerland) (*see* **Note 7**).
4. The glass slides with the PVDF membrane on which ganglioside antigens are spotted are blocked with 2% BSA in PBS for 60 min at room temperature.
5. After removing the blocked slide from the blocking solution, 1 mL of serum samples diluted 1:100 with 1% BSA in PBS is applied to a glass slide and incubated for 120 min at 4°C (*see* **Notes 3 and 8**).
6. After washing three times with 0.1% BSA in PBS for a total of 60 min at room temperature, Alexa Fluor 647 anti-human IgM and Alexa Fluor 555 anti-human IgG (Thermo Fisher Scientific, USA) diluted 1:1,000 with 1% BSA in PBS are applied as secondary antibodies for 60 min at 4°C.
7. After washing three times with 0.1% BSA in PBS for a total of 60 min at room temperature, the glass slides are washed with distilled water for 5 min.
8. The glass slides with the PVDF membrane are then dried at room temperature in the dark.
9. Fluorescence intensities are measured by scanning with a Typhoon 9200 (GE Health Care, Japan) (Fig. 3), and the quantification of fluorescence is analyzed by using Image Quant TL software (GE Health Care, Japan).

4 Notes

1. Draw a line 1.5 cm from the bottom of the TLC plate. The ganglioside antigen should be loaded on line 5 mm away from each antigen (Fig. 1).
2. The TLC plate can be dried quickly by cold air from a hair dryer after washing.

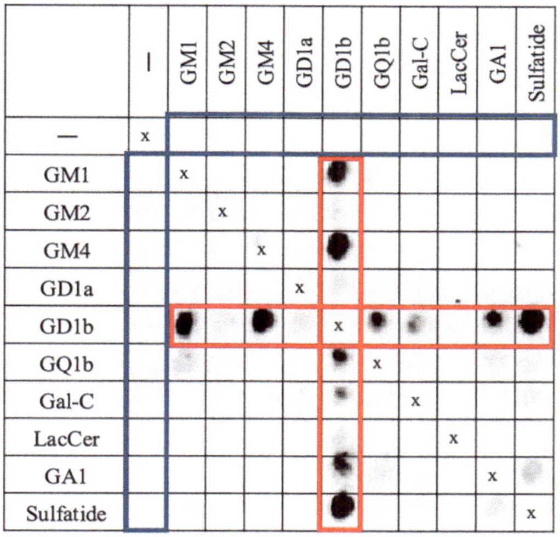

Fig. 3 Antibody detection by glycoarray. Spots on each lane (horizontal or vertical) contain ganglioside or glycolipid antigen written in the upper or left edge. For example, a spot on the horizontal GM1 lane and on the vertical GD1b lane contains GM1/GD1b complex. The leftmost lane and the uppermost lane are spotted with the individual antigens (marked by blue boxes). This serum sample from a patient with Guillain-Barré syndrome does not show any reactivity to individual glycolipids examined but shows strong reactivity to several GD1b-containing complexes (e.g., GM1/GD1b, GM4/GD1b, Sulfatide/GD1b, marked with red boxes). No antigen is spotted on the spaces marked with "X"

3. Cover the diluted serum or secondary antibody with parafilm during the primary or secondary antibody reaction.
4. The ganglioside complex antigens (Solution C and E) are mixed by vortexing before usage.
5. The ELISA plate incubation and washing of the glass slides are performed on the multishaker MMS (EYELA, Japan).
6. The leftmost well is a blank well in which no ganglioside antigen is coated. The corrected OD is calculated by subtracting the OD of the blank well from the OD of the well coated with ganglioside (Fig. 2).
7. The winCATS software can be programmed to spot 0.1 μL of each antigen at a 2-mm interval on the polyvinylidene difluoride (PVDF) membranes of 12 slides. The number and position of these spots are freely configurable.
8. Be careful not to allow the solution to flow through the glass slides and thus dry out the PVDF membrane during the primary and secondary antibody reactions.

Acknowledgments

This work was supported by JSPS KAKENHI Grant Number JP24K10672.

Conflict of Interest The authors have no conflicts of interest to declare that are relevant to the content of this chapter.

References

1. Ilyas AA, Willison HJ, Quarles RH et al (1988) Serum antibodies to gangliosides in Guillain-Barré syndrome. Ann Neurol 23:440–447
2. Chiba A, Kusunoki S, Shimizu T et al (1992) Serum IgG antibody to ganglioside GQ1b is a possible marker of Miller Fisher syndrome. Ann Neurol 31:677–679
3. Kusunoki S, Chiba A, Kon K et al (1994) N-acetylgalactosaminyl GD1a is a target molecule for serum antibody in Guillain-Barré syndrome. Ann Neurol 35:570–576
4. Ilyas AA, Quarles RH, Dalakas MC et al (1985) Monoclonal IgM in a patient with paraproteinemic polyneuropathy binds to gangliosides containing disialosyl groups. Ann Neurol 18:655–659
5. Willison HJ, O'Leary CP, Veitch J et al (2001) The clinical and laboratory features of chronic sensory ataxic neuropathy with anti-disialosyl IgM antibodies. Brain 124:1968–1977
6. Pestronk A, Cornblath DR, Ilyas AA et al (1988) A treatable multifocal motor neuropathy with antibodies to GM1 ganglioside. Ann Neurol 24:73–78
7. Kusunoki S, Willison HJ, Jacobs BC (2021) Antiglycolipid antibodies in Guillain-Barré and Fisher syndromes: discovery, current status and future perspective. J Neurol Neurosurg Psychiatry 92:311–318
8. Kaida K, Kusunoki S, Kanzaki M et al (2004) Anti-GQ1b antibody as a factor predictive of mechanical ventilation in Guillain-Barré syndrome. Neurology 62:821–824
9. Yamagishi Y, Kuwahara M, Suzuki H et al (2020) Serum IgG anti-GD1a antibody and mEGOS predict outcome in Guillain-Barré syndrome. J Neurol Neurosurg Psychiatry 91:1339–1342
10. Thomma RCM, Fokke C, Walgaard C et al (2023) High and persistent Anti-GM1 antibody titers are associated with poor clinical recovery in Guillain-Barré syndrome. Neurol Neuroimmunol Neuroinflamm 10:e200107
11. Kaida K, Morita D, Kanzaki M et al (2004) Ganglioside complexes as new target antigens in Guillain-Barré syndrome. Ann Neurol 56:567–571
12. Kaida K, Kanzaki M, Morita D et al (2006) Anti-ganglioside complex antibodies in Miller Fisher syndrome. J Neurol Neurosurg Psychiatry 77:1043–1046
13. Kuwahara M, Suzuki S, Takada K et al (2011) Antibodies to LM1 and LM1-containing ganglioside complexes in Guillain-Barré syndrome and chronic inflammatory demyelinating polyneuropathy. J Neuroimmunol 239:87–90
14. Galban-Horcajo F, Fitzpatrick AM, Hutton AJ et al (2013) Antibodies to heteromeric glycolipid complexes in multifocal motor neuropathy. Eur J Neurol 20:62–70
15. Nobile-Orazio E, Giannotta C, Briani C (2010) Anti-ganglioside complex IgM antibodies in multifocal motor neuropathy and chronic immune-mediated neuropathies. J Neuroimmunol 219:119–122
16. Rinaldi S, Brennan KM, Willison HJ (2012) Combinatorial glycoarray. Methods Mol Biol 808:413–423
17. Rinaldi S, Brennan KM, Kalna G et al (2013) Antibodies to heteromeric glycolipid complexes in Guillain-Barré syndrome. PLoS One 8:e82337
18. Morikawa M, Kuwahara M, Ueno R et al (2016) Serological study using glycoarray for detecting antibodies to glycolipids and glycolipid complexes in immune-mediated neuropathies. J Neuroimmunol 301:35–40

Check for updates

Chapter 29

Imaging and Characterization of Ganglioside-Induced Amyloid Fibril Formation

Teruhiko Matsubara

Abstract

High-density nanoclusters of gangliosides in membrane rafts interact with amyloid β protein (Aβ), which induces toxic Aβ aggregation. This chapter describes how to generate, detect, and assess the Aβ aggregates induced by ganglioside-containing liposomes and planar lipid bilayers.

Keywords Amyloid β protein, Ganglioside cluster, Membrane raft, Thioflavin T, Langmuir trough, Topography

1 Introduction

Neurotoxic protein aggregates have been implicated in the development of neurodegenerative diseases [1, 2]. Amyloid β protein (Aβ), a 39–43-amino acid residue polypeptide produced through the cleavage of amyloid precursor protein by secretases, oligomerizes and forms fibrils; these Aβ assemblies accelerate the development of Alzheimer's disease [3–5]. The Aβ assemblies are generally formed spontaneously [6], but their formation is enhanced by numerous factors, including gangliosides [7]. Gangliosides are abundant in the nervous system, and they exist at high densities in membrane (lipid) rafts comprising sphingomyelin (SM) and cholesterol [8]. Yanagisawa et al. reported that ganglioside-enriched microdomains in neuronal membranes are the key sites for the onset of Alzheimer's disease, in which a ganglioside-bound Aβ (GAβ) complex acts as an endogenous seed [9] (Fig. 1).

The ganglioside-induced Aβ assembly has been demonstrated on ganglioside-containing liposomes using a thioflavin T (ThT) assay, electron microscopy, and antibody assays [10, 11]. GM1, Galβ 1–3GalNAcβ1–4(Neu5Acα2–3)Galβ1–4Glc1–1'Cer, a representative ganglioside, is frequently used for generating ganglioside-containing lipid membranes; GD1a/b and GT1b also induce a large

Koichi Furukawa (ed.), *Gangliosides*, Methods in Molecular Biology, vol. 3035,
https://doi.org/10.1007/978-1-0716-5288-6_29,

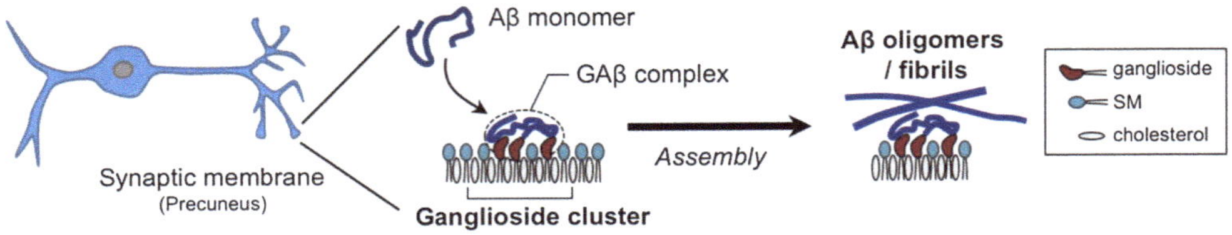

Fig. 1 Ganglioside clusters at synaptic membranes are responsible for ganglioside-bound Aβ (GAβ) complex, which acts as an endogenous seed for Aβ assemblies

amount of Aβ assemblies [12]. The typical lipid composition of the model membranes comprises GM1/sphingomyelin (SM)/cholesterol (0–10:45:45) [12], GM1/SM/cholesterol (20:40:40) [13], and GM1/SM/cholesterol/phospholipid (0–10:10:35:50) [14].

This chapter describes how to prepare ganglioside-containing lipid membranes, liposome and planar lipid bilayers for investigating Aβ aggregations. Liposome membranes are useful for ThT, circular dichroism (CD) [13], and cell viability studies, whereas planar lipid bilayers are useful for surface topographic imaging and Fourier transform infrared spectroscopic (FTIR)/Raman spectroscopic studies [15, 16]. Especially, in combination with atomic force microscopic (AFM) technology, the shapes of Aβ aggregates can be distinguished, and the elongation process of Aβ fibrils on the membrane can be visualized [12]. Finally, the chapter describes how to assess the toxicity of the Aβ assembly using rat and human cells [17, 18].

2 Materials

Aβ-sensitive ganglioside nanoclusters can be prepared with both liposomes and planar lipid bilayers composed of ganglioside, SM, and cholesterol (Fig. 2). These membranes are prepared from a lipid mixture. In addition, seed-free soluble Aβ is prepared without an organic solvent (e.g., dimethyl sulfoxide, DMSO), owing to the risk of membrane damage.

2.1 Lipids and Stock Solutions

1. Gangliosides and sphingomyelin of the highest purity without further purification. As the variety of ceramide composition depends on tissue, organ, and animal species (e.g., bovine brain and spinal cord), information about lipid extracts obtained is necessary.
2. Synthetic cholesterol and 1-palmitoyl-2-oleoyl-*sn*-glycero-3-phosphocholine (POPC) of >99% purity.
3. Ganglioside solution (1 mg/mL): Dissolve 1 mg of ganglioside in 1 mL of chloroform ($CHCl_3$)/methanol (MeOH) (2:1 or 4:1, *v/v*) in a 1-mL glass tube or volumetric flask. Store at –20°C. Alternatively, $CHCl_3$/ethanol (EtOH) (4:1, *v/v*) can be used.

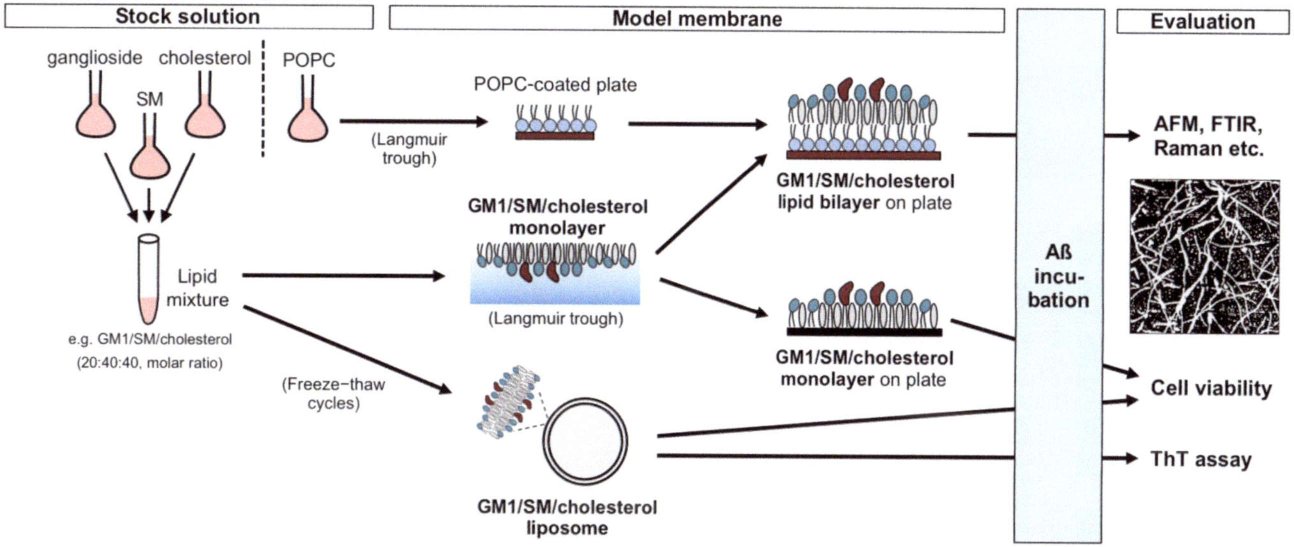

Fig. 2 Overview of experimental procedures of Aβ assemblies induced by ganglioside-containing membranes. Typical substrate plates are mica (AFM), a glass plate (spectroscopy), and a plastic plate (cell-based assay)

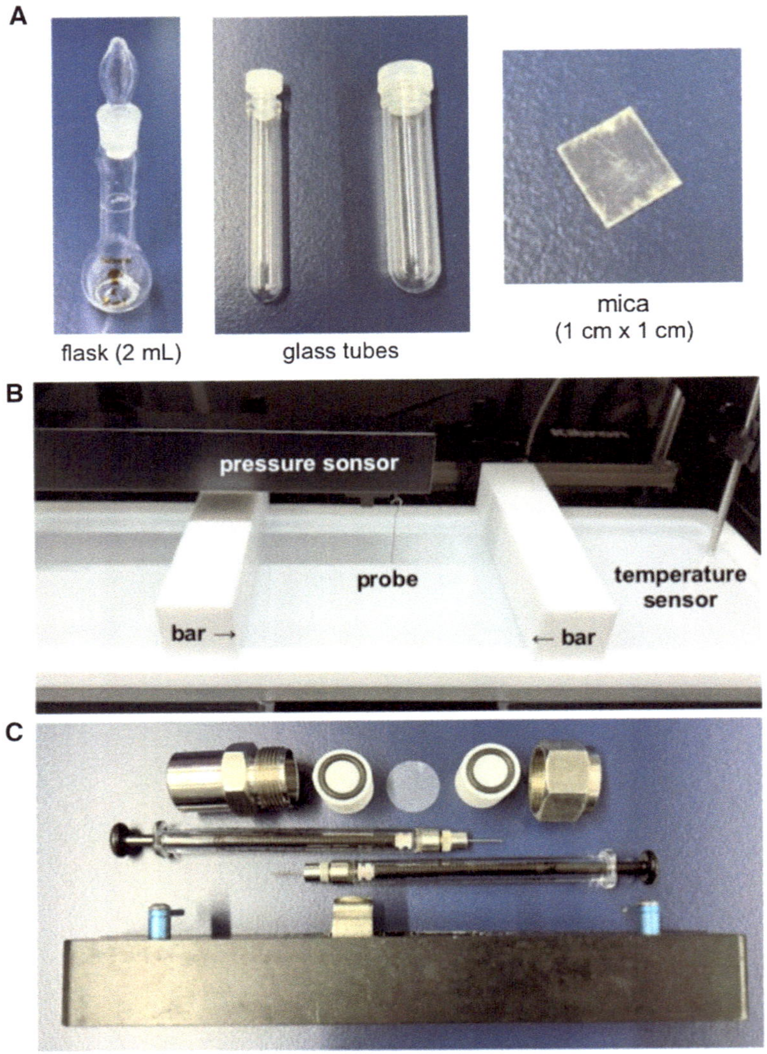

Fig. 3 Equipment pictures. (**A**) Voltametric flask, glass tubes, and mica. (**B**) Langmuir trough (Micro trough, Kibron Inc.). (**C**) Mini-extruder set (Avanti Polar Lipids, Inc.)

4. Sphingomyelin (SM) solution (5 mg/mL): Dissolve 5 mg of SM in 2 mL of chloroform in a 2-mL volumetric flask (Fig. 3A, left). Store at –20°C.
5. Cholesterol solution (5 mg/mL): Dissolve 10 mg of cholesterol in 2 mL of chloroform in a 2-mL volumetric flask. Store at –20°C.
6. POPC solution (10 mg/mL): Dissolve 10 mg of POPC in 2 mL of chloroform in a 2-mL volumetric flask. Store at –20°C.
7. Calculate the concentration of the lipid solution at molar base. If there is no information about the ceramide composition, the

molecular weight (Mw) of the ceramide portion of ganglioside and SM can be calculated based on the assumption that the fatty acid is stearic acid.

2.2 Lipid Mixture Solutions (Ganglioside/SM/Cholesterol)

1. Clean a 1-mL glass tube (internal diameter 5.7 mm, 50 mm length, with polyethylene cap) (Maruemu Corp., Osaka, Japan) (Fig. 3A, center) with nitric acid to avoid contamination.
2. Pour an appropriate volume of ganglioside (approximately 550 μL), SM (210 μL), and cholesterol solutions (50 μL) (ganglioside/SM/cholesterol, 20:40:40 by molar ratio) into the glass tube using a glass syringe with a stainless steel needle. Mix the tube by vortexing. Store the solution at –20°C until use.

2.3 Preparation of Ganglioside-Containing Liposomes

1. Pour the lipid mixture (ganglioside/SM/cholesterol) solution into ten glass tubes (internal diameter 13 mm, length 100 mm) 80 μL/tube.
2. Dry the solution under nitrogen gas at 40°C to evaporate the organic solvent. Optional: Vacuum desiccation for more than 1 h can be used.
3. Optional: Store at –20°C until use.
4. Suspend the resultant lipid film at the bottom of tube in PBS.
5. Freeze–thaw cycles: Freeze the solution in the test tube with liquid nitrogen (−196°C) for 1 min, and then thaw for 5 min. Repeat this cycle ten times.
6. Centrifuge the lipid suspension at 21,100 × *g* for 5 min, and resuspend the resultant pellet in PBS.
7. Sonicate the suspension for 20 min on ice to generate ganglioside-containing liposomes. The average size of the liposomes is 80 nm.
8. Optional: Passage the liposome solution through polycarbonate membranes of 100-nm pore size over ten times using a mini-extruder (Avanti Polar Lipids, Inc.).
9. Optional: Estimate the particle size of the liposomes using dynamic light scattering with a Zetasizer Nano ZS instrument (Malvern Instruments Co., Ltd., UK).

2.4 Preparation of POPC-Coated Mica (Fig. 4A)

1. Cut mica (The Nilaco Corp., Tokyo, Japan) into 1-cm squares (Fig. 3A, right). Remove the mica surface by peeling with a Scotch tape to obtain a freshly cleaved surface.
2. Fix the freshly cleaved mica on a glass slide using a double-stick tape, with the cleaved side facing up. Place the glass slide at the bottom of the trough.
3. Pour Milli-Q water (approximately 90 mL) into Langmuir trough.

4. Drop POPC solution (about 15 μL of 5 mg/mL) onto the water, and incubate the trough at 25°C for 15 min to prepare a POPC monolayer at the air–water interface.
5. Compress the POPC monolayer until 35 mN/m (*see* **Note 1**) of surface pressure at a speed of 0.2 mm/s.
6. Remove water slowly from the bottom of the trough at an edge using an aspirator to drop down the air–water interface. Immobilize the POPC monolayer on the mica surface.
7. Pick the mica surface and dry in an auto-dry desiccator overnight (POPC-coated mica). Use the POPC-coated mica within a week.

2.5 Preparation of Ganglioside-Containing Lipid Bilayer on Mica (Fig. 4B)

1. Pour Milli-Q water (110 mL) into Langmuir trough.
2. Drop the lipid mixure (ganglioside/SM/cholesterol) solution (about 15 μL) onto the water, and incubate the trough at 25°C for 15 min.
3. Compress the monolayer until 30 mN/m of surface pressure at a speed of 0.2 mm/s to prepare a ganglioside-containing lipid monolayer at the air–water interface.
4. Place the POPC-coated mica on a ganglioside-containing monolayer with the POPC-coated side down. Here, a lipid bilayer is reconstituted on mica.
5. Remove the excess lipid molecules of the monolayer surrounding the mica.
6. Pick the lipid bilayer-immobilized mica surface and keep it wet.

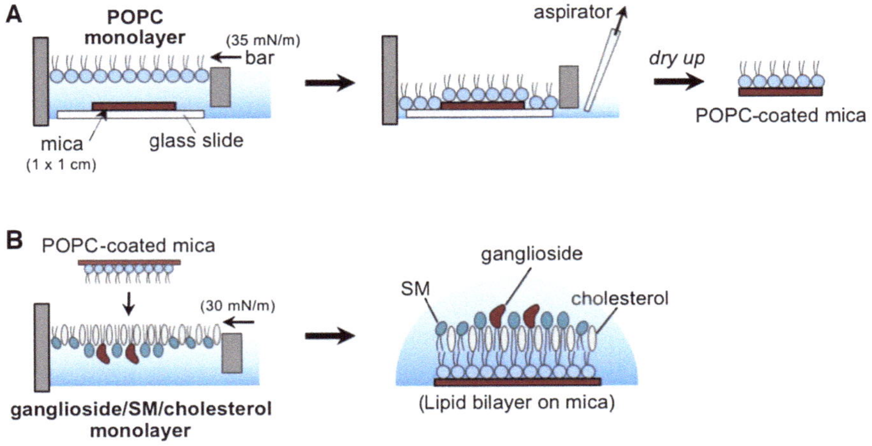

Fig. 4 Preparation of a planar lipid bilayer on a plate. (**A**) Preparation of POPC-coated mica. A monolayer of POPC is deposited on mica. (**B**) Preparation of a ganglioside-containing lipid bilayer on mica. A monolayer of ganglioside/SM/cholesterol is deposited on the POPC-coated mica to form a bilayer

2.6 Preparation Ganglioside-Containing Lipid Monolayer on a Plastic Disk

1. Prepare a ganglioside-containing lipid monolayer at the air–water interface (Sect. 2.5, Step 3).
2. Place a plastic disk (13 mm in diameter) (code 174950; Thermo Fisher Scientific Inc.) on the lipid monolayer. Remove the excess lipid molecules of the monolayer surrounding the disk.
3. Pick the lipid monolayer-immobilized disk and keep it wet.

2.7 Seed-Free Soluble Aβ

1. Dissolve synthetic $A\beta_{40}$ or $A\beta_{42}$ (Peptide Institute Inc., Osaka, Japan) in ice-cold 0.02% ammonia, and then ultracentrifuge the solution at 560,000 × g for 3 h at 4°C to remove undissolved peptide aggregates.
2. Collect the upper two-thirds ($A\beta_{40}$) or one-third ($A\beta_{42}$) fraction as a seed-free fraction. Estimate the concentration of the fractions using protein assays such as Bradford, bicinconic acid, and Lowry's methods.
3. Store aliquots of the seed-free fractions (80–120 μM) at −80°C until needed.
4. Prior to use, dilute each aliquot to make 1–20 μM solution in Dulbecco's PBS at pH 7.4.

2.8 ThT Solution

1. Dissolve 0.32 mg of ThT in 2 mL of Milli-Q water (500 μM stock solution), and store it at 4°C.
2. To prepare a 1:100 dilution (for 5 μM), add 1 μL of the stock solution to 99 μL of solution.

3 Methods

3.1 ThT Assay Using Ganglioside-Containing Liposomes (Total Amount of Aβ Aggregates)

1. Interact seed-free soluble $A\beta_{42}$ and ganglioside-containing liposomes (ganglioside/SM/cholesterol, 20:40:40) at 37°C in PBS containing ThT (5 μM) at a ganglioside/$A\beta_{42}$ molar ratio of 10:1.
2. Measure the fluorescence intensity indicating the amount of Aβ aggregates generated on a fluorescence spectrophotometer (FL-2500; Hitachi, Japan) at an excitation wavelength of 450 nm using a 3 mm path-length cuvette with a slit width of 5 nm at 25°C. Obtain emission spectra between 460 and 600 nm with the same slit width.
3. Alternatively, the fluorescence intensity can be measured in a black 96-well plate using a plate reader (Powerscan HT) (DS Pharma Biomedical Co., Ltd., Osaka, Japan). Record the intensity every hour in triplicate at excitation and emission wavelengths of 450 and 485 nm, respectively.

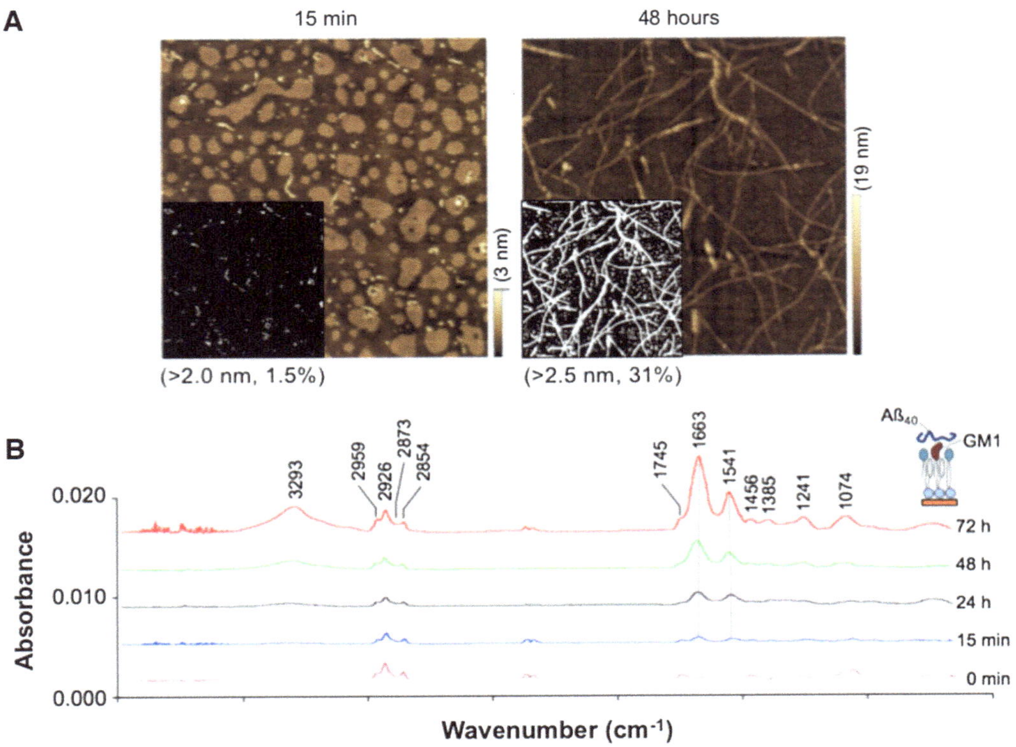

Fig. 5 Formation of Aβ fibrils on GM1/SM/cholesterol (20:40:40) [15]. (**A**) AFM images (4 μm × 4 μm) of GM1-containing membranes after incubation with 10 μM $A\beta_{40}$ at 37°C for 15 min and 48 h. Binarized AFM images were obtained by thresholding at 2.0–3.0 nm from the bottom (inset). (**B**) Fourier transform infrared reflection–absorption spectra of GM1-containing membranes after incubation with $A\beta_{40}$ for 0 min to 72 h. Amide I and amide II are around 1663–1668 and 1541–1542 cm^{-1}, respectively

3.2 AFM Observation of Aβ Assemblies on Planar Lipid Bilayer (Number and Length of Aβ Fibrils)

1. Incubate ganglioside-containing lipid bilayer (ganglioside/SM/cholesterol, 20:40:40) on mica with seed-free soluble $A\beta_{40}$ (10 μM) or $A\beta_{42}$ (1 μM) in PBS for 15 min to 72 h at 37°C.
2. After washing thrice with PBS, image the membrane in water at 25°C using an SPM-9600 atomic force microscope (Shimadzu Corp., Kyoto, Japan) and a 38-μm soft cantilever (BL-AC40TS-C2; Olympus, Tokyo, Japan) with integrated pyramidal silicon nitride tips with a spring constant of 0.1 $N m^{-1}$.
3. Acquire atomic force micrographs with a scan area of typically 4 μm^2 (2 μm × 2 μm) in a dynamic mode at 1–2 Hz. Perform AFM observations in at least triplicate, and use typical multiple images ($n \geq 3$) for further analyses. Indicate surface topography heights on a color scale (Fig. 5A).
4. Optional: To illuminate Aβ assemblies, atomic force micrographs can be binarised based on membrane heights (*see* **Note 2**) using image processing software (*see* **Note 3**).

5. Measure the lengths of the long and short axes of the Aβ assembly using ImageJ 1.46 (National Institutes of Health, Bethesda, MD).

3.3 Cell Viability Assay (Calcein-AM Assay)

1. Culture human neuroblastoma SH-SY5Y cells in Dulbecco's modified Eagle's medium with low glucose containing 10% foetal bovine serum, 10,000 U/mL penicillin, and 10 mg/mL streptomycin at 37°C under 5% CO_2.
2. Calcein-AM solution (1 mM): Dissolve 3',6'-di(*O*-acetyl)-4',5'-bis[*N*,*N*-bis(carboxymethyl) aminomethyl]fluorescein (Calcein-AM) (Dojindo Laboratories Co., Ltd., Kumamoto, Japan) in DMSO. Store the solution at −30°C.
3. Place a plastic disk coated with the ganglioside-containing lipid monolayer (Sect. 2.6, Step 3) into each well of a 24-well plate with the lipid-coated side facing up.
4. Add $A\beta_{42}$ solution (2 μM) to the well. After incubation at 37°C for 24 h, remove the solution from the wells.
5. Seed SH-SY5Y cells (8.0×10^4 cells/well) on the disk and incubate at 37°C for 48 h.
6. Wash the plastic disk twice with PBS and transfer onto a 24-well black plate.
7. Add 2 μM Calcein-AM solution (Step 2) in PBS into each well, and incubate the cells for 1 h at 37°C.
8. Measure the fluorescence intensities at excitation and emission wavelengths of 485 and 520 nm, respectively, with a slit width of 20 nm using a plate reader.

4 Notes

1. The surface pressure–molecular area data are observed for POPC in the closely packed state at 35 mN/m.
2. Binarize the micrographs based on the height and count the pixels. For example, the white area of a binarised image is identified as an Aβ-bound area by thresholding at 6 nm from the bottom.
3. For example, the GNU Image Manipulation Program (https://www.gimp.org).

Acknowledgments

This study was partially supported by JSPS KAKENHI (Grant Number JP23H04562).

Conflict of Interest The authors have no conflicts of interest to declare that are relevant to the content of this chapter.

References

1. Chiti F, Dobson CM (2017) Protein misfolding, amyloid formation, and human disease: a summary of progress over the last decade. Annu Rev Biochem 86:27–68
2. Sipe JD, Cohen AS (2000) Review: history of the amyloid fibril. J Struct Biol 130:88–98
3. Hardy J, Selkoe DJ (2002) The amyloid hypothesis of Alzheimer's disease: progress and problems on the road to therapeutics. Science 297:353–356
4. Duff K, Eckman C, Zehr C et al (1996) Increased amyloid-beta42(43) in brains of mice expressing mutant presenilin 1. Nature 383:710–713
5. Burdick D, Soreghan B, Kwon M et al (1992) Assembly and aggregation properties of synthetic Alzheimer's A4/beta amyloid peptide analogs. J Biol Chem 267:546–554
6. Wisniewski T, Ghiso J, Frangione B (1991) Peptides homologous to the amyloid protein of Alzheimer's disease containing a glutamine for glutamic acid substitution have accelerated amyloid fibril formation. Biochem Biophys Res Commun 179:1247–1254
7. Yanagisawa K (2011) Pathological significance of ganglioside clusters in Alzheimer's disease. J Neurochem 116:806–812
8. Simons K, Vaz WL (2004) Model systems, lipid rafts, and cell membranes. Annu Rev Biophys Biomol Struct 33:269–295
9. Yanagisawa K, Odaka A, Suzuki N et al (1995) GM1 ganglioside-bound amyloid beta-protein (A beta): a possible form of preamyloid in Alzheimer's disease. Nat Med 1:1062–1066
10. Yanagisawa K (2007) Role of gangliosides in Alzheimer's disease. Biochim Biophys Acta 1768:1943–1951
11. Yanagisawa K (2015) GM1 ganglioside and Alzheimer's disease. Glycoconj J 32:87–91
12. Matsubara T, Nishihara M, Yasumori H et al (2017) Size and shape of amyloid fibrils induced by ganglio- side nanoclusters: role of sialyl oligosaccharide in fibril formation. Langmuir 33:13874–13881
13. Kakio A, Nishimoto SI, Yanagisawa K et al (2001) Cholesterol-dependent formation of GM1 ganglioside-bound amyloid beta-protein, an endogenous seed for Alzheimer amyloid. J Biol Chem 276:24985–24990
14. Matsubara T, IIjima K, Kojima T et al (2021) Heterogeneous ganglioside-enriched nanoclusters with different densities in membrane rafts detected by a peptidyl molecular probe. Langmuir 37:646–654
15. Matsubara T, Yasumori H, Ito K et al (2018) Amyloid-β fibrils assembled on ganglioside-enriched membranes contain both parallel β-sheets and turns. J Biol Chem 293:14146–14154
16. Hu Z, Wang X, Wang W et al (2015) Raman spectroscopy for detecting supported planar lipid bilayers composed of ganglioside-GM1/sphingomyelin/cholesterol in the presence of amyloid-beta. Phys Chem Phys 17:22711–22720
17. Yamamoto N, Matsubara E, Maeda S et al (2007) A ganglioside-induced toxic soluble abeta assembly: Its enhanced formation from abeta bearing the arctic mutation. J Biol Chem 282:2646–2655
18. Evangelisti E, Cascella R, Becatti M et al (2016) Binding affinity of amyloid oligomers to cellular membranes is a generic indicator of cellular dysfunction in protein misfolding diseases. Sci Rep 6: 32721

Part VI

Application of Gangliosides

Chapter 30

Preparation of Ganglioside-Reconstituted Liposomes

Asako Shimoda and Kazunari Akiyoshi

Abstract

Liposomes are formed from phospholipids, which are the main components of biological membranes, and are widely used as models of biological membranes. Ganglioside-reconstituted liposomes can be prepared by mixing with phospholipids, providing a model system that closely mimics the cell surface environment. Liposome systems are used in various studies, including investigations of the biological functions of gangliosides and their applications in drug delivery systems (DDS). In this chapter, we describe the preparation and characterization of ganglioside-reconstituted liposomes.

Keywords Liposome, Reconstitution, Ganglioside, Extrusion, Dynamic light scattering

1 Introduction

Cell membrane lipids, including phospholipids, cholesterol, and glycolipids, play an important role in sustaining life. Gangliosides, which are major glycolipids, are involved in various physiological phenomena such as cell recognition, adhesion, and signaling [1, 2]. The characteristic of gangliosides as functional molecules includes their amphiphilic nature, allowing them to form various aggregates in aqueous environments. They closely relate to their functional expression of gangliosides in biomembrane [3, 4]. It is well known that cell membranes contain microdomains called lipid rafts composed of sphingolipids, sphingoglycolipids, and cholesterol, and that gangliosides are important in regulating the formation of lipid rafts [1]. Such microdomains are involved in membrane-mediated signal transduction [2] and also induce dynamic structural changes of the plasma membrane, such as the formation of dendrites [5].

Liposomes have been widely used as various models of biological membranes since they were first reported by Bangham in 1964, and they have greatly contributed to elucidating the functions of gangliosides [6].

In the 1980s, Hakomori et al. conducted extensive research on the recognition and adhesive functions of gangliosides on the cell

Koichi Furukawa (ed.), *Gangliosides*, Methods in Molecular Biology, vol. 3035,
https://doi.org/10.1007/978-1-0716-5288-6_30,

surface. During this period, liposomes were utilized as model systems [7, 8]. Because the density of glycans on the liposome surface can be modulated to a certain extent by altering the mixing ratio of gangliosides and phospholipids, liposomes serve as effective platforms for presenting glycans on their surface. For example, liposome-based systems have demonstrated that the state of ganglioside glycans on the cell surface plays an important role in intercellular adhesion. The interaction between GM3 and the mouse monoclonal antibody M2590, which is specific to GM3, was investigated using such liposomal systems. By varying the concentration of GM3 incorporated into the liposomes, it was shown that the binding characteristics with the antibody exhibited a threshold-like on-off response [9]. Furthermore, Sunamoto et al. investigated the interactions between various ganglioside-containing DPPC liposomes (with a particle size of approximately 150 nm) and immune cells, particularly T cells, which are known to possess glycan receptors. They found that especially GD1a and GT1b, when incorporated into liposomes, exhibited cell-stimulating activity at the nanomolar level, as evidenced by an increase in intracellular calcium concentration. In this case, the ganglioside concentration within the liposomes also showed a threshold-like on-off effect [10].

The phenomenon in which the function of glycans as ligands is enhanced through control of oligosaccharide topology or an increase in glycan density is generally referred to as the "glycan cluster effect" [11]. In biological membranes, glycolipids are known to form clusters, and these domain structures are considered crucial for the manifestation of their functions. In biological systems, the on-off regulation of cell functions by glycans is achieved through the precise control of glycan aggregation states on the cell membrane. This phenomenon has been successfully reproduced using liposome-based model systems. On the two-dimensional surface of liposomes, the conformation of ganglioside glycans becomes fixed. Additionally, the formation of domain structures through molecular association is thought to concentrate glycan ligands, thereby enhancing their binding to glycan receptors. In signal transduction pathways, multivalent interactions that induce receptor clustering are particularly important. Furthermore, at interfaces such as those of liposomes or cells, the microenvironment—including solvation conditions—differs significantly from the bulk solution. Molecular recognition processes at such interfaces, therefore, present intriguing challenges that should be explored at the molecular level.

Gangliosides-reconstituted liposomes have been widely employed as model systems to elucidate various biological phenomena. For example, the gangliosides-liposome system has revealed that GM1 ganglioside cluster domains play a critical role in the aggregation of amyloid β-protein (Aβ) on the cell surface, a process associated with the onset of Alzheimer's disease [12, 13].

Additionally, liposomes have long been used in DDS applications. In such contexts, blood stability and circulation time are crucial, and liposomes incorporating gangliosides have been shown to exhibit improved blood retention [14, 15]. Moreover, the addition of GM1 to polyethylene glycol-modified liposomes further enhances these properties [16].

More recently, GM1-incorporated liposomes have garnered attention for their antioxidant effects. This effect is attributed to liposomal exogenous GM1 activating the Nrf2 (nuclear factor erythroid 2-related factor 2) pathway, a key cellular defense mechanism against oxidative stress and other types of cellular damage [17].

GM3 is known to be overexpressed in cancer cells and recognized as a tumor-associated carbohydrate antigen (TACA) in melanoma tumors, making it a potential target for cancer vaccine immunotherapy. However, GM3 generally exhibits low immunogenicity and can easily induce immune tolerance, which has limited its effectiveness in cancer suppression. Recent studies have shown that GM3 interacts with CD169, a receptor expressed on antigen-presenting cells such as macrophages. This has led to the development of GM3-liposome-based cancer vaccines incorporating adjuvants [18]. Furthermore, immunotherapeutic approaches using ganglioside liposomes targeting $CD169^{+}$ macrophages—crucial for the early stages of viral infections such as herpes simplex keratitis (HSK)—have also been reported [19].

This article provides an overview of the preparation and characterization of reconstituted ganglioside liposomes.

2 Materials

1. Rotary evaporator.
2. Vortex mixer.
3. Vacuum desiccator.
4. Malvern Zetasizer Nano ZS (Malvern Instruments Ltd., Worcestershire, UK).
5. Avanti® Mini-Extruder (Avanti Polar Lipids, Alabaster, AL, USA).
6. 0.1, 0.2, 0.4, 0.6, and 1.0 μm polycarbonate membranes (Avanti Polar Lipids, Alabaster, AL, USA).
7. Nitrogen or Argon gas tank.
8. Round-bottom flask (50–250 mL).
9. Gangliosides: GM3, GD1a, GA1b, GQ1b, GT1b, GD3, etc. (Multiple Vendors, e.g., Sigma-Aldrich, Avanti Polar Lipids).

10. Phosphatidylcholines: e.g., 1,2-dipalmitoyl-sn-glycero-3-phosphocholine (DPPC), egg yolk PC, 1,2-distearoyl-sn-glycero-3-phosphocholine (DSPC), etc. (Avanti Polar Lipids).
11. Cholesterol (Multiple Vendors, e.g., Sigma-Aldrich, Avanti Polar Lipids, etc.).
12. Chloroform.
13. Methanol.
14. Phosphate-buffered saline (PBS).
15. Phospholipid assay kit (Multiple Vendors, e.g., FUJIFILM Wako, Sigma-Aldrich, Avanti Polar Lipids, CELL BIOLABS, INC.)

3 Methods

3.1 Preparation of Gangliosides-Reconstituted Liposomes Using a Thin-Film Hydration Method

1. A mixture of phospholipid (e.g. DPPC) and cholesterol (at a ratio of 10:1 in weight) and a ganglioside dissolved in chloroform/methanol (1:1, v/v) are added into a round-bottom flask (50–250 mL) (*see* **Notes 1 and 2**).
2. Remove the organic solvents using a rotary evaporator to prepare a lipid-thin film (*see* **Note 3**). The lipid-thin film is completely dried in a vacuum desiccator.
3. Add 3.0 mL of PBS without calcium and magnesium ions to the lipid-thin film and hydrate the film using a Vortex mixer.

3.2 Formation and Characterization of Small-Size Liposomes

1. Load the lipid suspension into a gas-tight syringe and extrude it several times by passing through polycarbonate membrane filters. The desired size of liposomes can be obtained by sequential extrusion with polycarbonate membranes with the pore sizes of 1.0, 0.6, 0.4, 0.2, and 0.1 μm using an Avanti® Mini-Extruder according to the manufacturer's instructions (*see* **Note 4**) (Fig. 1).
2. Filter the resulting solution through a 0.45 μm filter before size analysis.
3. Determine the mean size of the liposomes with dynamic light scattering by a Zetasizer Nano ZS.
4. Measure the liposomal phospholipids using a commercially available phospholipid assay kit. Adjust the final lipid concentration of the stock solution to 1.0×10^{-3} M. Store the liposome solution at 4°C and use it within a few days.

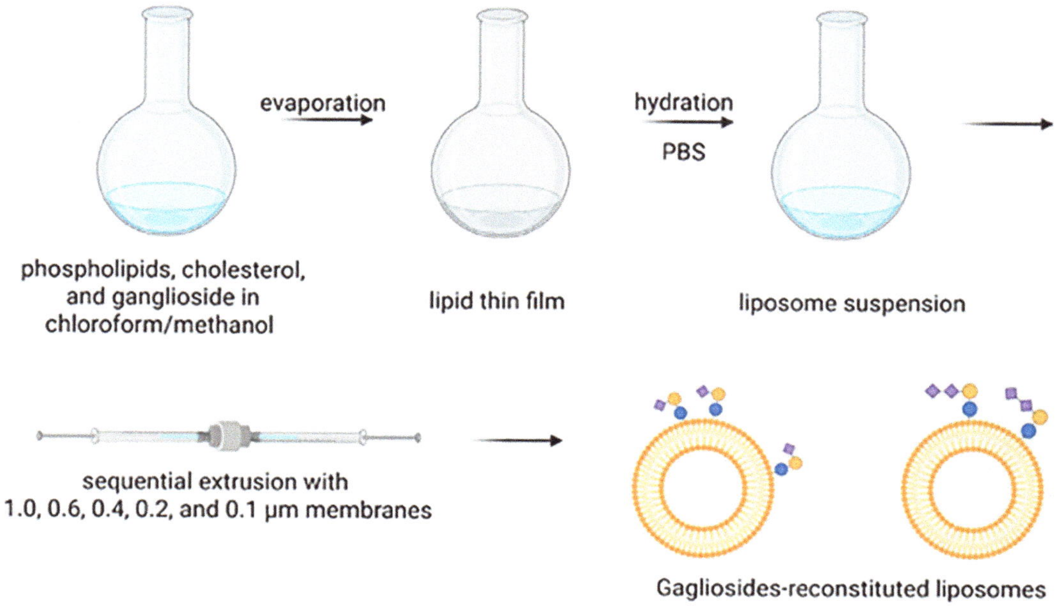

Fig. 1 Thin-film hydration method for preparation of gangliosides-reconstituted liposomes. (Figure created with BioRender.com)

4 Notes

1. The desired amount of ganglioside is 5.0–20.0 mol% of total liposomal lipids [20].
2. For preparing fluorescence-labeled liposomes, use fluorescent lipids (e.g., TopFluor™ TMR PC, C11 TopFluor™ GM3, TopFluor™ Cholesterol, 25-NBD Cholesterol, etc. from Avanti Polar Lipids). Fluorescent substances for cell membrane labeling (e.g., 1,1′-dioctadecyl-3,3,3′,3′tetra-methylindocarbocyanineperchlorate (DiI)) can also be used after preparing a liposome suspension.
3. For small-scale preparation, nitrogen or argon gas is used to remove the organic solvents.
4. Keep the temperature above the transition temperature of the lipid during hydration and extrusion.

Acknowledgments

This work was supported by JST FOREST Program (Grant Number JPMJFR225F, Japan), the Hakubi project at Kyoto University, and Grants-in-Aid for Scientific Research (JSPS KAKENHI, Grant Number 22H00585).

Confhealict of Interest The authors have no conflicts of interest to declare that are relevant to the content of this chapter.

References

1. Hakomori S, Handa K, Iwabuchi K et al (1998) Glycosignaling domain: a cell surface assembly of glycosphingolipids involved in cell adhesion and signaling. Glycobiology 8:xi–xviii
2. Hakomori S (2004) Carbohydrate-to-carbohydrate interaction in basic cell biology: a brief overview. Arch Biochem Biophys 426:173–181
3. Cantù L, Del Favero E, Sonnino S et al (2011) Gangliosides and the multiscale modulation of membrane structure. Chem Phys Lipids 164:796–807
4. Cantu' L, Del Favero E, Brocca P et al (2014) Multilevel structuring of ganglioside-containing aggregates. Adv Colloid Interface Sci 205:177–186
5. Akiyoshi K, Itaya A, Nomura SM et al (2002) Induction of neuron-like tubes and liposome networks by cooperative effect of gangliosides and phospholipids. FEBS Lett 532:283–287.
6. Bangham AD, Horne RW (1964) Negative staining of phospholipids and their structural modification by surface active agents as observed in the electron microscope. J Mol Biol 8:660–668
7. Otsuji E, Park YS, Tashiro K et al (1995) Inhibition of B16 melanoma metastasis by administration of GM3- or GG3-liposomes: blocking adhesion of melanoma cells to endothelial cells (anti-adhesion therapy) via inhibition of Ga-GG3Cer or Gn3-LacCer interaction. Int J Oncol 6:319–327
8. Song Y, Withers DA, Hakomori S (1998) Globoside-dependent adhesion of human embryonal carcinoma cells. J Biol Chem 273:2517–2525
9. Nores GA, Dohi T, Taniguchi M et al (1987) Density-dependent recognition of cell surface GM3 by a certain anti-melanoma antibody, and GM3 lactone as a possible immunogen: requirements for tumor-associated antigen and immunogen. J Immunol 139:3171–3176
10. Kato E, Akiyoshi K, Furuno T et al (1994) Interaction between ganglioside-containing liposome and rat T-lymphocyte: confocal fluorescence microscopic study. Biochem Biophys Res Commun 203:1750–1755
11. Yin XG, Lu J, Wang J et al (2021) Synthesis and evaluation of liposomal anti-GM3 cancer vaccine candidates covalently and noncovalently adjuvanted by αGalCer. J Med Chem 64:1951–1965
12. Yanagisawa K, Odaka A, Suzuki N et al (1995) GM1 ganglioside-bound amyloid β-protein (Aβ): a possible form of preamyloid in Alzheimer's disease. Nat Med 1:1062–1066
13. Matsuzaki K (2020) Aβ–ganglioside interactions in the pathogenesis of Alzheimer's disease. Biochim Biophys Acta, Biomembr 1862:183233
14. Gabizon A, Papahadjopoulos D (1988) Liposome formulations with prolonged circulation time in blood and enhanced uptake by tumors. Proc Natl Acad Sci USA 85:6949–6953
15. Allen TM, Hansen C, Rutledge J (1989) Liposomes with prolonged circulation times: factors affecting uptake by reticuloendothelial and other tissues. Biochim Biophys Acta 981:27–35
16. Yamauchi H, Yano T, Kato T et al (1995) Effects of sialic acid derivative on long circulation time and tumor concentration of liposomes. Int J Pharm 113:141–148
17. Weissig V, Joshi MD, Migrino RQ (2025) Cytoprotective effects of liposomal ganglioside GM1. J Liposome Res 35:212–217
18. Zhang J, Terreni M, Liu F et al (2024) Ganglioside GM3-based anticancer vaccines: reviewing the mechanism and current strategies. Biomed Pharmacother 176:116824
19. Shen W, Wang C, Jiang J et al (2024) Targeted delivery of herpes simplex virus glycoprotein D to CD169+ macrophages using ganglioside liposomes alleviates herpes simplex keratitis in mice. J Control Release 365:208–218
20. Kang E-C, Miyahara T, Akiyoshi K et al (2002) Partitioning of ganglioside-reconstituted liposomes in aqueous two-phase systems. J Bioact Compat Polym 17:87–104

Check for updates

Chapter 31

Intracerebroventricular Administration of Gangliosides for In Vivo Functional Studies in Mice

John Monyror, Melanie Horkey, Asifa K. Zaidi and Simonetta Sipione

Abstract

Levels of gangliosides are reduced in several neurological conditions, and their restoration has shown therapeutic potential in animal models of Huntington's disease and Parkinson's disease. However, the full range of functions of gangliosides in the brain, along with their molecular targets and mechanisms of action following exogenous administration, remains ill-defined. Due to their poor blood-brain barrier (BBB) permeability, gangliosides face significant challenges when delivered through non-invasive routes. Direct intracerebroventricular (ICV) infusion circumvents this issue by bypassing the BBB, ensuring a more precise delivery to brain structures and eliminating uncertainties related to brain penetration. Here, we describe a protocol for ICV infusion of gangliosides in mice that enables in vivo pre-clinical studies of their effects, while eliminating uncertainties about the extent of drug delivery to the brain through the BBB.

Keywords GM1, Huntington's disease, Neurodegeneration, Stereotaxic surgery, Osmotic minipumps

1 Introduction

Gangliosides are key molecules for the proper functioning of the brain. They play numerous roles in cell signaling, development, immunity, neuroprotection, and repair [1, 2]. In rare cases where genetic mutations impair ganglioside biosynthesis, severe and fatal neurodegenerative conditions develop, underscoring the important role of gangliosides in the brain. In more common neurodegenerative diseases, including Huntington's disease (HD) and Parkinson's disease (PD), deficient levels of gangliosides have been observed. This lack of gangliosides, particularly the lack of ganglioside GM1, may underpin some dysfunctions that contribute to the pathogenesis of these diseases [2–4]. Administration of exogenous GM1 modifies disease course in animal models of HD and PD, and rescues symptoms [5–10]. Excitingly, ganglioside administration has also shown promise in small clinical trials in PD patients [11–13]. The

Koichi Furukawa (ed.), *Gangliosides*, Methods in Molecular Biology, vol. 3035,
https://doi.org/10.1007/978-1-0716-5288-6_31,

mechanisms underlying the neuroprotective activity of endogenous as well as exogenously administered gangliosides remain to be fully elucidated.

A major challenge in studies that focus on the biology and pharmacology of gangliosides in vivo is the low permeability of the blood-brain barrier (BBB) to these glycosphingolipids [14–16]. In studies where radiolabeled ganglioside GM1 was administered intravenously, only 3.4% or 0.4% of the ganglioside was delivered to the brain of rats and cynomolgus monkeys, respectively [17, 18]. Efficient diffusion of therapeutics across the BBB typically requires a molecular weight below 400 Da, having fewer than eight chemical groups engaged in hydrogen bonds, and the absence of a negative charge, otherwise active transport mechanisms may be needed to cross the BBB [19–22]. Therefore, the poor brain permeability of GM1 is likely explained by its large size, high hydrogen bonding capacity, and negative charge. Nevertheless, a few clinical trials have shown beneficial effects of peripherally administered GM1 in PD patients [12, 23, 24]. Whether this is due to defects in BBB integrity in PD patients [25–27] that allow for increased brain accumulation of GM1, to the therapeutic effects of even a small fraction of the ganglioside reaching the brain, or to peripheral mechanisms remains to be determined. Furthermore, the accumulation of GM1 in off-target tissues must be considered when the ganglioside is administered peripherally. For instance, intravenous infusion of radiolabelled GM1 in non-human primates revealed significant accumulation in the heart, bone marrow, and lungs [17], which could affect ganglioside bioavailability and potentially result in undesirable side effects. Given the limitations of peripheral ganglioside administration, other delivery routes may be more suitable for pre-clinical studies in animal models, to eliminate the uncertainty surrounding ganglioside bioavailability and BBB penetration, enabling a more focused investigation into their biological and pharmacological effects.

By bypassing the BBB, direct intracerebroventricular (ICV) infusion provides the most reliable method for delivering gangliosides to the brain, ensuring maximum bioavailability. Although more invasive than subcutaneous, intravenous, or oral routes of administration, the certainty of effective delivery makes ICV infusion an ideal approach in studies that focus on mechanisms and molecular targets of gangliosides in vivo, at least in rodent models. However, ICV infusion may not be suitable to target all brain regions, as the infused molecules follow the cerebrospinal fluid (CSF) pathway and do not diffuse extensively into the brain parenchyma. Regions that are proximal to the ventricular system, such as the basal ganglia and the cortex, are preferentially targeted by ICV infusion of drugs [28]. Therefore, ICV infusion may be particularly appropriate in studies where the main target regions are adjacent to the ventricular system, such as the basal ganglia and cortex in models of HD and PD. Potential complications of ICV infusion include infection and the possibility of inaccurate targeting, both of which can be minimized through proper sterile technique, post-surgical antibiotic administration, and adequate surgical training.

In this chapter, we present a protocol for continuous ICV infusion of ganglioside GM1 into the mouse brain using mini-osmotic pumps. This approach allows for chronic drug administration in animals for up to 42 days, minimizes animal handling, and allows animals to move freely post-surgery, facilitating the analysis of long-term effects of drug administration, functional studies, and behavioral testing. This protocol can be adapted to administer other compounds with limited or uncertain BBB penetration.

2 Materials

2.1 ALZET Mini-Osmotic Pump Preparation

- Artificial CSF (Harvard Apparatus #59-7316).
- Monosialoganglioside GM1 (from bovine brain, SIGMA # G9652, or any other suitable source).
- ALZET mini-osmotic pumps:
  - Model 2004—28 days infusion (200 μL capacity, average infusion rate of 0.25 μL/h).
  - Model 2006—42 days infusion (200 μL capacity, average infusion rate of 0.15 μL/h).
- ALZET Brain Infusion Kit 3 (ALZET #0008851). Kit components:
  - 3 mm brain infusion cannulas.
  - Catheter tubing.
  - Depth-adjustment spacers.
- Sterile Dulbecco's phosphate-buffered saline solution (DPBS).
- Sterile polypropylene conical tubes.
- Sterile 100 mm dishes.
- 1 cc syringes.
- 21-gauge needles.
- Sterile gloves.
- Sterile forceps.
- Biological safety cabinet (biosafety level 2).
- 37°C incubator.

2.2 Isoflurane Anesthesia and Surgical Preparation

- Isoflurane anesthetic.
- Ohmeda Isotec 4 Isoflurane Precision Vaporizer.
- Bain Circuit. This is a specialized coaxial breathing system designed to deliver anesthetic gases while preventing the rebreathing of exhaled gases.
- Mouse Anesthetic Mask.
- Oxygen source (e.g., compressed oxygen cylinder).
- Scavenger system (e.g., building exhaust connection or carbon filter unit, e.g., Omnicon #80120).
- Towels, drapes, and cloths (to cover the animal during and post-operation).
- Insulating material between the animal and the work surface (e.g., water circulating heating pad and paper towel).
- Sealed induction chamber connected to scavenger system (e.g., a lidded polycarbonate chamber with an inlet for anesthetic gas and outlet for the scavenger system).
- Liposic® sterile lubricating ophthalmic ointment.
- Fur trimmer (e.g., Wahl #8040-100).
- Lint roller (e.g., 3M #7100136596).
- Stanhexidine (4% w/v chlorhexidine gluconate solution, Omega Laboratories L0000008).

2.3 Surgery

- Drill (Dremel Mini-Mite Drill, Model 750).
- Drill bit (Dremel 105 Engraving Cutter).
- Sutures—5-0 PDS*II (polydioxanone) RB-1; 17 mm ½ c taper needle (Ethicon #Z303H).
- Stereotaxic frame (Kopf model 940).
- Povidone-iodine (Ricca Chemical Company #3955).
- Vetbond Tissue Adhesive (3M #1469SB).
- Sterile surgical tools:
  - Thumb dressing standard forceps ×2.
  - Mosquito forceps, curved.
  - Converse-angled scissors.
  - Spring scissors.
  - Needle holder, large (>6.5″).
  - Scalpel blades and a scalpel.
- Hydrogen peroxide.
- Cotton-tipped applicators to clean the surgical site (Puritan #806-WC).

- Permanent marker (e.g., ultra-fine tipped blue Sharpie #37113).
- 26-gauge needle (e.g., Air-Tite Products Co. Inc. # N2612).
- Normal sterile saline (0.9% sodium chloride).
- Solutions for administration:
  - 0.2 mg/mL Meloxicam in sterile saline. Use 5mg/kg dose (or 250 µL/10 g mouse weight).
  - 0.6 mg/mL Slow-release Buprenorphine in sterile saline. Use 0.5 mg/kg .

2.4 Post-Operative Care

- Disposable cage with lid.
- Incubator set at 38°C.
- Antibiotics: 3.1 mM of sulfamethoxazole and 540.3 µM of trimethoprim in mouse drinking water.

3 Methods

3.1 ALZET Pump Preparation

1. **Pump priming.** Two to three days before surgery, prepare pumps under sterile conditions in a laminar flow hood, as per manufacturer instructions (*see* **Note 1**). Pumps must be 'primed' as described below, to ensure immediate and reliable drug delivery once implanted in the animals. The length of priming depends on the pump model. For ALZET 2004 and 2006 models, the minimum priming time is 40 h and 60 h, respectively. Priming length can be increased by up to 24 h without affecting the duration of drug delivery in mice.
2. Prepare two sterile 50 mL polypropylene conical tubes by labeling one "CSF" and the other "GM1." Fill each tube with 30 mL of sterile DPBS. These are the tubes that will hold the pumps during priming.
3. **Dissolve GM1 in CSF** at the appropriate concentration (*see* Box 1).
   - For Model 2004 pumps, prepare 3.6 mM GM1 solution in artificial CSF (infusion rate = 1.41 µg GM1/h).
   - For Model 2006 pumps, prepare 6 mM GM1 solution in artificial CSF (infusion rate = 1.41 µg GM1/h).

   Prepare 300 µL solution/pump (this includes ~20% more volume than needed to account for losses in filling needles and syringes). For example, to prepare 300 µL 6 mM GM1 solution, add 300 µL of sterile CSF to 2.77 mg of bovine GM1. Pipette gently to mix and leave the solution at room

BOX 1. Calculating drug concentration in the pump for ICV delivery.

The cerebrospinal fluid (CSF) is produced by the cells of the choroid plexus in the wall of the lateral and third ventricles, and flows with pulsatile movement in the direction: lateral ventricles → third ventricle → fourth ventricle → subarachnoid space, where it is reabsorbed.

The total volume of mouse CSF is 35μl
The rate of CSF synthesis is 18 μl/h [30]

To simplify the model of drug distribution for infusion calculations, we can use a first-order approximation, assuming the CSF flows in a fixed-volume cylindrical space. Every hour more than half of the total CSF volume is replaced, and within two hours, the entire CSF is expected to be renewed.

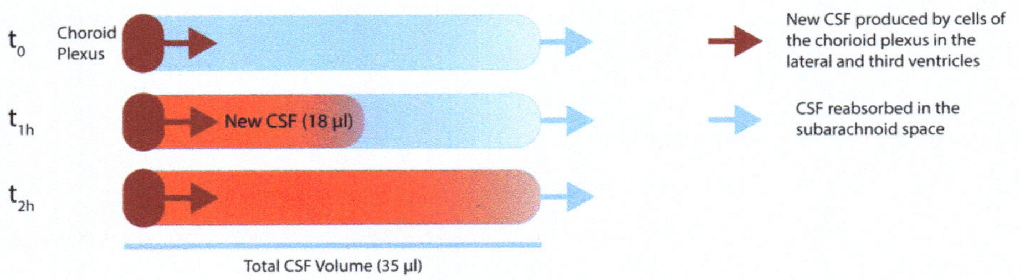

In this model, we assume that the infused drug mixes primarily with the newly synthesized CSF and is carried along with the CSF flow through the ventricular system. After 2h, when the CSF has been replaced once, the drug concentration in the CSF should approximate the desired target concentration. This is a simplification, as the drug is not directly infused into the newly secreted CSF, but rather into the lateral ventricle at some distance from the site of secretion, and drug distribution in the CSF can be affected by many factors, including drug molecular weight, pulsatile flow, diffusion, potential interaction with CSF proteins etc.

Nevertheless, this first-order approximation provides a reasonable estimation of the relationship between the concentration of the drug in the infusion pump and the final concentration in the CSF. The formula below models the relationship between the drug concentration in the infusion pump and the final concentration in the CSF, factoring in the rate of CSF synthesis and the pump infusion rate:

$$\textbf{Drug Concentration in the Pump} = \textbf{Desired Final Drug Concentration in the CSF} \times \frac{\textbf{Rate of CSF Synthesis}}{\textbf{Pump Flow Rate}}$$

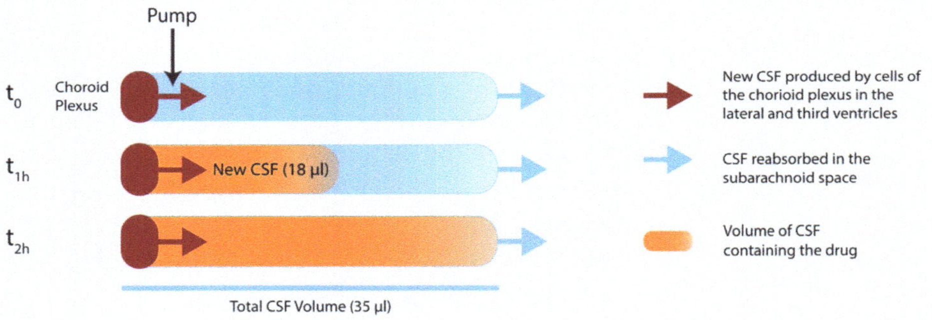

temperature until ready to fill the pumps. GM1 will form micelles in aqueous solution.

4. **Pump filling**. Fill control CSF pumps first to avoid carryover of the drug into the control condition as each infusion kit provides only one blunt-tipped 27-gauge filling tube (*see* **Note 2**).
5. Unpack pumps and flow moderators and place them in a 100 mm dish under sterile conditions in a laminar flow hood. Remove the plastic flanges from all flow moderators.
6. From the infusion kits, take one piece of tubing for two pumps. This protocol assumes eight pumps will be prepared

(four filled with CSF and four with GM1), but it can be scaled up or down as needed. Cut the tubes in half using sterile scissors and place them in the dish with the pumps and the flow moderators.

7. Draw 1 mL of CSF into a 1 cc syringe using a 21 g needle. Remove any air bubbles (*see* **Note 3**).
8. Insert the 21 g needle into each of the four pieces of tubing to be filled with CSF, one at a time. Proceed to fill each tube. Recap the needle and remove it.
9. Attach the blunt-tipped 27-gauge filling tube to the syringe and remove any air bubbles.
10. Holding one of the pumps in an upright position, insert the filling tube through the opening at the top until it can go no further (*see* **Note 4**). Record the volume of solution in the syringe.
11. Push the plunger of the syringe *slowly* while continuing to hold the pump upright. This will minimize any air bubbles being introduced into the pump. When the solution bubbles out of the opening at the top, stop filling and carefully remove the filling tube. Record the volume of solution remaining in the syringe—the difference in volume from the previous step gives the approximate fill volume for the pumps. If the latter is less than the mean fill volume provided by the manufacturer, air may be trapped in the pump. If so, empty the incompletely filled pump and refill.
12. Insert a flow moderator into the top of the pump until ~3 mm of the moderator is left sticking out. A drop of solution should appear at the top of the moderator.
13. Attach a pre-filled tube from Step 8 to the moderator, pushing it down on the moderator as far as possible (*see* **Note 5**). Place the pump in the "CSF" conical tube, immersing it completely in the sterile DPBS, with the tubing sticking out to prevent DPBS from entering the tubing. Cap the tube to maintain sterility (*see* **Note 6**).
14. Repeat the filling procedure (Steps 8–13) for all remaining CSF-loaded pumps. Place the conical tube in an incubator at 37°C until surgery.
15. Start filling GM1-loaded pumps, filling the tubing and the pumps with the GM1 solution prepared earlier, using the same procedure illustrated for CSF pump loading in Steps 7–14.
16. Leave the tubes containing the pumps in a 37°C incubator for 2–3 days prior to use, depending on the pump model.

3.2 Mice Preparation for Surgery and Isoflurane Anesthesia

1. **Weigh each mouse and record their weight.** This is required to properly dose animals with analgesic drugs.
2. Confirm that you have adequate levels of oxygen and isoflurane for all mice undergoing surgery.
3. Connect a vaporizer to an oxygen source and a Bain Circuit with a nose cone. Ensure that the scavenger system is set up properly.
4. **Pre-anesthetic administration of slow-release buprenorphine.** Inject the mouse with 0.5 mg/kg buprenorphine subcutaneously, before transferring it to the induction chamber. This step induces partial sedation, reduces animal stress, and lowers the minimum alveolar concentration of isoflurane needed to maintain anesthesia.
5. **Mouse anesthesia.** Transfer the mouse into the induction chamber and seal it. Turn on the oxygen. The flow rate depends on the size of the induction chamber. For example, a 1L-chamber typically requires an oxygen flow rate of 1 L/min, and a 2L-chamber requires a 2 L/min flow rate. Larger induction chambers with a capacity above 3 L require a 2 L/min flow rate to prevent rapid cooling of the mouse.
6. Pre-oxygenate the chamber by delivering only oxygen for 4–5 mouse breaths (5–10 s).
7. To prevent breath holding and to provide smooth induction of anesthesia, turn on the isoflurane gauge and slowly increase the flow starting with 0.5% isoflurane and increasing by 0.5% increments to a maximum of 4%. If the mouse is highly agitated, quickly deliver 3%–5% isoflurane.
8. Once the righting reflex is lost and the mouse is breathing deeply, slowly turn off the isoflurane and flush the chamber for 2–3 s with pure oxygen to remove waste gas.
9. Open the induction chamber. Supporting the whole body of the mouse and keeping the mouse level, gently transfer the mouse to the nose cone.
10. Turn the oxygen flow to 0.5–1 L/min (*see* **Note 7**). Turn the isoflurane vaporizer up to 2%–3% (*see* **Note 8**).
11. Apply sterile lubricating eye ointment to the mouse (*see* **Note 9**).
12. Perform a toe pinch *on all four limbs* to assess whether the mouse is at the surgical plane. Heart rate should be rapid and regular (300–600 bpm), respiration rate should involve both thorax and abdomen and should have a regular pattern (90–200 bpm), and extremities should remain pink, not pale, gray, or blue-tinged. If these requirements are met, you may proceed with surgery (*see* **Note 10**).

13. **Prepare the infusion kit** while the mouse is placed under anesthesia. Open the infusion kit and insert it into the cannula holder of the stereotaxic apparatus with the part that will be attached to the pump pointing towards the mouse tail.
14. Attach a piece of empty tubing provided with the infusion kits to the end of a 21 g needle attached to a syringe filled with CSF (or GM1). Attach the other end of the tubing to the part of the infusion kit that will be attached to the pump, and inject some CSF or GM1 until a small droplet of solution appears on the infusion end of the kit. This will remove any air in the infusion kit and prepare it for immediate pumping (*see* **Note 11**).
15. **Prepare the mouse for surgery**. Confirm the mouse is at or near the surgical plane.
16. Remove the fur on the top of the mouse head using fur trimmers. Ensure the fur is removed from ear to ear and from the eyes to the rear of the skull (*see* **Note 12**).
17. Prepare the surgical site. Clean the exposed skin on top of the head with Stanhexidine using a cotton-tipped applicator. Repeat two more times. Disinfect with 70% ethanol once and let the skin air dry.

3.3 Surgery

1. **Position the mouse in the stereotaxic frame**. Place the mouse onto a paper towel covering a circulating water heater pad to maintain body temperature. Cover the mouse with a sterile drape to ensure a sterile environment during surgery. Open the mouse's mouth slightly and place its incisors into the hole of the lower bar of the stereotaxic frame. This secures the mouse head in place. Cover the snout with the gas mask connected to the isoflurane supply, but do not tighten the mask yet, so that adjustments to the mouse head position can be made.
2. Maintain the oxygen flow between 0.5 and 1 L/min and isoflurane between 2% and 3% to ensure that the mouse remains in the surgical plane. Regularly monitor the mouse's heart rate, respiratory rate, pedal reflexes, and the color of its extremities to ensure the correct depth of anesthesia. Adjust the level of isoflurane as necessary.
3. Insert the left ear bar into the mouse's ear canal. Ensure that the bar does not apply pressure to the skull, as this can lead to head injury and asymmetric positioning of the mouse head. Center the mouse's head as much as possible before tightening the ear bar.
4. Insert the right ear bar in the same manner as the left, until the skull is held firmly in place without excessive pressure that could damage it. Tighten the ear bar. The typical position of

ear bars is ~0.4 on both the left and right sides of the stereotaxic frame, but individual adjustments may be necessary.

5. Tighten the gas mask gently once the mouse is properly centered and its head immobilized.
6. Confirm that the skull is secured by gently pressing sideward and downward on the head. If there is any movement, the skull is not properly secured. In such a case, release the mouse and try again until the skull is firmly secured (*see* **Note 13**).
7. **Perform final preparation of the surgical area**: disinfect the top of the skull three times by applying povidone-iodine with a cotton-tipped applicator.
8. **Prepare for pump installation on the back of the mouse.** Using a scalpel blade or spring scissors, make an approximately 1.2 cm longitudinal incision from just anterior to Bregma to the back of the skull.
9. Open the incision and use the scalpel to carefully scrape away any connective tissue adhering to the skull. This will ensure a clean surface for the next steps and will allow precise identification of anatomical landmarks.
10. Create a subcutaneous pocket for the pump on the back of the mouse. Lift the skin at the posterior end of the incision with a set of forceps and gently insert the large needle holder under the skin along the spine. Insert the needle holder approximately 4 cm along the left shoulder. Once the needle holder is inserted, gently open and close it a couple of times to create a loose subcutaneous pocket. The pocket should be large enough to accommodate the ALZET pump without excessive tension or compression.
11. **Insert the pump**. Holding the base of the pump where the tubing is attached, gently insert it all the way into the subcutaneous pocket. The entire pump should be positioned under the skin, including the portion where the pump attaches to the tubing, which should be covered by the skin. Tuck the tubing out of the way, wrapped under the left ear bar until it is connected to the infusion kit.
12. **Level the skull**. Ensure the skull is free of connective tissue. If needed, scrape it again with a scalpel. This is essential for the accurate determination of the infusion site.
13. Visualize sutures with hydrogen peroxide. Soak a cotton swab in hydrogen peroxide and swab the skull. Remove excess solutions with the dry end of the swab. This will make the cranial sutures appear white, which helps locate Bregma and Lambda.

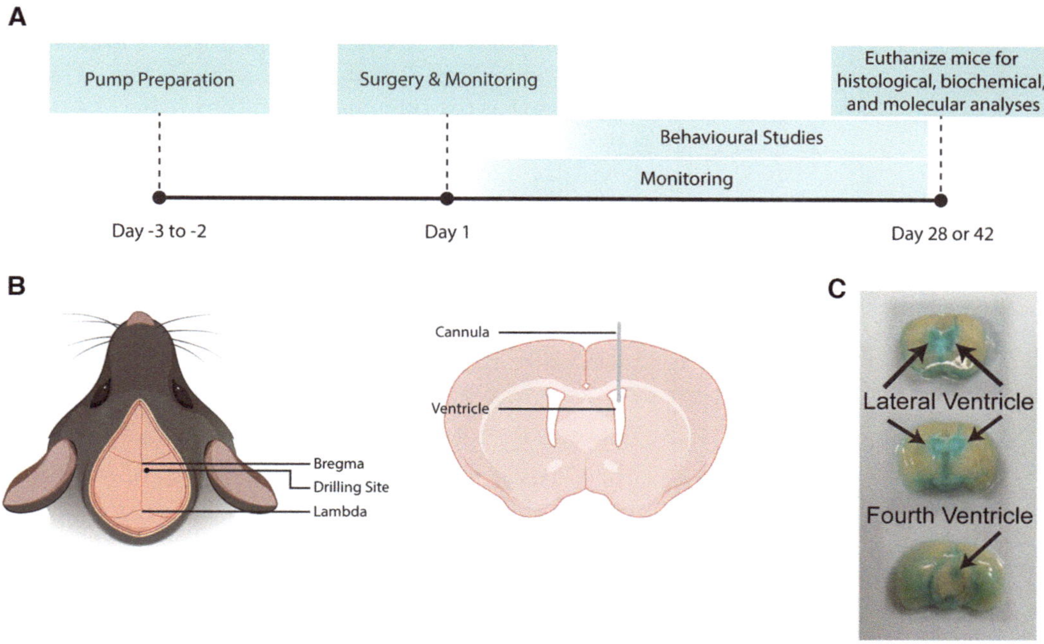

Fig. 1 (**A**) Timeline of experiments involving ICV infusion of gangliosides. Two to three days prior to surgery, the osmotic pumps must be primed with CSF or GM1. Surgery, which includes induction of anesthesia, surgical preparation on the mouse, and stereotaxic surgery, is performed on day 1. Mouse welfare must be monitored after the surgery and in the following days. GM1 infusion can be performed up to day 28 or 42, depending on the size of the pump that is used. Mice should be allowed to recover for a few days before starting any behavioral studies. At day 28 or 42, mice may be euthanized for histological, biochemical, and molecular analyses. (**B**) (Left) Diagram showing the approximate location of the drilling site on the mouse skull for infusion into the right ventricle. Bregma and Lambda are indicated. The drilling site is located 1.2 mm posterior and 0.6 mm to the right of Bregma. (Right) Coronal view of the mouse brain showing the cannula in the right ventricle. (**C**) Representative images of 2.5% Fast Blue dye distribution throughout the mouse brain ventricular system and adjacent brain parenchyma. Following insertion of the infusion kit, 20–40 μL of Fast Blue dye was injected into the ventricles. After 2–3 min, the mouse was perfused with 0.3% heparin in 1× PBS, and the brain was removed and cut coronally. The injection site is clearly visible in the top two sections. (Panel **C** is adapted from DeVos and Miller [29] with permission)

14. Locate Lambda and position the infusion kit directly over it, just touching the skull. Record the Z-axis reading from the vernier scale of the stereotaxic apparatus.
15. Locate and record Bregma. Raise the infusion kit slightly and reposition it over Bregma. Lower it until it touches the skull, then record the Z-axis reading on the vernier scale. If the skull is level, the difference in Z-axis values between Bregma and Lambda should be within 0.1 mm of each other. If not, adjust the nose of the mouse until the Z-axis readings are within 0.1 mm and the skull is level.

16. **Determine the coordinates for the drill site**. With the infusion kit positioned over Bregma, set the X and Y axes to zero on the electronic display of the stereotaxic frame. This will establish Bregma as the reference point for further coordinate adjustments.
17. Record X- and Y-axis vernier scale values. These will be different from the electronic display, where the coordinates have been set to zero.
18. Subtract 0.06 from the vernier Y-axis (the larger number) to calculate the rostro-caudal Y-coordinate of the drilling site.
19. Subtract 0.12 from the vernier X-axis (the smaller number) to calculate the lateral X-coordinate of the drilling site.
20. Raise the cannula holder but leave it in place over Bregma.
21. Move the Y-axis arm backward (toward the tail of the mouse) until the digital display reads "1.2."
22. Move the X-axis arm laterally (to the right) until the display reads "−0.6."
23. Check the vernier scales on the stereotaxic frame to confirm that the X- and Y-axis values match the differences (target coordinates) calculated in Steps 18 and 19. If there is a discrepancy between the digital display and the vernier scale, use the calculated values on the vernier scale, as they are more accurate for positioning.
24. Lower the infusion kit until it is just touching the skull. This is the exact location to drill (Fig. 1B). Raise the cannula holder and mark the site using a marker.
25. Lower the cannula holder and compare the mark to the position of the infusion kit. If the marker is slightly off-center, either remark or compensate accordingly during the drilling step.
26. **Drill the hole**. Securely insert the engraving cutter into the drill. Hold the drill with your dominant hand, and gently begin drilling at the marked site. Drill lightly and carefully, until just through the skull. Avoid damaging underlying tissue.
27. Verify skull penetration. Check that the drill has penetrated the skull by using a 90° bent-tipped 26-gauge needle. Use the needle to clear any bone burrs around the edge of the hole and ensure that the meninges have been punctured.
28. **Prepare the infusion kit with adhesive**. Liberally coat the underside of the infusion kit with Vetbond, taking care not to block the cannula.
29. **Attach the infusion kit to the skull**. Slowly lower the cannula holder until the infusion kit touches the skull, pulling

the skin to the side to avoid gluing it. Allow the adhesive to cure for a few seconds. Add more Vetbond if necessary. The length of the infusion kit (3 mm) is optimized to reach the mouse lateral ventricle (Fig. 1B, *see* **Note 14**).

30. **Connect the pump to the infusion kit**. Trim the tubing on the pump so it is approximately 1.5 cm in length (long enough to reach the infusion kit without restricting the mouse's head movement). Ensure the tubing is completely filled with solution (*see* **Note 15**).
31. Supporting the cannula holder from behind, attach the tubing from the pump to the infusion kit until it is flush with the end of the infusion kit, to prevent leaks or disconnection.

Table 1

Animal welfare score sheet. Use one table per mouse to track their progress every hour after surgery. Administer butorphanol and meloxicam if mice are showing severe signs of pain

Animal welfare score sheet	
Mouse I.D.	**Date of Operation**
Pre-study bodyweight	**Surgeon**
Date	
Time	
Body weight & B.A.R. (bright, alert, responsive) SCORE (normal is 0, score 1, 2, 3 for ↑ severity)	
Body wt yesterday	
Body wt today	
Body wt change	
Body wt score	
B.A.R.	
General clinical signs (score normal animal as 0, score 1, 2, 3 for ↑ severity)	
Breathing	
Posture	
Gait	
Hair coat	
Wound	
Behavioral signs of pain (score normal animals as 0, score 1, 2, 3 for ↑ severity)	
Vocalization (when handled)	
Nest	
Total	
INITIALS	

32. **Administer post-surgery analgesic**. Inject 5 mg/kg meloxicam (250 μL/10 g of mouse) subcutaneously, along with 1 mL of saline solution to aid recovery. Inject the solutions in the lower back of the mouse, one solution on either side of the spine (*see* **Note 16**).
33. **Release the infusion kit** by loosening the cannula screw. Using forceps to hold the infusion kit in place, slowly raise the cannula to completely release the kit. Use bent scissors to trim the top of the infusion kit.
34. **Close the incision**. Pull the skin closed over the infusion kit. Make a single interrupted suture anterior to the infusion kit. Make a second suture just posterior to the infusion kit. Make three more interrupted sutures posterior to the infusion kit. Secure each suture with 4 knots (*see* **Note 17**).

3.4 Post-Operative Recovery and Care

1. Transfer the mouse to a clean disposable cage with a paper towel at the bottom and cover with a lid. Transfer the cage to an incubator set at 38°C connected to an oxygen supply for optimal recovery (*see* **Note 18**).
2. Once the mouse begins to move around after surgery, return it to its home cage (*see* **Note 19**) and place two to three pieces of food at the bottom for easy access (*see* **Note 20**).
3. Monitor mice every hour, recording their progress on the Animal Welfare Score Sheet (Table 1). If, by 8 h after surgery, the mice show signs of pain (e.g., decreased activity, hunching, piloerection), inject them with meloxicam (5 mg/kg) subcutaneously.
4. Return cages to the animal room once the mice have recovered and are stable.
5. To prevent post-surgical infections, ensure the mice are given antibiotic water containing sulfamethoxazole and trimethoprim.
6. Check the mice twice daily for the first 2 days after surgery, then once daily for the next 7 days. If signs of pain are present in the first 72 hours, administer meloxicam. If the mice still show signs of pain beyond 72 hours, give either meloxicam or another dose of slow-release buprenorphine.
7. Once the mice have fully recovered, proceed to any planned behavioral studies or further experimental procedures (*see* **Note 21**).

4 Notes

1. It is critical to dissolve GM1 and fill the pumps in a biosafety cabinet under laminar flow to maintain aseptic conditions. This helps to prevent contamination that could lead to infections or degradation of the compound during delivery.
2. Always use sterilized forceps when handling the pumps during filling and priming.
3. A 1cc syringe must be used to fill the pumps. Larger syringes are not recommended, as they may fill the pump too quickly, increasing the risk of air bubbles remaining trapped within the pumps, which would disrupt drug delivery.
4. It is essential that the syringe and attached filling tube are completely free of air bubbles. Air bubbles in the system will prevent the solution from flowing, leading to inconsistent dosing or delayed delivery.
5. When connecting the flow moderator to the tubing, it is essential to maintain the continuity of liquid without any air bubbles between the flow moderator and the tubing. Even small air bubbles will disrupt or delay drug delivery, affecting experimental outcomes.
6. The pump must be entirely submerged in sterile DPBS and maintained in an upright position. The end of the tubing must remain above the surface of DPBS to avoid dilution of the solution in the tubing.
7. The oxygen flow rate during anesthesia depends on the calibration of the specific vaporizer used. Refer to the manufacturer's specifications to ensure accurate delivery.
8. The isoflurane level must be adjusted based on the mouse's heart rate, respiration rate, color of extremities, and pedal reflexes. Close monitoring of these parameters helps to maintain the appropriate level of anesthesia during surgery.
9. Due to the lack of blinking under anesthesia, regular application of eye ointment throughout the surgery is necessary to prevent the cornea from drying out and causing injury to the mouse's eyes.
10. If a mouse is exhibiting signs of pain and inadequate depth of anesthesia (e.g., blinking, increased heart and respiratory rate, body movement, and/or pedal reflexes), isoflurane should be increased by 0.5% increments. After waiting one minute, reassess if the mouse has reached the surgical plane. If the anesthesia appears too profound, which can be indicated by a decreased respiration rate with increased abdominal involvement, a decrease in heart rate, and pale or grey extremities,

isoflurane should be decreased by 0.5% increments. After waiting one minute, reassess if the mouse is in the surgical plane and has a normal respiratory rate, heart rate, and pink extremities.

11. Ensure that the infusion kit is fully primed with CSF or GM1 solution before attaching the pump. Any air bubbles or incomplete priming will cause interruptions in drug delivery.
12. After shaving the surgical site, any remaining loose shaved hair can be removed using a lint roller. This helps maintain a clean area and minimizes the risk of infections during surgery.
13. The skull must be held firmly in place before drilling. This is critical for accurate targeting of brain structures, since a loose or unstable skull could lead to improper placement of the infusion kit.
14. Correct placement of the cannula can only be determined at the end of the experiment by immunohistochemistry, which will show the cannula track surrounded by scar tissue. Therefore, it is advisable to practice and verify proper cannula insertion in advance, using test mice. To verify proper cannula placement, 20–40 μL of a dye solution (e.g., trypan blue, fast blue, or Evans blue) can be injected after the infusion kit is attached. Following the injection, the mouse can be euthanized, and the brain can be cut coronally adjacent to the cannula to confirm the presence of the dye in the brain ventricle. Alternatively, the dye injection can be performed at the end of the experiment and GM1 administration, following euthanasia, after disconnecting the pump from the infusion kit.
15. If correctly primed and no air bubbles are trapped in the connecting tubing, the pump should start infusing its contents immediately. There is no way to determine the correct operation of the pump at this stage. It is advisable to remove the pump from the animal at the end of the experiment and measure the actual residual volume of solution inside the pump. The latter should not be more than 10%–15% different from the expected residual volume (calculated based on pump capacity, flow rate, and overall duration of infusion). In our experience, only <1%–2% of implanted pumps did not work or had a significantly lower flow rate than indicated by the manufacturer.
16. Inject meloxicam and saline on opposite sides of the spine to ensure the whole volume of both solutions can be accommodated under the skin without leaking.
17. In total, you will make five sutures, each of them secured with 4 knots. To prevent the skin from tearing, do not pull the skin too tightly during suturing. If a rupture occurs, mice must be re-anesthetized and re-sutured to close the wound. It is

acceptable to leave the top of the infusion kit uncovered by the skin.

18. After surgery, it is critical to let the mouse recover in a warm incubator (38°C). The cage must have a lid to prevent escape as the mouse regains motor functions.
19. Following surgery, mice must be housed individually to prevent cage mates from undoing sutures.
20. It is important to leave two to three pieces of food on the floor of the cage, as mice recovering from surgery may not be able to reach food placed on the wire lid.
21. All procedures performed on mice must receive approval from and conform to the ethical standards set by the Animal Care Committee of the respective institution. All animal procedures described in this protocol were approved by the University of Alberta's Animal Care and Use Committee and are in accordance with the guidelines of the Canadian Council on Animal Care.

Acknowledgments

This study was funded by grants from the Canadian Institutes of Health Research (PS 175090) and the Natural Sciences and Engineering Research Council of Canada (RGPIN-2024-04875). Figure 1B was created in BioRender. Monyror J (2024) BioRender.com/d86w843 and BioRender.com/w61f029.

Conflict of Interest

Simonetta Sipione is a member of the scientific advisory board of Zulia Inc. (USA) and received consultant honoraria from TRB Chemedica International SA (Switzerland).

References

1. Ledeen RW, Wu G (2015) The multi-tasked life of GM1 ganglioside, a true factotum of nature. Trends Biochem Sci 40:407–418
2. Sipione S, Monyror J, Galleguillos D et al (2020) Gangliosides in the brain: physiology, pathophysiology and therapeutic applications. Front Neurosci 14:572965
3. Galleguillos D, Wang Q, Steinberg N et al (2022) Anti-inflammatory role of GM1 and other gangliosides on microglia. J Neuroinflammation 19:9
4. Maglione V, Marchi P, Di Pardo A et al (2010) Impaired ganglioside metabolism in Huntington's disease and neuroprotective role of GM1. J Neurosci 30:4072–4080
5. Alpaugh M, Galleguillos D, Forero J et al (2017) Disease-modifying effects of ganglioside GM1 in Huntington's disease models. EMBO Mol Med 9:1537–1557
6. Di Pardo A, Maglione V, Alpaugh M et al (2012) Ganglioside GM1 induces phosphorylation of mutant huntingtin and restores normal

motor behavior in Huntington disease mice. Proc Natl Acad Sci U S A 109:3528–3533
7. Itokazu Y, Fuchigami T, Morgan JC et al (2021) Intranasal infusion of GD3 and GM1 gangliosides downregulates alpha-synuclein and controls tyrosine hydroxylase gene in a PD model mouse. Mol Ther 29:3059–3071
8. Schneider JS, Aras R, Williams CK et al (2019) GM1 ganglioside modifies alpha-synuclein toxicity and is neuroprotective in a rat alpha-synuclein model of Parkinson's disease. Sci Rep 9: 8362
9. Schneider JS, Kean A, DiStefano L (1995) GM1 ganglioside rescues substantia nigra pars compacta neurons and increases dopamine synthesis in residual nigrostriatal dopaminergic neurons in MPTP-treated mice. J Neurosci Res 42:117–123
10. Schneider JS, Pope A, Simpson K et al (1992) Recovery from experimental parkinsonism in primates with GM1 ganglioside treatment. Science 256:843–846
11. Magistretti PJ, Geisler FH, Schneider JS et al (2019) Gangliosides: treatment avenues in neurodegenerative disease. Front Neurol 10:859
12. Schneider JS, Sendek S, Daskalakis C et al (2010) GM1 ganglioside in Parkinson's disease: results of a five year open study. J Neurol Sci 292:45–51
13. Schneider JS, Cambi F, Gollomp SM et al (2015) GM1 ganglioside in Parkinson's disease: pilot study of effects on dopamine transporter binding. J Neurol Sci 356:118–123
14. Ghidoni R, Trinchera M, Venerando B et al (1986) Incorporation and metabolism of exogenous GM1 ganglioside in rat liver. Biochem J 237:147–155
15. Saulino MF, Schengrund CL (1994) Differential accumulation of gangliosides by the brains of MPTP-lesioned mice. J Neurosci Res 37:384–391
16. Polo A, Kirschne G, Guidotti A et al (1994) Brain content of glycosphingolipids after oral administration of monosialogangliosides GM1 and LIGA20 to rats. Mol Chem Neuropathol 21:41–53
17. Revunov E, Johnstrom P, Arakawa R et al (2020) First radiolabeling of a ganglioside with a positron emitting radionuclide: in vivo PET demonstrates low exposure of radiofluorinated GM1 in non-human primate brain. ACS Chem Neurosci 11:1245–1249
18. Dumontet C, Rebbaa A, Portoukalian J (1992) Kinetics and organ distribution of [14C]-sialic acid-GM3 and [3H]-sphingosine-GM1 after intravenous injection in rats. Biochem Biophys Res Commun 189:1410–1416
19. Fischer H, Gottschlich R, Seelig A (1998) Blood-brain barrier permeation: molecular parameters governing passive diffusion. J Membr Biol 165:201–211
20. van de Waterbeemd H, Camenisch G, Folkers G et al (1998) Estimation of blood-brain barrier crossing of drugs using molecular size and shape, and H-bonding descriptors. J Drug Target 6:151–165
21. Pardridge WM (2005) The blood-brain barrier: bottleneck in brain drug development. NeuroRx 2:3–14
22. Wu D, Chen Q, Chen X et al (2023) The blood-brain barrier: structure, regulation, and drug delivery. Signal Transduct Target Ther 8:217
23. Schneider JS, Gollomp SM, Sendek S et al (2013) A randomized, controlled, delayed start trial of GM1 ganglioside in treated Parkinson's disease patients. J Neurol Sci 324:140–148
24. Schneider JS, Roeltgen DP, Mancall EL et al (1998) Parkinson's disease: improved function with GM1 ganglioside treatment in a randomized placebo-controlled study. Neurology 50:1630–1636
25. Gray MT, Woulfe JM (2015) Striatal blood-brain barrier permeability in Parkinson's disease. J Cereb Blood Flow Metab 35:747–750
26. Al-Bachari S, Naish JH, Parker GJM et al (2020) Blood-brain barrier leakage is increased in Parkinson's disease. Front Physiol 11:593026
27. Kortekaas R, Leenders KL, van Oostrom JC et al. (2005) Blood-brain barrier dysfunction in parkinsonian midbrain in vivo. Ann Neurol 57:176–179
28. Pardridge WM (2016) CSF, blood-brain barrier, and brain drug delivery. Expert Opin Drug Deliv 13:963–975
29. DeVos SL, Miller TM (2013) Direct intraventricular delivery of drugs to the rodent central nervous system. J Vis Exp 75:e50326
30. Pardridge WM (1991) Peptide drug delivery to the brain/William M. Pardridge. Raven Press, New York

Check for updates

Chapter 32

Gangliosides in Neural Stem Cell Fate Determination and Nerve Cell Specification: Preparation and Administration

Yutaka Itokazu, Toshio Ariga, Takahiro Fuchigami and Dongpei Li

Abstract

Gangliosides are sialylated glycosphingolipids with essential yet enigmatic functions in both healthy and disease brains. Among them, GD3 represents the predominant glycosphingolipid species in neural stem cells (NSCs), comprising over 80% of the total. On the other hand, the expression of GM1 is upregulated through a GM1-modulated epigenetic gene regulation mechanism involving GM2 synthase. GM1, in turn, influences the transcriptional activity of neuronal genes to maintain neuronal functions. To elucidate the functional roles of gangliosides, we introduce methods to isolate GD3 and GM1 and administer them into the mouse brain to investigate their functions on NSC fate determination and nerve cell specification.

Keywords Carbohydrate, Ganglioside, GD3, GM1, Ganglioside purification, Lipid membrane, Microdomain, Neural stem cell, Neurogenesis, Regeneration

1 Introduction

Glycosphingolipid composition, including gangliosides [1], is dynamically changed during neural differentiation (GD3 => GM1), thereby regulating distinct stages of cell fate determination. Our and others' studies suggest that GD3 maintains neural stem cell (NSC) characteristics and that GM1 epigenetically promotes neuronal differentiation and maintains neuronal functions [2–17] (Fig. 1). GD3 is the predominant species in NSCs and GD3-synthase (sialyltransferase II; *St8Sia1*) knockout (GD3S-KO) revealed reduction of postnatal NSC pools with severe behavioral deficits including cognitive impairment [13], depression-like phenotypes [16], and olfactory dysfunction [3, 4]. Exogenous administration of GD3 significantly restored the NSC pools and enhanced the stemness of NSCs with multipotency and self-renewal, followed by restored neuronal functions [3, 8]. Our group discovered that GD3 is involved in the maintenance

Yutaka Itokazu and Toshio Ariga contributed equally to this work.

Koichi Furukawa (ed.), *Gangliosides*, Methods in Molecular Biology, vol. 3035,
https://doi.org/10.1007/978-1-0716-5288-6_32,

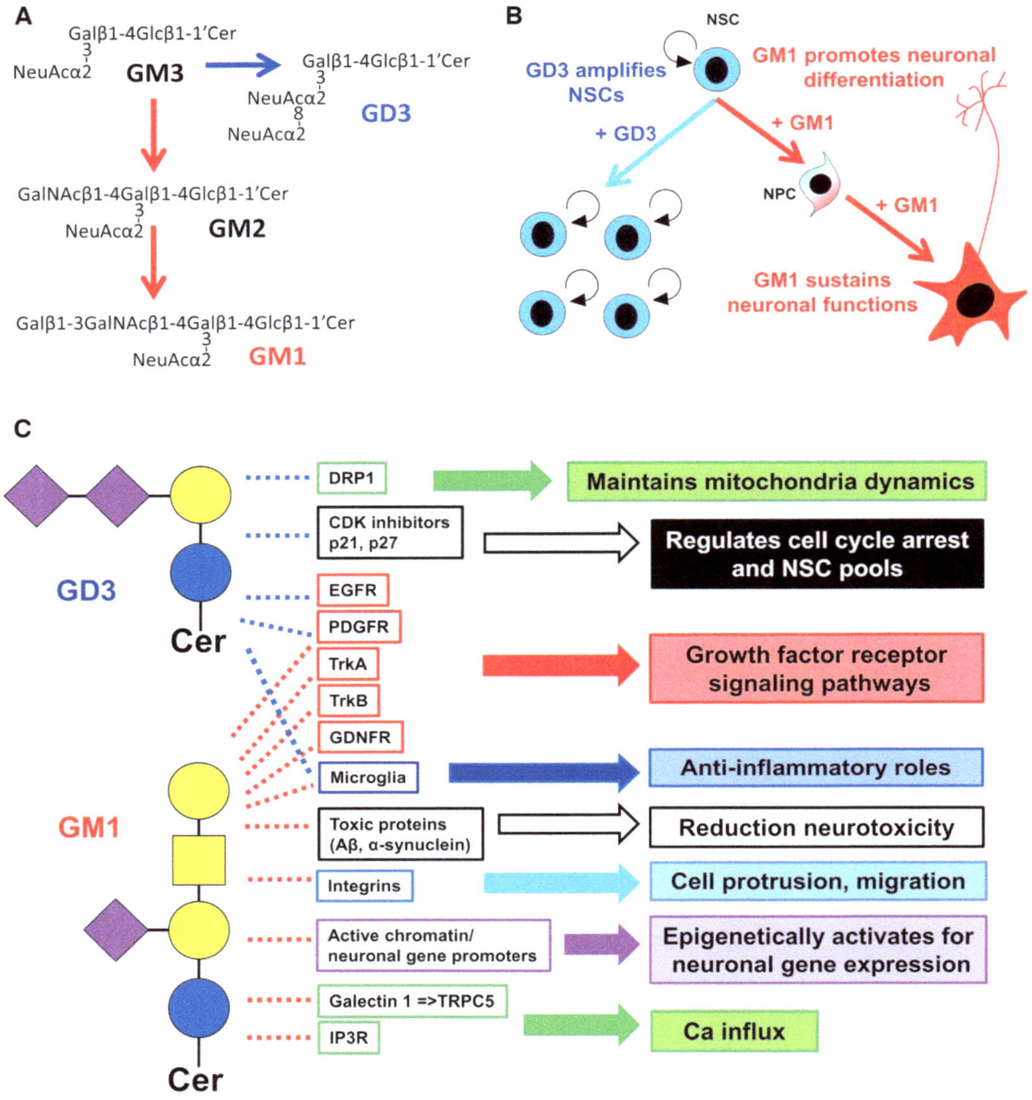

Fig. 1 Metabolic pathways and structure of GD3 and GM1 gangliosides. (**A**) The nomenclature for gangliosides and their components is based on that of Svennerholm and the IUPAC–IUBMB Joint Commission on Biochemical Nomenclature [32, 33]. (**B**) GD3 restores the number of neural stem cells (NSCs), and GM1 restores neuronal precursor cells (NPCs) and neuronal functions. (**C**) Examples of multifunctionality of GD3 and GM1: Without GD3, increased Drp1 levels induce aberrant mitochondrial fragmentation. GD3 suppresses p21 expression and maintains p27 expression in NSCs. Gangliosides bind to the neurotrophic factor receptors to regulate their signaling. Gangliosides decrease inflammatory microglia and increase phagocytosis. Gangliosides prevent and even reduce the accumulation of toxic proteins [e.g., amyloid-beta peptides (Aβs) and alpha-synuclein (α-syn)] in neurodegenerative diseases. GM1 promotes integrin signaling for cell protrusion, including neurite outgrowth and cell migration. Nuclear GM1 epigenetically promotes neuronal gene expression to sustain healthy neuronal functions. GM1 induces Ca influx

of NSC fate determination by interacting with epidermal growth factor receptors (EGFRs) [17], by modulating the expression of cyclin-dependent kinase (CDK) inhibitors p27 and p21 [3, 4], and by regulating mitochondrial dynamics via associating with a mitochondrial fission protein, the dynamin-related protein-1 (Drp1) [13]. Furthermore, we discovered that nuclear GM1 promotes neuronal differentiation by an epigenetic regulatory mechanism. GM1 binds with acetylated histones on the promoter of *N-acetylgalactosaminyltransferase (GalNAcT; GM2 synthase (GM2S); B4galnt1)* as well as on the *NeuroD1* in differentiated neurons [14]. In addition, epigenetic activation of the GM2S gene was detected as accompanied by an apparent induction of neuronal differentiation in NSCs responding to an exogenous supplement of GM1 [9, 15]. Interestingly, GM1 induced epigenetic activation of the *tyrosine hydroxylase (TH)* gene, with recruitment of Nurr1 and PITX3, dopaminergic neuron-associated transcription factors, to the *TH* promoter region [7]. In this way, GM1 epigenetically regulates dopaminergic neuron-specific gene expression, and it would modify Parkinson's disease. Multifunctional gangliosides significantly modulate lipid microdomains to regulate functions of important molecules on multiple sites: the plasma membrane, mitochondrial membrane, and nuclear membrane. Versatile gangliosides regulate functional neurons as well as sustain NSC functions via modulating protein and gene activities on ganglioside microdomains [18]. Maintaining proper ganglioside microdomains benefits healthy neuronal development and millions of senior citizens with neurodegenerative diseases [19].

Functional roles of gangliosides have been studied with ganglioside knockout (KO) mice (i.e., glycosyltransferase KO mice). However, KO mice cannot exclude the effect of ganglioside metabolic changes (e.g., upregulated or downregulated other than interesting molecules). To elucidate the functional roles of gangliosides, exogenous administration of ganglioside(s) is a critical and desirable experiment. In previous studies, intracerebroventricular administration was the most reliable method of delivering gangliosides into the brain [8, 20, 21]. We have successfully developed a more convenient non-invasive delivery procedure to the brain by intranasal infusion of gangliosides [3, 7, 19, 22]. In the present work, we review the laboratory methods that one can use to study the ganglioside function on NSC fate determinations in the mouse brain.

2 Materials

1. Chloroform.
2. Methanol.
3. Sodium acetate.

4. Sodium acetate in methanol (0.05 and 0.3 M)
5. Ammonia.
6. *n*-propanol (1-propanol).
7. Calcium chloride.
8. DEAE-Sephadex A-25.
9. Iatrobeads.
10. Glass chromatography columns (5 cm inner diameter (i.d.) × 150 cm; 2.5 cm i.d. × 200 cm; 2.2 cm i.d. × 100 cm; 1.5 cm i.d. × 100 cm; 1.2 cm i.d. × 56 cm; 2 cm i.d. × 100 cm; 0.4 cm i.d. × 110 cm).
11. Bovine, porcine, ovine, or human brains (for GM1). Bovine buttermilk powder (for GD3).
12. Saline (0.9% sodium chloride).
13. Ganglioside solution in Saline.
14. Anesthetic agent (e.g., ketamine hydrochloride and xylazine hydrochloride).
15. Stereotaxic apparatus.
16. Mini-osmotic pumps (100 μl capacity for 7-day infusion).
17. Gel loading capillary tips.

3 Methods

3.1 Purification of GM1

The procedure of GM1 isolation is based on Ando and Yu [23] with minor modifications of Ariga and Yu [24]. *See* **Note 1** for chemical synthesis of GM1.

1. Homogenize the brain (about 2.3 kg) with ten volumes each of chloroform/methanol 2:1, 1:1, and 1:2 (v/v).
2. Combine all lipid extracts.
3. Evaporate solvent.
4. Dissolve the residue in 12 l of chloroform/methanol/water 30:60:8 (v/v/v) (Solvent A).
5. Apply to a DEAE-Sephadex A-25 column chromatography (acetate form).
6. Elute the neutral lipids with 15 l of chloroform/methanol/water 30:60:8 (v/v/v) and 2 l of methanol.
7. Add 13 l of 0.2 M sodium acetate in methanol to recover the acidic lipids.
8. Evaporate solvent.

9. Dissolve the residue in 1 l of water.
10. Dialyze against 3 l of distilled water for 3 days, changing the distilled water daily.
11. Lyophilize the retentate.
12. Dissolve the residue in 500 ml of chloroform/methanol/water 30:60:8 (v/v/v)
13. Apply to a DEAE-Sephadex A-25 column chromatography.
14. At this stage, there are still lesser amounts of residual neutral lipids. Elute the neutral lipids with 1 l of chloroform/methanol/water 30:60:8 (v/v/v) and 500 ml of methanol.
15. Elute the acidic glycolipids with 4 l of a linear gradient system prepared from sodium acetate in methanol (0.05 and 0.3 M).
16. Combine the monosialoganglioside fractions.
17. Evaporate solvent.
18. Dissolve the residue in 200 ml of distilled water.
19. Dialyze against 3 l of distilled water.
20. Lyophilize the retentate.
21. Dissolve the residue in 20 ml of chloroform/methanol/water 70:30:1 (v/v/v).
22. Apply to an Iatrobeads column chromatography.
23. Elute from column with 2 l of a linear gradient system of chloroform/methanol/water 65:35:4 and 45:55:5 (v/v/v).
24. Combine the fraction containing GM1.
25. Evaporate solvent.
26. Dissolve the residue in 20 ml of chloroform/methanol/water 70:30:1 (v/v/v).
27. Apply again to another Iatrobeads column chromatography.
28. Elute from column with 1 l of a linear gradient system of chloroform/methanol/water 65:35:4 and 45:55:5 (v/v/v).
29. Evaporate solvent.
30. Dissolve the residue in 20 ml of chloroform/methanol/water 70:30:1 (v/v/v).
31. Apply the final Iatrobeads column chromatography.
32. Elute from column with 500 ml of *n*-propanol/water 85:15 (v/v).
33. Confirm the purity of GM1 by high-performance thin-layer chromatography (HPTLC) with the following solvent system: (a) chloroform/methanol/0.02% calcium chloride 50:45:10 (v/v/v) or (b) chloroform-methanol/5 M ammonia/0.4% calcium chloride 60:45:4:5 (v/v/v); and/or by Mass spectrometry (MS)-based ganglioside profiling.

3.2 Purification of GD3

The procedure of GD3 isolation is based on previous publications [25, 26]. *See* **Note 2** for chemical synthesis of GD3.

1. Suspend the bovine buttermilk powder (1.7 kg, containing 32% protein, 8.5% mineral, 49% carbohydrates, and 3.5% other small molecular weight compounds) in 1 l of water.
2. Dialyze against 3 l of distilled water for 3 days to remove small oligosaccharides, minerals, and other small molecular weight materials.
3. Lyophilize the retentate.
4. Extract total lipids with ten volumes each of chloroform/methanol 2:1, 1:1, and 1:2 (v/v).
5. Combine all lipid extracts.
6. Adjust to a ratio of chloroform/methanol/water 2:1:0.6 (v/v/v).
7. Allow the mixture to stand, then collect the upper phase. Add an equal volume of chloroform/methanol/water 3:48:47(v/v/v) to the mixture. Repeat this step and partition three times in total. This partitioning procedure follows the method described by Folch et al. [27]. Combine the ganglioside fractions from the upper phase.
8. Combine the ganglioside fractions from the upper phase.
9. Evaporate solvent.
10. Dialyze the residue against 3 l of distilled water for 3 days with frequent changing of water.
11. Lyophilize the retentate.
12. Adjust the residue with chloroform/methanol/water 30:60:8 (v/v/v).
13. Apply to a DEAE-Sephadex A-25 column chromatography.
14. Remove neutral lipids with ten volumes of chloroform/methanol/water 30:60:8 (v/v/v).
15. Elute the crude gangliosides by stepwise elution with a solvent system of 1 l each of 0.05 M, 0.2 M, 0.8 M sodium acetate in chloroform/methanol-/water 30:60:8 (v/v/v).
16. Collect each 15 ml of effluent fractions.
17. Pool the fractions 1-80 as monosialogangliosides, 80-110 as disialogangliosides, and 110-140 as trisialo- gangliosides.
18. Evaporate solvent.
19. Dissolve the residue in 20 ml of chloroform/methanol/water 70:30:1 (v/v/v).
20. Apply to an Iatrobeads column chromatography.

21. Elute from column with 2 l of a linear gradient elution using 1 l each of chloroform/methanol/water 70:30:3 and 35:65:5 (v/v/v) to yield GD3.
22. Evaporate solvent.
23. Dissolve the residue in 20 ml of chloroform/methanol/water 70:30:1 (v/v/v).
24. Apply the final Iatrobeads column chromatography.
25. Elute from column with 1 l of a linear gradient elution using 1 l each of chloroform/methanol/water 70:30:3 and 35:65:5 (v/v/v) to yield GD3.
26. Confirm the purity of GD3 by high-performance thin-layer chromatography (HPTLC) with the following solvent system: (a) *n*-propanol/water 80:20 (v/v), or (b) chloroform/methanol/0.5% calcium chloride 55:45:10 (v/v/v); and/or by Mass spectrometry (MS)-based ganglioside profiling. Store the purified lipids (*see* **Note 3**).

3.3 Intracerebroventricular Administration of Gangliosides

The procedure of intracerebroventricular administration of gangliosides (Fig. 2) is based on Itokazu et al. [8] (*see* **Note 4**).

1. Prepare ganglioside solution in (5 mg/kg-weight/day, 12 μl for a daily infusion).
2. Load ganglioside solution into the Mini-osmotic pumps (ALZET, Cupertino, CA) and connect the catheter tube and cannula as instructed in the manual.

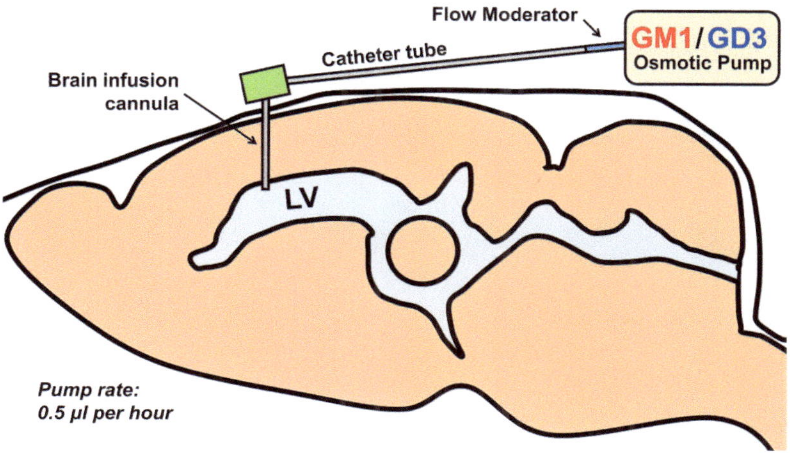

Fig. 2 Intracerebroventricular administration of gangliosides using a mini-osmotic pump. Implant the brain infusion cannula into the lateral ventricle, and the pump (filled with GD3 or GM1) attached by tubing to the brain infusion cannula subcutaneously in the neck region of the back

3. Anesthetize the mouse with ketamine hydrochloride (40 mg/kg body weight) and xylazine hydrochloride (4 mg/kg body weight).
4. Fix the mouse on a stereotaxic apparatus.
5. Shave and disinfect the scalp; make a midline sagittal incision and expose the skull; and prepare a subcutaneous pocket in the back of the mice to hold the osmotic pump.
6. Identify the bregma as the reference point. Move the tip of the stereotaxic coordinates to AP: 0.4 mm, ML: 1.5 mm. Drill a hole in the skull. Adjust the cannula to DV: 3 mm and apply glue to the cannula holder. Insert the cannula into the desired place, and direct pressure should be applied to make sure the cannula holder is attached to the skull tightly.
7. Place the osmotic pump in the subcutaneous pocket.
8. Close the scalp and the skin of the back with a suture or skin glue. Analyze mice (*see* **Note 5**).

3.4 Intranasal Administration of Gangliosides

The procedure of intranasal administration of gangliosides (Fig. 3) is based on Itokazu et al. [7] and Fuchigami et al. [3] (*see* **Note 6**).

1. Prepare ganglioside solution in saline (5 mg/kg-weight/day, 24 µl for a daily infusion). If the weight of a mouse is 20 g, a solution of 4.17 mg/ml of ganglioside in saline will be prepared. Twenty-four µl is the amount needed for a dosage of 5 mg/kg of body weight per day.

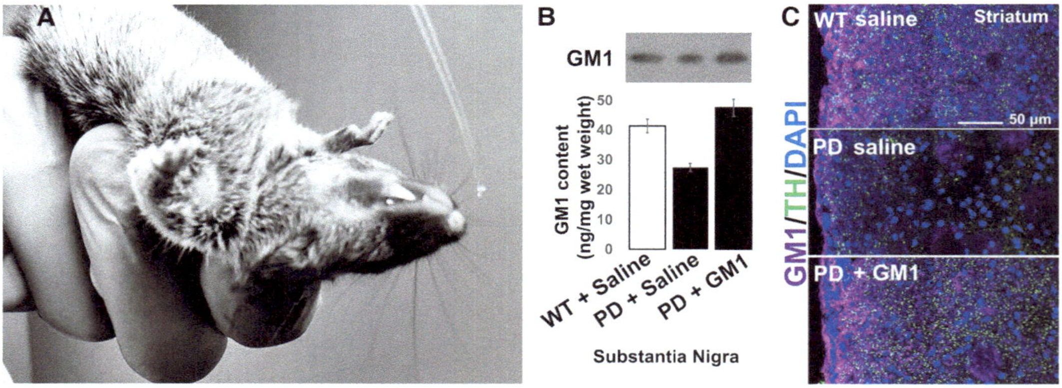

Fig. 3 Intranasal administration of gangliosides. (**A**) Intranasal drug delivery is an alternative administration, using the direct anatomic pathway between the olfactory/trigeminal neuroepithelium of the nasal mucosa and the brain. (**B**) Nasal GM1 (5 mg/kg/day for 28 days) restored GM1 levels in the Parkinson's disease (PD) model (A53T) mouse brain (by HPTLC). (**C**) Intranasally infused GM1 (5 mg/kg/day for 28 days) recovered the tyrosine hydroxylase (TH)+ axons (green) in the striatum of PD model mice. This suggests that chronic dopamine depletion seen in the PD-model mice and possibly in PD patients might be restored by intranasal GM1

2. Hold the mouse on the back of the neck using the thumb and pointer finger.
3. Invert with the ventral side facing up towards the ceiling.
4. Infuse 6 μl of ganglioside solution or saline into the naris of the mouse using capillary tips (Fig. 3A).
5. Keep holding the mouse in this position for 30 s.
6. Repeat the procedure for an additional three times. Infuse each 6 μl of ganglioside solution into the right and left nares twice (total 24 μl infusion).
7. Intranasally infuse daily for the desired days. Analyze mice (*see* **Note 5**).

4 Notes

1. Momoi et al. [28] reported the purification of a large amount of GM1 from bovine brain. Sugimoto et al. reported the chemical synthesis of GM1 [29].
2. Chemical synthesis of GD3 was reported by Ito et al. [30] and Ishida et al. [31].
3. Purified lipids can be stored in the freezer.
4. All animal experiments were approved by the Institutional Animal Care and Use Committee (IACUC) at Augusta University (AU) according to the National Institutes of Health (NIH) guidelines and were performed with approved animal protocols (references AUP 2009-0240 and 2014-0694).
5. Ganglioside-treated mice will be sacrificed for further NSC analyses.
6. In our experiments, no detectable adverse effects occurred during or after intranasal administration of gangliosides. Tissue sections were analyzed with hematoxylin and eosin (HE) staining, and no abnormal structures were found in the olfactory epithelium, brain, eye, spinal cord, tongue, thymus, lung, liver, spleen, kidney, stomach, intestine, colon, bone, or bone marrow after intranasal GD3 or GM1 infusion. Neither histological damage nor tumor formation was observed. Enzyme-linked immunosorbent assay (ELISA) revealed that intranasal ganglioside infusion (28 days) did not elevate the titer of ganglioside antibodies in the serum. Thus, intranasally administered gangliosides are safe and show no toxicity.

Acknowledgments

This work was partly supported by a National Institute of Neurological Disorders and Stroke grant (Y.I.: R01NS100839), a Sheffield Memorial Grant of the Central Savannah River Area (CSRA) Parkinson Support Group (Y.I.), and the excellent infrastructural support of the Department of Pharmacology and Toxicology (Chair: Dr. Alvin Terry), Medical College of Georgia at Augusta University.

Conflict of Interest The authors have no conflicts of interest to declare that are relevant to the content of this chapter.

References

1. Yu RK, Tsai YT, Ariga T et al (2011) Structures, biosynthesis, and functions of gangliosides—an overview. J Oleo Sci 60:537–544
2. Fantini J, Barrantes FJ (2009) Sphingolipid/cholesterol regulation of neurotransmitter receptor conformation and function. Biochim Biophys Acta 1788:2345–2361
3. Fuchigami T, Itokazu Y, Morgan JC et al (2023) Restoration of adult neurogenesis by intranasal administration of gangliosides GD3 and GM1 in the olfactory bulb of A53T alpha-synuclein-expressing Parkinson's-disease model mice. Mol Neurobiol 60:3329–3344
4. Fuchigami T, Itokazu Y, Yu RK et al (2024) Ganglioside GD3 regulates neural stem cell quiescence and controls postnatal neurogenesis. GLIA 72:167–183
5. Furukawa K, Ohmi Y, Ohkawa Y et al (2019) New era of research on cancer-associated glycosphingolipids. Cancer Sci 110:1544–1551
6. Galleguillos D, Wang Q, Steinberg N et al (2022) Anti-inflammatory role of GM1 and other gangliosides on microglia. J Neuroinflammation 19:9
7. Itokazu Y, Fuchigami T, Morgan JC et al (2021) Intranasal infusion of GD3 and GM1 gangliosides downregulates alpha-synuclein and controls tyrosine hydroxylase gene in a PD model mouse. Mol Ther 29:3059–3071
8. Itokazu Y, Li D, Yu RK (2019) Intracerebroventricular infusion of gangliosides augments the adult neural stem cell pool in mouse brain. ASN Neuro 11:1759091419884859
9. Itokazu Y, Tsai YT, Yu RK (2017) Epigenetic regulation of ganglioside expression in neural stem cells and neuronal cells. Glycoconj J 34:749–756
10. Itokazu Y, Wang J, Yu RK (2018) Gangliosides in nerve cell specification. Prog Mol Biol Transl Sci 156:241–263
11. Ledeen RW, Wu G (2015) The multi-tasked life of GM1 ganglioside, a true factotum of nature. Trends Biochem Sci 40:407–418
12. Schneider JS, Singh G, Williams CK et al (2022) GM1 ganglioside modifies microglial and neuroinflammatory responses to alpha-synuclein in the rat AAV-A53T alpha-synuclein model of Parkinson's disease. Mol Cell Neurosci 120:103729
13. Tang FL, Wang J, Itokazu Y et al (2021) Ganglioside GD3 regulates dendritic growth in newborn neurons in adult mouse hippocampus via modulation of mitochondrial dynamics. J Neurochem 156:819–833
14. Tsai YT, Itokazu Y, Yu RK (2016) GM1 ganglioside is involved in epigenetic activation loci of neuronal cells. Neurochem Res 41:107–115
15. Tsai YT, Yu RK (2014) Epigenetic activation of mouse ganglioside synthase genes: implications for neurogenesis. J Neurochem 128:101–110
16. Wang J, Cheng A, Wakade C et al (2014) Ganglioside GD3 is required for neurogenesis and long-term maintenance of neural stem cells in the postnatal mouse brain. J Neurosci 34:13790–13800
17. Wang J, Yu RK (2013) Interaction of ganglioside GD3 with an EGF receptor sustains the self-renewal ability of mouse neural stem cells in vitro. Proc Natl Acad Sci U S A 110:19137–19142
18. Itokazu Y, Yu RK (2023) Ganglioside microdomains on cellular and intracellular membranes regulate neuronal cell fate determination. Adv Neurobiol 29:281–304
19. Itokazu Y, Fuchigami T, Yu RK (2023) Functional impairment of the nervous system with glycolipid deficiencies. Adv Neurobiol 29:419–448
20. Di Pardo A, Maglione V, Alpaugh M et al (2012) Ganglioside GM1 induces phosphorylation of mutant huntingtin and restores normal

motor behavior in Huntington disease mice. Proc Natl Acad Sci U S A 109:3528–3533
21. Svennerholm L, Bråne G, Karlsson I et al (2002) Alzheimer disease—effect of continuous intracerebroventricular treatment with GM1 ganglioside and a systematic activation programme. Dement Geriatr Cogn Disord 14:128–136
22. Itokazu Y (2024) Multifunctional glycolipids as multi-targeting therapeutics for neural regeneration. Neural Regen Res 19:707–708
23. Ando S, Yu RK (1977) Isolation and characterization of a novel trisialoganglioside, GT1a, from human brain. J Biol Chem 252:6247–6250
24. Ariga T, Yu RK (1987) Isolation and characterization of ganglioside GM1b from normal human brain. J Lipid Res 28:285–291
25. Ledeen RW, Yu RK (1982) Gangliosides: structure, isolation, and analysis. Methods Enzymol 83:139–191
26. Ren S, Scarsdale JN, Ariga T et al (1992) O-acetylated gangliosides in bovine buttermilk. Characterization of 7-O-acetyl, 9-O-acetyl, and 7,9-di-O-acetyl GD3. J Biol Chem 267:12632–12638
27. Folch J, Lees M, Sloane Stanley GH (1957) A simple method for the isolation and purification of total lipides from animal tissues. J Biol Chem 226:497–509
28. Momoi T, Ando S, Magai Y (1976) High resolution preparative column chromatographic system for gangliosides using DEAE-Sephadex and a new porus silica, Iatrobeads. Biochim Biophys Acta 441:488–497
29. Sugimoto M, Numata M, Koike K et al (1986) Total synthesis of gangliosides GM1 and GM2. Carbohydr Res 156:C1–C5
30. Ito Y, Numata M, Sugimoto M et al (1989) Highly stereoselective synthesis of ganglioside Gd3 .65. J Am Chem Soc 111: 8508–8510
31. Ishida H, Ohta Y, Tsukada Y et al (1993) A synthetic approach to polysialogangliosides containing alpha-sialyl-(2–>8)-sialic acid: total synthesis of ganglioside GD3. Carbohydr Res 246:75–88
32. Svennerholm L (1963) Chromatographic Separation of human brain gangliosides. J Neurochem 10:613–623
33. No authors listed (1977) The nomenclature of lipids. Recommendations (1976) IUPAC-IUB Commission on Biochemical Nomenclature. Lipids 12:455–468

Index

Koichi Furukawa (ed.), *Gangliosides*, Methods in Molecular Biology, vol. 3035,
https://doi.org/10.1007/978-1-0716-5288-6,

S

T

U

MIX
Papier aus verantwortungsvollen Quellen
Paper from responsible sources
FSC® C105338

FSC
www.fsc.org

If you have any concerns about our products,
you can contact us on
ProductSafety@springernature.com

In case Publisher is established outside the EU,
the EU authorized representative is:
Springer Nature Customer Service Center GmbH
Europaplatz 3, 69115 Heidelberg, Germany

Printed by Libri Plureos GmbH
in Hamburg, Germany